STUDIES OF APPALACHIAN GEOLOGY: CENTRAL AND SOUTHERN

STUDIES OF APPALACHIAN GEOLOGY:

Central and Southern

Edited by

GEORGE W. FISHER

JOHNS HOPKINS UNIVERSITY
BALTIMORE, MARYLAND

F. J. PETTIJOHN

JOHNS HOPKINS UNIVERSITY
BALTIMORE, MARYLAND

J. C. REED, JR.

BRANCH OF ROCKY MOUNTAIN ENVIRONMENTAL GEOLOGY
DENVER FEDERAL CENTER
DENVER, COLORADO

and

KENNETH N. WEAVER

MARYLAND GEOLOGICAL SURVEY
BALTIMORE, MARYLAND

INTERSCIENCE PUBLISHERS

A DIVISION OF JOHN WILEY & SONS
NEW YORK   LONDON   SYDNEY   TORONTO

280196

ERNST CLOOS

TO

Ernst Cloos

WHO REKINDLED A SPIRIT
OF INQUIRY INTO
APPALACHIAN GEOLOGY

THIS VOLUME IS AFFECTIONATELY
DEDICATED BY HIS FRIENDS AND
FORMER STUDENTS

List of Contributors

ADAMS, Robert W., Department of Geology and Earth Science, State University College at Brockport, Brockport, New York 14420.

BERGIN, M. J., Branch of Organic Fuels, U. S. Geological Survey, Washington, D. C. 20242.

BROWN, William R., Department of Geology, University of Kentucky, Lexington, Kentucky 40506.

BRYANT, Bruce, U. S. Geological Survey, Room 2474, Bldg. 25, Federal Center, Denver, Colorado 80225.

COLTON, G. W., Branch of Astrogeology, U. S. Geological Survey, Flagstaff, Arizona 86001.

COOPER, Byron N., Department of Geological Sciences, Virginia Polytechnic Institute, Blacksburg, Virginia 24061.

DOE, Bruce, Isotope Geology Branch, U. S. Geological Survey, Federal Center, Denver, Colorado 80225.

DRAKE, A. A., Jr., U. S. Geological Survey, Washington, D. C. 20242.

ESPENSHADE, G. H., Eastern States Branch, U. S. Geological Survey, Washington, D. C. 20242.

FISHER, George W., Department of Geology, Johns Hopkins University, Baltimore, Maryland 21218.

GWINN, Vinton E., Department of Geology, University of South Carolina, Columbia, South Carolina 29208 (deceased).

HADLEY, Jarvis B., U. S. Geological Survey, Eastern States Branch, Agricultural Research Center, Beltsville, Maryland 20242.

HARRIS, Leonard D., U. S. Geological Survey, Room 11—Post Office Building, Knoxville, Tennessee.

HOPSON, C. A., Department of Geology, University of California, Santa Barbara, California 93106.

*HUNTER, Ralph E., Illinois State Geological Survey, Urbana, Illinois 61801.

HURST, Vernon J., Department of Geology, University of Georgia, Athens, Georgia 30601.

KING, Phillip B., U. S. Geological Survey, 345 Middlefield Road, Menlo Park, California 94025.

* Present Address: U. S. Geological Survey, 345 Middlefield Road, Menlo Park, California 94025.

McIVER, Norman L., P. O. Box 60775, New Orleans, Louisiana 70160.

MECKEL, Lawrence D., Jr., Shell Oil Co., 1700 Broadway, Denver, Colorado.

MYERS, W. B., U. S. Geological Survey, Federal Center, Denver, Colorado 80225.

OVERSTREET, William C., Foreign Geology Branch U. S. Geological Survey, Washington, D. C. 20242.

OWENS, James P., 14528 Bauer Drive, Rockville, Maryland 20853.

PETTIJOHN, F. J., Department of Geology, Johns Hopkins University, Baltimore, Maryland 21218.

RANKIN, Douglas W., Eastern States Branch, U. S. Geological Survey, Washington, D. C. 20242.

REED, J. C., Jr., U. S. Geological Survey Branch of Rocky Mountain Environmental Geology, Denver Federal Center, Denver, Colorado 80225

RODGERS, John, Department of Geology, Yale University, New Haven, Connecticut 06520.

SOUTHWICK, David L., Department of Geology, MacAlester College, St. Paul Minnesota 55101.

SUNDELIUS, Harold W., Department of Geology, Wittenberg University, Springfield, Ohio 45501.

TILTON, George R., Department of Geology, University of California, Santa Barbara, California 93106.

WATERS, Aaron C., Division of Natural Sciences, University of California, Santa Cruz, California 95060.

WEAVER, Kenneth N., Maryland Geological Survey, Johns Hopkins University, Baltimore, Maryland 21218.

WISE, Donald U., Department of Geology, Franklin and Marshall College, Lancaster, Pennsylvania 17604.

WOOD, Gordon H., Jr., Branch of Organic Fuels, U. S. Geological Survey, Washington, D. C. 20242.

Preface

The Appalachians are one of the classic mountain ranges of the world. Many concepts fundamental to geology were born and nurtured among these long ridges, narrow valleys, and gentle summits. Generations of geologists have studied them and pondered the problems they pose. Many of these problems are still unresolved, and mapping is continuing at an accelerated pace. Some of this recent work has buttressed ideas long held. Some has required modification of older ideas. And some has revealed new problems, hitherto unrecognized.

The classical interpretations of Appalachian geology, and the evolution of these ideas, are thoroughly treated in a number of excellent books. However, the results of most of the recent work are scattered through a vast number of papers, state geological survey reports, and unpublished sources. Many students of the Appalachians have for some time felt the need for a stocktaking. In what direction is Appalachian geology trending? Where are the major problem areas? What facts have been established, and what questions remain? Where might the answers to these questions be found? This book is an attempt to answer these questions for the southern and central Appalachians. It is not an exhaustive synthesis of Appalachian Geology. Written against a background of well-known, classical ideas, it represents an attempt to summarize the results of current work in the Appalachians, to make them more readily available to geologists in other areas, and to provide a reformulation of problems for geologists working in the Appalachians.

In a book of this sort, it is essential to provide room for alternate interpretations, and to avoid excessive editorializing. This consideration dictated the format, a collection of papers by active workers. Some of the papers present regional summaries of recent work; others describe more limited areas, and attempt to relate local features to regional problems. In order to provide a loose, but coherent framework, the papers are grouped in four sections, one dealing with the stratigraphy and sedimentation of the Valley and Ridge and Appalachian Plateau; another with the structure and tectonics of the same regions; a third with the geology of the Blue Ridge and the Reading Prong; and a last section on the Piedmont. In order to relate the individual papers more closely, and to provide a regional perspective, we have written short introductory comments for each section. As is almost inevitable in a volume of this sort, there are some gaps. We have indicated these omissions and, in part, tried to fill them in the introductions.

Ernst Cloos has played a major role in Appalachian studies. He has spent nearly thirty-seven years in the field, has introduced innumerable students to its problems, and has conducted many Appalachian field excursions for professional colleagues. It seemed to some of us, therefore, that an Appalachian volume would be a fitting tribute to the man who had contributed so much to our understanding of this classic area. C. A. Hopson was one of the original group which conceived of this project. He helped greatly with formulating the philosophy of the volume, making preliminary investigation of possible publication arrangements and solicited many of the papers. When, because of the press of other duties he had to withdraw from the project, G. W. Fisher consented to take his place. We are indebted to Hopson for his assistance during the formative stages. We are also indebted to many others for their cooperation and especially to the contributors, many of whom were students of Ernst Cloos. We are particularly grateful to R. N. Ginsburg for his help in editing the papers on stratigraphy and sedimentation.

The publication of this book was assisted by a grant from the Geology Alumni Publications Fund of the Johns Hopkins University.

THE EDITORS

Ernst Cloos—An Appreciation

"Will you write an 'appreciation' of Ernst Cloos for the Appalachian volume?" "Why yes, of course."

So it was agreed. Now, with pen in hand, how can I epitomize in a few sentences the amazing career of this big, energetic, and warm-hearted man who made such an impact on Appalachian geology, brought his department at Johns Hopkins University to first-class stature shortly after being named to its chairmanship, and sparked a period of refreshing new ideas and constructive leadership as President of the Geological Society of America?

I believe that I know his secret for success, and it is very simple: early in life Ernst Cloos learned the invaluable lesson of making the most of what was available. Long ago when both of us were young, we climbed together among the peaks of the High Sierra. One day years later, I followed him along a tortuous and smelly ravine in Ellicott City and asked "Ernst, how could you give up the Sierra to work here?" His answer was direct and gentle, yet spoken with a touch of asperity: "It was depression days and there were many starving people in California. Johns Hopkins has been very good to me, and the structure of these rocks is magnificent. Come a little farther and I will show you." Sure enough, around a few more bends we came upon an amazing set of exposures wherein the Ellicott City "granite" bared most of its secrets. Lineations were clear and easy to measure. Quartz was strained and bluish from internal defects. Hornblendes with tails appeared to be swimming madly upstream (in the direction of tectonic transport) like a flock of tadpoles frightened by a voracious kingfisher. And the huge poikiloblastic microclines tantalized you with complex helicitic structures, as well as evidence of crushing and recrystallization. Having just studied somewhat similar rocks in Washington and in Norway, I was entranced, and must have chattered like a ten-year-old schoolboy. Cloos watched the buildup of my enthusiasm with interest, pointing to more and more interesting features until he brought me to the base of a low bluff swept free of vegetation by the swiftly moving stream. We agreed this was the best place of all! Then, with a bit of sly amusement showing in his deep baritone, he boomed out: "You see geology makes beautiful things for us to study everywhere. You just have to search, that is all." As he spoke he stretched his arm to full length, pointing to the top of the cliff behind me, and he kept his arm outstretched until finally I turned and gazed up at the top of the cliff over my head. There, hanging over the edge so as to easily disgorge their contents, were five open privies.

Are you surprised that such a guy is a winner? And that even as this is being written he is out hammering on the metabasalts of the Blue Ridge?

AARON C. WATERS

Vita: Ernst Cloos

b. May 17, 1898 at Saarbrücken, Germany
m. December 27, 1923 to Margret Spemann
Children—Mrs. W. R. Evitt
    Mrs. F. C. Evering

EDUCATION:

Univ. Freiburg i.Br., Göttingen, and Breslau; Ph.D. Breslau 1923.

PROFESSIONAL RECORD:

Geophysicist, Seismos G.M.B.M. (Hannover) 1924-1930; 1924-1926 in Louisiana and Texas; 1926-1929 in England, Norway, Sweden, Italy, Holland, Iran and Iraq; 1928-1930 at Hannover, Germany.

Investigated batholiths in California under grant from Notgemeinschaft der Deutschen Wissenschaften, 1930-1931.

POSITIONS WITH THE JOHNS HOPKINS UNIVERSITY:

Res. Lect. Structural Geology, 1931-33; Lect. 1933-37; Assoc. Prof. 1937-1941; Prof. 1941-1968; Chairman 1951-1963; Prof. Emeritus 1968: Academic Council 1948-53, 54-59, 60-65.

Consultant, Esso Production Research Co., Thomasville Stone and Lime Co., Harry T. Campbell Corp.

PUBLIC AND PROFESSIONAL SERVICE:

Chairman, Div. Geol. and Geography, National Research Council, 1950-53.
Chairman, Maryland State Commission for Maryland Geological Survey, 1962-.

Acting Director, Maryland Geological Survey, 1962-63.
Vice-President, Geological Soc. Washington, 1947.
Councillor, Geological Soc. of America, 1947-49, Vice-Pres. 1953; Pres. 1954; Past-Pres. 1955.

HONORS

Phi Beta Kappa 1938.
Amer. Phil. Soc. 1954.
Nat. Acad. Science 1953.
Represented The Johns Hopkins Univ. at 100 anniversary of Nat. Acad. of Science in 1963.
Represented The Johns Hopkins Univ. at 500 anniversary Univ. Basel in 1960.
Recipient Guggenheim Fellowship 1956-57.
Elected to Finnish Academy of Sciences.
Distinguished Lecturer—American Assoc. of Petroleum Geologists in 1942.
Gustav Steinmann Medal of Geologischen Vereinigung, 1968.

PROFESSIONAL SOCIETY MEMBERSHIPS:

Geological Soc. of America.
National Assoc. Geology Teachers.
American Geophysical Union.
Geological Soc. of London.
Geological Assoc. of Canada.
Geological Soc. of Finland.
The Geologist's Assoc.
Geologische Vereinigung.
American Assoc. for the Advancement of Science.
Geological Soc. of Washington.

MILITARY RECORD:

German Air Force in first World War; shot down over Switzerland; interned for duration.

Bibliography of Ernst Cloos, to 1968

1922, Tektonik des Granits von Gorkau (Kr. Nimptsch) in Schlesien. Preussiche Geologische Landesanstalt. Abhandlungen Vol. 89, pt. V, p. 93–102.

——, Tektonik und Parallelgefüge im Granit und Granitporphür des nördlichen Schwarzwaldes. Abhandlungen Preussische Geologische Landesanstalt. Vol. 89, pt. IX, p. 137–141.

1927, with Hans Cloos, Die Quellkuppe des Drachenfels am Rhein. Ihre Tektonik und Bildungsweise. Zeitschrift für Vulkanologie. Vol. 11, Heft 1, p. 33–40.

——, with Hans Cloos, Das Strömungsbild der Wolkenburg im Siebengebirge. Zeitschrift für Vulkanologie. Vol. 11, Geft 2, p. 93–95.

1931, Der Sierra Nevada–Pluton. Geol. Rundschau, Band 22, Heft 6, p. 372–384.

——, Mechanism of the intrusion of the granite masses between Mono Lake and the Mother Lode (abstracts); Pan-Am. Geologists, vol. 55, no. 5, p. 373; Geol. Soc. America Bull., vol. 43, no. 1, p. 236, 1932; Washington Acad. Sci. Jour., vol. 22, no. 11, pp. 319–320, 1932.

1932, Structural survey of the granodiorite south of Mariposa, Calif.: Am. Jour. Sci. 5th Ser., vol. 23, pp. 280–304.

——, Feather joints as indicators of the direction of movement on faults, thrusts, joints, and magmatic contacts. Natl. Acad. Sci. Proc., vol. 19, no. 5, pp. 387–395.

——, Motion pictures of geologic events (abstract). Geol. Soc. America Bull., vol. 43, no. 1, p. 172; Pan-Am. Geologists, vol. 57, no. 1, p. 80.

1933, Structure of the Sierra Nevada batholith. Guidebook 16, Middle California and Western Nevada, XVI International Geological Congress, pp. 40–45.

——, Structure of the Ellicott City granite, Md.; Natl. Acad. Sci. Proc., vol. 19, no. 1, pp. 130–138.

1934, Auto radio—an aid in geologic mapping. Am. Jour. Sci. 5th Ser., vol. 28, no. 166, pp. 255–268; abstract, Zeitschr. Geophysik, Jahrg. 10, Heft ⅚, 1934.

——, Auto radio als Hilfsmittel geologiocher Kartierung. Zeitschr. f. Geophysik, vol. 10, p. 252–258, 1934.

——, The Loon Lake pluton, Bancroft area, Ontario, Canada. Jour. Geology, vol. 42, no. 4, pp. 393–399.

——, with Hans Cloos, Pre-Cambrian structure of the Beartooth, the Big Horn and the Black Hills uplifts and its coincidence with Tertiary uplifting. Abstract. Geol. Soc. America Proc. 1933–34, p. 56.

1935, Mother Lode and Sierra Nevada batholiths. Jour. Geology, vol. 43, no. 3, pp. 225–249. Abstract. Geol. Soc. America Bull., vol. 44, pp. 79–80, 1933.

——, (Review of) Geologic structures, by Bailey Willis and Robin Willis, 3d ed. rev., 1934; Econ. Geology, vol. 30, no. 8, pp. 936–939, 1935.

1936, Der Sierra–Nevada–Pluton in Californien: Neues Jahrb. Beilage-Band Union Trans. 16 Ann. Mtg., p. 274, Natl. Research Council, 1935.

——, with Howard Garland Hershey, Structural age determination of Piedmont intrusives in Maryland. Natl. Acad. Sci. Proc., vol. 22, no. 1, pp. 71–80. Abstract, Washington Acad. Sci. Jour., vol. 26, no. 9, p. 383, 1936.

1937, The application of recent structural methods in the interpretation of the crystalline rocks of Maryland. Maryland Geol. Survey (Rept.), vol. 13, pp. 27–105.

1940, and Carl Huntington Broedel, Geologic map of Howard and adjacent parts of Montgomery and Baltimore Counties (Md.), Maryland Geol. Survey.

——, Crustal shortening and axial divergence in the Applachians of southeastern Pennsylvania and Maryland. Geol. Soc. Am. Bull., v. 51, no. 6, p. 845–872.

1941, Flowage and cleavage in Appalachian folding N.Y. Acad. Sci. Trans., ser. 2, v. 3, no. 7, pp. 185–190.

——, and Anna Martha Hietanen, Geology of the Martic overthrust and the Glenarm series in Pennsylvania and Maryland. Geol. Soc. Am. Spec. Paper 35, xiii, 207 p.

1942, Distortion of stratigraphic thicknesses due to folding. Natl. Acad. Sci. Proc., v. 28, no. 10, pp. 401–407, 1942. Abstract, Tulsa Geol. Soc. Digest, v. 11, 1942–43, p. 49, 1943.

——, Fabric analyses of rock-flowage (abs). Am. Geophys. Union Trans. 23rd Ann. Mtg. Pt. 2, p. 707–708.

1943, and Carl Huntington Broedel, Reverse faulting north of Harrisburg, Pennsylvania. Geol. Soc. Am. Bull., v. 54, no. 9, pp. 1375–1397.

——, Method of measuring changes of stratigraphic thicknesses due to flowage and folding. Am. Geophys. Union Trans. 24th Ann. Mtg., Pt. 1, pp. 273–280.

1945, Correlation of lineation with rock-movement (summary). Am. Geophys. Union Trans. 25th Ann. Mtg., Pt. 4, pp. 660–662.

——, Memorial of Edward Bennett Mathews (1869–1944). Am. Mineralogist, v. 30, nos. 3–4, pp. 135–141.

——, History of geology in graphical representation. Geol. Soc. Am. Bull., v. 56, no. 4, pp. 385–388.

1946, Lineation, a critical review and annotated bibliography. Geol. Soc. Am. Mem. 18, vi, 122 p.

1947, Boudinage. Am. Geophys. Union Trans., v. 28, no. 4, pp. 626–632.

——, Tectonic transport and fabric in a Maryland granite. Comm. geol. Finlande Bull. no. 140, pp. 1–14.

——, Oölite deformation in the South Mountain fold, Maryland. Geol. Soc. Am. Bull., vol. 58, no. 9, pp. 843–917.

1949, Structures of the basement rocks of Pennsylvania and Maryland and their effect on overlying structures (abs). Oil and Gas Jour., v. 47, no. 24, p. 90, 1948; Am. Assoc. Petrol. Geol. Bull., v. 32, no. 11, p. 2162.

1950, The geology of the South Mountain anticlinorium, Maryland, Guidebook 1. Johns Hopkins Univ. Studies in Geology, no. 16, pt. 1.

——, and Judson Lowell Anderson, The geology of Bear Island, Potomac River, Maryland, Guidebook 2.

——, and John Calvin Reed, Memorial to Robert Ellsworth Fellows (1915–1949). Geol. Soc. America Proc. 1949, p. 159–162.

1951, History and geography of Washington County, in The physical features of Washington County. Md. Geol. Survey, Washington County (Rept. 14), p. 1–16.

——, Stratigraphy of sedimentary rocks, in The physical features of Washington County. Md. Geol. Survey Washington County (Rept. 14), 17–94.

——, Igneous rocks, in The physical features of Washington County. Md. Geol. Survey, Washington County (Rept. 14), p. 95–97.

——, Structural geology of Washington County, in The physical features of Washington County. Md. Geol. Survey, Washington County (Rept. 14), p. 124–163.

——, Mineral resources of Washington County, in The physical features of Washington County. Md. Geol. Survey, Washington (Rept. 14), p. 164–178; with a section on marl by J. T. Singewald, Jr.

——, Ground water resources, in The physical features of Washington County. Md. Geol. Survey, Washington (Rept. 14), p. 179–193.

1953, Appalachenprofil in Maryland, Geol. Rundschau, Band 41, p. 145–160, Stuttgart, Germany.

——, and Charles Wythe Cooke, Geologic map of Montgomery County and the District of Columbia. Md. Geol. Survey.

——, Lineation—review of literature 1942–1952. Geol. Soc. America Mem. 18, Supp., 14 p.

1955, Experimental analysis of fracture patterns. Geol. Soc. America Bull., v. 66, no. 3, p. 241–256.

1956, Fabric at granodiorite–schist contact, Bear Island, Maryland. Tschermaks Mineralog, u. Petrog. Mitt., Folge 3, Band 4, Heft 1–4, p. 81–89, Vienna.

——, Memorial to Robert Balk (1899–1955). Geol. Soc. America Proc. 1955, p. 93–100.

1957, Cost of educating one geologist. A.A.P.G. Bull., v. 41, no. 10, p. 2364–2368.

——, Blue Ridge tectonics between Harrisburg, Pennsylvania, and Ashville, North Carolina. Natl. Acad. Sci. Proc., v. 43, no. 9, p. 834–839.

1958, Structural geology of South Mountain and Appalachians in Maryland, Guidebooks 4–5. Johns Hopkins Univ. Studies in Geology, no. 17, 85 p. Includes a section by T. D. Murphy, which is cited individually.

——, Lineation und Bewegung, eine Diskussionsbemerkung. Geologie, Jahrg. 7, Heft 3–6, p. 307–311, Berlin.

1959, Memorial to Robert Milton Overbeck (1887–1958). Geol. Soc. America Proc. 1958, p. 161–164.

1960, and D. U. Wise, The Martic Problem and the New Providence Railroad Cut, in Some tectonic and structural problems of the Appalachian Piedmont along the Susquehanna River. Pennsylvania Geologists, 25th Ann. Field Conf., Oct. 1960, Guidebook, p. 39–48, 51–52.

1961, Bedding slips, wedges, and folding in layered sequences. Comptes Rendus. Geol. Soc. of Finland, v. 33, pp. 106–122.

1963, Review of Turner and Weiss, Structural analysis of metamorphic tectonites. McGraw Hill Book Co., N. Y., 1963. Trans. Am. Geophysical Union, v. 44, no. 3, 1963.

1964, History and Geography of Howard and Montgomery Counties; Review of the Post-Triassic Rocks; Structural Geology of Howard and Montgomery Counties. The Geology of Howard and Montgomery Counties, Md. Geol. Survey. pp. 1–10, 18–26, 216–259.

——, Appalachenprofil 1964, Geol. Rundschau, Bd. 54, p. 812–834.

——, Wedging, bedding plane slips, and gravity tectonics in the Appalachians, in Tectonics of the southern Appalachians, VPI Dept. Geol. Sci. Mem. p. 64–70.

——, Memorial to Joseph T. Singewald, Jr., State Geologists Journal, v. 16, no. 1, p. 314, 1964.

1968, Slickensides, striae, and mineral growth as tectonic indicators. Geol. Soc. America. (Abs) Regional Metting, Washington, February 1968.

——, Experimental analysis of Gulf Coast fracture patterns. A.A.P.G. Bulletin, March 1968, 28 p.

Contents

THE VALLEY AND RIDGE AND APPALACHIAN PLATEAU— STRATIGRAPHY AND SEDIMENTATION

Chapter 1

Introduction

F. J. PETTIJOHN

THE APPALACHIAN MOUNTAINS consist of a series of linear, level-topped ridges and intervening narrow valleys (the "Valley and Ridge Province") resulting from the differential erosion of a thick sequence of unmetamorphosed Paleozoic sedimentary rocks. These ridges extend from Pennsylvania southwestward into Georgia and Alabama where the belt widens somewhat and is amalgamated with prominent highlands upheld by late Precambrian strata in the Great Smoky Mountains. The Appalachians continue northeastward through the Hudson Valley and western New England into the Canadian Maritime Provinces where they are known under various local names.

The greater Appalachian province, however, is broader and more comprehensive than the Appalachian Mountains proper. In a general way this larger deformed belt can be divided into two parts which King has called the "sedimentary Appalachians" confined to the Appalachians proper, and the "crystalline Appalachians" which characterize the Piedmont (King, 1959, p. 44). Between the two stands the Blue Ridge which partakes of something of the character of each of the two areas.

The papers in this section deal exclusively with the sedimentary Appalachians; those of the later sections are concerned with the crystalline Appalachians and deal more specifically with the age of the metasediments and their relation to the Paleozoic strata of the Valley and Ridge Province. Colton's paper reviews the general stratigraphic framework. Colton makes several significant points. Most surprising is that, contrary to common belief, the rate of deposition of the carbonates, which form the bulk of the lower half of the section, was no different from that of the clastic sediments so prominent in the upper half

of the section. A second significant observation is that the centers of greatest thickness of the various clastic sequences which lie near the easternmost limits of the basin shifted widely from time to time. The maximum depocenter in the late Devonian, for example, lay in Pennsylvania whereas that for the Pennsylvanian was in Alabama.

Meckel discusses the central Appalachian Paleozoic alluvial deposits of which there are three, each of which culminates in a quartzitic, conglomeratic formation (the Silurian Tuscarora, the Mississippian Pocono, and the Pennsylvanian Pottsville). It is to these three formations that the central Appalachians owe their distinctive topography for these beds constitute the prominent ridges in the folded belt or Valley and Ridge Province.

McIver deals with the turbidite facies found in the central Appalachian province. Three such are known though the oldest, in the late Precambrian Glenarm, is confined to the Piedmont and is less amenable to study than the others (Ordovician Martinsburg and Devonian "Portage") because of metamorphism.

Regretfully the Appalachian carbonate rocks are not adequately treated. The carbonate sequences are likewise three in number. Only the uppermost and least impressive, the Greenbriar–Loyalhanna of Mississippian age, is here reported on by Adams. Each of the carbonate sequences records the culmination of a major marine transgression from the west just as each alluvial conglomeratic sandstone is the culmination of the intervening regressive phase. In upward sequence the carbonates decline in thickness and importance as the conglomeratic beds conversely increase in coarseness and conglomeratic character. The Loyalhanna, however, differs radically in character from the other carbonates and presents an interesting problem in carbonate deposition and paleocurrent analysis. Unlike the lowest Paleozoic carbonates of Cambro-Ordovician age most of which bear the earmarks of intertidal deposition, the Loyalhanna is strongly cross-bedded. The cross-bedding in this formation presents a paradox. It seemingly records a transport direction up the paleoslope and in a direction opposite to that from which the sediment apparently came.

Ralph Hunter's contribution on the Silurian Clinton is a study in iron sedimentation. He is able to recognize a succession of facies, wholly continental in the east and southeast and fully marine in the northwest, with a series of mixed brackish water facies in between. The combination of faunal, mineralogical, and paleocurrent data make an integrated interpretation of the Clinton possible. Of interest to the students of iron-bearing formations are the relations of the iron minerals to these facies.

The picture that emerges from these papers and from earlier work is that of a great elongate sedimentary basin—a miogeosyncline—being filled in the central and western parts mainly by carbonates of tidal flat and supratidal origin during the early Paleozoic and a poorly known eugeosynclinal tract on the east in which slates and other clastics, with perhaps volcanics in some places (Carolina), accumulated at the same time. The nature of the boundary between the western miogeosyncline and the eastern eugeosyncline is unknown. Was it, as in New England, a geanticlinal ridge? Or was the geosynclinal basin one with a great carbonate platform on the west flanking a deep basin on the east in which only thin muds accumulated? Was it open to the ocean on the east or was it flanked by a narrow tectonic land or even a substantial borderland?

Deformation and metamorphism of the eugeosynclinal tract in Ordovician time was followed by uplift and erosion of these early Paleozoic rocks and the Precambrian on which they were deposited. This uplift is recorded in the miogeosyncline by the late Ordovician–early Silurian clastic wedge consisting of black shales and marine turbidites (Martinsburg) overlain by continental deposits culminating in the conglomeratic Tuscarora and Shawangunk. This Taconic "cycle" was followed by a marine transgression and a second epoch of carbonate deposition which, in turn, was brought to an end by a second uplift in the source land and a late Devonian—early Mississippian clastic wedge, again beginning with black shales and marine turbidites ("Portage") fol-

lowed by continental deposits culminating in the conglomeratic Pocono. This Acadian cycle was followed by a third but aborted marine transgression which produced the Greenbriar–Loyalhanna carbonates. The final uplift in the source land led to a flood of continental clastics of which the Pottsville conglomerates were the culmination.

This, in brief, is the story of the sedimentary Appalachians—or the central Appalachians, to be more precise. But it is not the whole story. There was a long history prior to the Cambrian and there were thick accumulations of late Precambrian sediments in both the central and southern Appalachians. But this story is presented in these sections of this book dealing with the Blue Ridge and Piedmont Provinces.

Two major problems that have long confronted the student of the Appalachians are still not wholly resolved. One has to do with the source of the sediments which filled the Appalachian geosyncline from which the Appalachian Mountains arose. To what extent were these materials derived from the continental interior—from the Canadian Shield or craton in particular—and to what extent were they derived from sources exterior to the continent such as the once-postulated borderland of Appalachia? The second major question has to do with the relations between the sedimentary Appalachians and the crystalline Appalachians. Are the rocks of the latter, found mainly in the Piedmont province, in fact the Precambrian basement on which the Paleozoic sediments were deposited or are they the metamorphic derivatives of a former eastward extension of the Paleozoic fill or both.

What new evidence do we now have that bears on these old questions? First, we have the internal evidence of the sedimentary rocks themselves. As was shown by Krynine (1940) some time ago, and by Pelletier (1958), Folk (1960), McIver (1961), McBride (1962), Yeakel (1962), and Meckel (1967) more recently, careful petrographic study of the sandstones of late Ordovician and younger systems shows that they contain very little feldspar, are devoid of volcanic debris, are quartz-rich, and contain particles of metaargillaceous rocks such as slate, phyllites, and metaquartzite. From these observations we infer a source land with few, if any, deep-seated plutonic granites and gneisses, little or no volcanic rocks, and composed largely of sedimentary and metasedimentary rocks. Such conclusions seem to rule out either the Canadian Shield with its extensive granites and related rocks or peripheral island arcs of volcanic construction.

A second line of new evidence is derived from recent paleocurrent studies. The eastern or southeastern source inferred by earlier workers based on facies considerations is fully confirmed by extensive detailed mapping of current-produced sedimentary structures—chiefly cross-bedding in the Paleozoic alluvial sandstones. Various studies (Pelletier, 1958; McIver, 1961; Yeakel, 1962; Meckel, 1967) have shown a remarkably persistent and widespread paleocurrent transport from southeast to northwest beginning in late Ordovician and continuing to the close of the Paleozoic. Only in the Cambrian is there evidence of movement from the west or northwest. The widespread paleocurrent pattern and the persistence of the current systems through time leave little doubt of the reality of a southeast source land from late Ordovician on.

Another line of evidence has been the infrequent but undeniable discovery of Paleozoic fossils in Piedmont rocks once thought to be unfossiliferous Precambrian. It is significant, perhaps, that all of the fossils have been found in slates (Peach Bottom in Maryland, Arvonia and Quantico in Virginia, and in the Carolina slate belt) and are Ordovician or earlier in age.

Another important line of new evidence is that of the radiometric ages. These seem to suggest that the metamorphism and deformation of the Piedmont was completed by late Ordovician (Wetherill et al., 1966).

The evidence from the fossils coupled with that of the radiometric dates suggests that the crystalline Appalachians were metamorphosed, deformed, and uplifted by the close of the Ordovician and may well have been the principal source of all the sediments in the sedimentary Appalachians of late Ordovician age and younger. This is consistent with the petrographic evidence mentioned above.

If these conjectures are correct, then several interesting questions arise. From whence came the clastic materials which comprise the early Paleozoic sediments of the Piedmont? The earlier Paleozoic rocks in the sedimentary Appalachians are largely carbonates—forming nearly half the total Paleozoic record. These carbonate sediments extend far into the interior of the continent, where they thin somewhat and overlap on the craton and may at times have covered it completely. This carbonate armor essentially precludes the craton as a significant source of clastic materials. Was there then a source land still further southeast?

There is very little evidence bearing on the source of either the early Paleozoic or late Pre-

cambrian metasediments within the Piedmont province. Facies considerations suggest an eastern or southeastern source for the Snowbird Group and a northeastern source for the overlying Great Smoky Group (Hadley and Goldsmith, p. 47, 68) of the late Precambrian Ocoee Series. Meager data on cross-bedding in metasediments of similar age in Georgia indicate sediment transport from northeast to southwest (Mellen, 1956). The late Precambrian Glenarm Series of Maryland is believed to have been derived from an eastern source (Hopson, 1964, p. 130) as was much of the Lynchburg (Brown, Chapter 23, this volume). It is clear, therefore, that in early Paleozoic and even in Precambrian time, sediments were supplied mainly from an eastern source and that with the exception of some Cambrian sands, they could not have been derived from the continental interior or craton.

The nature of the eastern source land has been disputed: its existence even denied (Kay, 1951, p. 31; Dietz, 1963, p. 330). Some have reduced it to a geanticlinal welt arising from the geosyncline; others describe it as an island arc in part volcanic. As we have seen the character of the debris is, in general, incompatible with a volcanic source land. Furthermore, as noted by Kuenen (1948, p. 333), mountainous island tracts produce little sediment. That a narrow tectonic ridge will not do was also demonstrated by material balance calculations in Naylor and Boucot (1965, p. 157) for the Silurian sediments of Ludlow age in the New England–Maritime region. As Barrell noted long ago (1914, p. 248), the great volume of Devonian deltaic sediments presumes a major land area. Viva Appalachia!

REFERENCES

Barrell, J., 1913, The Upper Devonian delta and the Appalachian geosyncline: Am. Jour. Sci., v. 36, p. 429–472; v. 37, p. 87–109, 225–253.

Dietz, R. S., 1963, An actualistic concept of geosynclines and mountain building: Jour. Geology, v. 71, p. 314–343.

Folk, R. L., 1960, Petrography and origin of the Tuscarora, Rose Hill, and Keefer formations, Lower and Middle Silurian of eastern West Virginia: Jour. Sed. Petrology, v. 30, p. 1–58.

Hadley, J. B., and Goldsmith, R., 1963, Geology of the eastern Great Smoky Mountains North Carolina and Tennessee: U. S. Geol. Survey Prof. Paper 349-B, 118 pages.

Hopson, C. A., 1964, The crystalline rocks of Howard and Montgomery Counties in The Geology of Howard and Montgomery Counties: Maryland Geol. Survey, 359 pages.

Kay, Marshall, 1951, North American Geosynclines: Geol. Soc. America Memoir 48, 143 pages.

King, P. B., 1959, The Evolution of North America: Princeton, N.J., Princeton Univ. Press, 190 pages (esp. p. 44).

Krynine, P. D., 1940, Petrology and genesis of the Third Bradford Sand: Pennsylvania State Min. Industries Exp. Station Bull. 29, 134 p.

Kuenen, Ph. H., 1958, Problems concerning source and transportation of flysch sediments: Geol. en Mijnbouw (n.s.), v. 20e, p. 329–339.

McBride, E. F., 1962, Flysch and associated beds of the Martinsburg Formation (Ordovician), central Appalachians: Jour. Sed. Petrology, v. 32, p. 39–91.

McIver, N. L., 1961, Sedimentation of the Upper Devonian marine sediments of the central Appalachians: Unpublished Ph.D. dissertation, The Johns Hopkins Univ., Baltimore.

Meckel, L. D. Jr., 1967, Origin of Pottsville conglomerates (Pennsylvanian) in the central Appalachians: Geol. Soc. America, Bull., v. 78, p. 223–258.

Mellen, James, 1956, Pre-Cambrian sedimentation in the northeast part of the Cohutta Mountain Quadrangle, Georgia: Georgia Mineral Newsletter, v. 9, p. 46–61.

Naylor, R. S. and Boucot, A. J., 1965, Origin and distribution of rocks of Ludlow age (Late Silurian) in the northern Appalachians: Amer. Jour. Sci., v. 263, p. 153–169.

Pelletier, B. R., 1958, Pocono paleocurrents in Pennsylvania and Maryland: Geol. Soc. America, Bull., v. 69, p. 1033–1064.

Wetherill, G. W., Tilton, G. R., Davis, G. L., Hart, S. R., and Hopson, C. A., 1966, Age measurements in the Maryland Piedmont: Jour. Geophysical Research, v. 71, p. 2139–2155.

Yeakel, L. S. Jr., 1962, Tuscarora, Juniata, and Bald Eagle paleocurrents and paleogeography in the central Appalachians: Geol. Soc. America, Bull., v. 73, p. 1515–1540.

Chapter 2

The Appalachian Basin—Its Depositional Sequences and Their Geologic Relationships*

GEORGE W. COLTON

INTRODUCTION

THIS PAPER is an outgrowth of a report originally prepared in 1960 for the Division of Reactor Development, U. S. Atomic Energy Commission, to aid in appraising the potential of the Appalachian basin for the underground disposal of radioactive waste solutions (Colton, 1961). In the present paper the stratigraphic relationships of the predominantly miogeosynclinal suite of rocks that fill the Appalachian basin are emphasized; all references to their waste disposal potential and most references to their structural environment have been deleted.

Many new structural and stratigraphic data have been obtained from deep test wells drilled since the original study was completed seven years ago. The geology of many areas has since been mapped for the first time or remapped, new sections have been measured in detail and previously described sections have been remeasured, often with different results, and many regional or subregional summary papers presenting new conclusions have appeared. So much new material has become available that the maps in the original report had to be recompiled for this presentation. The elapsed period of time also enabled the writer to compile and present additional quantitative data, much of it volumetric.

The writer is aware of the danger of an oversimplified treatment of the geology of a large, complex area. However, if his purpose has been accomplished, this summary study will furnish a convenient stratigraphic and geographic frame-

work within which many of the detailed papers in this volume can be viewed.

Organization of Report

For convenience of description the stratigraphic column has been subdivided into nine parts on the basis of gross lithologic composition. The subdivisions are referred to as "sequences," and word prefixes indicate their general age and gross lithologic composition. For example, throughout most of the basin the Upper Ordovician clastic sequence consists largely of noncarbonate clastic rocks (shale, mudstone, siltstone, and sandstone) predominantly of Late Ordovician age. The lithologic composition, thickness, nomenclature, and stratigraphic relationships of the major units composing the individual sequences are summarized by means of representative stratigraphic sections in diagrammatic form. The extent, thickness, and highly generalized outcrop of most of the sequences are shown on individual maps, and the distribution of selected rock types or stratigraphic units included in the sequence is also shown on several of the isopach maps. The structural attitude of the sequences is summarized briefly with the aid of two structure contour maps which show the elevation of the top of the Precambrian basement complex and of a horizon near the middle of the Paleozoic with regard to mean sea level.

Location and Extent of Study Area

The Appalachian basin (Fig. 1) is an oblong sedimentary basin in the eastern United States which extends from the Canadian shield in southern Quebec and Ontario Provinces, Canada,

* Publication authorized by the Director, U. S. Geological Survey.

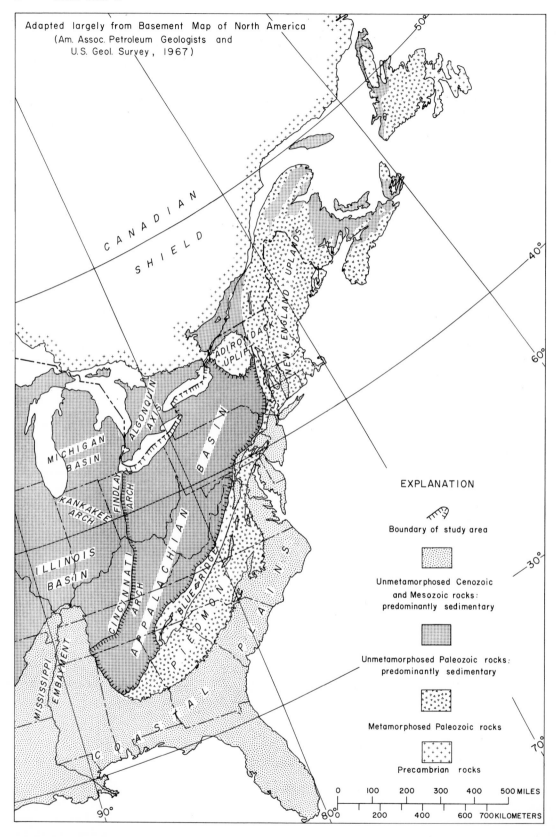

FIGURE 1. Map of eastern North America relating major geologic and physiographic features to the study area.

southwestward to central Alabama, approximately parallel to the Atlantic coastline. It is not a physiographic basin; on the contrary, much of it is occupied by mountains and plateaus. For the purposes of this study the Canadian part of the basin and the part in New York State north and east of the Adirondack uplift are excluded.

The west edge of the study area is here defined as a sinuous line that extends from the southern end of the Algonquin axis near Lake Erie to southwestern Tennessee along the crests of the Findlay and Cincinnati arches. The south edge of the study area is defined to coincide with the boundary between Paleozoic rocks of the Appalachian Plateaus province and the overlapping Cretaceous strata of the Mississippi embayment of the Gulf Coastal Plain province.

The eastern boundary of the study area was adapted from the "Basement map of North America" (Am. Assoc. Petroleum Geologists and U. S. Geol. Survey, 1967). Throughout much of its extent the boundary is the surface contact between unmetamorphosed or slightly metamorphosed rocks of Paleozoic age on the west with metasedimentary, metavolcanic, and intrusive rocks of Precambrian and Paleozoic age on the east. It coincides with the west flank of the Blue Ridge in the south and with the west edge of the New England Uplands in the north. The boundary as thus drawn is convenient for most purposes of this study, but it excludes a thick sequence of stratified rocks of Precambrian age (not shown in Fig. 1) that some workers might consider as part of the basin fill underlying parts of Pennsylvania, Maryland, Virginia, Tennessee, North Carolina, and Georgia. Across much of southeastern Pennsylvania where Precambrian rocks are not present along the strike of the Blue Ridge, the downfaulted west edges of the Triassic sedimentary basins mark the eastern boundary of the study area. In the Hudson Valley region of New York the west edge of the Taconic klippe is arbitrarily designated as the eastern boundary. From northwestern Georgia to the south end of the basin, the eastern boundary of the study area is marked by the contact of unmetamorphosed Paleozoic rocks on the west with metamorphosed Paleozoic rocks on the east. In this area the boundary of the study area coincides with the boundary between the Valley and Ridge and the Piedmont physiographic provinces.

The terms "study area" and "Appalachian basin" will be used interchangeably in this paper, although the study area is smaller by about 15 percent than the Appalachian basin itself. The 15 percent excluded consists of shallow peripheral

parts of the basin in southern Canada and that part of northern New York State north and east of the Adirondack uplift; the volume of sediment thus eliminated from further discussion is less than 5 percent of the total sedimentary basin fill. Outlined in this manner, the Appalachian basin is about 1,030 mi (1,657 km) long and about 330 mi (530 km) wide at its widest point. It covers an area of about 206,900 sq mi (536,000 sq km), including all of West Virginia, large parts of New York, Pennsylvania, Maryland, Ohio, Kentucky, Virginia, Tennessee, Georgia, and Alabama, and small areas in New Jersey and North Carolina.

ACKNOWLEDGMENTS

This study was facilitated by the help of many people. Special appreciation is expressed to Gail M. Everhart of the U. S. Geological Survey who did much of the preliminary work by reviewing the literature and compiling published information on base maps. Many of the maps included in this report evolved from her compilations. Phoebe E. Bernat assisted greatly by performing some of the volumetric calculations, and Peter R. Margolin made many improvements in the text. R. D. Carrol and R. E. Sabala graciously interrupted a busy schedule to draft the figures.

Advice and much helpful information were freely supplied by Wallace de Witt, Jr., J. F. Pepper (deceased), G. H. Wood, Jr., H. H. Arndt, L. D. Harris, K. E. Englund, and J. B. Epstein—all of the U.S. Geological Survey. The writer thanks the above persons and the following, whose critical reviews have improved the original manuscript considerably: Wallace de Witt, Jr., M. H. Hait, J. F. McCauley, and R. S. Saunders—all of the U. S. Geological Survey.

GEOLOGIC FRAMEWORK

Depositional Framework

The Appalachian basin as delineated in preceding paragraphs is an elongate downwarped segment of the earth's crust in which a great thickness of sediment accumulated. Most of the sediment was deposited in shallow seas that occupied the downwarped area for long periods throughout the Paleozoic era. To some extent concurrently with deposition, but largely after the bulk of the sediments had been deposited, the mass of sedimentary rock was uplifted and deformed. Finally, erosional processes, which are still active, created the present topography.

Subsidence and deposition commenced in late Precambrian time in the eastern part of the basin and continued intermittently throughout most of the basin until late Paleozoic or early Mesozoic time. As the basin subsided, sediments derived largely from source areas to the east—and partly from emergent areas to the north and west—accumulated in the basin. Contributions from the north and west were greatest during Cambrian and Early Ordovician times when the bulk of the carbonate basin fill was deposited. During most of the middle and latter parts of the Paleozoic Era, sediments from the east formed the thick clastic wedges that constitute more than one-half the volume of the total basin fill. Again, in late Paleozoic time, especially during the Mississippian and Pennsylvanian Periods, appreciable clastic material entered the basin from the north and was deposited in the northwest part.

Subsidence as well as deposition was most rapid along the east edge of the basin close to the main source area of most of the sediments. The resulting asymmetry of the basin is shown in Figure 2, a stratigraphic section drawn perpendicular to the long axis of the basin. This marked asymmetry is also shown on Figure 3, a structure contour map of the top of the Precambrian basement complex beneath the Paleozoic basin fill.

Stratigraphic and sedimentologic studies suggest that most of the sediments were deposited in shallow water, indicating that the rate of deposition closely balanced the rate of subsidence of the sea floor. At times deposition was more rapid than subsidence, and predominantly red, brown, or tan sediments accumulated above sea level. Rocks of these colors are most common near the eastern periphery of the basin, especially in the northeast part. During some periods subsidence, at least locally, may have been much more rapid than deposition, and sediments may have accumulated in water of considerable depth. Localized subsidence of this type probably occurred during part of Middle and Late Ordovician time in eastern Pennsylvania. At other times subsidence halted or uplift began, resulting in periods of nondeposition or in the erosion of previously deposited sediments. Hiatuses in the stratigraphic section due to nondeposition or to erosion are most pronounced along the western, northern, and eastern margins of the basin.

In general most of the sediments in the eastern part of the basin—especially in the northeastern part—accumulated in a deltaic environment; most of the sediments in the central part accumulated offshore in the trough of the marine basin; and most of the sediments in the western part accumulated on the shallow peripheral platform or shelf. The areal distribution of the principal rock types in the Appalachian basin (Fig. 4) reflects the geographic location of the predominant environments of deposition. Sandstone, siltstone, and red beds make up a relatively high percentage of the stratigraphic column in the eastern and northeastern parts of the basin, where deltaic environments prevailed throughout much of the history of the basin. In the central part of the basin, or the trough of the basin of accumulation, argillaceous rocks constitute a higher percentage of the rock sequence than elsewhere. Carbonate rocks predominate in the western part of the basin where shelf or platform environments prevailed.

The relative abundance of the principal rock types in the Appalachian basin is compared with that in the Michigan basin and with calculated and measured worldwide abundances (Fig. 5). Of particular interest are the facts that both basins are relatively deficient in shale and, compared with worldwide abundances, uncommonly rich in carbonate rocks. Of the two basins, the Michigan basin contains the least shale and (relatively) the most carbonates and evaporites.

The intermediate standing of the Appalachian basin (Fig. 5) suggests that it received or trapped a relatively greater amount of argillaceous terrigenous debris than the smaller more isolated intracratonic Michigan basin—classified as an autogeosyncline by Kay (1951, p. 20)—but only about half the argillaceous material that geochemical calculations suggest is available for deposition. Presumably most of this material accumulates in eugeosynclines and in the deeper parts of the permanent ocean basins. Consequently, the deficiency of shale and the overabundance of carbonates in the Appalachian basin are indicative of its predominantly miogeosynclinal nature, and hence of prolonged deposition in a predominantly shallow-water environment. The sandstone–shale–carbonate ratios of the metamorphosed Paleozoic stratified rocks of the Piedmont and New England uplands, which presumably accumulated in a predominantly eugeosynclinal environment, probably more nearly approximate worldwide ratios than do those of the Appalachian basin.

There is, perhaps not surprisingly, general agreement in the abundances of sandstone calculated by different methods in different areas. Both the Appalachian and Michigan basins contain 23 percent sandstone. Other reported values are 32 percent (Leith and Mead, 1915), 15 percent (Clarke, 1924), 14 percent (Kuenen, 1941),

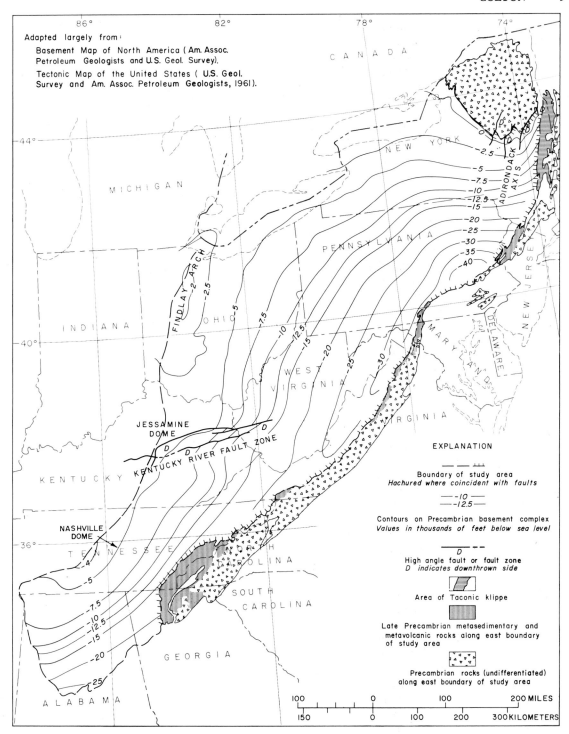

FIGURE 3. Configuration of the Precambrian basement floor in the Appalachian basin.

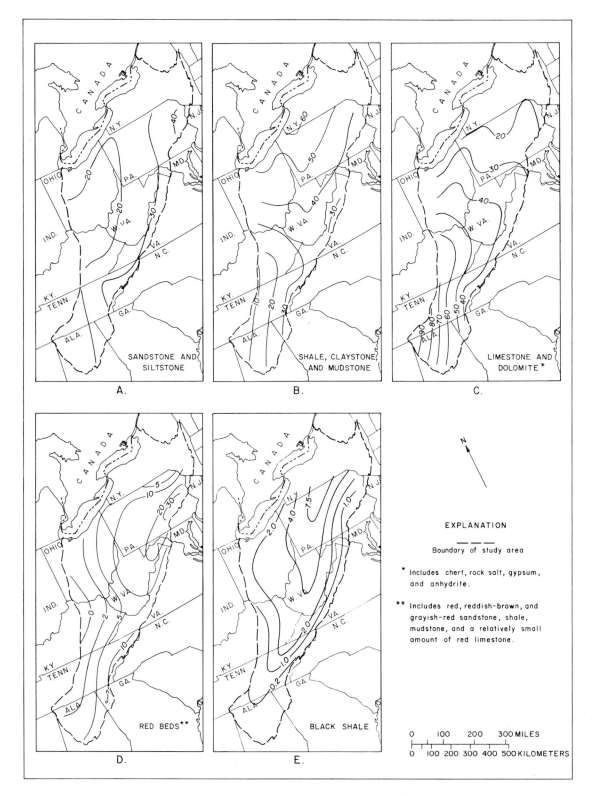

FIGURE 4. A, B, and C, distribution of principal lithologies composing the Paleozoic basin fill shown by lines of equal percentage; some restoration of younger Paleozoic rocks toward peripheral parts of basin. D and E include parts of the rock types shown in A–C.

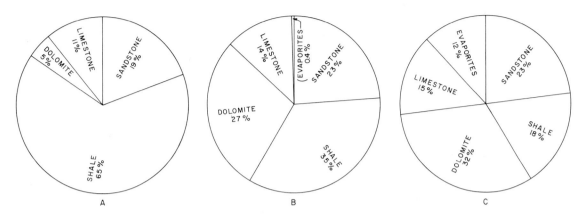

LITHOLOGIC PERCENTAGES

FIGURE 5. Relative abundance of the principal lithologies. (A) Worldwide; (B) Appalachian basin; (C) Michigan basin. The worldwide diagram is average of six determinations based on geochemical calculations and measured sections in various parts of the world (Leith and Mead (2), 1915; Clarke, 1924; Goldschmidt, 1933, recalculated 1954; Kuenen, 1941; Krynine, 1948, recalculated by Pettijohn, 1957). Appalachian basin diagram based on analysis of published well-log descriptions and outcrop measurements. Michigan basin diagram (Cohee and Landes, 1958, slightly modified by present writer) based on well-log descriptions.

and 22 percent (Krynine, 1948, recalculated by Pettijohn, 1957).

The Paleozoic rocks that compose the basin fill can be divided into eight gross lithologic units. Each unit—designated a sequence* in this report—accumulated while a rather uniform depositional environment or a suite of closely related environments prevailed throughout much of the basin. The contacts between units are determined primarily on the basis of vertical change in lithology; precise age relationships are of secondary importance. Consequently, some contacts do not coincide with the generally accepted boundaries between geologic periods and eras in many parts of the basin. The eight sequences are listed in Table 1 and are compared with the major divisions of the time scale.

A ninth bedded sequence—the very thick Precambrian stratified sequence—is shown in Table 1 for reference. It crops out in the western part of the Blue Ridge province and may not extend

* The sequences defined here are distinguished primarily by their bulk lithologic composition and are meant to apply only to the Appalachian basin. They do not coincide with the "sequences" of Sloss and others (1949, p. 110–112) and Sloss (1963), which are distinguished primarily by bounding surfaces that constitute interregional unconformities and are of near-continental extent.

TABLE 1.

Sequences (this report)	Geologic period
Pennsylvanian sequence	Permian
	Pennsylvanian
Mississippian sequence	Mississippian
Devonian clastic sequence	Devonian
Silurian–Devonian carbonate sequence	
Silurian clastic sequence	Silurian
Upper Ordovician clastic sequence	Ordovician
Cambrian–Ordovician carbonate sequence	
Lower Cambrian clastic sequence	Cambrian
(Upper Precambrian stratified sequence)	Precambrian
(Precambrian basement complex)	

appreciably westward into the study area. Although some workers have shown that it may be conformable in some areas with the earliest Paleozoic rocks of the basin, it is not here considered as part of the basin fill. These Precambrian strata contain much volcanic material—both flows and pyroclastic debris. They may represent a sequence that originally accumulated well to the east and was subsequently tectonically displaced westward to its present site.

Although two sequences—the Upper Ordovician clastic sequence and the Silurian clastic sequence—are somewhat similar lithologically, they are separated for two reasons. First, a pronounced structural and stratigraphic discontinuity divides them in some places. Second, the lower sequence is at present not significantly productive of oil and gas, whereas the upper sequence is highly productive in many areas. The Mississippian and Pennsylvanian sequences are also similar in many respects, but they are separated because Mississippian rocks, in contrast to Pennsylvanian and Permian rocks, are highly productive of gas and oil but contain very little mineable coal or clay.

The volumetric relations of the sequences are shown diagrammatically in Figure 6, which includes estimates of the lengths of time during which the sequences accumulated. This figure and Figure 5B show that: (1) the two intervals

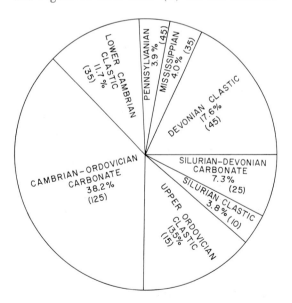

FIGURE 6. Relative volumes of the Paleozoic sequences recognized in this report expressed as percent of total basin fill (510,000 cu mi, 2,126,000 cu km). Figures in parentheses are approximate length of time in millions of years (adapted by Kulp, 1961) during which sequence accumulated where sequence is most typically and fully developed.

of carbonate deposition lasted approximately 150 million years, and the rocks compose about 42 percent of the Paleozoic basin fill; and (2) deposition of noncarbonate rocks lasted about 185 million years, and these rocks compose about 58 percent of the fill. If the compaction rate of carbonate sediment is assumed to equal the average rate for noncarbonate mud and sand combined, then the relative rates of deposition can be easily calculated from these data, provided two other factors are considered: (1) the amount of material of one type (carbonate or noncarbonate) included in the sequences of the other type, and (2) the amount and type of rock missing beneath unconformities. The two carbonate sequences do contain a large amount of sandstone and a smaller amount of shale. The clastic sequences—especially the Mississippian sequence—include an appreciable amount of limestone. The writer believes, as a first approximation, that the volumes of extraneous rocks present in the two types of sequence are about equal. A first consideration of the strata missing beneath unconformities suggests that somewhat more carbonate rock is missing in carbonate sequences than noncarbonate rock associated with unconformities within or between clastic sequences.

From the above the writer estimates that the overall rate of carbonate deposition during the 335 million years of the Paleozoic Era was slightly higher than the rate of noncarbonate deposition; a ratio of approximately 1.17:1.0 seems reasonable in the light of the preliminary data presently available.

These calculations are of course approximate and at best apply only to the material now present within the study area. If, in these calculations, a significant volume of rock that originally constituted part of the basin fill has been excluded, then the estimates of lithologic abundances and comparative rates of deposition given cannot be validly applied to the basin of accumulation.

A study of Figure 2 strongly indicates that part of the Paleozoic succession may indeed be missing. In this cross section, which extends southeastward across the basin from northwestern Ohio to the west edge of the Blue Ridge, the very thick east-thickening wedge of Cambrian–Ordovician carbonate rocks terminates abruptly at the Blue Ridge front. This apparently anomalous condition suggests to the writer that the eastern part of the sequence is missing. Presumably the carbonate wedge present within the study area intertongued farther eastward with an even thicker succession of noncarbonate rocks composed predominantly of shale and secondarily of sandstone. If it did exist, the succession may be missing because of erosion after uplift of the Precambrian Blue Ridge complex or, more likely, because of the westward displacement of the Blue Ridge complex over the eastern part of the Cambrian–Ordovician basin. If such a noncarbonate mass had been included in the calculations of lithologic abundances, the results would have been different. A lower percentage of carbonate rocks and a higher percentage of noncarbonate rocks would have resulted, and the abundances for the Appalachian basin would have more closely approached the worldwide abundances. Additionally, the calculated rate of carbonate relative to noncarbonate deposition would have been lower.

Significant errors in calculation may also have been introduced by use of conventional equal area projection maps. If palinspastic maps had been used the large amount of crustal shortening along the eastern side of the basin would have been corrected for and the calculated abundances and depositional rates might have been different. Construction of palinspastic maps for this study was not practical, however, particularly in view of the large size of the basin and the structural complexity of the eastern part of the basin.

The sediments that accumulated during late Precambrian and Paleozoic time attained a composite maximum total thickness of at least 63,000 ft (19,200 m). However, the greatest thickness preserved at any one place in the basin is probably between 35,000 ft (10,670 m) and 40,000 ft (12,200 m) (G. H. Wood, Jr., oral commun., 1960). This thick section, which occurs in east-central Pennsylvania, comprises rocks ranging in age from Early Cambrian to Late Pennsylvanian. In addition to being very thick, the Paleozoic column in the Appalachian basin is remarkably complete from Cambrian to Lower Permian. With one possible exception, no significant stratigraphic gap or hiatus is known to extend the full area of the basin. The exception may be an unconformity between the Lower and Middle Ordovician Series, but throughout most of the eastern part of the basin at least, the gap seems to be small, in that appreciably less than a series is absent.

A large volume of rock has been removed by erosion, largely since late Paleozoic time. Preliminary calculations based on measurement of restored cross sections extending the full width of the basin indicate that about 24 percent of the rocks deposited before the middle of Permian time has since been removed. This is probably a conservative estimate of the entire volume of rock eroded from the basin because in the present absence of all rocks younger than middle Permian the writer has based his calculations on the assumption that none were deposited. It seems certain, however, that Mesozoic rocks at the southern end of the basin once extended northeast of their present erosional limit, and hence occupied part —an unknown part—of the study area. The recent report (Pierce, 1965) of an isolated occurrence of unconsolidated Cretaceous sediments in southern Pennsylvania suggests that some Mesozoic rocks may have been present in the northern part of the area. Finally, the location of the Triassic basins in Pennsylvania adjacent to the boundary of the study area suggests that the northeast part of the area may have been covered by Triassic rocks if the present basins are erosional remnants of much larger centers of accumulation.

The volume of Paleozoic rocks present in the basin today is approximately 510,000 cu mi (2,126,000 cu km)—a figure obtained by measuring the volume of the basin below sea level from Figure 2, and by adding a volume that would occupy the area of the basin to a height of 1,800 ft or 549 m, the approximate average elevation of the land surface. The volume of Precambrian stratified sedimentary and volcanic rocks present in the basin could not be measured because subsurface data are lacking; however, it is probably less than 10,000 cu mi (41,700 cu km).

Structural Framework

The structural configuration of the Appalachian basin is primarily the result of two factors: (1) the original shape of the downwarped segment of the earth's crust; (2) diastrophic processes that deformed the Precambrian crystalline floor and the mass of younger sedimentary rock in the basin. The basin was compressed perpendicular to its long axis by stresses, the resultants of which were directed from the east and southeast. Compression was most severe in the eastern part of the basin. Although the intensity of structural deformation decreases progressively to the west and north, the basin can be divided into two segments on the basis of the type and frequency of occurrence of the resulting structures. The boundary between the more intensely deformed eastern part and the less intensely deformed western part is marked by the Appalachian structural front (Price, 1931). The trend of the structural front (Fig. 7) closely coincides with topographic features known as the Catskill

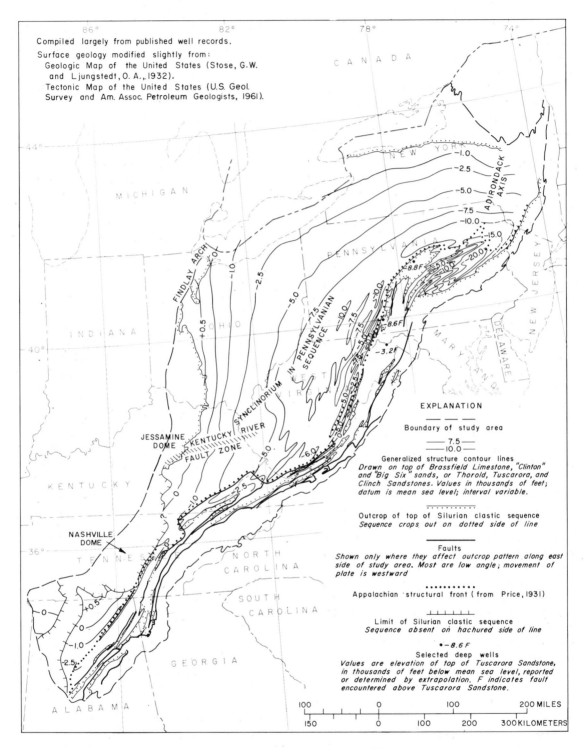

FIGURE 7. Generalized structural configuration of a horizon in the Silurian clastic sequence.

Escarpment, the Allegheny Front, and the Cumberland Escarpment.

The rocks in the eastern segment—the area generally included in the Valley and Ridge physiographic province—are characterized by numerous closely spaced folds and many low-angle thrust faults. The axes of folds and the traces of faults closely parallel the long axis of the basin. Many of the folds are oversteepened or overturned to the west on their west flanks. Movement of the overthrust blocks along the thrust faults was toward the northwest. Tear faults and high-angle reverse faults are also present in the Valley and Ridge province. Axial plane cleavage is present in some incompetent argillaceous rocks along the eastern margin of the basin. Cleavage is well developed in Precambrian stratified rocks along the west side of the Blue Ridge. Within the Valley and Ridge province the folds become progressivly gentler and thrust faults become progressively less numerous and more widely spaced toward the west. Structure contours are not shown in Figure 7 in most of the area east of the Appalachian structural front because of structural complexities, insufficient subsurface data, and limitations imposed by the small scale of the map. Where structure contour lines are drawn in the eastern part of the basin, they are to be regarded only as an attempt to portray a highly generalized concept of the configuration of the rocks at the scale of the map. The details of structure are, of course, much more complex.

The rocks in the western segment of the basin —the area included in the Appalachian Plateaus physiographic province and the easternmost part of the Central Lowlands province—are characterized by the general absence of thrust faults intersecting the surface and by gentle, approximately symmetric surface folds. The folds are steeper, more closely spaced, and more commonly parallel to the regional trend near the structural front. Additionally, some anticlines near the front are asymmetric to the southeast, and many have high-angle reverse faults near their crests. Westward the folds become flatter, more widely spaced, more varied in trend, and are less commonly faulted. In the westernmost part of the basin deformation is restricted largely to very broad gentle arches and elongate domes. Among these are the Findlay and Cincinnati arches and the Jessamine and Nashville domes. A major fault zone or declivity—the Kentucky River fault zone—crosses eastern Kentucky oblique to the regional trend of the Appalachian basin (Figs. 3 and 7). The zone comprises many fault slices bounded by nearly vertical fault planes. The

apparent displacement is obviously downward on the south, although this may be, in part, the result of right-lateral slip-strike movement. Some of the accompanying isopach maps suggest that movement was indeed right-lateral and that it was greatest during Cambrian time. The writer believes that this fault zone may extend east across West Virginia to southern Page County, Va., although Woodward (1964, p. 345–346) indicated that it extends northeast into Maryland and across the southeast corner of Pennsylvania.

Although the Appalachian basin was subjected to deformation many times throughout its history, several rather distinct episodes of diastrophic and epeirogenic activity have been recognized within the basin.

The first episode that resulted in a widespread hiatus in the geologic record occurred in late Precambrian time and probably continued into Early Ordovician time. As a result of uplift and erosion of the early Precambrian metamorphic and igneous terrane prior to deposition of younger strata, a marked angular unconformity is present between the crystalline basement complex and the stratified rocks that comprise the basin fill.

The next major disturbance—the Blountian of some workers, especially Rodgers (1953, p. 124) —was a gentle epeirogenic uplift that occurred in early Middle Ordovician time (Fig. 8E). It resulted in an erosional disconformity of basinwide extent in the upper part of the Cambrian–Ordovician carbonate sequence.

In parts of the basin there is conclusive evidence of severe diastrophic activity in Late Ordovician time. This orogeny—the Taconic— was strongest in the northeastern part of the Appalachian basin. In parts of eastern Pennsylvania and northwestern New Jersey a marked angular unconformity (Fig. 8D) exists between highly deformed fine-grained Upper Ordovician rocks and slightly deformed coarse-grained Lower Silurian rocks. Along the southern flank of the Adirondack uplift, where rocks of late Middle and early Late Ordovician age are discomformably overlain by rocks of late Early Silurian age, the orogeny apparently lasted longer than elsewhere.

A widespread but gentle epeirogenic uplift occurred before Late Devonian time. It was strongest in the southern part of the area and is marked by an erosional disconformity (Fig. 8C) at the base of the predominantly Upper Devonian Chattanooga Shale in much of Tennessee, northwestern Georgia, and northern Alabama. The disconformity increases in magnitude from the trough of the basin northwestward toward the

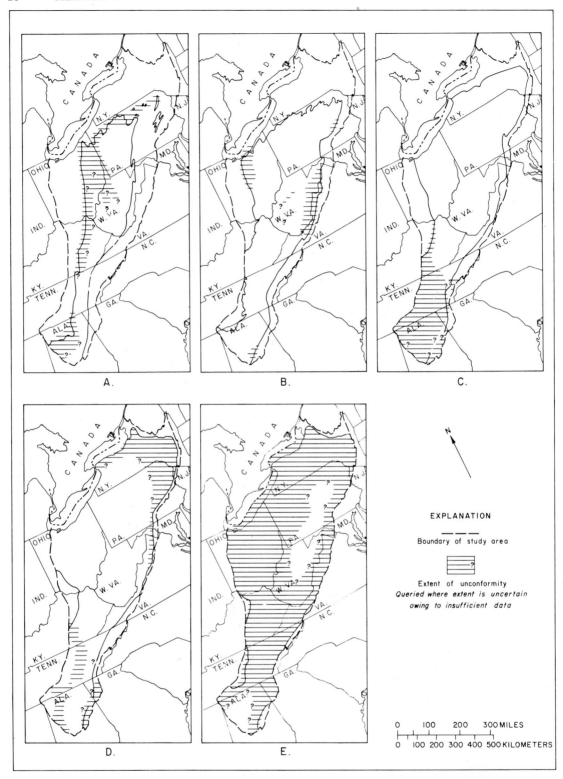

FIGURE 8.   Sketch maps of major unconformities (A) at base of Pennsylvanian
System, (B) at or near base of Mississippian System, (C) at or near base of
Devonian clastic sequence, (D) at base of Silurian System, (E) at base of Middle
Ordovician Series.

Nashville dome and southeastward toward the east edge of the basin. In parts of central Tennessee and northeastern Alabama, rocks of late Middle Devonian age rest on rocks of Middle Ordovician age.

In latest Devonian time or in Early Mississipian time gentle uplift occurred along the northern margin of the basin. Basal Mississippian conglomeratic sandstones, in part derived from the north, disconformably overlie fine-grained Upper Devonian rocks (Fig. 8B).

Finally, in Late Mississippian time or in very Early Pennsylvanian time, gentle uplift in the northwest part of the basin resulted in an erosional disconformity (Fig. 8A) between the Mississippian and Pennsylvanian Systems. The disconformity is most pronounced along a narrow belt trending south across eastern Ohio into West Virginia and possibly into southern West Virginia. In the northern part of the belt, rocks of Early Pennsylvanian age are locally channeled into rocks of Early Mississippian age. In some areas in the southeastern part of the basin also, especially in northern Alabama, Mississippian and Pennsylvanian rocks are separated by a disconformity.

In late Paleozoic time and probably in early Mesozoic time the Appalachian basin was profoundly affected by deformation commonly referred to as the Appalachian orogeny. It may have occurred in two phases, the earlier phase dominated by compressional forces and the later phase by tensional forces. The system of parallel folds that is characteristic of much of the area, as well as most of the longitudinal thrust faults, tear faults, and other high-angle faults were formed at this time. The duration of the orogeny cannot be accurately determined because of a large gap in the geologic record within the basin. No Paleozoic rocks younger than Early Permian, no Mesozoic rocks, and no Cenozoic strata except a thin spotty veneer of Quaternary sediments are present. Diastrophism associated with the Appalachian orogeny began no earlier than Early Permian time and may have extended at least into Late Triassic time. Downfaulting along parts of the east edge of the Appalachian basin, the rapid accumulation of Late Triassic red beds in the downfaulted basins, and the subsequent intrusion of many basaltic dikes into the Triassic rocks and into Paleozoic rocks in the eastern and northern parts of the Appalachian basin may represent the last phase of the revolution. However, these phenomena of Late Triassic age may represent a later disturbance—the Palisades disturbance of some workers.

Since the end of the Appalachian orogeny deformation apparently has been restricted to episodes of relatively mild epeirogenic uplift followed by periods of erosional planation.

STRATIGRAPHY

Upper Precambrian Stratified Sequence

Rocks of Precambrian age bordering the study area crop out (Figs. 1 and 3) in the Adirondack Mountains at the north end of the Appalachian basin and along much of the eastern margin of the basin. They are divided into two categories on the basis of stratigraphic position and lithology: the Precambrian basement complex and the Upper Precambrian stratified sequence. As the name implies, the Precambrian basement complex forms the floor or the basement of the Appalachian basin. The generalized configuration of the basement surface is shown on Figure 3. Contours in most of the central part of the basin are hypothetical and were drawn in part by extrapolation from the northern and western flanks of the basin where data were available, and in part by determination of the maximum stratigraphic thicknesses of Paleozoic rocks reported in the central and eastern parts. Where exposed, the complex consists largely of schists, gneisses, and a wide range of intrusive rocks such as granite, granodiorite, quartz syenite, gabbro, and anorthosite. Because a study of the crystalline rocks of the basement complex is not within the scope of this report, they will not be discussed further.

The stratigraphically higher Precambrian stratified sequence is a thick succession of interbedded metasedimentary and metavolcanic rocks unconformably overlying the basement complex. The gross relationships of the major units of the sequence are shown diagrammatically in Figure 9. Along the outcrop, the Precambrian stratified sequence ranges in thickness from a feather edge in eastern Pennsylvania, to probably more than 30,000 ft (9,150 m) in east-central Tennessee (King and others, 1958, p. 947).

The westward extent of the Late Precambrian stratified rocks along the floor of the Appalachian basin is unknown. However, King (1950, p. 13) presents evidence that the west edge of the Catoctin Greenstone coincides closely with the west side of the Catoctin–Blue Ridge anticlinorium. He attributes the rapid pinchout of the Catoctin to nondeposition and overlap as the Catoctin accumulated, and to truncation by erosion in post-Catoctin time. Bloomer and Werner (1955, Fig. 4) suggests that the Swift Run Formation, as well as the Catoctin, pinches out in the vicinity of the west flank of the Blue Ridge Mountains.

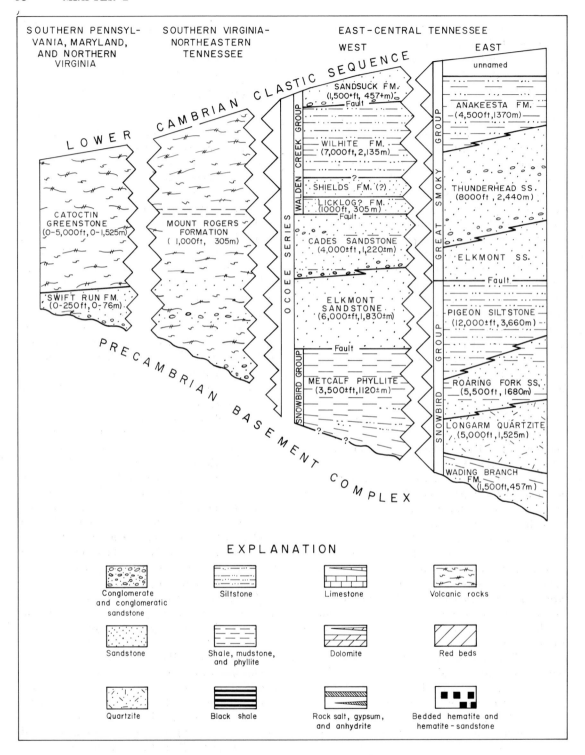

FIGURE 9. Schematic diagram showing relations of units included in Precambrian stratified sequence. Breaks between columns indicate uncertainty of lateral relationships. Approximate or average thicknesses in parentheses. The stratigraphic nomenclature used in this illustration is from many sources and does not necessarily conform to U. S. Geological Survey usage.

Lower Cambrian Clastic Sequence

The Lower Cambrian clastic sequence is here defined to include the oldest predominantly clastic materials of Cambrian age throughout the basin (Fig. 10).

In much of the basin the boundaries of the sequence delineate a blanket of quartzose and arkosic detritus laid down in the shallow waters of a northwest-transgressing Cambrian sea. It is a wedge-shaped mass (Fig. 11) that is thickest along the east margin of the basin, where its rocks are predominantly Early Cambrian in age, and thinnest along the west and north margins, where its rocks are predominantly Late Cambrian in age. An exception to the above age relationship occurs along the east side of the basin, from southern Virginia southward to central Alabama. In this area the sequence is defined to include in its upper part all the Middle Cambrian Series and the lowest formation of the Upper Cambrian Series. These rocks, which compose most of the Conasauga Group, consist of alternating formations of shale and limestone. They are included in the Lower Cambrian clastic sequence because shale is slightly preponderant over limestone. Shortly west of this belt the Conasauga Group is included in the overlying Cambrian–Ordovician sequence because the proportion of limestone to shale is reversed. This change in the upper boundary of the Lower Cambrian clastic sequence accounts in part for the high gradient of the thickness lines in the southeast part of the basin in Figure 11.

Along the west flank of the Blue Ridge, where the Lower Cambrian sequence is underlain by the upper Precambrian stratified sequence, the contact between the two is commonly conformable (Bloomer and Werner, 1955, p. 589–599; Cloos, 1958, p. 8). Locally, however, the contact is reportedly unconformable (King, 1950, p. 13–14). Where the Lower Cambrian sequence rests on the Precambrian basement complex—as it does throughout most of the basin—the contact is markedly unconformable.

Relationships between Lower Cambrian rocks, upper Precambrian stratified rocks, and rocks of the basement complex are very uncertain in the southeast part of the study area. Stose and Stose (1944, p. 412–413) state that the upper Precambrian Ocoee Series extends southwestward across Georgia into Alabama and is equivalent to the Talladega Series, which continues across Alabama to the southern end of the Piedmont region. However, the Tectonic map of the United States (U. S. Geol. Survey and Am. Assoc. Petroleum Geologists, 1961) shows metamorphic rocks of Paleozoic age occupying the southern part of the belt designated as Ocoee and Talladega by the Stoses. Griffin (1951, p. 43–48) summarized the data regarding the age of the Talladega Series in eastern Alabama and concluded that the upper part of the Talladega was of late Paleozoic age.

Thickness. The Lower Cambrian sequence is a wedge-shaped body (Fig. 11). The greatest thickness reported is in southwestern Virginia where Miller (1944, Table 1) showed approximately 10,000 ft (3,050 m) of Lower Cambrian strata. The sequence thins rapidly from the east edge of the basin to the northwest. In eastern Kentucky, however, the regional thickness gradient is noticeably altered; abrupt changes occur along the dislocation marked at the surface by the Kentucky River fault system. In deep wells in eastern Kentucky a thin sandstone sequence is present north of the fault zone and a thick sequence of shale, sandstone, and interbedded carbonate rock is present south of the fault zone. Although subsurface control is not sufficiently dense to be unequivocal, the thickness pattern suggests the possibility of right-lateral strike-slip displacement along the fault zone as suggested by Summerson (1962, p. 5) from a study of the configuration of the basement surface.

Cambrian–Ordovician Carbonate Sequence

A sequence composed largely of dolomite and limestone and a lesser volume of quartzose sandstone conformably overlies the Lower Cambrian clastic sequence in most of the Appalachian basin (Figs. 12 and 13). It is composed predominantly of rocks of Middle and Late Cambrian, and Early and Middle Ordovician age, and is here designated the Cambrian–Ordovician carbonate sequence.

The lower part of the sequence consists largely of interbedded dolomite, sandy dolomite, dolomitic quartz sandstone, and very pure quartz sandstone. The upper part of the carbonate sequence consists largely of limestone and impure argillaceous limestone. A thin unit of greenish shale—the Glenwood Shale—separates the lower dolomitic part of the sequence from the upper nondolomitic parts. The Glenwood marks the approximate position of the St. Peter Sandstone, which occurs in places along the east flank of the Cincinnati arch and extensively in areas west of the study area. Both of these thin units are associated with an unconformity, nearly basinwide in extent, at the base of the Middle Ordovician Series (Fig. 8E). The magnitude of the unconformity, which increases northwestward across

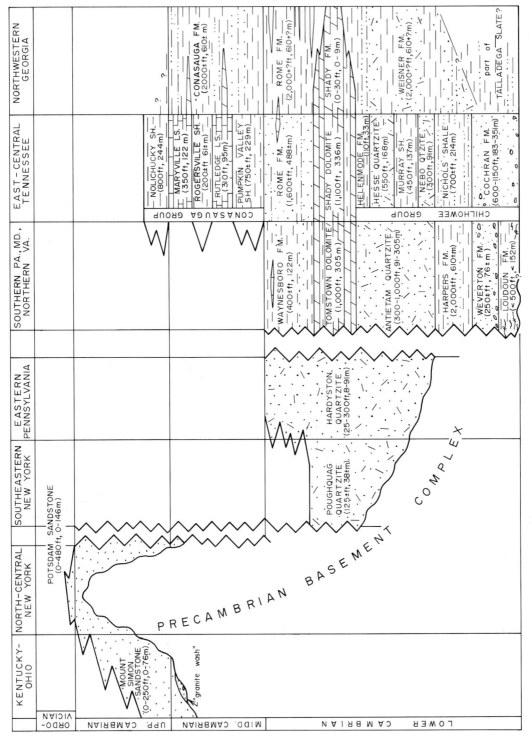

FIGURE 10. Schematic diagram showing relations of units included in Lower Cambrian clastic sequence. Breaks between columns indicate uncertainty of lateral relationships. Time boundaries apply only to Paleozoic rocks. Patterns are explained in Figure 9. The stratigraphic nomenclature used in this illustration is from many sources and does not necessarily conform to U. S. Geological Survey usage.

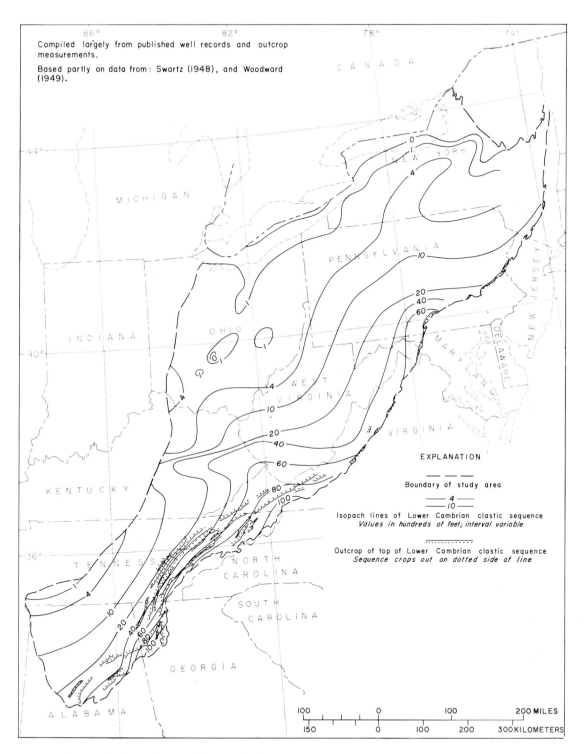

FIGURE 11. Isopach map of Lower Cambrian clastic sequence.

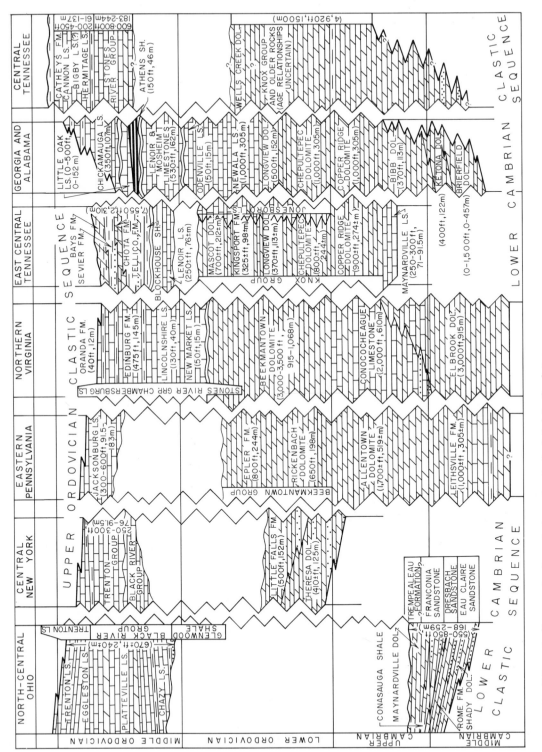

FIGURE 12. Schematic diagram showing relations of units included in Cambrian–Ordovician carbonate sequence. Breaks between columns indicate uncertainty of lateral relationships. Patterns are explained in Figure 9. The stratigraphic nomenclature used in this illustration is from many sources and does not necessarily conform to U. S. Geological Survey usage.

the basin, has recently been thoroughly documented in Ohio by Calvert (1962, 1963a, and 1964). On the crest of the Waverly arch of Woodward (1961, p. 1645–1646) in north-central Ohio, Middle Ordovician beds rest on Upper Cambrian beds. Recent studies in Pennsylvania by Wagner (1966) and in New York by Flagler (1966) show a similar truncation of successively older units beneath the same unconformity northward across these two states.

In eastern Tennessee and parts of contiguous states, thick wedges of noncarbonate clastic rock (largely the Tellico, Sevier, and Bays Formations) are included in the upper part of the Cambrian–Ordovician carbonate sequence.

Thickness. The Cambrian–Ordovician carbonate sequence ranges in thickness from about 600 ft (183 m) in northern New York to slightly more than 10,000 ft (3,050 m) in southeastern Tennessee. The axis of greatest thickness extends from southeast New York to northern Alabama approximately parallel to the east edge of the Appalachian basin (Fig. 13). It was suggested earlier (p. 13) that the juxtaposition of the axis of greatest thickness and the east edge of the basin, at least in Maryland, parts of Virginia, and eastern Tennessee, may have resulted from the northwestward displacement of the Blue Ridge complex over rocks, now hidden, that are laterally equivalent to the Cambrian–Ordovician carbonate sequence. From the axis of greatest thickness the sequence thins gradually westward toward the Findlay arch, Cincinnati arch, and Nashville dome. In eastern Pennsylvania, where the westward displacement of Precambrian rocks may have been much less, the sequence thins markedly eastward from the belt of thickest carbonate rocks.

Upper Ordovician Clastic Sequence

A sequence of predominantly noncalcareous clastic rocks conformably overlies the Cambrian–Ordovician carbonate sequence throughout most of the basin (Figs. 14, 15). It consists largely of shale (black to red), siltstone and sandstone (gray to red), and smaller amounts of limestone and quartz–pebble conglomerate. The bulk of the clastic sequence is Late Ordovician in age. In parts of the Appalachian basin, however, especially along the northeastern and eastern edges, some rocks of Middle Ordovician age are included, so that the boundary between the two sequences will coincide with major change in lithology from carbonate rocks below to noncarbonate rocks above. The Upper Ordovician clastic sequence thins rather uniformly to the

north, west, and southwest from its area of greatest thickness along the northeastern edge of the basin (Fig. 16). The most noticeable exception to this pattern is in northeastern Pennsylvania and southeastern New York where the sequence thins rapidly westward against the southwestern extension of the Adirondack axis (Fig. 3) then thickens slightly west of the axis. On Figure 16 the thickness lines in this area were drawn so that they abut because the writer could not connect them on the basis of the thickness data presently available. If an anomaly actually exists there it may be due to the lateral displacement of a large mass of pre-Silurian strata westward toward the axis.

The boundary between the Upper Ordovician clastic sequence and the underlying Cambrian–Ordovician carbonate sequence is conformable, and normally gradational, throughout the basin. The boundary with the overlying Silurian clastic sequence, however, is disconformable in the northeast and southwest parts of the basin (Fig. 8D), and is sharply defined in most other parts. Stratigraphically the disconformity is most pronounced south of the Adirondack uplift in eastern New York, where all of the Upper Ordovician and part of the Middle Ordovician Series are absent. Volumetrically the unconformity is greatest in eastern Pennsylvania and contiguous parts of New Jersey. Here the Upper Ordovician clastic sequence attains its greatest thickness, although the upper part of the Martinsburg Formation and two younger thick coarse-grained clastic units (the Oswego Sandstone and the Juniata Formation of central Pennsylvania) are absent.

The absence of the younger Upper Ordovician rocks along the eastern outcrop belts and the sharp angular unconformity at the contact of the Martinsburg with the next younger sequence are evidence of diastrophic activity during Late Ordovician or Early Silurian time, and suggest that the original thickness of the Upper Ordovician clastic sequence may have been appreciably greater than the 9,800–12,800 ft (2,987–3,900 m) (Drake and Epstein, 1967) still preserved.

The percentage of calcium carbonate in the Upper Ordovician sequence increases southwestward from eastern Tennessee across Georgia into northeastern Alabama. Near the south end of the study area, Butts (1926, p. 119–133) included all rocks of Late Ordovician age in the Chickamauga Limestone. The Chickamauga includes rocks of Middle and Late Ordovician age, as well as several internal unconformities. It is composed

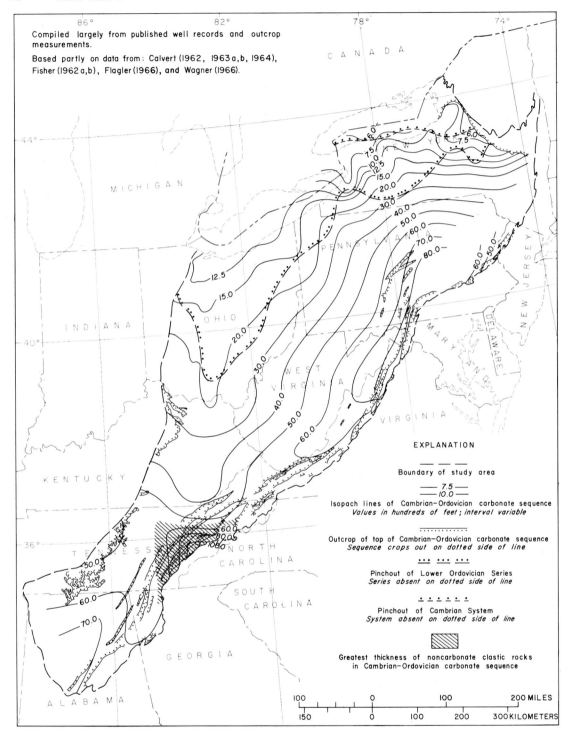

FIGURE 13. Isopach map of Cambrian–Ordovician carbonate sequence.

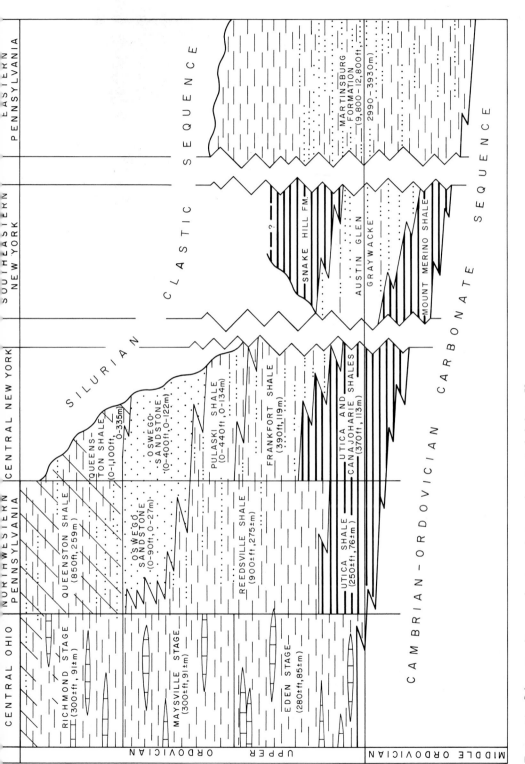

FIGURE 14. Schematic diagram showing relations of units included in Upper Ordovician clastic sequence in northern part of study area. Breaks between columns indicate uncertainty of lateral relationships. Patterns are explained in Figure 9. The stratigraphic nomenclature used in this illustration is from many sources and does not necessarily conform to U. S. Geological Survey usage.

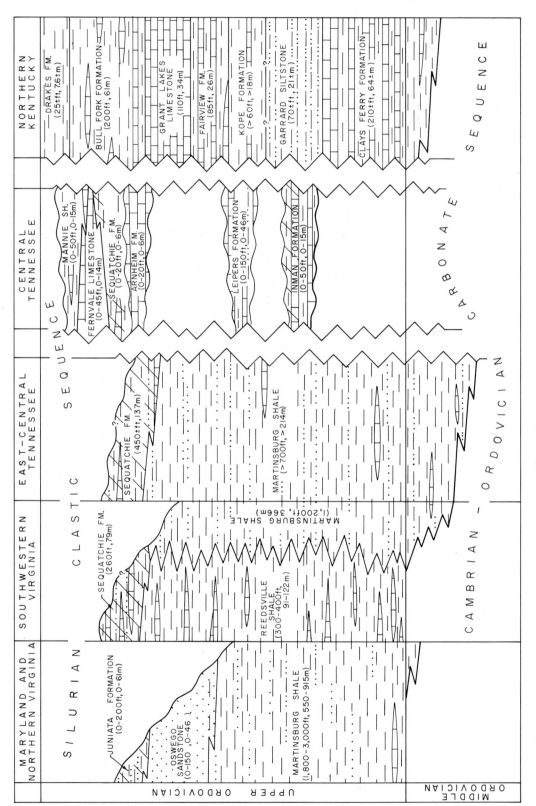

FIGURE 15. Schematic diagram showing relations of units included in Upper Ordovician clastic sequence in southern part of study area. Breaks between columns indicate uncertainty of lateral relationships. Patterns are explained in Figure 9. The stratigraphic nomenclature used in this illustration is from many sources and does not necessarily conform to U. S. Geological Survey usage.

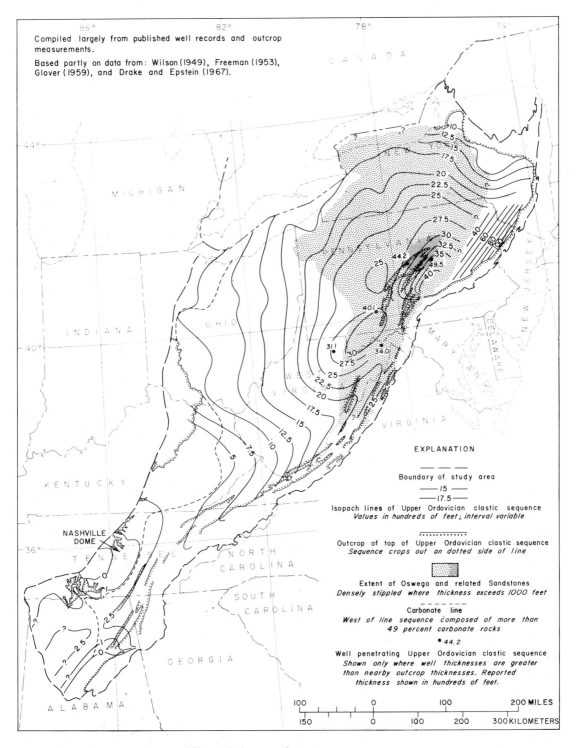

Compiled largely from published well records and outcrop measurements.

Based partly on data from: Wilson (1949), Freeman (1953), Glover (1959), and Drake and Epstein (1967).

EXPLANATION

— — — —
Boundary of study area

——— 15 ———
———17.5———
Isopach lines of Upper Ordovician clastic sequence
Values in hundreds of feet; interval variable

....
Outcrop of top of Upper Ordovician clastic sequence
Sequence crops out on dotted side of line

Extent of Oswego and related Sandstones
Densely stippled where thickness exceeds 1000 feet

— — — —
Carbonate line
West of line sequence composed of more than 49 percent carbonate rocks

• 44.2
Well penetrating Upper Ordovician clastic sequence
Shown only where well thicknesses are greater than nearby outcrop thicknesses. Reported thickness shown in hundreds of feet.

FIGURE 16. Isopach map of Upper Ordovician clastic sequence.

largely of medium- to thick-bedded fine-grained blue limestone. An unconformity at the top of the Chickamauga results in the absence of rocks equivalent to the Oswego, Sequatchie, Juniata, and Queenston. The unconformity increases in magnitude to the southeast. Data (Butts, 1940) near the south end of the basin indicate that rocks of late Middle Ordovician, Late Ordovician, Silurian, and Early Devonian ages are locally absent near the east edge of the study area (Fig. 16). In this region the Middle Ordovician Little Oak Limestone is unconformably overlain by sandstone of Early or Middle Devonian age.

The Upper Ordovician strata are difficult to trace from the outcrop belt in northeast Alabama into the subsurface in northwest Alabama. Sample studies of wells drilled for oil and gas in the northwest part of Alabama (McGlamery, 1955) indicate that the Chattanooga Shale of Devonian age is underlain by 100–400 ft (30.5–122 m) of interbedded greenish gray to red limestone and sandy limestone, green and red shale, and some fine-grained sandstone. This sequence was considered by McGlamery to be Silurian in age. The Silurian rocks are underlain by a thick sequence of gray limestone designated by the same writer as Ordovician in age. If no unconformity separates these strata, part of the gray limestone sequence may be Late Ordovician in age. If an unconformity is present, none of the gray limestone may be Late Ordovician in age. Because the exact relationships of these rocks are not known, the isopach lines in northern Alabama (Fig. 16) are generalized and largely speculative.

Silurian Clastic Sequence

The Late Ordovician clastic sequence is overlain by a thin sequence of predominantly clastic rocks (Fig. 17), mainly of Early Silurian age, that extends throughout most of the Appalachian basin (Fig. 18). Like the underlying Late Ordovician sequence, the Early Silurian clastic sequence is thickest (about 2,600 ft, 793 m) and coarsest grained in the northeastern part of the basin where it consists largely of sandstone and conglomerate. It thins to the north, west, and southwest and wedges out entirely against the east side of the Nashville dome. In the central part of the basin it is composed largely of shale, siltstone, and sandstone; along the western part of the Appalachian basin the sequence contains much sandy limestone and dolomite. Bedded oölitic hematite and beds of hematitic sandstone or quartzite are characteristic of the sequence throughout much of its extent.

The Silurian clastic sequence is noticeably absent in eastern New York and in several areas in the southwestern part of the basin. In New York its abrupt pinchout south of the Adirondack uplift is well illustrated by Fisher (1959). It apparently was caused by depositional thinning against the northern part of the Adirondack axis, and by many intervals of erosion during the Silurian Period. Similar conditions apparently existed in northern Alabama, northwestern Georgia, and central Tennessee, although erosion may have continued well into Devonian time.

Except in the northeast and southwest parts of the basin the sequence is conformable on the underlying sequence, and in most of the basin it is conformable with the overlying Silurian–Devonian carbonate sequence. However, in eastern New York the Early Silurian clastic sequence is disconformably overlain by the Silurian–Devonian carbonate sequence in areas peripheral to its pinchout (Fig. 18) against the Adirondack axis. In parts of northeastern Alabama, northwestern Georgia, and southern Tennessee the Silurian clastic sequence is sharply disconformable with the younger Devonian clastic sequence, and the intervening Silurian–Devonian carbonate sequence is absent.

Some of the most productive oil and gas sands in the Appalachian basin occur in the Silurian clastic sequence. Among them are the "Clinton sands" of eastern Ohio, the "Medina sands" of the Albion Group in the western half of New York and the northwest corner of Pennsylvania, the "Big Six" sand in parts of eastern Kentucky and southwestern West Virginia, and the Tuscarora Sandstone in parts of western West Virginia. The Tuscarora and equivalent rocks are largely untested potential reservoirs of gas in much of Pennsylvania, south-central New York, southeastern Ohio, and much of West Virginia.

The thickness and stratigraphic relations of the sequence in the subsurface are uncertain in some areas in the southern half of the basin. For example, wells drilled in northern Alabama (McGlamery, 1955) penetrated interbedded shale, siltstone, sandstone, and limestone overlying limestone of Ordovician age and underlying black shale of Devonian age. The content of limestone increases northward toward the Alabama–Tennessee border. Part of the heterogeneous sequence encountered in north-central Alabama probably is equivalent to the Red Mountain Formation, and consequently is included in the Early Silurian sequence of this report. As a result some of the thickness lines (Fig. 18) in the southern part of the study area are to be regarded only as an

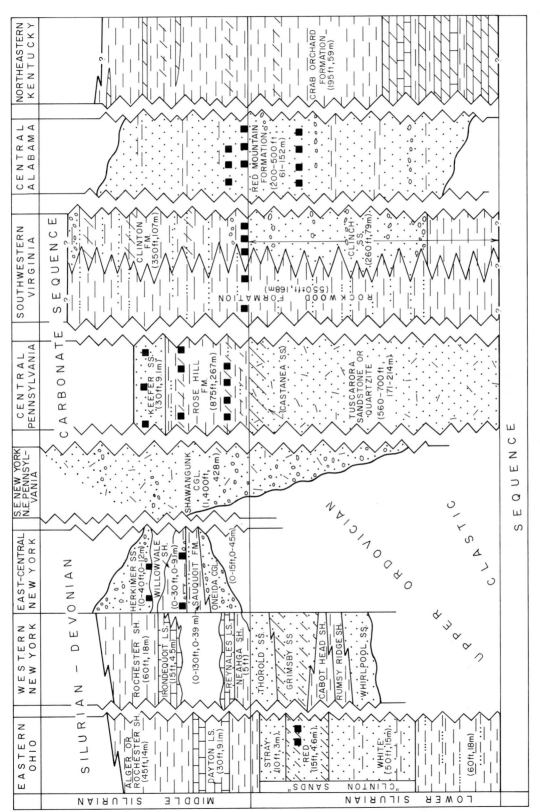

FIGURE 17. Schematic diagram showing relations of units included in Silurian clastic sequence. Breaks between columns indicate uncertainty of lateral relationships. Time boundaries apply only to Silurian rocks. Patterns are explained in figure 9. The stratigraphic nomenclature used in this illustration is from many sources and does not necessarily conform to U. S. Geological Survey usage.

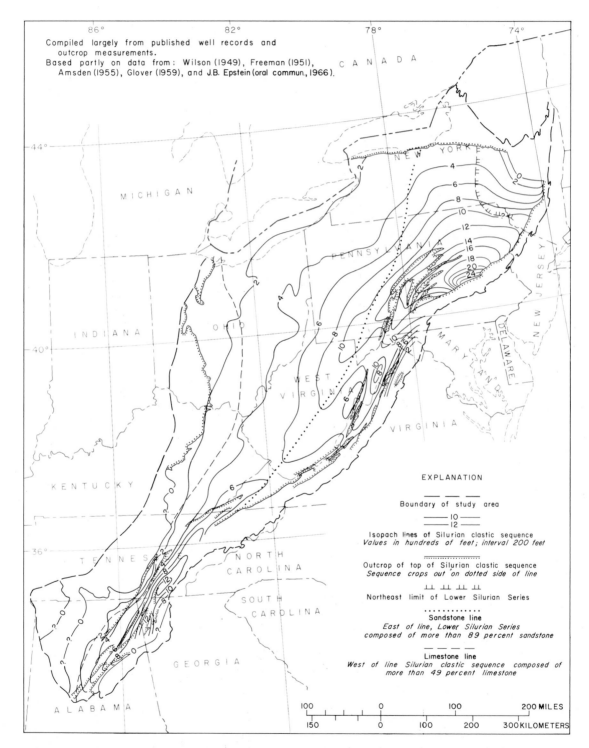

FIGURE 18. Isopach map of Silurian clastic sequence.

approximation of the configuration of the Silurian clastic sequence.

Silurian–Devonian Carbonate Sequence

A moderately thick succession of carbonate rocks of Middle and Late Silurian age, Early Devonian age, and some of early Middle Devonian age (Fig. 19) underlies much of the Appalachian basin except the periphery and much of the southern part. Like the underlying sequences it is wedge-shaped, thick on the east and thin on the west (Fig. 20). It is thickest—approximately 3,300 ft (1,005 m)—in northeastern Pennsylvania. The carbonate sequence thins toward the southern part of the Appalachian basin, and is absent in part of Kentucky, most of Tennessee, all of Georgia, and much of Alabama. In northeastern Pennsylvania and adjacent parts of New Jersey and New York the lower half of the Silurian–Devonian carbonate sequence is a thick wedge of red sandstone, siltstone, and shale, constituting the Bloomsburg Red Beds and the High Falls Formation. To the north, west, and southwest the red-bed facies grades into an alternating succession of shale and siltstone (gray, red, and brown), limestone and dolomite (gray and tan), and rock salt, gypsum, and anhydrite. Toward the margins of the sequence dolomite and limestone take the place of most of the noncarbonate rocks.

Over more than half the geographic extent of the sequence, a thin unit of orthoquartzite—the Oriskany or Ridgeley Sandstone—is present in the upper part.

The Silurian–Devonian carbonate sequence conformably overlies the Silurian clastic sequence. In turn it is overlain by a predominantly clastic sequence of Middle and Upper Devonian rocks. Normally the upper boundary is conformable, but locally in northeastern Kentucky it may be disconformable.

Of particular interest in the isopach map (Fig. 20) are the two axes of thin strata extending from southeastern New York to south-central Pennsylvania, and from westernmost New York to central Maryland. Both are generally substantiated by surface and subsurface data. The first, which coincides with the Adirondack axis, has been previously traced by Woodward (1957, p. 1431) by the wedgeout of the Oriskany Sandstone along its flanks. Except in the northern part of the first axis, both axes seem to be caused by the gradual thinning of many of the constituent units of the sequence, and not by a single large hiatus in the column.

The disappearance of the sequence toward the southwest is due in part to thinning and wedging out of some of the formations in the sequence, and to the intertonguing of thick-bedded facies with thin-bedded facies. In addition, the exclusion of a relatively small volume of clastic rocks equivalent in age to the upper part of the carbonate sequence elsewhere accentuates the southwestward pinchout of the sequence. For reasons of convenience discussed later, these clastic rocks are included in the overlying Devonian clastic sequence.

From western Maryland (section shown in Fig. 19), the sequence thins southwestward along the outcrop belts in West Virginia and Virginia. In the northern part of these two states the sequence includes virtually the same units present in Maryland, and ranges from 500 to 2,000 ft (153–610 m) in thickness. In the southern part of these states and in the contiguous part of Tennessee the sequence consists of no more than two formations. The lower, the Hancock Dolomite or Limestone, consists largely of dolomite to the west (Miller and Fuller, 1954, p. 155–158), and interbedded limestone, dolomite, siltstone, and sandstone to the east (Harris and Miller, 1958). In general it ranges in thickness from 19 to 190 ft (5.8–58 m). Locally the Hancock is overlain by the Wildcat Valley Sandstone (Miller and others, 1964) which consists primarily of fossiliferous calcareous quartz sandstone as much as 45 ft (14 m) thick.

Locally in northwestern Georgia and eastern Alabama a thin succession of sandstone and some chert, equivalent in age to the upper part of the carbonate sequence elsewhere, is present. Included are the Clear Branch Sandstone, Frog Mountain Sandstone, Ragland Sandstone, and the Armuchee Chert. These units crop out in isolated areas and have not been traced laterally into the other units of the Silurian–Devonian carbonate sequence. In addition they are noncarbonate rocks, so they are more conveniently included in the overlying Devonian clastic sequence.

A thin unit of argillaceous limestone and calcareous shale is locally present in northwestern Alabama and in contiguous counties in southern Tennessee. The rocks occupy the interval between the top of the Osgood Shale of McGlamery (1955) (the upper unit of the underlying Silurian clastic sequence) and the base of the Chattanooga Shale (the basal unit of the Devonian clastic sequence). In this area rocks of the carbonate sequence may total as much as 225 ft (69 m) in thickness, but generally are much

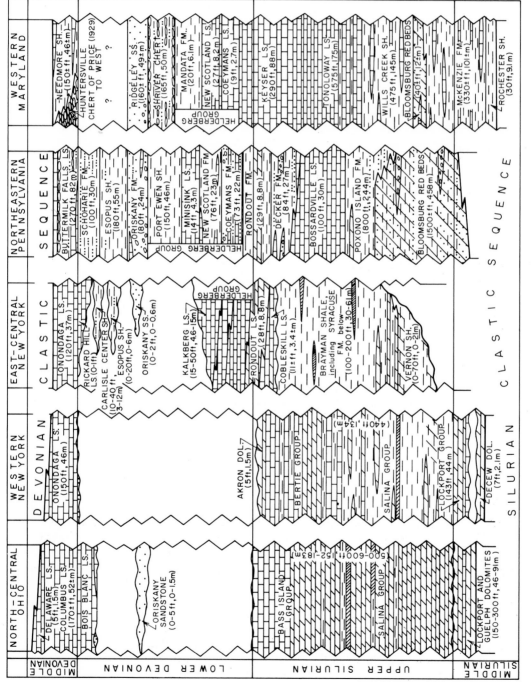

FIGURE 19. Schematic diagram showing relations of units included in Silurian–Devonian carbonate sequence. Breaks between columns indicate uncertainty of lateral relationships. Patterns are explained in Figure 9. The stratigraphic nomenclature used in this illustration is from many sources and does not necessarily conform to U. S. Geological Survey usage.

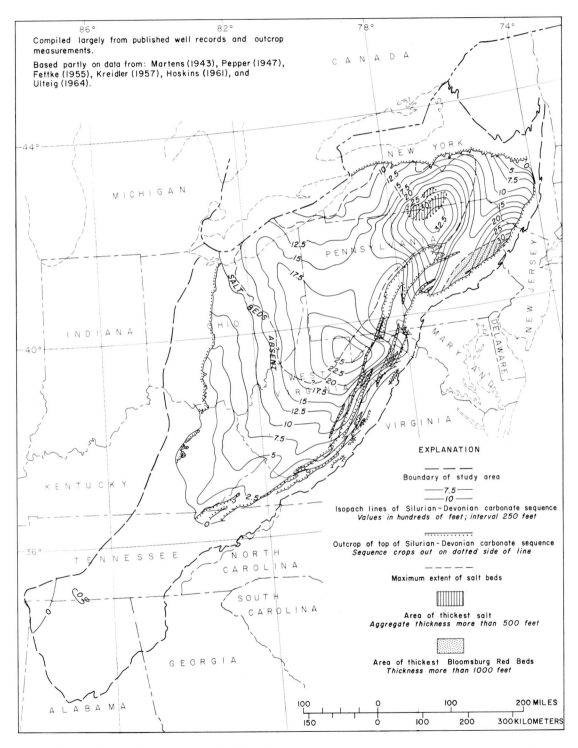

FIGURE 20. Isopach map of Silurian–Devonian carbonate sequence.

thinner. They may represent an accumulation of predominantly fine-grained calcareous rocks preserved within embayments on the south flank of the Nashville dome. With these exceptions, the Silurian–Devonian carbonate sequence is absent elsewhere in the southern part of the Appalachian basin.

Devonian Clastic Sequence

The Devonian clastic sequence is a moderately thick sequence composed largely of shale, mudrock, siltstone, and sandstone, primarily of Middle and Late Devonian age (Fig. 21). It extends throughout most of the Appalachian basin, but is absent in the vicinity of Jefferson County, Ala., and in a narrow belt in part of north-central Alabama and south-central Tennessee (Fig. 22). In all but the southern part of the basin, where locally it is in unconformable contact with the Silurian clastic sequence or the older Upper Ordovician and Cambrian–Ordovician sequences, the Devonian clastic sequence overlies the Silurian–Devonian carbonate sequence. The contact is conformable in most areas, but is commonly sharply defined. Throughout almost all of its extent the Devonian sequence is overlain by rocks of the Mississippian system. Local exceptions, however, occur in New York and Pennsylvania along the western half of their common border that closely follows the 42nd parallel. In the northern part of the basin the upper boundary is placed at the base of the lowest Mississippian formation that is noticeably coarser grained than the bulk of the rocks below. In various areas this formation is the Berea, Cussewago, or Corry Sandstone, the Pocono Formation or Group, or the Price Sandstone or Formation. The contact between the two sequences is normally either conformable but abrupt, or slightly disconformable (Fig. 8B). In the southern part of the basin the upper boundary is placed at the top of the black Chattanooga Shale and at the base of the lowest formation entirely of Mississippian age. In various areas this formation is the Maury Formation (gray calcareous shale), the Fort Payne Chert or Formation (gray limestone and chert), or the Grainger Shale (gray shale and siltstone). The contact between the Devonian and Mississippian sequences is conformable in most of the southern Appalachians.

Only in the subsurface in eastern Kentucky is there appreciable difficulty in recognizing and defining the upper boundary of the sequence. Consequently, the thickness lines on Figure 22 in this area should be considered only an approximation of the true configuration of the sequence. Most of the data used are from well-sample studies by Freeman (1951). In the shallow subsurface on the east flank of the Cincinnati arch, the sequence is easily defined by the top and bottom of the New Albany or Chattanooga Shale. Farther east the sequence was arbitrarily defined to extend from the base of the New Albany to the base of the lowest siltstone or sandstone reported by Freeman in the succession designated by her as New Providence (Mississippian) or as Lower Mississippian. In the easternmost corner of Kentucky the top of the Devonian clastic sequence is marked by the base of the siltstone or sandstone designated as the Berea by Freeman.

The Devonian clastic sequence is wedge-shaped, being thickest near the east margin of the Appalachian basin and thinnest near the west margin; its thickness diminishes to a feather edge in parts of Alabama and Tennessee. Relatively coarse-grained sedimentary rocks, including red rocks deposited in a subaerial environment, predominate in the northeastern part of the basin where the sequence is thickest, more than 10,000 ft (3,050 m). Medium-grained gray rocks deposited in a marine environment predominate where the sequence is of intermediate thickness. Fine-grained black shale and calcareous gray shale predominate in the southwestern part of the Appalachian basin where the sequence is thinnest.

Mississippian Sequence

The Devonian clastic sequence is overlain by a relatively thin succession of rocks entirely of Mississippian age (Fig. 23)—the Mississippian sequence of this paper. The contact between the two is conformable in most of the study area, but is slightly disconformable (Fig. 8B) in the northwestern part and along much of the eastern margin. The Mississippian sequence is, in turn, overlain by rocks of Pennsylvanian age. The contact between Mississippian and Pennsylvanian rocks is commonly disconformable, especially in the western and northern parts of the area (Fig. 8A).

Like the underlying sequences, the Mississippian sequence is grossly wedge-shaped (Fig. 24), but areas of anomalously thin rock in eastern Ohio and western West Virginia and an abrupt wedgeout in northern Pennsylvania significantly modify the wedge shape. Anomalous thicknesses in both areas are largely due to erosion in Late Mississippian or Early Pennsylvanian time. The greatest reported thickness known to the writer is in southwestern Virginia, where Averitt (1941) reported more than 6,800 ft (2,075 m) of Mississippian strata. A thickness of more than 6,000 ft (1,830 m) in eastern Pennsylvania has also been reported by Wood and others (1962, p. C39).

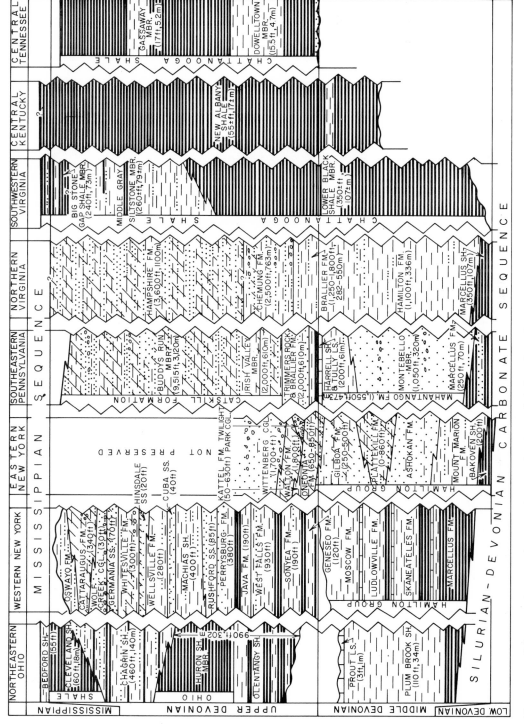

FIGURE 21. Schematic diagram showing relations of units included in Devonian clastic sequence. Breaks between columns indicate uncertainty of lateral relationships. Patterns are explained in Figure 9. The stratigraphic nomenclature used in this illustration is from many sources and does not necessarily conform to U. S. Geological Survey usage.

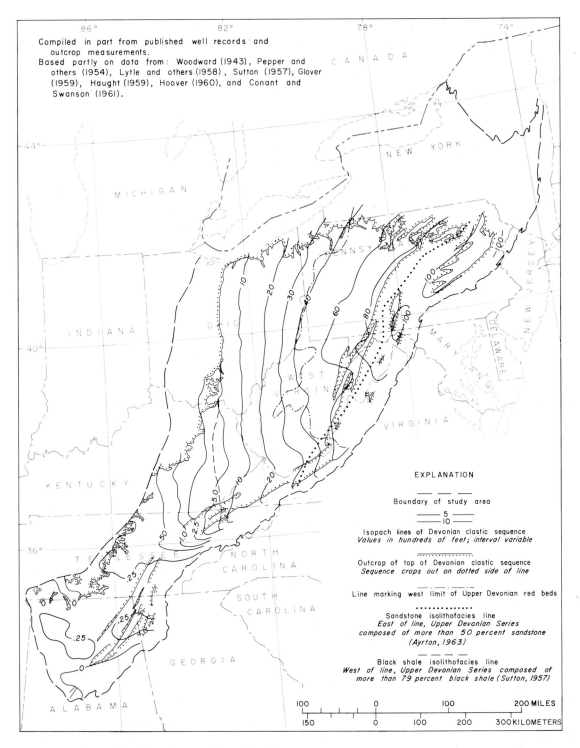

FIGURE 22. Isopach map of Devonian clastic sequence.

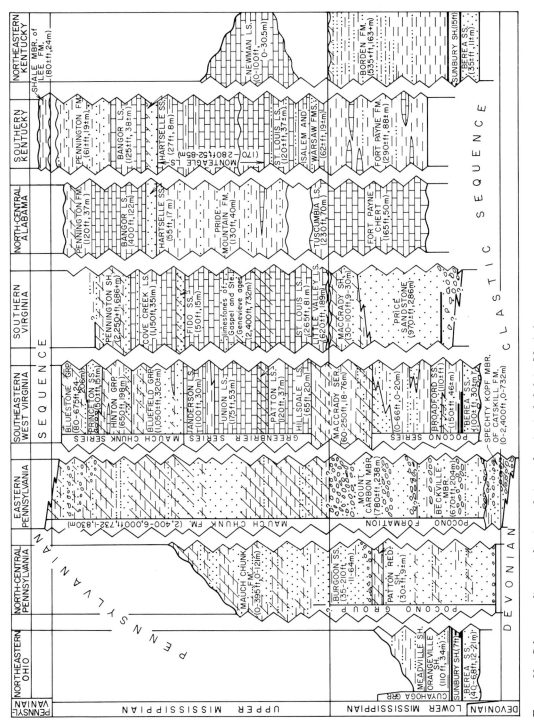

FIGURE 23. Schematic diagram showing relations of units included in Mississippian sequence. Breaks between columns indicate uncertainty of lateral relationships. Patterns are explained in figure 9. The stratigraphic nomenclature used in this illustration is from many sources and does not necessarily conform to U. S. Geological Survey usage.

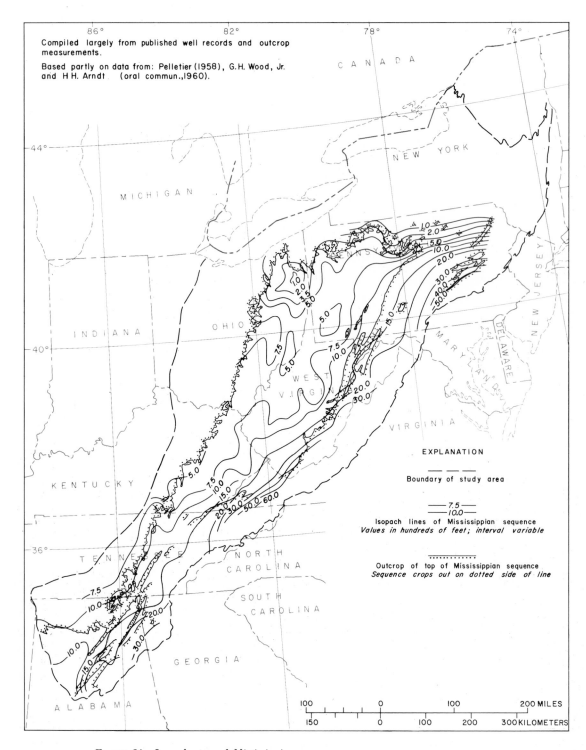

Compiled largely from published well records and outcrop measurements.

Based partly on data from: Pelletier (1958), G.H. Wood, Jr. and H.H. Arndt (oral commun.,1960).

EXPLANATION

———————
Boundary of study area

——— 7.5 ———
——— 10.0 ———
Isopach lines of Mississippian sequence
Values in hundreds of feet; interval variable

·················
Outcrop of top of Mississippian sequence
Sequence crops out on dotted side of line

FIGURE 24. Isopach map of Mississippian sequence.

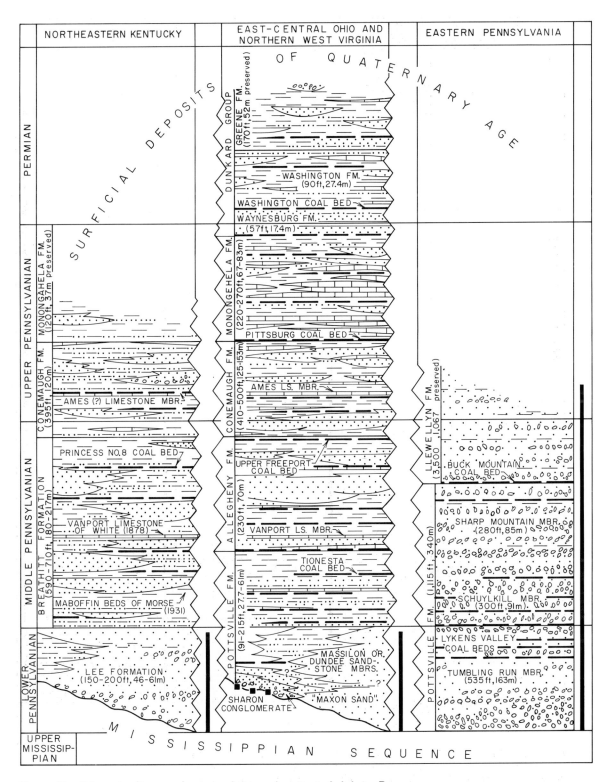

FIGURE 25. Schematic diagram showing relations of units included in Pennsylvanian sequence in northern part of study area. Breaks between columns indicate uncertainty of lateral relationships; solid bars indicate relative thickness of sections. Patterns are explained in Figure 9. The stratigraphic nomenclature used in this illustration is from many sources and does not necessarily conform to U. S. Geological Survey usage.

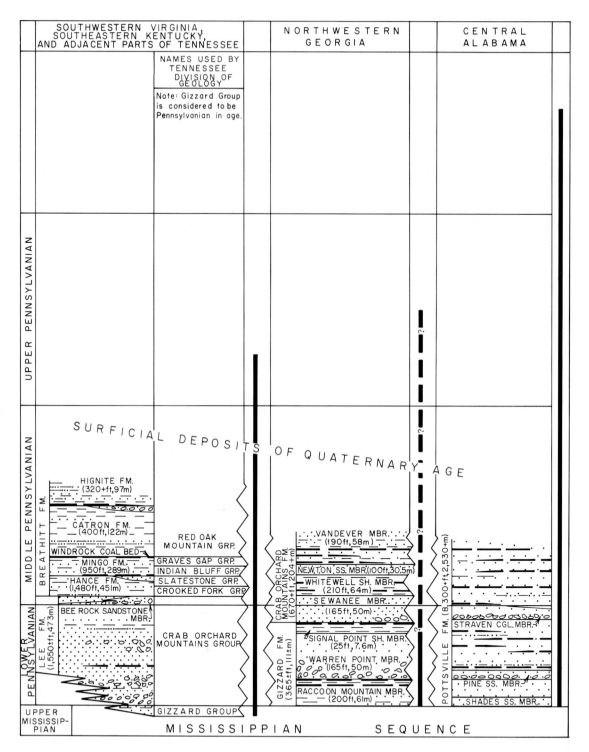

FIGURE 26. Schematic diagram showing relations of units included in Pennsylvanian sequence in southern part of study area. Breaks between columns indicate uncertainty of lateral relationships; solid bars indicate relative thickness of sections, scale same as Figure 25. Patterns are explained in Figure 9. The stratigraphic nomenclature used in this illustration is from many sources and does not necessarily conform to U. S. Geological Survey usage.

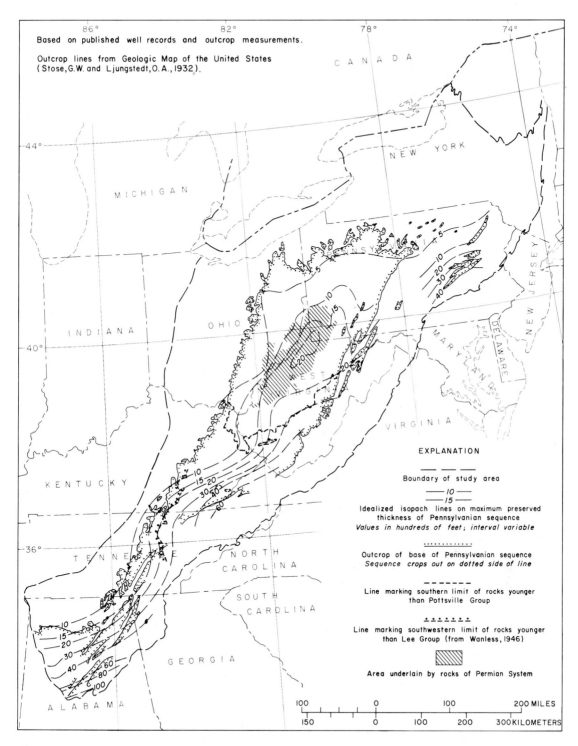

FIGURE 27. Generalized isopach map of maximum preserved thickness of Pennsylvanian sequence. Sequence portrayed as if thickness were unaffected by stream
erosion.

In general the constituent units are coarsest grained along the eastern margin, where the sequence is thickest. Red and tan sandstone, conglomerate, and shale predominate in the northeast part of the study area; gray shale and siltstone in the northwest part; interbedded sandstone and limestone in the east-central and southeast parts; and limestone and shale in the southwest part.

Many of the rocks in the Mississippian sequence are highly productive of oil and gas. Among the better known productive units are the Berea Sand, the Weir, Big Injun, and Squaw sands, the Murrysville or "Gas" sand and the Greenbriar or "Big Lime" in the northern half of the basin, and the Fort Payne Chert and Princeton Sandstone in the southern half.

Pennsylvanian Sequence

The Mississippian sequence is overlain by the Pennsylvanian sequence—here defined to include all Paleozoic rocks of post-Mississippian age preserved in the basin today (Figs. 25 and 26). In much of the basin the basal boundary is marked by an abrupt change from thin-bedded relatively fine-grained rocks below to massively bedded conglomeratic (quartz–pebble) sandstones above. Most of the sequence consists of rocks of the Pennsylvanian System, but it also includes, in the north-central part of the study area (Fig. 27), remnants of the Permian System. The two systems are conveniently grouped together because there is no marked lithologic change at their common boundary, and because there are gross lithologic similarities. Both are composed largely of relatively thin, commonly lenticular, alternating units of sandstone, conglomerate, siltstone, shale, and mudstone; smaller volumes and thinner beds of claystone, limestone, and coal are also present. In addition, both systems contain some red beds.

The Pennsylvanian sequence originally constituted a much thicker and more extensive wedge of clastic rocks, nearly three-quarters of which has been removed since Paleozoic time. At no point in the basin is the complete original thickness of the sequence preserved. The preserved lower parts of the sequence are restricted to the central part of the Appalachian basin from Pennsylvania to Alabama, and to many of the deeper synclines of the Valley and Ridge province along the east side of the basin (Fig. 26). Calculations based on published geologic sections extending the width of the basin, from western Ohio to eastern Pennsylvania, indicate 72 percent of the sequence originally present in these states

has since been removed. Highly preliminary estimates suggest that this figure may also be approximately correct for the southern part of the study area. For these measurements the writer assumed that (1) the sequence originally extended across the full width of the basin, (2) thickness gradients established for the preserved lower part of the sequence are proportional to those of the entire original sequence, and (3) no rocks younger than the youngest Permian rocks now preserved were ever present.

Disregarding local variations, the Pennsylvanian sequence is thickest and coarsest grained along the eastern side of the basin. In eastern Pennsylvania, where the upper half of the Pennsylvanian System and all of the Permian System are absent, Wood and others (1962) report a thickness of about 4,600 ft (1,400 m). Sandstone, conglomeratic sandstone, and conglomerate predominate. In central Alabama, where only the Lower Pennsylvanian Series is preserved, Butts (1940) reported a thickness of nearly 9,200 ft (2,800 m). Most of the section is composed of sandstone. In the western part of the basin the sequence is normally less than 1,000 ft (305 m) thick and is composed on the average of about 30 percent sandstone (including conglomerate), 60 percent shale (including claystone), and 10 percent limestone and coal. The above relations between thickness and composition are grossly similar in type to those of several of the older clastic sequences. Unlike the other clastic sequences, however, the Pennsylvanian sequence contains some coarse-grained conglomerates along its northern and northwestern outcrop belts. The areal distribution of these are convincing evidence of a northern source area—probably the Canadian shield and possibly the area of the Adirondack uplift.

REFERENCES

American Association of Petroleum Geologists and U. S. Geological Survey, 1967, Basement map of North America between latitudes 24 degrees and 60 degrees N: Washington, D. C., U. S. Geol. Survey, scale 1:5,000,000.

Amsden, T. W., 1955, Lithofacies map of Lower Silurian deposits in central and eastern United States and Canada: Am. Assoc. Petroleum Geologists Bull., v. 39, no. 1, p. 60–74.

Averitt, Paul, 1941, The Early Grove gas field, Scott and Washington Counties, Virginia: Virginia Geol. Survey Bull. 56, 50 p.

Ayrton, W. G., 1963, Isopach and lithofacies map of the Upper Devonian of northeastern United States, *in* Shepps, V. C., ed., Symposium on Middle and Upper Devonian stratigraphy of Pennsylvania and adjacent states: Pennsylvania Geol. Survey, 4th ser., Bull. G–39, p. 3–6.

Bloomer, R. O., and Werner, H. J., 1955, Geology of the Blue Ridge region in central Virginia: Geol. Soc. America Bull., v. 66, no. 5, p. 579–606.

Butts, Charles, 1926, The Paleozoic rocks, *in* Geology of Alabama: Alabama Geol. Survey Spec. Rept. 14, p. 40–230.

————, 1940, Description of the Montevallo and Columbiana quadrangles [Alabama]: U. S. Geol. Survey Geol. Atlas, Folio 226, 20 p.

Calvert, W. L., 1962, Sub-Trenton rocks from Lee County, Virginia, to Fayette County, Ohio: Ohio Geol. Survey Rept. Inv. 45, 57 p.

————, 1963a, A cross section of sub-Trenton rocks from Wood County, West Virginia to Fayette County, Illinois: Ohio Div. of Geol. Survey Rept. Inv. 48, 33 p.

————, 1963b, Sub-Trenton rocks of Ohio in cross sections from West Virginia and Pennsylvania to Michigan: Ohio Div. Geol. Survey Rept. Inv. 49, 5 p.

————, 1964, Sub-Trenton rocks from Fayette County, Ohio to Brant County, Ontario: Ohio Div. Geol. Survey Rept. Inv. 52, 7 p.

Clarke, F. W., 1924, The data of geochemistry: 5th ed., U. S. Geol. Survey Bull. 770, 841 p.

Cloos, Ernst, 1958, Structural geology of South Mountain and Appalachians in Maryland, Guidebooks 4–5: Johns Hopkins Univ. Studies in Geology, no. 17, 85 p.

Cohee, G. V., and Landes, K. K., 1958, Oil in the Michigan basin, *in* Weeks, L. G., ed., Habitat of oil—a symposium: Tulsa, Okla., Am. Assoc. Petroleum Geologists, p. 473–493.

Colton, G. W., 1961, Geologic summary of the Appalachian basin: U. S. Geol. Survey open-file report TEI–791, 121 p.

Conant, L. C., and Swanson, V. E., 1961, Chattanooga shale and related rocks of central Tennessee and nearby areas: U. S. Geol. Survey Prof. Paper 357, 91 p.

Cooper, B. N., 1939, Geology of the Draper Mountain area, Va.: Virginia Geol. Survey Bull. 55, 98 p.

Drake, A. A., Jr., and Epstein, J. B., 1967, The Martinsburg Formation (Middle and Upper Ordovician) in the Delaware Valley, Pennsylvania–New Jersey: U. S. Geol. Survey Bull. 1244–H, p. H1–H16.

Fettke, C. R., 1955, Preliminary report, occurrence of rock salt in Pennsylvania: Pennsylvania Geol. Survey, 4th ser., Progr. Rept. 145.

Fisher, D. W., 1959, Correlation of the Silurian rocks in New York State: New York State Mus., Geol. Survey Map and Chart ser. [no. 1], with text.

————, 1962a, Correlation of the Cambrian rocks in New York State: New York State Mus. Map and Chart ser., no. 2.

————, 1962b, Correlation of the Ordovician rocks in New York State: New York State Mus. Map and Chart ser., no. 2.

Flagler, C. W., 1966, Subsurface Cambrian and Ordovician stratigraphy of the Trenton Group—Precambrian interval in New York State: New York State Mus. Map and Chart ser., no. 8, 57 p.

Freeman, L. B., 1951, Regional aspects of Silurian and Devonian stratigraphy in Kentucky: Kentucky Geol. Survey, ser. 9, Bull. 6, 565 p.

————, 1953, Regional subsurface stratigraphy of the Cambrian and Ordovician in Kentucky and vicinity: Kentucky Geol. Survey, ser. 9, Bull., no. 12, 352 p.

Glover, Lynn, 1959, Stratigraphy and uranium content of the Chattanooga shale in northeastern Alabama, northwestern Georgia, and eastern Tennessee: U. S. Geol. Survey Bull. 1087–E, p. 133–168.

Goldschmidt, V. M., 1933, Grundlagen der quantitativen Geochemie: Fortschrift Mineral. Krist. Petrog., v. 17, p. 112–156.

————, 1954, Geochemistry: Oxford, Clarendon Press, 730 p.

Griffin, R. H., 1951, Structure and petrography of the Hillabee sill and associated metamorphics of Alabama: Alabama Geol. Survey Bull. 63, 74 p.

Harris, L. D., and Miller, R. L., 1958, Geology of the Duffield quadrangle, Virginia: U. S. Geol. Survey Geol. Quad. Map GQ–111.

Haught, O. L., 1959, Oil and gas in southern West Virginia: West Virginia Geol. Survey Bull., no. 16, 39 p.

Hoover, K. V., 1960, Devonian–Mississippian shale sequence in Ohio: Ohio Div. Geol. Survey Inf. Circ. 27, 154 p.

Hoskins, D. M., 1961, Stratigraphy and paleontology of the Bloomsburg Formation of Pennsylvania and adjacent states: Pennsylvania Geol. Survey, 4th ser., Bull. G–36, 125 p.

Kay, G. M., 1951, North American geosynclines: Geol. Soc. America Mem. 48, 143 p.

King, P. B., 1950, Geology of the Elkton area, Virginia: U. S. Geol. Survey Prof. Paper 230, 82 p.

————, Hadley, J. B., Neuman, R. B., and Hamilton, W. B., 1958, Stratigraphy of the Ocoee Series, Great Smoky Mountains, Tennessee and North Carolina: Geol. Soc. America Bull., v. 69, p. 947–966.

Kreidler, W. L., 1957, Occurrence of Silurian salt in New York State: New York State Mus. and Sci. Service Bull., no. 361, 56 p.

Krynine, P. D., 1948, The megascopic study and field classification of sedimentary rocks: Jour. Geol. v. 56, p. 156.

Kuenen, P. H., 1941, Geochemical calculations concerning the total mass of sediments in the earth: Am. Jour. Sci., v. 239, no. 3, p. 161–190.

Kulp, J. L., 1961, Geologic time scale: Science v. 133, no. 3459, p. 1105–1114.

Leith, C. K., and Mead, W. J., 1915, Metamorphic geology: New York, Henry Holt and Co., 337 p.

Lytle, W. S., Bergsten, J. M., Cate, A. S., and Wagner, W. R., 1958, Oil and gas developments in Pennsylvania in 1957: Pennsylvania Geol. Survey, Prog. Rept. 154, 46 p.

McGlamery, Winnie, 1955, Subsurface stratigraphy of northwest Alabama: Alabama Geol. Survey, Bull. 64, 503 p.

Martens, J. H. C., 1943, Rock salt deposits of West Virginia: West Virginia Geol. Survey Bull. 7, 67 p.

Miller, R. L., 1944, Geology of manganese deposits of the Glade Mountain district, Virginia: Virginia Geol. Survey Bull. 61, 150 p.

————, and Fuller, J. O., 1954, Geology and oil resources of the Rose Hill district—the fenster area of the Cumberland overthrust block, Lee County, Virginia: Virginia Geol. Survey Bull. 71, 383 p.

————, Harris, L. D., and Roen, J. B., 1964, The Wildcat Valley Sandstone (Devonian) of southwest Virginia, *in* U. S. Geol. Survey Prof. Paper 501–B, p. B49–B52.

Morse, W. C., 1931, Pennsylvanian invertebrate fauna: Kentucky Geol. Survey ser. 6, v. 36, p. 293–348.

Pelletier, B. R., 1958, Pocono paleocurrents in Pennsylvania and Maryland: Geol. Soc. America Bull., v. 69, no. 8, p. 1033–1063.

Pepper, J. F., 1947, Areal extent and thickness of the salt deposits of Ohio: Ohio Jour. Sci., v. 47, p. 225–239.

_____, de Witt, Wallace, Jr., and Demarest, D. F., 1954, Geology of the Bedford shale and Berea sandstone in the Appalachian basin: U. S. Geol. Survey Prof. Paper 259, 111 p.

Pettijohn, F. J., 1957, Sedimentary rocks: 2d ed., New York, Harper and Bros., 718 p.

Pierce, K. L., 1965, Geomorphic significance of a Cretaceous deposit in the Great Valley of southern Pennsylvania: U. S. Geol. Survey Prof. Paper 525–C, p. C152–C156.

Price, P. H., 1931, The Appalachian structural front: Jour. Geol., V. 39, no. 1, p. 24–44.

Rickard, L. V., 1964, Correlation of the Devonian rocks in New York State: New York State Mus. and Sci. Service, Geol. Survey Map and Chart ser., no. 4.

Rodgers, John, 1953, Geologic map of East Tennessee with explanatory text: Tennessee Div. Geology Bull. 58, pt. 2, 168 p.

Sloss, L. L., 1963, Sequences in the cratonic interior of North America: Geol. Soc. America Bull., v. 74, p. 93–113.

_____, Krumbein, W. C., and Dapples, E. C., 1949, Integrated facies analysis, in Longwell, C. R., chm., Sedimentary facies in geologic history: Geol. Soc. America Mem. 39, p. 91–123.

Stose, G. W., and Ljungstedt, O. A., comps., 1932, Geologic map of the United States: Washington, D. C., U. S. Geol. Survey, 4 sheets, scale 1:2,500,000.

_____, and Stose, A. J., 1944, The Chilhowee group and Ocoee series of the southern Appalachians: Am. Jour. Sci., v. 242, no. 7, p. 367–390; no. 8, p. 401–416.

Summerson, C. H., 1962, Precambrian in Ohio and adjoining areas: Ohio Geol. Survey Rept. Inv. 44, 16 p.

Sutton, R. G., 1957, Lithofacies map of Upper Devonian in eastern United States: Am. Assoc. Petroleum Geologists Bull., v. 41, no. 4, p. 750–755.

Swartz, F. M., 1948, Trenton and sub-Trenton of outcrop areas in New York, Pennsylvania, and Maryland, in Galey, J. T., ed., Appalachian Basin Ordovician symposium: Am. Assoc. Petroleum Geologists Bull., v. 32, no. 8, p. 1493–1595.

Ulteig, J. R., 1964, Upper Niagaran and Cayugan stratigraphy of northeastern Ohio and adjacent areas: Ohio Div. Geol. Survey Rept. Inv. 51, 48 p.

U. S. Geological Survey and American Association of Petroleum Geologists, 1961, Tectonic map of the United States, exclusive of Alaska and Hawaii: 2 sheets, scale 1:2,500,000 [1962].

Wagner, W. R., 1966, Stratigraphy of the Cambrian to Middle Ordovician rocks of central and western Pennsylvania: Pennsylvania Geol. Survey Gen. Geol. Rept. G49, 156 p.

Wanless, H. R., 1946, Pennsylvania geology of a part of the southern Appalachian coal field: Geol. Soc. America Mem. 13, 162 p.

Wilson, C. W., Jr., 1949, Pre-Chattanooga stratigraphy in Tennessee: Tennessee Div. Geology Bull. 56, 407 p.

Wood, G. H., Jr., and others, 1962, Pennsylvanian rocks of the southern part of the anthracite region of eastern Pennsylvania: U. S. Geol. Survey Prof. Paper 450–C, p. C39–C42.

Woodward, H. P., 1943, Devonian system of West Virginia: West Virginia Geol. Survey [Rept.] v. 15, 665 p.

_____, 1949, The Cambrian system of West Virginia: West Virginia Geol. Survey [Rept.], v. 20, 317 p.

_____, 1957, Structural elements of northeastern Appalachians: Bull. Am. Assoc. Petroleum Geologists, v. 41, no. 7, p. 1429–1440.

_____, 1961, Preliminary subsurface study of southeastern Appalachian Interior Plateau: Am. Assoc. Petroleum Geologists Bull., v. 45, no. 10, p. 1634–1655.

_____, 1964, Central Appalachian tectonics and the deep basin: Am. Assoc. Petroleum Geologists Bull., v. 48, no. 3, p. 338–356.

The sources of information for the stratigraphic names, age relationships, thicknesses, and lithologies shown on the schematic stratigraphic diagrams (figures 9, 10, 12, 14, 15, 17, 19, 21, 23, 25, and 26) follow:

Figure 9

Cloos, Ernst, 1958, Structural geology of South Mountain and Appalachians in Maryland, in Pennsylvania Geologists Guidebooks (4,5) 23d Ann. Field Conf., 1958; The Johns Hopkins Univ. Studies in Geology, no. 17, F. J. Pettijohn, ed., 85 p.

Hadley, J. B., and Goldsmith, Richard, 1963, Geology of the eastern Great Smoky Mountains, North Carolina and Tennessee: U. S. Geol. Survey Prof. Paper 349–B, p. B1–B118.

Neuman, R. B., and Nelson, W. H., 1965, Geology of the western Great Smoky Mountains, Tennessee: U. S. Geol. Survey Prof. Paper 349–D, p. D1–D81.

Stose, G. W., and Stose, A. J., 1944, The Chilhowee group and Ocoee Series of the Southern Appalachians: Am. Jour. Sci., v. 242, no. 7, p. 367–390; no. 8, p. 401–416.

Figure 10

Cloos, Ernst, 1958, Structural Geology of South Mountain and Appalachians in Maryland, in Pennsylvania Geologists Guidebooks (4,5) 23d Ann. Field Conf., 1958: The Johns Hopkins Univ. Studies in Geology, no. 17, F. J. Pettijohn, ed., 85 p.

Fisher, D. W., 1962, Correlation of the Cambrian rocks in New York State: New York State Mus. Map and Chart ser., no. 2.

Flagler, C. W., 1966, Subsurface Cambrian and Ordovician stratigraphy of the Trenton Group—Precambrian interval in New York State: New York State Mus. Map and Chart ser., no. 8, 57 p.

Howell, B. F., chm., and others, 1944, Correlation of the Cambrian formations of North America [Chart no. 1]: Geol. Soc. America Bull., v. 55, p. 993–1003.

Kesler, T. L., 1950, Geology and mineral deposits of the Cartersville district, Georgia: U. S. Geol. Survey Prof. Paper 224, p. 1–97.

King, P. B., 1950, Geology of the Elkton area: U. S. Geol. Survey Prof. Paper 230, 82 p.

Neuman, R. B., and Nelson, W. H., 1965, Geology of the western Great Smoky Mountains, Tennessee: U. S. Geol. Survey Prof. Paper 349–D, p. D1–D81.

Willard, Bradford, 1955, Cambrian contacts in eastern Pennsylvania: Geol. Soc. America Bull., v. 66, p. 819–834.

Figure 12

Butts, Charles, 1926, The Paleozoic rocks in Alabama, *in* Geology of Alabama: Alabama Geol. Survey Spec. Rept. 14, p. 40–230.

————, and Edmundson, R. S., 1966, Geology and mineral resources of Frederick County: Virginia Div. of Mineral Resources Bull. 80, 142 p.

Calvert, W. L., 1964, Sub-Trenton rocks from Fayette County, Ohio, to Brant County, Ontario: Ohio Div. Geol. Survey Rept. Inv. no. 52, 7 p.

Drake, A. A., Jr., 1967, Geologic map of the Easton quadrangle, New Jersey–Pennsylvania: U. S. Geol. Survey Geol. Quad. Map GQ–594.

Fisher, D. W., 1962, Correlation of the Cambrian rocks in New York State: New York State Mus. Map and Chart ser., no. 2.

————, 1962b, Correlation of the Ordovician rocks in New York State: New York State Mus. Map and Chart ser., no. 3.

Flagler, D. W., 1966, Subsurface Cambrian and Ordovician stratigraphy of the Trenton Group—Precambrian interval in New York State: New York State Mus. Map and Chart ser., no. 8, 57 p.

King, P. B., 1950, Geology of the Elkton area, Virginia: U. S. Geol. Survey Prof. Paper 230, 82 p.

Neuman, R. B., and Nelson, W. H., 1965, Geology of the western Great Smoky Mountains, Tennessee: U. S. Geol. Survey Prof. Paper 349–D, p. D1–D81.

————, and Wilson, R. L., 1960, Geology of the Blockhouse quadrangle, Tennessee: U. S. Geol. Survey Geol. Quad. Map GQ–131.

Twenhofel, W. H., chm., and others, 1954, Correlation of the Ordovician formations of North America [Chart no. 2]: Geol. Soc. America Bull., v. 65, p. 247–298.

Wilson, C. W., Jr., 1949, Pre-Chattanooga stratigraphy in Tennessee: Tennessee Dept. Conserv., Div. Geol. Bull. 56, 407 p.

Figure 14

Drake, A. A., Jr., and Epstein, J. B., 1967, The Martinsburg Formation (Middle and Upper Ordovician) in the Delaware Valley, Pennsylvania–New Jersey: U. S. Geol. Survey Bull. 1244–H, p. H1–H16.

Donnerstag, Phillip, and others, 1950, Sample study and correlation of C. C. Lobdell no. 1 well: New York State Mus. Circ. 28, 15 p.

Fisher, D. W., 1962, Correlation of the Ordovician rocks in New York State: New York State Mus. Map and Chart Ser., no. 3.

Stout, W. E., and Lemay, C. A., 1940, Paleozoic and pre-Cambrian rocks of Vance Well, Delaware County, Ohio: Am. Assoc. Petroleum Geologists Bull., v. 24, no. 4, p. 672–692.

Twenhofel, W. H., chm., and others, 1954, Correlation of the Ordovician formations of North America [Chart no. 2]: Geol. Soc. America Bull., v. 65, p. 247–298.

Wagner, W. R., 1958, Emma McKnight no. 1 well. Pymatuning Township, Mercer County: Pennsylvania Geol. Survey, 4th ser., Well–Sample Record no. 30, 36 p.

Figure 15

Cattermole, J. M., 1966, Geologic map of the Fountain City quadrangle, Knox County, Tennessee: U. S. Geol. Survey Geol. Quad. Map GQ–513.

————, 1966, Geologic map of the John Sevier quadrangle, Knox County, Tennessee: U. S. Geol. Survey Geol. Quad. map GQ–514.

Cloos, Ernst, 1951, Stratigraphy of sedimentary rocks of Washington County, *in* The physical features of Washington County: Maryland Dept. Geol., Mines and Water Resources, p. 17–94.

Greene, R. C., 1965, Geologic map of the Kirksville quadrangle, Garrard and Madison Counties, Kentucky: U. S. Geol. Survey Geol. Quad. Map GQ–452.

Harris, L. D., and Miller, R. L., 1958, Geology of the Duffield quadrangle, Virginia: U. S. Geol. Survey Geol. Quad. Map GQ–111.

Schilling, F. A., Jr., and Peck, J. H., 1967, Geologic map of the Orangeburg quadrangle, northeastern Kentucky: U. S. Geol. Survey Geol. Quad. Map GQ–588.

Twenhofel, W. H., chm., and others, 1954, Correlation of the Ordovician formations of North America [Chart no. 2]: Geol. Soc. America Bull., v. 65, p. 247–298.

Wilson, C. W., Jr., 1949, Pre-Chattanooga stratigraphy in Tennessee: Tennessee Dept. of Conserv., Div. of Geology Bull. 56, 407 p.

Figure 17

Butts, Charles, 1926, The Paleozoic rocks in Alabama, *in* Geology of Alabama: Alabama Geol. Survey Spec. Rept. 14, p. 40–230.

Conlin, R. R., and Hoskins, D. M., 1962, Geology and mineral resources of the Mifflintown quadrangle, Pennsylvania: Pennsylvania Geol. Survey, 4th ser., Atlas A–126, 46 p.

Englund, K. J., 1964, Geology of the Middlesboro South quadrangle, Tennessee–Kentucky–Virginia: U. S. Geol. Survey Geol. Quad. Map GQ–301.

Fisher, D. W., 1959, Correlation of the Silurian rocks in New York State: New York State Mus., Geol. Survey Map and Chart Ser. [no. 1], with text.

Harris, L. D., and Miller, R. L., 1963, Geology of the Stickleyville quadrangle, Virginia: U. S. Geol. Survey Geol. Quad. Map GQ–238.

Kindle, E. M., and Taylor, F. B., 1913, Description of the Niagara quadrangle [New York]: U. S. Geol. Survey Geol. Atlas, Folio 190.

Peck, J. H., and Pierce, K. L., 1966, Geologic map of part of the Manchester Islands quadrangle, Lewis County, Kentucky: U. S. Geol. Survey Geol. Quad. Map GQ–581.

Pepper, J. F., de Witt, W., Jr., and Everhart, G. M., 1953, The "Clinton" sands in Canton, Dover, Massillon, and Navarre quadrangles, Ohio: U. S. Geol. Survey Bull. 1003–A, 15 p.

Rickard, L. V., and Zenger, D. H., 1964, Stratigraphy and paleontology of the Richfield Springs and Cooperstown quadrangles, New York: New York State Mus. Bull. no. 396, 101 p.

Swartz, C. K., chm., and others, 1942, Correlation of the Silurian formations of North America [Chart no. 3]: Geol. Soc. America Bull., v. 53, p. 533–538.

Figure 19

de Witt, Wallace, Jr., and Colton, G. W., 1964, Bedrock geology of the Evitts Creek and Pattersons Creek quadrangles, Maryland, Pennsylvania, and West Virginia: U. S. Geol. Survey Bull. 1173, 90 p.

Dow, J. W., 1962, Lower and Middle Devonian limestones in northeastern Ohio and adjacent areas: Ohio Div. Geol. Survey Rept. Inv. no. 42, 67 p.

Fisher, D. W., 1959, Correlation of the Silurian rocks in New York State: New York State Mus., Geol. Survey Map and Chart ser. [no. 1], with text.

Kindle, E. M., and Taylor, F. B., 1913, Description of the Niagara quadrangle [New York]: U. S. Geol. Survey Geol. Atlas, Folio 190.

Rickard, L. V., and Zenger, D. H., 1964, Stratigraphy and paleontology of the Richfield Springs and Cooperstown quadrangles, New York: New York State Mus. Bull. no. 396, 101 p.

Woodward, H. P., 1943, Devonian System of West Virginia: West Virginia Geol. Survey vol. XV, 655 p.

Figure 21

Butts, Charles, and Edmundson, R. S., 1966, Geology and mineral resources of Frederick County [Virginia]: Virginia Div. of Mineral Resources Bull. 80, 142 p.

Chadwick, G. H., 1944, Geology of the Catskill and Kaaterskill quadrangles, pt. II, Silurian and Devonian geology, with a chapter on glacial geology: New York State Mus. Bull. 336, 251 p.

Cooper, G. A., chm., and others, 1942, Correlation of the Devonian formations of North America [Chart no. 4]: Geol. Soc. America Bull., no. 12, pt. I, p. 1729–1794.

Fletcher, F. W., 1963, Regional stratigraphy of Middle and Upper Devonian non-marine rocks in southeastern New York, in Shepps, V. C., Symposium on Middle and Upper Devonian stratigraphy of Pennsylvania and adjacent states: Pennsylvania Geol. Survey, 4th ser., General Geol. Rept. G–39, p. 25–41.

Freeman, L. B., 1951, Regional aspects of Silurian and Devonian stratigraphy in Kentucky: Kentucky Geol. Survey, ser. 9, Bull. 6, 565 p.

Gualtieri, J. L., 1967, Geologic map of the Crab Orchard quadrangle, Lincoln County, Kentucky: U. S. Geol. Survey Geol. Quad. Map GQ–571.

Hass, W. H., 1956, Age and correlation of the Chattanooga shale and the Maury formation: U. S. Geol. Survey Prof. Paper 286, 47 p.

Hoover, K. V., 1960, Devonian–Mississippian shale sequence in Ohio: Ohio Div. of Geol. Survey Inf. Circ. no. 27, 154 p.

Rickard, L. V., 1964, Correlation of the Devonian rocks in New York State: New York State Mus. and Sci. Service, Geol. Survey, Map and Chart ser., no. 4.

Roen, J. B., Miller, R. L., and Huddle, J. W., 1964, The Chattanooga Shale (Devonian and Mississippian) in the vicinity of Big Stone Gap, Virginia, in Geological Survey Research 1964: U. S. Geol. Survey Prof. Paper 501–B, p. B43–B48.

Figure 23

Averitt, Paul, 1941, The Early Grove gas field: Virginia Geol. Survey Bull. 56, 50 p.

Colton, G. W., 1967, Subsurface structure map of parts of Lycoming, Clinton, Tioga, and Potter Counties, Pennsylvania: Pennsylvania Geol. Survey, 4th ser., Map 14.

Ebright, J. R., 1952, The Hyner and Ferney anticlines and adjacent areas, Centre, Clinton and Lycoming Counties, Pennsylvania: Pennsylvania Geol. Survey, 4th ser., Bull. M35, 32 p.

Erickson, R. L., 1966, Geologic map of part of the Friendship quadrangle, Lewis and Greenup Counties, Kentucky: U. S. Geol. Survey Geol. Quad. Map GQ–526.

Lewis, R. Q., Sr., and Thaden, R. E., 1965, Geologic map of the Cumberland City quadrangle, southern Kentucky: U. S. Geol. Survey Geol. Quad. Map GQ–475.

Morris, R. H., Geologic map of parts of the Concord and Buena Vista quadrangles, Lewis County, Kentucky: U. S. Geol. Survey Geol. Quad. Map GQ–525.

Pepper, J. F., de Witt, Wallace, Jr., and Demarest, D. F., 1954, Geology of the Bedford shale and Berea sandstone on the Appalachian basin: U. S. Geol. Survey Prof. Paper 259, 111 p.

Price, P. H., and Heck, E. T., 1939, Greenbrier County [West Virginia]: West Virginia Geol. Survey [Rept.], 846 p.

Trexler, J. P., Wood, G. H., Jr., and Arndt, H. H., 1962, Uppermost Devonian and Lower Mississippian rocks of the western part of the Anthracite region of eastern Pennsylvania: Art. 73 in U. S. Geol. Survey Prof. Paper 450–C, p. C36–C39.

Welch, Stewart W., 1959, Mississippian rocks of the northern part of the Black Warrior basin, Alabama and Mississippi: U. S. Geol. Survey Oil and Gas Inv. Chart OC–62.

Weller, J. M., chm., and others, 1948, Correlation of the Mississippian formations of North America [Chart no. 5]: Geol. Soc. America Bull., v. 59, p. 91–106.

Figure 25

Berryhill, H. L., Jr., 1963, Geology and coal resources of Belmont County, Ohio: U. S. Geol. Survey Prof. Paper 380, 113 p.

————, Jr., and Swanson, V. E., 1962, Revised stratigraphic nomenclature for Upper Pennsylvanian and Lower Permian rocks Washington County, Pennsylvania: Art. 75 in Geological Survey Research 1962: U. S. Geol. Survey Prof. Paper 450–C, p. C43–C46.

Dobrovolny, E., Sharps, J. A., and Ferm, J. C., 1963, Geology of the Ashland quadrangle, Kentucky–Ohio, and the Catlettsburg quadrangle in Kentucky: U. S. Geol. Survey Geol. Quad. Map GQ–196.

Englund, K. J., and DeLaney, A. O., 1966, Geologic map of the Sandy Hook quadrangle, Elliott and Morgan Counties, Kentucky: U. S. Geol. Survey Geol. Quad. Map GQ–521.

Lamborn, R. E., 1956, Geology of Tuscarawas County: Ohio Div. of Geol. Survey Bull. 55, 269 p.

Moore, R. C., chm., and others 1944, Correlation of Pennsylvanian formations of North America [Chart no. 6]: Geol. Soc. America Bull., v. 55, no. 6, p. 657–706.

Spencer, F. D., 1964, Geology of the Boltsfork quadrangle and part of the Burnaugh quadrangle, Kentucky: U. S. Geol. Survey Geol. Quad. Map GQ–316.

Wood, G. H., Jr., Trexler, J. P., and Arndt, H. H., 1962, Pennsylvanian rocks of the southern part of the Anthracite region of eastern Pennsylvania: Art. 74 in Geological Survey Research 1962: U. S. Geol. Survey Prof. Paper 450–C, p. C39–C42.

Figure 26

Butts, Charles, 1926, The Paleozoic rocks, *in* Geology of Alabama: Alabama Geol. Survey Spec. Rept. 14, p. 40–230.

Culbertson, W. C., 1963a, Correlation of the Parkwood Formation and the lower members of the Pottsville Formation in Alabama: Art. 193 *in* Geological Survey Research 1962: U. S. Geol. Survey Prof. Paper 450–E, p. E47–E50.

————, 1963b, Pennsylvanian nomenclature in northwest Georgia: Art. 194 *in* Geological Survey Research 1962: U. S. Geol. Survey Prof. Paper 450–E, p. E51–E57.

Englund, K. J., 1964, Geology of the Middlesboro South quadrangle, Tennessee–Kentucky–Virginia: U. S. Geol. Survey Geol. Quad. Map GQ–301.

————, and others, 1961, Geology of the Ewing quadrangle, Kentucky and Virginia: U. S. Geol. Survey Geol. Quad. Map GQ–172.

Moore, R. C., chm., and others, 1944, Correlation of the Pennsylvanian formations of North America [Chart no. 6]: Geol. Soc. America Bull., v. 55, no. 6, p. 657–706.

Wanless, H. R., 1946, Pennsylvanian geology of a part of the southern Appalachian coal field: Geol. Soc. America Mem. 13, 162 p.

Chapter 3

Paleozoic Alluvial Deposition in the Central Appalachians: A Summary

L. D. MECKEL

INTRODUCTION

THICK WEDGES OF nonmarine clastics were deposited at three times during the filling of the Appalachian geosyncline in the central Appalachians: once in the late Ordovician and early Silurian and twice in the late Devonian through Pennsylvanian. Each sequence is associated with a major source area orogeny and terminated a period of widespread marine deposition. These thick, dominantly alluvial accumulations of lithologies, ranging from conglomerate to coal, thin markedly away from the source and intertongue basinward with shallow marine sections.

These alluvial deposits, the subject of many studies, display similar facies and depositional patterns. Their common attributes are the focal point of this paper; the associated marine deposits are considered only where pertinent to understanding the complete depositional framework.

My objectives are: (1) to summarize the stratigraphic and genetic relationships of these alluvial strata in the central Appalachians; (2) to document the recurrent patterns of alluvial deposition; (3) to provide a depositional model for alluvial deposition in a geosynclinal trough; and (4) to suggest areas where additional data and studies are needed.

Much of this summary comes from three published studies: Yeakel's* (1962) on the Bald

* To whose memory this paper is dedicated. A professional colleague and personal friend, Lloyd S. Yeakel made significant contributions to Appalachian geology and his work is drawn on heavily for this summary article.

Eagle, Juniata, and Tuscarora; Pelletier's (1958) on the Pocono; and Meckel's (1967) on the Pottsville. Each is a regional stratigraphic–petrologic–paleocurrent analysis of these units within the central Appalachians; in their original forms these studies were doctoral dissertations. Both unpublished data from these theses together with material obtained from other sources were utilized for this study.

This paper is in three parts. The first describes the general basin framework and stratigraphy of the alluvial sediments. The second considers the current and dispersal patterns within the nonmarine wedges. The last discusses the recognition of upward-fining fluvial cycles. These stratigraphic and sedimentologic aspects are then brought together to outline a depositional model for Paleozoic alluvial deposition in the central Appalachians.

GENERAL BASIN FRAMEWORK

The central Appalachians lie south and southeast of the Canadian Shield and southwest of the Adirondack Uplift. The area is bordered on the east and southeast by the crystalline complex of the Piedmont. To the west, two broad upwarps—the Findlay and Cincinnati arches—flank the basin; these arches separate the more mobile Appalachian Basin from the stable intracratonic basins to the west. The main northeast-southwest trending structural features of the central Appalachians continue southward as the southern Appalachians. The relations of the central Appalachians to these major structural features of the northeastern United States are shown in Figure 1.

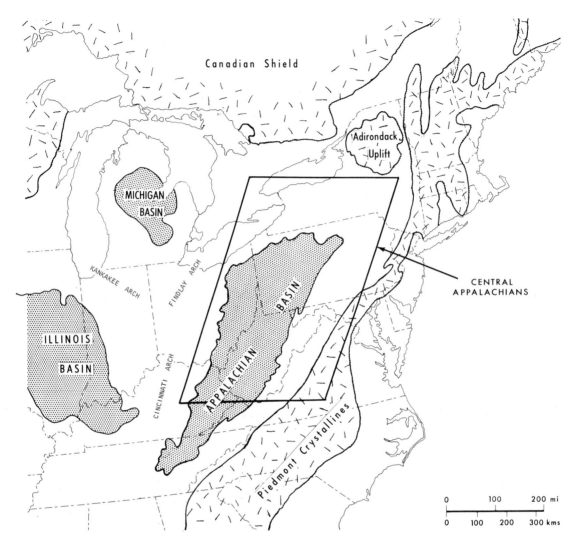

FIGURE 1. Relations of central Appalachian study area to major structural features of the northeastern United States.

The area of interest includes parts of New York, New Jersey, Pennsylvania, Ohio, Maryland, West Virginia, and Virginia. The area comprises the folded Valley and Ridge Province on the east and the gently deformed Appalachian Plateau Province to the west.

The central Appalachian structural basin is an elongate trough which consists of a narrow zone of maximum subsidence (the geosyncline) along the southeastern margin which is transitional with a broad stable platform area (the foreland) to the north and west. The axis of the structural trough (line of greatest downwarp) lies near the southeastern margin of the basin and trends northeast-southwest. This asymmetric structural basin was flanked by an active borderland

(Appalachia) on the southeast and a stable cratonic source (Canadian Shield) on the north; it widens to the southwest. Virtually all of the clastic material is supplied by the borderland; the northern source makes only very small and intermittent contributions. Proximity to the major southeastern source serves to define the proximal (eastern) and distal (western) portions of the basin. The bathymetric axis of the basin (the line of greatest water depth) lies several hundred miles west of the structural axis.

In the tectonic borderland flanking the basin, there were three major episodes of orogenic activity: the Taconic Orogeny in late Ordovician time, the Acadian Orogeny in mid-Devonian time, and the Appalachian Revolution which occurred intermittently throughout late Mississippian, Pennsylvanian, and Permian times. Associated with each orogeny were periods of regional regression characterized by widespread alluvial deposition. Strata associated with the Taconic Orogeny form a mid-Paleozoic nonmarine wedge; those associated with the Acadian and Appalachian events constitute two late Paleozoic wedges which tend to merge, being separated only by thin marine strata in the distal areas. The two older nonmarine accumulations terminate periods of widespread marine (turbidite) deposition within the basin (see Chapter 4 by N. L. McIver in this volume).

Each wedge comprises several formations, the boundaries of which are arbitrarily defined on the basis of color or lithologic change or both. Many of the formational boundaries correspond with facies boundaries and therefore cross time—stratigraphic boundaries. The Taconic wedge consists of the following formations from bottom to top: the Bald Eagle (Oswego),* Juniata (Queenstown), Tuscarora (Medina, Clinch, Shawangunk), and Castanae (Grimsby). The Acadian wedge includes the Catskill, Pocono (Price), and lower part of the Mauch Chunk Formations; the youngest wedge includes the upper Mauch Chunk, Pottsville, and Llewelyn Formations. Figure 2 shows the stratigraphic positions and descriptions of these formations in the eastern part of the basin, where they are thickest.

These sediments represent a variety of depositional environments, all of which are elements of a broad coastal plain flanking a marine embayment. The coastal plain is composed of high-gradient alluvial fans in the east which coalesce into a broad alluvial apron which emerges basinward (westward) into low-gradient deltaic and intertidal plains. In those parts of the basin where the nonmarine intertongues with coastal and marine deposits, the latter may predominate without a change in formation name. For example, most of the Tuscarora in eastern West Virginia has been interpreted as beach deposits flanking the coastal plain which occurs farther to the east (Folk, 1960). Herein lies the problem of using formation names to discuss the alluvial deposits. Therefore, in this paper formation names are used sparingly and lithologic facies terms are used to describe the alluvial sediments.

* The formation names used in this report are those established in eastern and central Pennsylvania. The names in parentheses are lateral equivalents of these formations elsewhere in the central Appalachians.

STRATIGRAPHY

Facies

The alluvial deposits are in general characterized by two lithofacies. One is coarse-grained, consisting primarily of conglomerate and sandstone. These coarse deposits are typically white, gray, olive gray, or tan. Relatively thin units of gray shale, which contain coal beds in the Carboniferous sections, can occur in this facies but are most common in the distal part of the basin. The sandstones of this facies are predominantly subgraywackes and protoquartzites. The dominant sedimentary structure is cross-bedding; other common features include cut-and-fill structures, intraformational shale–pebble conglomerates, plant fragments (restricted to late Paleozoic sections), parting lineation, and *Arthophycus*.

The other lithofacies is composed predominantly of red shale and siltstone which contain some interbedded red sandstones. The fine-grained red-bed sediments are most commonly massive in appearance but do contain mud cracks, ripple marks, small-scale cross-bedding, and flaser bedding. The sandstones are similar, except for color, to those described for the other facies.

In many parts of the basin the two facies are transitional with one another. The transition zones, ranging from 50 to 2,000 ft (150–600 m) thick, consist of interbedded strata of both facies.

Because formation boundaries in these nonmarine sections are picked on the basis of color and/or lithology, the alluvial formations are largely, but not entirely, composed of one lithofacies. For instance, the mid-Paleozoic Juniata and Castanae Formations and the late Paleozoic Catskill and Mauch Chunk Formations consist predominantly of the fine-grained, red-bed facies.* The mid-Paleozoic Tuscarora and late Paleozoic Pocono, Pottsville, and Llewelyn Formations are examples of the coarse-grained (conglomeratic), nonred facies. For purposes of stratigraphic mapping the transition zones are included as members of the red-bed formations. Because of this, the red formations contain intercalated nonred sandstone and conglomerate units, which are most common near the upper and lower contacts; whereas interbeds of red strata are rare or absent in the nonred formations.

* The Ordovician Bald Eagle is a dubious unit of this facies. Though mottled green and grayish green, it has been interpreted on the basis of stratigraphy and petrology to have been originally a red unit which was leached green by pore water expelled from the underlying marine Reedsville Formation (Horowitz, 1965).

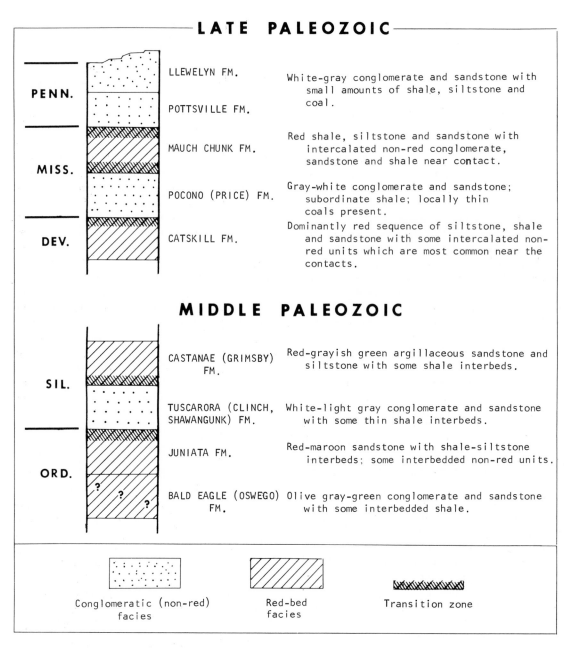

FIGURE 2. Generalized stratigraphic column showing positions and descriptions of Paleozoic alluvial formations, central Appalachians.

In this paper, the terms "red-bed facies" and "conglomeratic facies" are used to refer to the two lithofacies for these are the most striking attributes of these deposits as seen in the field in the proximal part of the basin.

Both lithofacies occur in each of the three alluvial wedges and show a recurrent stratigraphic arrangement with respect to one another.

Where complete, a wedge comprises three parts: (1) a lower red-bed facies; (2) a middle conglomeratic facies; and (3) an upper red-bed facies. Table 1 shows the formations which make up each alluvial wedge and relates these to the lithofacies. Of the three wedges, the mid-Paleozoic one is most complete in its preservation. The youngest is incomplete due to regional uplift

TABLE 1. Relation of Formation Names to Lithologic Facies within Paleozoic Alluvial Wedges of the Central Appalachians

	Taconic wedge	Acadian wedge	Appalachian wedge
Upper red-bed facies	Castanae	Mauch Chunk (lower part)	—
Conglomeratic facies	Tuscarora	Pocono	Pottsville, Llewellyn
Lower red-bed facies	Juniata, Bald Eagle(?)	Catskill	Mauch Chunk (upper part)

and folding of the Appalachian Basin at the end of the Paleozoic prior to deposition of an upper red-bed facies.

Genetically the alluvial sediments consist predominantly of two facies: channel and non-channel. The channel deposits are coarse-grained (conglomerate and sandstone) and are typically nonred in color; they are red only where locally interbedded with thick red-bed sequences such as in the Juniata and Catskill Formations. The nonchannel deposits are fine-grained (shale and siltstone) and include sediments of flood plain, lacustrine (playa), and tidal flat environments. The fine-grained facies consists of two color types, red and gray, which seem to differ genetically. The red deposits, by far the most abundant, occur in thick sections such as in the Catskill and Mauch Chunk; they are probably lacustrine or tidal flat in origin. The gray nonchannel deposits are much thinner and occur interbedded with thick channel sequences, as in the Pocono and Pottsville. These gray sediments are predominantly flood basin in origin and contain coals which are autochthonous swamp deposits. Therefore the red color seems to correlate more with environment of deposition than with grain size.

Though the origin of the red-bed color is beyond the scope of this paper, some field observations in the Mauch Chunk–Pottsville transition zone bear on this complex problem. The red color either (1) may be inherited from the source area, or (2) may have originated in the environment during or shortly after deposition. If the red coloration were due to erosion of red soils in the source area, the matrix clays of the interbedded conglomerates would also be red (Edwards, 1955). Because the interbedded coarse clastics in the transition zones contain fine-grained, gray-green matrix similar to that in the red beds except for color, either (a) there were two sources, one furnishing red material to the areas depositing fine-grained material and one furnishing gray material to the channels, or (b) all the material was derived from a red source, the channel deposits being differentially reduced *in situ* during or after deposition. If the red and gray deposits represent material from two

sources, it is hard to envision, from the distribution of the deposits, a reasonable depositional model. The red shale–pebble conglomerates and paper-thin red shale laminae in the coarse non-red units suggest that these units were not differentially reduced. Therefore by elimination, it seems that the red color was probably produced at the site of deposition as has been envisioned for other red-bed units (Clark, 1962; Walker, 1967).

In summary, the dominantly red-bed formations (Juniata, Catskill, Mauch Chunk) consist of mixed nonchannel and channel deposits, the latter (both red and nonred) being sandwiched between thick, fine-grained nonchannel sections which predominate in most parts of the basin. The conglomeratic formations (Tuscarora, Pocono, Pottsville) consist primarily of coalescing coarse channel deposits; the finer nonchannel sediments, though locally present in many areas, are most common in the distal parts of the basin where the marine and nonmarine intertongue.

Facies Relations

Stratigraphic relations are most evident in sections normal to the structural axis, i.e., in northwest–southeast sections. Figure 3, which summarizes these relations, though drawn for Pennsylvania, applies in general throughout the basin. The sections show two relations which are important for understanding the stratigraphic framework of these alluvial deposits: (1) that of nonmarine to marine strata; and (2) that of the alluvial red-bed facies to the conglomeratic facies. Many of these relations have been established or suggested only through recent detailed field mapping by both the federal and state surveys. Summary Figure 11 may be helpful for visualizing these relationships.

(1) *Nonmarine–Marine Relations.* The broad depositional pattern is one of thick alluvial clastics in the east which thin and intertongue westward with thinner time-equivalent marine strata. The geographic position of this nonmarine to marine transition within the basin varies for each deposit. At times of maximum regression its average position is in western

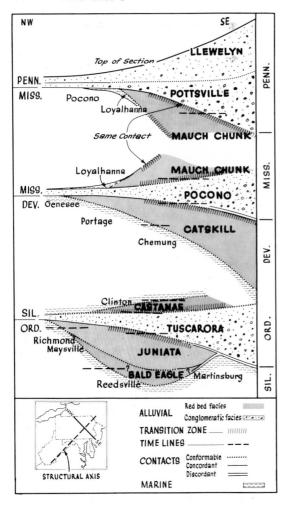

FIGURE 3. Transverse cross sections of the three alluvial wedges. Time lines are essentially horizontal.

Pennsylvania and West Virginia; at times of maximum transgression it moves eastward into eastern Pennsylvania.

Each alluvial wedge was formed in three stages: (a) an initial period of regional regression during which continental environments displace marine environments; (b) followed by a period of maximum nonmarine regression; and (c) a final period of regional transgression during which the marine encroaches back across the basin. Refer to Figure 3.

(a) PERIODS OF REGIONAL REGRESSION. During these periods the rate of deposition exceeds the rate of subsidence and there is a westward progradation of the coastal plain; therefore, the transitional marine–nonmarine contact rises in stratigraphic section to the west.* During these regressions the alluvial intertongues with rapidly

deposited shallow marine deposits which overlie turbidites. A horizontal time line passes through three facies; from east to west they are an alluvial conglomeratic facies, a red-bed facies, and a marine facies. During these periods it is typically the red-bed facies which intertongues with the shallow marine.

(b) PERIODS OF MAXIMUM REGRESSION. During these periods alluvial deposition has its maximum areal extent into the basin from the southeast and occurs in the distal parts of the basin. These periods are generally characterized by an absence of the alluvial red-bed facies; the nonred facies intertongues directly with marine facies. These periods represent times when the rates of deposition and subsidence are approximately equal. Local transgressions and regressions occur at this time but are epeirogenically produced and are limited to the distal part of the basin.

(c) PERIODS OF REGIONAL TRANSGRESSION. During these periods the rate of subsidence exceeds that of deposition and marine deposition advances eastward; the marine–nonmarine contact therefore rises in section from west to east.* During regional transgressions alluvial beds typically intertongue with slowly deposited shallow marine clastics and carbonates. From east to west, a time line passes from a conglomeratic facies to a red-bed facies to a marine facies. Therefore, as during periods of regression, it is typically the red-bed facies which intertongues westward with the shallow marine during transgressions.

Thus the red-bed facies has a restricted spatial distribution within the basin. Areally it is essentially restricted to the trough area along the southeastern margin of the basin and laps only as thin tongues onto the foreland (e.g., the Mauch Chunk). Stratigraphically (with respect to time) it occurs only during periods of regional transgressions and regressions.

The red beds deposited during regressions and transgressions are lithologically and genetically similar. However, stratigraphically they differ in several respects, suggesting that the tectonic framework of the basin is different during times of red-bed deposition. During transgressions the basin seems to be more stable and receives less clastic material than during regressions as evidenced by:

1. The regressive red units intertongue basinward with rapidly deposited marine clas-

---

* Therefore, in a regressive stratigraphic section, nonmarine units overlie marine units.

* Therefore, in a transgressive stratigraphic section, marine units overlie nonmarine units.

tics; the transgressive units typically intertongue basinward with slowly deposited
marine clastics, chemical sediments, and
carbonates.

2. Sourceward the regressive units intertongue
with coarser (more conglomeratic), less
mature channel deposits whereas the transgressive units intertongue with finer
grained (less conglomeratic), more mature
channel deposits.

3. The transition zone with the conglomeratic
facies appears to be thicker during periods
of regression.

**(2) Red-Bed–Conglomeratic Facies Re-
lations.** Contact relations between the two alluvial facies are variable across the basin. Within
the axial (trough) part of the basin the two facies
are typically transitional with one another. Where
conformable the contacts are characterized by a
pronounced transition zone in which strata of
the red-bed and conglomeratic facies are interbedded. These transitional intertonguing boundaries cross time lines (see Fig. 3) and in each
case fine-grained red clastics occur basinward of
time-equivalent coarse, nonred clastics with which
they intertongue. Therefore these transitional
contacts change stratigraphic position across the
basin during periods of transgressions and regressions.

Laterally the contacts not only change stratigraphic position but change character as well.
Toward the distal part of the basin the contacts
become unconformable and no transition zone is
present. The strata on each side of the epeirogenic unconformity (disconformity) are typically
concordant. The time hiatus represented by the
unconformity commonly increases toward the
more stable cratonic areas. The paleogeologic
map (Fig. 4) of the pre-Pennsylvanian surface
illustrates this well; progressively older Mississippian strata are truncated to the west and
northwest. In this figure the time interval represented by the Mississippian–Pennsylvanian unconformity increases away from the area of
maximum subsidence where deposition was continuous from the Mississippian to the Pennsylvanian.

Locally within and east of the zone of maximum downwarp the transitional contacts become
unconformable, commonly displaying an angular
discordance. These unconformities are produced
by tilting and autocannibalism of sediment flanking the tectonically active margin of the basin.
Best documented at the Ordovician–Silurian and
Devonian–Mississippian contacts, they are not
common elsewhere because all or most of the

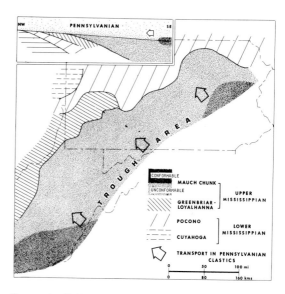

FIGURE 4. Paleogeologic map of the pre-Pennsylvanian
surface in the central Appalachians.

eastern side of the structural basin has been
eroded.

In addition to these major unconformities, the
continental sediments are characterized by numerous local diastems and erosional unconformities at the base of channels.

Thickness, Lithofacies, and Fauna

Each nonmarine formation is markedly wedge-
shaped, being thickest at or near its southeastern
outcrop limit and thinning markedly to the west
and north. Figures 3 and 5 depict these pronounced changes normal to the basin axis. The
isopachs are commonly irregularly arcuate, being
convex to the north and northwest.

In considering regional thickness variations,
one is permanently hampered by lack of data east
of the present outcrop limit. Only in the Bald
Eagle and Juniata (Ordovician), Catskill (Devonian), and Pottsville (Pennsylvanian) Formations can a northeast–southwest trending axis of
maximum subsidence be established. Based on this
limited control the Ordovician structural axis was
located in central Pennsylvania; in younger units
the axis shifted eastward and was positioned in
eastern Pennsylvania.

The pronounced westward thinning of each
unit can be attributed to a combination of subsidence, depositional thinning, and postdepositional erosion prior to deposition of overlying
units. Most of the thinning appears to be related
to subsidence and deposition for it is accompanied by westward intertonguing of beds of one
formation with time equivalent strata of a different formation. For example, the Mauch Chunk

MID-PALEOZOIC   LATE-PALEOZOIC

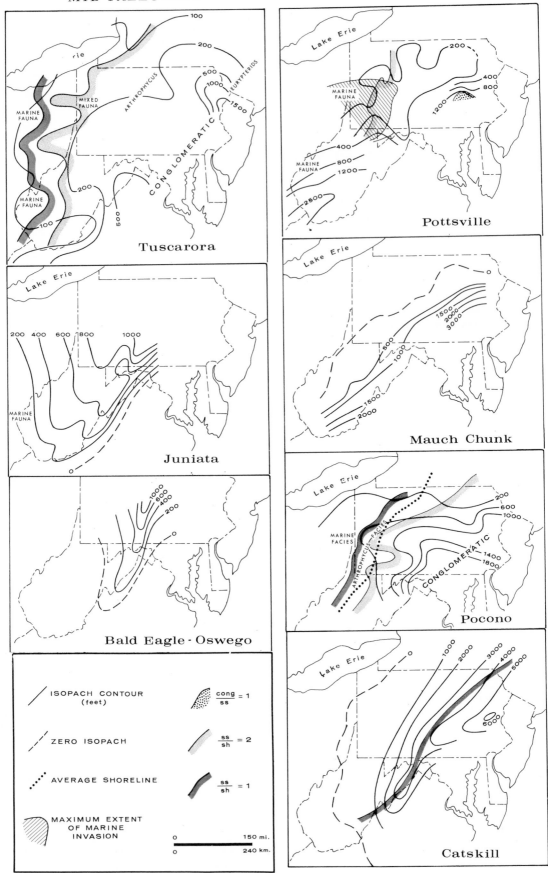

Tuscarora

Pottsville

Juniata

Mauch Chunk

Bald Eagle · Oswego

Pocono

ISOPACH CONTOUR
(feet)

$\frac{cong}{ss} = 1$

ZERO ISOPACH

$\frac{ss}{sh} = 2$

AVERAGE SHORELINE

$\frac{ss}{sh} = 1$

MAXIMUM EXTENT
OF MARINE
INVASION

0   150 mi.

0   240 km.

Catskill

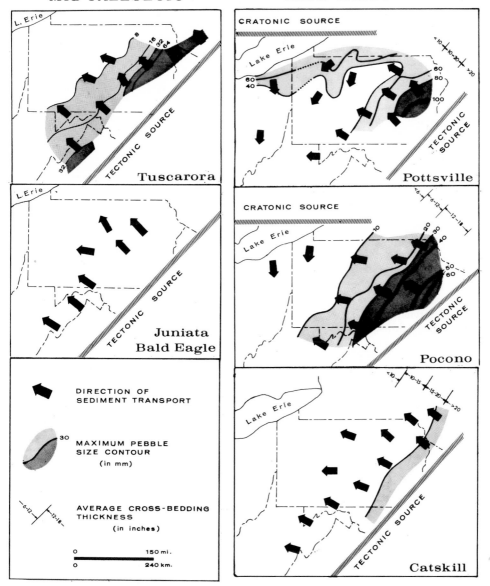

MID-PALEOZOIC   LATE-PALEOZOIC

FIGURE 6. Summary of current and dispersal patterns of Paleozoic alluvial formations, central Appalachians. Scale same for each map. Sources of data: Juniata, Bald Eagle, Tuscarora (Yeakel, 1962); Catskill (McIver, 1961; Burtner, 1963); Pocono (de Witt, 1951; Pelletier, 1958); Pottsville (Meckel, 1964, 1967).

FIGURE 5. Summary of isopach and lithofacies data for Paleozoic alluvial formations, central Appalachians. Scale same for each map. Sources of data: Bald Eagle, Juniata (Woodward, 1951); Tuscarora (Amsden, 1955; Yeakel, 1962); Catskill (McIver, 1961); Pocono (Pelletier, 1958); Mauch Chunk (Meckel, 1964); and Pottsville (Branson, 1962; Meckel, 1967).

in eastern Pennsylvania thins approximately 3,200 ft (1,100 m) within 60 mi (100 km); of this, more than 2,000 ft (660 m) takes place across an area where the Mauch Chunk contacts are conformable and intertonguing. Some of the stratigraphic thinning can be attributed to the artificial and sometimes subjective methods of stratigraphic subdivision. Because relatively homogeneous coarse clastics (conglomerate and sandstone) become intercalated westward with well-defined and laterally extensive shale, coal, and limestone, subdivision of strata in the western areas is common. Thus one formation in the east may become two or more time-equivalent formations to the west, only one of which bears the name used to the east.

Lithologically each formation shows a similar regional pattern: the clastics are coarsest in the eastern areas and become progressively finer to the west, northwest, and north. In general the lithofacies trends parallel the isopachs. These trends are irregularly arcuate and tend to be convex to the north and northwest.

Lithofacies variations are best documented for formations dominantly composed of the conglomeratic facies (Tuscarora, Pocono, and Pottsville Formations; Fig. 5). Each shows a pronounced westward change in lithologic proportions. Conglomerate and sandstone predominate in the eastern areas; both, particularly the conglomerate, decrease in relative abundance to the west where the finer clastics (shale, siltstone) become progressively more abundant. Only in the westernmost areas does shale predominate over sandstone. For example, the Pocono is a predominantly conglomerate–sandstone lithology in eastern Pennsylvania and changes to a pebbly sandstone–shale association and then a sandstone–shale facies to the west. The contours having a sand/shale ratio of 2 have a similar position (western Pennsylvania and central West Virginia) and trend (northeast–southwest) for all the conglomeratic facies.

Lithofacies data for one formation composed dominantly of red-bed facies (the Catskill) suggest a similar westward change from coarse to fine grain size. The Catskill, however, contains less coarse material and therefore is characterized by a higher proportion of finer clastics. For instance, the transition from predominantly sandstone to predominantly shale occurs in central Pennsylvania, more than 100 mi east of the similar change for formations composed mainly of nonred strata.

Marine fauna begin to appear west of the 2:1 sand to shale contour; they are common near the 1:1 contour. The strata containing marine fauna increase in both number and total thickness to the west into Ohio and to the southeast into West Virginia and Kentucky. The thickest marine sections occur still farther west on the stable interior platform.

An *Arthophycus* facies occurs landward of the marine fauna in both the Tuscarora and Pocono. This facies appears to characterize nearshore environments along the edge of the coastal plain. Landward of the *Arthophycus* facies there is a general absence of fauna except for the Tuscarora where fresh-water *Eurypterids* occur in the easternmost areas. Refer to Figure 5.

CURRENT AND DISPERSAL PATTERNS

Stratigraphic relations define the broad facies patterns within the basin. However, it remains to relate these patterns to the dispersal patterns of the clastic material. Regional mapping of both directional structures (principally cross-bedding) and variations in scalar properties (principally lithology and grain size) within the nonmarine section corroborates the stratigraphic interpretation and provides information on the way in which the central Appalachian basin was filled.

The most useful structure for reconstructing paleocurrents in these alluvial deposits is cross-bedding. However, many scalar properties—lithologic proportions, composition, grain size, cross-bedding thickness—which are controlled totally or in part by the current system show systematic relationships to the paleocurrents and are thus also useful in reconstructing dispersal patterns. Figure 6 summarizes the dispersal patterns and locations of source areas for several of the nonmarine formations.

Cross-Bedding

For each formation, the principal direction of sediment transport, as indicated by cross-bedding, was uniformly from the southeast to the northwest. See Figure 6. Figure 7 depicts even more clearly the remarkable stability of these nonmarine currents with time. The nonmarine paleocurrents therefore display a consistent relationship to basin architecture during the Paleozoic: the direction of infilling was transverse to the structural axis of the basin.

The depositional slope (paleoslope) reconstructed from the current azimuths dipped from the principal southeastern source area toward the shallow marine environments to the west and northwest. The gradient along this depositional surface gradually decreases basinward as it does in modern coastal plains. This basinward decrease

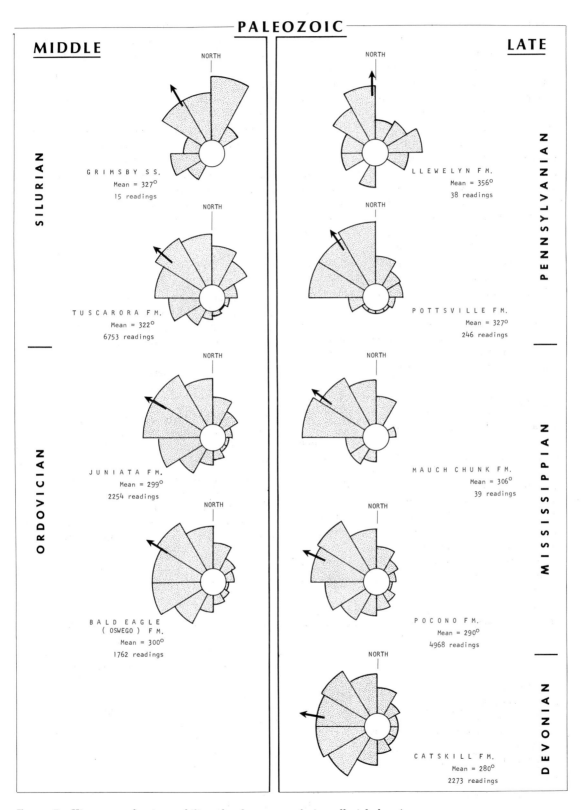

FIGURE 7. Histograms showing stability of paleocurrents during alluvial deposition in the central Appalachians. Sources of data: Bald Eagle, Juniata, Tuscarora, Pocono (after Pettijohn, 1962); Grimsby (Hunter, 1960); Catskill (Burtner, 1963); Mauch Chunk, Pottsville, Llewelyn (Meckel, 1964, 1967).

in gradient is reflected in the increase in scatter of cross-bedding azimuths from east to west, i.e., from the alluvial upper part of the plain to the deltaic lower part where channels branch and bifurcate in a seaward direction. Part of this scatter in the distal areas may be produced by the inclusion of cross-bedding observations of inter-tonguing coastal-marine sands with those of the alluvial channel deposits. Current directions in the distal parts of many of the Paleozoic non-marine systems also show a more westerly or even southwesterly orientation with respect to those in the proximal areas. This suggests a deflection of currents toward the major marine areas to the southwest, a factor which is quite evident during Pottsville deposition.

A second dispersal system, intermittent during the Mississippian and Pennsylvanian times, contributed small volumes of material to the basin. This system originated in a stable northern (cratonic) source which bounded the basin to the north. Fluvial material was dispersed southward and southwestward across the relatively stable, slowly subsiding northern margin of the depositional basin.

The existence of this second source has been well documented in both Pocono and Pottsville strata in northwestern Pennsylvania and eastern Ohio. This cratonic source may have contributed small amounts of material at other times but either (1) the material was mixed and diluted with that from the tectonic source and is now not recognizable, or (2) it exists but has not yet been recognized in the subsurface, or (3) strata with a northern source have been removed by erosion along the northern margin of the basin.

The clastic material entering the basin from the north is petrographically different from that en-tering from the southeastern tectonic source and reflects the more stable nature of the northern source. These petrographic differences, best demonstrated in the Pottsville, are compared in Table 2. This northern cratonic source probably comprised the Canadian Shield and Adirondack Uplift (see Fig. 1).

Therefore, during at least the terminal phase of infilling, the central Appalachian Basin was bounded on both the north and southeast by coastal plains. It seems very likely that these merged to the northeast to close the basin in that direction. The distribution and maximum extent of marine environments during Pennsylvanian time corroborates this (see Fig. 5).

Scalar Properties

Areal variations of lithologic proportions, maximum pebble size, mean cross-bedding thickness, and composition bear a consistent relation to the paleocurrents and are therefore useful in reconstructing dispersal patterns within these alluvial clastics.

Figure 5 best shows the westward lithofacies change. The proportion of shale increases; that of sand and conglomerate decreases. The lithofacies contours generally trend normal to the direction of sediment transport.

A systematic downcurrent decrease in maximum pebble size* has been well demonstrated for the Bald Eagle, Tuscarora, Pocono, upper Mauch Chunk, and Pottsville Formations (Fig. 6). For each formation the maximum pebble size averages 100 mm or larger along the southeastern margin of the outcrop area and decreases very

* Average maximum diameter of the ten largest vein quartz pebbles.

TABLE 2.   Petrographic Differences in Material Supplied by Tectonic and Cratonic Source Areas, Pottsville Formation

	SE tectonic source	N cratonic source
Mineralogy:		
$\frac{\text{Monocrystalline}}{\text{Polycrystalline}}$ quartz ratio	Low (<4)	High (>10)
Characteristic nonopaque heavy minerals*	Z T m	T Z r
$\frac{\text{Zircon}}{\text{Tourmaline}}$ ratio	High (>1)	Low (<1)
Sandstone:		
Mineralogical maturity	Mature	Supermature
Textile maturity	Immature–Submature	Mature
Dominant type	Protoquartzite	Orthoquartzite

* Listed in order of decreasing abundance. Capitalization denotes >5 percent; lower case, <5 percent.
Z = zircon; T = tourmaline; r = rutile; m = metamorphic suite (kyanite, staurolite, sillimanite, garnet, edipote).

rapidly to the northwest. Contours of equal grain size essentially parallel isopach and lithofacies contours. Compare Figures 5 and 6.

In those units which display a systematic downcurrent decrease in pebble size it is theoretically possible to calculate an approximate distance to the upcurrent margin of the depositional basin by assuming what average maximum pebble size accumulates at the basin's edge. This has been done for the Tuscarora (Yeakel, 1962), Pocono Pelletier, 1958), and Pottsville (Meckel, 1967) Formations using Sternberg's Law which relates the downcurrent decline of mean (also maximum) pebble size to the distance of transport. For each formation the basin's edge was estimated to be southeast of or near Philadelphia, Pennsylvania.

In some units (Catskill, Pocono, and Pottsville), the contours of equal cross-bedding thickness are essentially normal to the regional current pattern. The mean thickness averages approximately 20 in. (51 cm) in the eastern-most areas and decreases to 5–10 in. (10–20 cm) in the distal parts of the basin (Fig. 6).

The composition of both the conglomerates and sandstones typically becomes more mature in the downcurrent direction. However, pebble* roundness and sphericity show little or no regional change. Both of these parameters change most rapidly early in a pebble's abrasion history and more slowly later as they approach a limiting value (Pettijohn, 1957, p. 545, 550). The high values of both roundness and sphericity in these deposits coupled with the regional uniformity of these values suggest the limiting values for vein quartz pebbles had been reached in each of the major alluvial conglomerates.

FLUVIAL CYCLES

The criteria used by most workers in interpreting an alluvial origin for these sediments have been general and include the trends and interrelations of cross-bedding dip azimuth, maximum pebble size, isopach, and lithofacies. Supporting evidence has included the occurrence of specific types of sedimentary structures, the absence of marine fauna, and the presence of extensive coals. In recent years the vertical sequences (i.e., internal organization) within coarse deposits have been used to refine interpretations within alluvial deposits (Bernard and Major, 1963; Visher, 1965a,b; Allen, 1965a,b; Potter, 1967). Such data are scarce for alluvial sections in the Appala-

* Again restricted to vein quartz pebbles which are most common.

chian Basin, mostly because they have not been looked for, not because they do not exist.

Allen (1965b) has pointed out that probably one of the most distinctive criteria of alluvial deposits are fining-upward cycles. Such cycles have been reported in the Catskill (Allen, 1965b), Mauch Chunk, and Pottsville (Meckel, 1964) Formations in Pennsylvania. Similar sequences have been observed by the author in the Pocono of western Maryland and eastern Pennsylvania and are to be expected in the Bald Eagle, Juniata, and Tuscarora Formations.

The upward-fining sequences in the Mauch Chunk and Pottsville Formations are characterized by:

(1) Contact relations: disconformable (erosional) basal contact which sharply truncates underlying strata; the upper contact is commonly the sharp erosional base of the next cycle.

(2) Grain size: an upward-fining of both average and maximum grain size; pebbles, if present, are largest and most common in the lower part.

(3) Sedimentary structures: the lower beds tend to be massive or horizontally bedded; cross-bedding is common in the lower and middle parts; small-scale cross-bedding, ripple bedding, and intercalated thin sand beds and shale are most common in the upper part.

(4) Bedding thickness: bedding is thickest (2–15 ft) in the lower part and becomes much thinner (< 6 in.) toward the top of the section.

These upward-fining cycles (Figs. 8 and 9) are interpreted as fluvial channel deposits. The systematic stratigraphic changes reflect decreasing energy in the channel system, either as the result of an energy gradient along a depositional slope of a laterally meandering point bar or from a decrease in energy as a channel deteriorates. Individual channel deposits in the Appalachians range from 10 to over 100 ft (3–30 m) thick. Overlying fine-grained deposits are usually levee or flood basin deposits.

Typically individual channel deposits coalesce stratigraphically to form a complex channel deposit. Where this occurs the individual sequences are commonly modified by postdepositional scour and erosion which leave only partial channel sequences. These multiple channel sections consist primarily of the lower parts of the individual channel deposits, the upper parts having been completely or partially removed by erosion. If enough has been removed, the recognition of these compounded partial channel deposits becomes difficult.

FIGURE 8. Conglomeratic alluvial channel deposit (approximately 40 ft thick), Mauch Chunk Formation near Ashland, Pennsylvania. Note three subcycles within channel deposit. Bedding is vertical and top is to the left. Legend for lithologic log on Figure 10.

FIGURE 9.   Channel sandstone (approximately 50 ft thick) on the lower part of
the alluvial plain, Pottsville Group near Coalmont, Pennsylvania. Hat for scale
near base of channel. Legend for lithologic log on Figure 10.

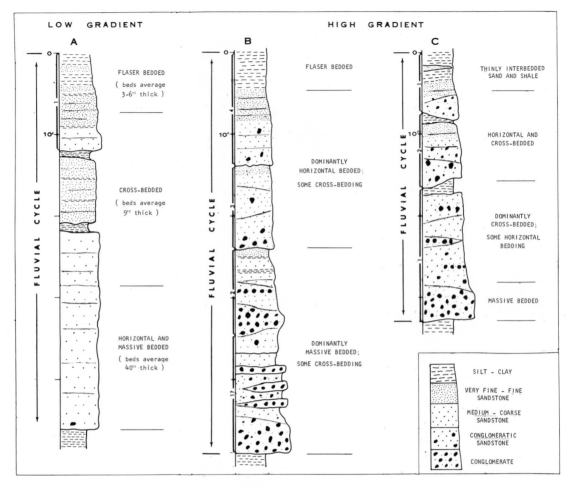

FIGURE 10. Vertical profiles showing three upward-fining fluvial cycles. Profile A is a low-gradient channel deposit in the Pottsville Group near Reynoldsville, Pennsylvania. Profiles B and C are high-gradient channel deposits in the Mauch Chunk Formation near Pottsville, Pennsylvania. Individual channel deposits may contain one or more subcycles (numbered along left margin of each section) representing major flood deposits. Scale is the same for all profiles.

Both high-gradient and low-gradient channel deposits can be identified. The high-gradient deposits are characterized by a coarse lithology (dominantly conglomerate and conglomeratic sandstone), an irregular upward change in various sedimentary attributes, and the common occurrence of irregular and lenticular beds. In contrast the low-gradient deposits are commonly much finer grained (dominantly fine to medium sandstone), show a more systematic stratigraphic variation in many of the sedimentary features, and display less lenticular bedding. Vertical sections typical of these two channel types are compared in Figure 10.

Individual channel deposits, 30–80 ft thick, commonly contain thinner subcycles (Fig. 10). These subcycles, 3–20 ft, are characterized by a sharp lower contact and an upward decrease in

grain size. Each subcycle suggests a period of scour followed by a period of deposition in which the energy of the moving water decreases with time, finally culminating in a period of slack water represented by the finer grain sizes (commonly clay) at the top of the subcycle. Scour followed by deposition of sand and then clay occurs during floods in many modern channels; therefore, these subcycles are probably flood cycles within a channel. These flood cycles can occur either individually or be repeated, one stacked on top of the other. One subcycle commonly removes part of the underlying one. Where repeated within a single channel section, successive subcycles tend to be thinner and have a finer average grain size. Refer to Figure 10.

Geographically the high-gradient channel deposits are largely restricted to the eastern part of the basin (paleogeographically the upper part of the coastal plain) where they coalesce to form thick sections. Toward the distal part of the basin (lower coastal plain) the sections thin and separate into a number of low-gradient channel sands (sheets and belts) which are separated by finer grained deposits, e.g., in the Pottsville. In the Pennsylvanian these interbedded channel deposits, fine nonchannel sediments, and coals have been termed nonmarine cyclothems. Genetically these cyclothems are upward-fining channel deposits overlain by flood basin sediment. In many instances these cycles are produced as fluvial environments shift laterally over a subsiding coastal plain.

The channel types are but part of a continuous spectrum which ranges from alluvial fan to deltaic channels. In many cases, sedimentologic criteria from the channels alone are not sufficient to distinguish the various genetic types. For this, studies of the immediately adjacent stratigraphic framework are necessary.

DEPOSITIONAL MODEL

A depositional model for alluvial sediments based on stratigraphic and paleocurrent data depicts these strata as a thick wedge which extends into the basin from the tectonically active source. This wedge represents a balance between the rate of basinal subsidence and the rate of deposition

which is related to the tectonic activity in the source area. Table 3 describes the major elements of this model which are summarized in a transverse cross section in Figure 11. This cross section shows that at a given time the following relationships exist from east to west. During both regional transgressions and regressions coarse channel deposits (nonred) are transitional westward with finer red-bed deposits which in turn are transitional with marine deposits. During periods of maximum regression, however, conglomeratic nonmarine units fine to the west where they intertongue with marine deposits; the red-bed facies is absent.

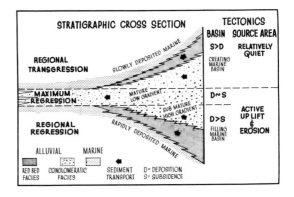

FIGURE 11. Schematic section across the basin which relates nonmarine deposition to basin tectonics. Time lines are essentially horizontal.

TABLE 3. Depositional Model for Paleozoic Alluvial Deposition in Central Appalachians

Basin Geometry	Elongate (trends NE–SW); widens to the SW; asymmetric in cross section, composed of a narrow trough area on the SE which is transitional with broad stable platform area to N and W.
Source Areas	Tectonically active area along SE margin of basin; stable cratonic area along N margin. Both composed dominantly of sedimentary and metamorphic rocks.
Recurrent Nonmarine Unit	A thick wedge of terrigenous sediment produced by a major orogeny in the tectonic source. Each wedge comprises strata deposited during a regional regression overlain by strata deposited during a regional transgression.
Sedimentary Associations	Wedge composed of three lithologic parts: (1) a lower red-bed facies, (2) a middle conglomeratic facies, and (3) an upper red-bed facies. Each is markedly wedge-shaped and thins away from tectonically active margin of basis. Red strata thin out completely in the distal parts of the basin.
Depositional Pattern	Dominantly transverse fill from tectonic source; depositional strike parallels basin axis and is normal to paleoslope; clastic ratios (sand to shale, conglomerate to sand), maximum grain size, and cross-bedding thickness decrease in downcurrent direction (i.e., SE to NW); compositional maturity increases downslope. Sandstones dominantly protoquartzites. Small intermittent contributions from northern cratonic source. Sandstones dominantly orthoquartzites.
Facies Relations	Coastal plain environment in which high-gradient alluvial fans along eastern margin of basin are transitional downslope with low-gradient alluvial, lacustrine, and deltaic deposits which in turn are transitional westward with time-equivalent marine units. Coarse channel deposits characterized by upward-fining fluvial sequences are the most common genetic unit.

Other workers (Potter and Pettijohn, 1963; Pettijohn, Potter, and Siever, 1965) have also used the similarity of attributes of some of these predominantly nonmarine deposits to develop a sedimentary or depositional model for Appalachian molasse.

AREAS OF FUTURE WORK

Though studied for over 100 years, the alluvial sections of the central Appalachians still pose many interesting and unanswered questions. At present our recognition of the various genetic units (both fine and coarse grained) which comprise these alluvial deposits is primitive. Additional data are needed to establish criteria for recognition of these genetic units and to understand their organization. For instance, what characterizes the deltaic environments of the coastal plain? What features distinguish channel deposits of braided streams from those of meandering streams?

The depositional, pinchout edge of the nonmarine with the marine is not well understood. In this zone where the marine and nonmarine intertongue, what are the stratigraphic and genetic relations of the various depositional units? Many of the low-gradient nonmarine facies occur in this prism of sediment.

Still largely unanswered is the Appalachian red-bed problem which is ripe for restudy in light of new geochemical information and better geologic mapping. What are the origins and significance of these red clastics?

The sedimentology of some Paleozoic alluvial deposits (such as the Llewelyn Formation and its time equivalents) has yet to be studied regionally in detail.

These areas of future work point out that the depositional model suggested in this paper is based on incomplete data. More detailed studies of both the stratigraphic and sedimentologic characteristics of these genetically related deposits will certainly permit considerable refinement and elaboration of this model.

REFERENCES

Allen, J. R. L., 1965a, A review of the origin and characteristics of Recent alluvial sediments: Sedimentology (Special Issue), v. 5, 191 p.

————, 1965b, Fining upwards cycles in alluvial successions: Liverpool and Manchester Geol. Jour., v. 4, p. 229–246.

Amsden, T. W., 1955, Lithofacies map of Lower Silurian deposits in the central and eastern United States and Canada: Am. Assoc. Petr. Geol. Bull., v. 39, p. 60–74.

Bernard, H. A., and Major, C. F., 1963, Recent meander belt deposits of the Brazos River: An alluvial "sand" model: Am. Assoc. Petr. Geol. Bull., v. 47, p. 350 (abstract).

Branson, C. C., 1962, Pennsylvanian System of central Appalachians, p. 97–116: in Pennsylvanian System in the United States—a symposium: Am. Assoc. Petr. Geol. Special Publication 1.

Burtner, R. L., 1963, Sediment dispersal patterns within the Catskill facies of southeastern New York and northeastern Pennsylvania: Pennsylvania Geol. Survey, General Geology Report G39, p. 7–23.

Clark, J., 1962, Field interpretation of red beds: Geol. Soc. America Bull., v. 73, p. 423–428.

de Witt, W., Jr., 1961, Stratigraphy of the Berea sandstone and associated rocks in northeastern Ohio and northwestern Pennsylvania: Geol. Soc. America Bull., v. 62, p. 1347–1370.

Edwards, J. D., 1955, Studies of some early Tertiary red conglomerates of central Mexico: U. S. Geol. Survey Prof. Paper 264H, p. 153–183.

Folk, R. L., 1960, Petrography and origin of the Tuscarora, Rose Hill, and Keefer Formations, Lower and Middle Silurian of eastern West Virginia: Jour. Sed. Pet., v. 30, p. 1–58.

Horowitz, D. H., 1965, Petrology of the Upper Ordovician and Lower Silurian rocks in the central Appalachians: Unpublished Ph.D. Thesis: The Pennsylvania State University, 221 p.

Hunter, R. E., 1960, Iron sedimentation in the Clinton Group of the central Applachian Basin: Ph.D. Thesis: Johns Hopkins University, 242 p.

McIver, N. L., 1961, Upper Devonian marine sedimentation in the central Appalachians: Unpublished Ph.D. Thesis: The Johns Hopkins University, 347 p.

Meckel, L. D., 1964, Pottsville sedimentology central Appalachians: Ph.D. Thesis: The Johns Hopkins University, 411 p.

————, 1967, Origin of Pottsville conglomerates (Pennsylvanian) in the central Appalachians: Geol. Soc. America Bull., v. 78, p. 223–258.

Pelletier, B. R., 1957, Pocono paleocurrents: Ph.D. Thesis: The Johns Hopkins University, 285 p.

————, 1958, Pocono paleocurrents in Pennsylvania and Maryland: Geol. Soc. America Bull., v. 79, p. 1033–1064.

Pettijohn, F. J., 1957, Sedimentary Rocks: 2nd ed., New York, Harper and Brothers, 718 p.

————, 1962, Paleocurrents and paleogeography: Am. Assoc. Petr. Geol. Bull., v. 46, p. 1468–1493.

————, and others, 1965, Geology of sand and sandstone: Indiana Geol. Survey, Bloomington, Indiana, 225 p.

Potter, P. E., 1967, Sand bodies and sedimentary environments: A review: Am. Assoc. Petr. Geol. Bull., v. 51, p. 337–365.

————, and Pettijohn, F. J., 1963, Paleocurrents and basin analysis: Heidelberg, Springer–Verlag, 296 p.

Visher, G. S., 1965a, Fluvial processes as interpreted from ancient and Recent fluvial deposits: in Primary sedimentary structures and their hydrodynamic interpretation, Soc. Econ. Paleon. and Miner., Special Publication No. 12, p. 116–132.

_____, 1965b, Use of vertical profile in environment reconstruction: Am. Assoc. Petr. Geol. Bull., v. 49, p. 41–61.

Walker, T. R., 1967, Formation of red beds in modern and ancient deserts: Geol. Soc. America Bull., v. 78, p. 353–368.

Woodward, H. P., 1951, Ordovician System of West Virginia: West Virginia Geol. Survey Bull. 21, 627 p.

Yeakel, L. S., 1959, Tuscarora, Juniata, and Bald Eagle paleocurrents and paleogeography in the central Appalachians, Ph.D. Thesis: The Johns Hopkins University, 454 p.

_____, 1962, Tuscarora, Juniata, and Bald Eagle paleocurrents and paleogeography in the central Appalachians: Geol. Soc. America Bull., v. 73, p. 1515–1540.

Chapter 4

Appalachian Turbidites

N. L. McIVER

INTRODUCTION

THE WORD "TURBIDITE" has come to designate a sedimentary facies characterized by graded sandstones. Other features common to this facies include sandstones which are moderate to poorly sorted, continuous bedding of even thickness, thick sequence of sandstone–shale interbeds, current bottom marks on sandstone beds, absence of megafauna in shale and siltstone beds, and absence of medium- to large-scale cross-bedding. These sandstones are interpreted to be deposits of turbidity or density currents in relatively deep water (below normal wave base).

In attempting to establish an interpretation of deep water sand deposition, sedimentologists in the 1950's tended to emphasize the similarities common to turbidite sequences. The similarities are now generally recognized and the density current origin is widely accepted. However, as experience with these sediments has broadened, one is made more aware of facies differences within the group. Deep water sediments show a complete gradation from shale with thin sand interbeds to sand with little or no shale, and from siltstone to coarse conglomerates. Acknowledging that such variations exist, the problem is to interpret them relative to source areas, basin geometry, and tectonic history.

The purpose of this paper will be to summarize two Appalachian turbidite examples, note their differences as well as similarities, and interpret the variations. Three turbidite sequences are reported in the central Appalachian stratigraphic column—the Wissahickon Formation of the Glenarm Series, Precambrian to Early Cambrian (Hopson, 1964); the Martinsburg Formation, Upper Ordovician (McBride, 1962); and the "Portage" rocks, known by various stratigraphic

names in this area, of Upper Devonian age (McIver, 1961). Primary depositional features and basin setting of the latter two formations will be compared and contrasted in this paper. Adequate comparison with the Wissahickon example is not possible because primary structures and textures are too greatly altered by metamorphic recrystallization and deformation. Descriptions by Hopson, however, indicate a close resemblance to the Martinsburg Formation.

MARTINSBURG FORMATION

General

Sandstones of the Martinsburg Formation have been interpreted as turbidites. The evidence bearing on this conclusion has been published by McBride (1962). The Martinsburg Formation crops out in a continuous belt 3–18 mi (5–30 km) wide in the Appalachian Great Valley. Generalized stratigraphic relations of the Martinsburg and the Reedsville facies are summarized in Figure 1. The Martinsburg turbidites grade into underlying dark gray to black shale which in turn are conformable with underlying argillaceous Trenton limestones. The upper contact is generally truncated by the Bald Eagle (Oswego) Formation and by the basal Silurian Tuscarora Formation in eastern Pennsylvania.

A local exposure of fossiliferous sandstone and interbedded shale, the Shochary Sandstone, is thought to represent the youngest Martinsburg in this area. The Reedville shale lithofacies crops out in the folded belt of central Pennsylvania. The Reedsville shales become silty in the upper half and culminate in a 50-ft (15 m) fossiliferous sandstone at the top. The Reedsville sediments are more distal (from the source area) and ap-

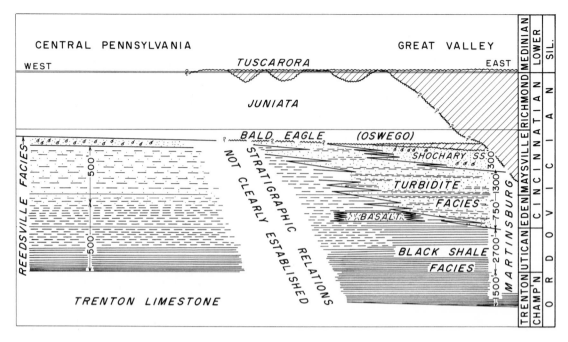

FIGURE 1. Generalized stratigraphic relations of the Martinsburg Formation and Reedsville facies in Pennsylvania. Thicknesses are only approximate.

pear to be somewhat shallower water deposits than the Martinsburg turbidites.

True thickness of the Martinsburg Formation is not known due to intense folding and faulting. Estimates range from 2,250 to 4,000 ft (700–1,200 m) for the main outcrop belt (McBride, 1962, p. 43). The preserved section includes approximately one-third sand-shale (turbidite) facies and two-thirds shale facies.

Petrography

Shale makes up from 30 to 60 percent of the turbidite units. The shales are generally dark gray and carbonaceous with rare graptolites on parting planes. These shales are commonly interbedded with graded siltstones and, higher in the section, with graywackes. The gray and black shales are considered to be "normal pelagic deposits representing a very slow rate of deposition below wave base" (McBride, 1962, p. 46). Siliceous shales and cherts occur locally in the lower Martinsburg. Radiolaria and graptolites have been reported in these shales.

The Martinsburg sandstones contain greater than 15 percent detrital matrix (finer than 0.02 mm), thus are classified as graywackes (Pettijohn, 1957, p. 291). McBride (1962, p. 61) reports that matrix makes up 22–50 percent of the rock for the 55 thin sections studied. The sands are described as fine to coarse grain having fair to poor sorting. Calcite cement, occurring

chiefly as replacement intergrowths, makes up an average of 8 percent of these graywackes. The principal framework constituents are quartz, feldspar, and lithic grains (Fig. 2). The relative proportions of each are shown in Figure 3. The Martinsburg sandstones are mostly lithic graywackes and those from the Reedsville are subgraywackes.

Lenticular conglomerates and pebbly sandstones are an important but minor component of the Martinsburg. They occur as thin (generally less than 3 or 4 ft thick) lenses in the sandstones. McBride (1962, p. 74–76) describes two types— a quartz pebble conglomerate and a polymictic conglomerate. The quartz conglomerates consist of subrounded to well-rounded vein quartz with less than 10 percent metaquartzite and shale pebbles, cemented by a graywacke matrix. The polymictic conglomerates contain 20–60 percent pebbles of limestone, shale, chert, graywacke, siltstone, quartzite, volcanics, and fossil debris "floating" in a sandy matrix.

The variety of detrital minerals which occur in the Martinsburg led McBride (1962, p. 65) to conclude that the sedimentary material was derived from "terrigenous and carbonate sedimentary rocks, low-grade metamorphic, and acid plutonic rocks; and trace contributions from extrusive volcanic rocks." Regional variations suggested a more metamorphic or igneous source to the northeast and a dominantly sedimentary source to the southwest.

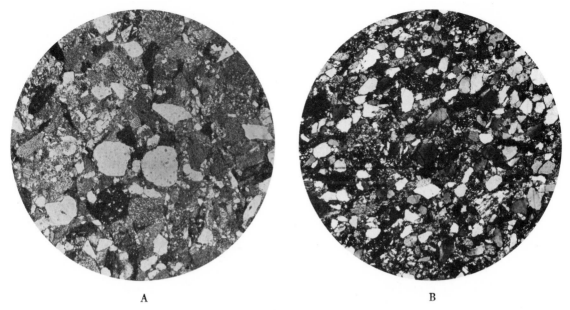

FIGURE 2. Photomicrographs of Martinsburg graywackes (A) Lithic graywacke, crossed nicols, 12 ×. (B) Graywacke in which grain imbrication is evident. Bedding is parallel to dark line. Photos courtesy of E. McBride.

Sedimentary Structures

The Martinsburg sandstones and interbedded shales are characteristically very well bedded. The beds are even and laterally continuous (Fig. 4). Bedding thickness varies widely up to several feet. According to McBride (1962, p. 44) over 90 percent of the graywacke beds are less than 7 in. (17.8 cm) thick. Internal bedding structures include graded bedding, horizontal lamination, small-scale cross lamination, convolute bedding, and massive bedding. Graded bedding is very common and particularly well developed in the coarser sandstones. Some thicker beds contain several graded sediment units. Various types of laminated bedding are most common in the finer grain beds, or in the upper, silty portion of graded beds. Structureless beds are relatively rare.

Martinsburg sole markings include all of the primary structures commonly associated with graded sandstone sequences. Most common are groove casts and flute casts (Fig. 4). In the Martinsburg both types show great variation in form, size, and abundance. Less common sole structures include load casts, scour- and channel-fill, and organic markings.

Directional Features

Preferred grain orientation in the bedding plane is most pronounced in the Martinsburg lithic types which contain a high proportion of elongate grains. McBride also found generally good agreement between the orientation of fabric and sole markings. Furthermore, in the plane normal to bedding and parallel to the current direction, the grains have a preferred imbricate orientation, dipping in the up-current direction an average of 20 degrees.

Current directional data from 124 locations are summarized in rose diagrams for 13 subregional divisions of the main outcrop belt (McBride, 1962). These data show considerable scatter, both at single field locations and in the subregional groupings. Although some scatter may have been introduced by grouping locations, by mixing various types of data, and by tectonic movements, that which occurs at individual outcrops indicates that the dispersion of currents is real and complex.

This scatter makes interpretation difficult. Of the 13 subregional groupings, 9 clearly show a single major current trend, 4 show a significant secondary trend, and in 2 cases there is no obvious primary mode (see McBride, 1962, Figs. 28–30). Using both types of data in combination (that which gives sense of direction and direction only) the primary and secondary current trends were intepreted and plotted on an outcrop base map (Fig. 5). This simplified version of McBride's paleocurrent map clearly shows a tendency for the currents to be either parallel or at a high angle to the outcrop belt. The secondary mode at Subdivision 10 may not be significant or may have the reverse sense of direction. At the

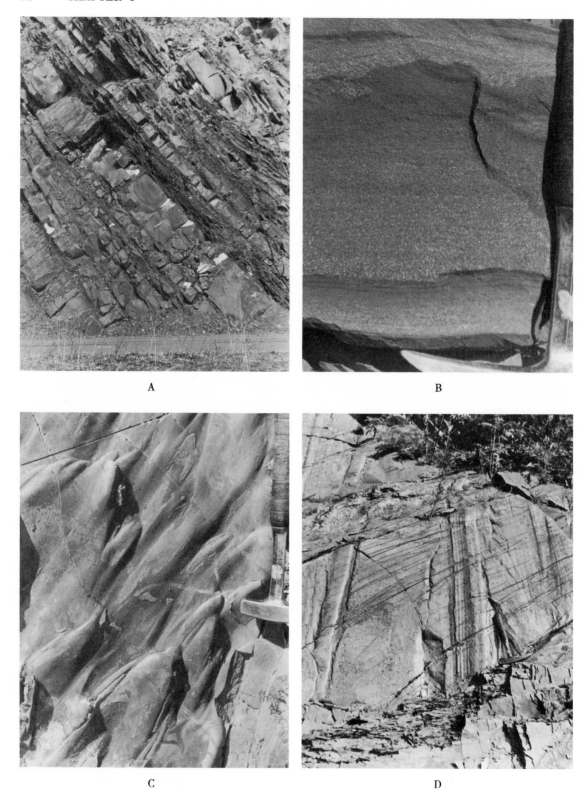

A

B

C

D

FIGURE 3. Primary structures which characterize the Martinsburg sandstones. (A) Medium to thick interbeds of sandstone and shale, Hamburg, Pennsylvania. (B) Graded graywacke. (C) Flute casts. (D) Large groove casts. Photos courtesy of E. McBride. (From McBride, 1962.)

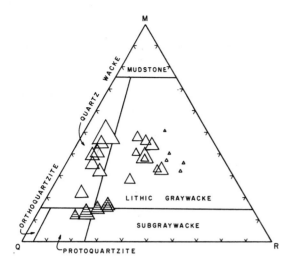

FIGURE 4. Four component composition diagram of Martinsburg (open triangles) and Reedsville (hatchured triangles) sandstones. Feldspar percent is indicated by the length of the base of each triangle. Modified after McBride (1962).

southern end of the area currents indicated in Subdivisions 1 and 2 flowed northward, parallel to the outcrop. In Subdivisions 11, 12, and 13 flute casts indicate southerly flow, also parallel to the outcrop.

Paleogeography

On the basis of facies distribution, McBride (1962, p. 81–83) concluded that the Martinsburg sandstones were deposited in a linear trough essentially parallel to the present tectonic strike.

Opposing paleocurrent directions parallel to the strike support the interpretation of a linear trough. They further indicate a bathymetric low in the trough in east-central Pennsylvania.

Detrital material was apparently supplied from the southeast side of the trough as well as the ends. There is no clear evidence from facies relations, sandstone bedding thickness variations, grain size, or mineralogy to indicate the major supply at either end of the basin. The only apparent regional facies relations indicate that the western Reedsville facies was distal relative to the sand source. On the basis of the scattered current directions McBride suggests that the trough had a flat basin floor without significant regional dip. The scatter could also be accounted for by complex, divergent distributary channel systems on submarine fans. Interfan areas could have opposed directional features produced.

Estimates of the paleobathymetry during Martinsburg deposition range from bathyal to abyssal. In the absence of diagnostic fauna the depth can never be established. However, the following lines of indirect evidence are cited in support of a deep water interpretation: (1) abundant sedimentary structures and textures in common with other known deep-water deposits, (2) absence of shallow water structures or organisms, (3) presence of red shales, the color of which was considered primary (Moseley, 1948, p. 22) and analogous to present-day deep-water red oozes, and (4) the occurrence of radiolarian cherts.

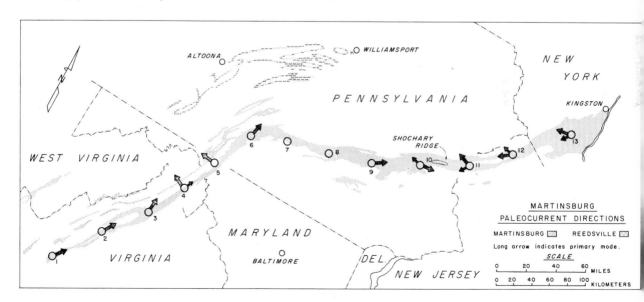

FIGURE 5. Interpreted average current directions for 13 subregional groups of data, Martinsburg Formation. The averages are based on 732 readings by McBride (1962).

UPPER DEVONIAN TURBIDITES

General

To the west and northwest of the Catskill deltaic complex occur some interbedded shales, siltstones, and fine sandstones which are recognized as turbidites. These rocks are designated the "Portage" lithofacies of the Upper Devonian regressive sequence (Woodward, 1943; McIver, 1961; Sutton, 1963). The name "Portage" is used because, in describing the original Portage Group of western New York, James Hall (1843) noted many features of these rocks which characterize them as turbidites, including the first account of bottom markings. As stratigraphic names multiplied in this area the name "Portage" fell into disuse (Chadwick, 1935). The more common stratigraphic units comprising the "Portage" facies, as well as the subjacent and superjacent facies in West Virginia, Maryland, and Pennsylvania, are shown in Figure 6.

The Upper Devonian in this area is entirely composed of a clastic wedge of sediments related to the Catskill delta system. The wedge trends parallel to the folded belt, and extends from a preserved maximum thickness over 8,000 ft (2,450 m) in eastern Pennsylvania to a pinchout edge on the Cincinnati Arch in central Ohio. The sand: shale ratio systematically decreases from greater than 75 percent in the east to less than 10 percent in Ohio (Ayrton, 1963).

The "Portage" is underlain by up to 300 ft (90+ m) of black and dark to medium gray shales of the "Genesee" lithofacies. The "Genesee" is represented by the Genesee Shale of Western Maryland, the Burket and Harrell shales of West Virginia and Pennsylvania.

The "Portage" is overlain by the shallow marine or shelf deposits known as "Chemung" lithofacies. The "Chemung" is best seen in the

type area of the Chemung Formation, on the Chemung River in southeastern New York. Other examples of this lithofacies include the Chemung Formation exposed at Keyser, West Virginia, and the upper Trimmers Rock sandstone in the Susquehanna Valley. The "Chemung" lithofacies is characterized by thin to medium, uneven beds of siltstone and sandstone, interbedded with shales and mudstones, generally containing lenses of megafauna and abundant organic markings. At a few localities lenses of rounded, vein quartz pebbles are present. In most areas there is a general increase in sand from less than 50 percent in the lower half to 60–75 percent in the upper 100 ft (30 m). The "Chemung" siltstones and sandstones are massive, laminated, rippled, and cross-bedded. Minor channeling and ball-and-pillow structures are also common. Throughout the folded belt the "Chemung" facies is between 1,500 and 3,000 ft thick (450–900 m).

Contact between "Portage" and "Chemung" facies in the Valley and Ridge Province is gradational through 100–200 ft (30–60 m), depending on one's interpretation as well as the nature of the transition. In New York these facies are more difficult to separate. In the Finger Lakes district rocks of "Portage" and "Chemung" lithology occur in alternating repetition, generally with interspersed black shale zones. These stratigraphic units, separated by black shales, have been mapped in detail and demonstrate the lateral equivalency of "Chemung" and "Portage" facies (de Witt and Colton, 1959; Sutton, 1963). Such correlation is not well established in the folded belt but the same regional relationships are thought to exist between the gross units (Woodward, 1963, p. 286). These relationships are schematically shown in Figure 7 for Pennsylvania.

"Portage" Lithofacies

The "Portage" lithofacies as defined here is typified by the Brallier Shale in Pennsylvania and West Virginia and the Woodmont Shale in Maryland. It consists of even interbeds of shale, siltstone, and very fine sandstone. Sandy units range from 20 to 50 percent sandstone. Characteristic primary structures include sole markings, horizontal lamination, small-scale cross lamination, and convolute lamination. Throughout Pennsylvania and Maryland these rocks have an average thickness of 1,500–1,800 ft (450–550 m). Poor exposures, structural complications, and difficulty in defining the contacts preclude the possibility of satisfactorily isopaching this unit.

The "Portage" is generally agreed to be sparsely fossiliferous. These beds contain a so-

LITHOFACIES	WEST VIRGINIA	MARYLAND		PENNSYLVANIA
"CATSKILL"	HAMPSHIRE FM.	HAMPSHIRE FM.		CATSKILL "MAGNAFACIES"
"CHEMUNG"	CHEMUNG FM.	CHEMUNG MBR. PARKHEAD SS.		CHEMUNG FM. U. TRIMMERS ROCK
"PORTAGE"	BRALLIER SH. U. HARRELL SH.	WOODMONT MBR.		L. TRIMMERS ROCK BRALLIER SH.
"GENESEE"	L. HARRELL SH.	BURKET SH. GENESEE SH.		HARRELL SH. BURKET SH.
		TULLY LST.		TULLY LST.

FIGURE 6. Correlation chart of principal stratigraphic units comprising four lithofacies of the Upper Devonian in the central Appalachians.

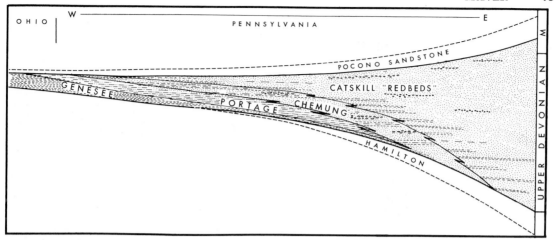

FIGURE 7. Generalized cross section of lithofacies in the Upper Devonian of Pennsylvania.

called "Naples" fauna of thin shelled pelecypods, gastropods, brachiopods, and a characteristic track, *Pteridichnites bisereatus.*

Sandstone Petrography

Because of fine texture, petrographic study of the "Portage" sandstones is difficult. Of 15 "Portage" samples examined in thin section, 7 are fine to very fine grain sandstones and the remainder are siltstones. The frameworks average 69 percent of the rock. Of the whole rock, quartz (+ chert) makes up 60 percent, feldspar 3 percent, rock fragments 5 percent, micas and other heavy minerals 1 percent, and the remainder is matrix. Rock fragments of shale and siltstone may exceed 5 percent but in a fine grain, highly compacted rock they are difficult to distinguish from matrix.

Heavy minerals were separated after treating crushed samples with concentrated hydrofluoric acid. Opaques, including magnetite, hematite, leucoxene, pyrite, and hydrous iron oxide (?), make up 60–75 percent of the separates. The remainder is a mixture of round to angular zircon, tourmaline, chlorite and traces of pyroxene, hornblende, rutile, and epidote.

The matrix is a dense mat of illitic clay, chlorite, sericite, silica (or quartz?), and opaques or semiopaques (carbonaceous material?). Locally the matrix and adjacent shales contain finely disseminated calcite.

Seven sandstones are plotted on a triangular classification diagram in Figure 8. They are in the category of quartz wackes and protoquartzites.

These sandstones could not be disaggregated for size analysis. Furthermore, point counting was not found satisfactory for very fine sand and silt. Size estimates were made of the framework

fractions by comparing them with artificially packed and impregnated mixtures. The median grain size of the framework ranges up to fine sand (maximum Md = 0.15 mm). The sorting of the framework is good to moderate (Trask sorting coefficient of 1.30–1.70).

The "Portage" lithofacies is thought to be derived from a source of sedimentary and low-rank metamorphic rocks. This was also the conclusion of Krynine (1940) after studying "Chemung" facies Third Bradford Sand in northwestern Pennsylvania. This conclusion is based on the high ratio of quartz + chert to feldspar + rock fragments, the sedimentary origin of the rock fragments, the rounded to angular mixture of stable heavy minerals, and lack of abundant mafics of typical high-rank metamorphic minerals.

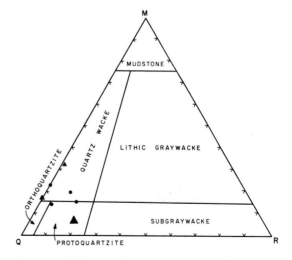

FIGURE 8. Four component diagram of Portage sandstones—quartz (Q), rock fragments (R), matrix (M), and feldspar (indicated by the length of the triangle base or by a dot for less than 1 percent).

FIGURE 9. Interbedded shales, siltstones and fine sandstones of the "Portage" facies, Sunbury, Penna.

Sedimentary Structures

For a complete description of sedimentary structures the reader is referred to other sources (Clarke, 1917; McIver, 1961; Potter and Pettijohn, 1963; Pettijohn and Potter, 1964; Frakes, 1967). Bedding of the "Portage" sandstones and siltstones is typical of most fine-grain turbidite deposits—they are remarkably even, rhythmic interbeds with shale (Fig. 9). Some statistics on the nature of internal bedding structure and bedding thickness at two localities of typical "Portage" are given in Table 1. Graded bedding was generally not apparent in the field, nor could grading be detected by examination of cut and polished samples under the binocular microscope. Frakes (1967) was able to demonstrate slight grading in thin sections of six "Portage" siltstones cut normal to bedding. In laminated beds there is commonly a distinct graded upper contact in contrast to a sharp lower contact (Fig. 10). The lack of typical grading of sand beds can best be explained by the limited range of grain size available at the site of deposition.

Small-scale cross lamination (generally less than 1 in. thick) is very common in "Portage" siltstones and sandstones. It usually occurs in the upper part of laminated beds or at the top of massive beds and may be associated with convolute bedding.

Sole marks are abundant locally, particularly in the Finger Lakes district, New York. Flute casts and a variety of substratal grooves and striations are present (Figs. 11 and 12). Flute casts are relatively uncommon compared to groove casts.

FIGURE 10. Typical sequence of internal structures in a "Portage" siltstone. Note the sharp basal contact with dark shale and gradational upper contact. A thin zone of fossil fragments (mostly crinoidal) occurs at top of the horizontal laminations.

FIGURE 11.   Flute and groove casts on the base of a "Portage" sandstone bed. (From Pettijohn and Potter, 1964.)

FIGURE 12.   Groove casts, small prod casts, striations and chevron casts on the base of Portage sandstone bed. (From Pettijohn and Potter, 1964.)

Directional Features

Fifty-six oriented samples were collected at 16 outcrops in New York and Pennsylvania. Most of the samples are from relatively undeformed sections and all beds sampled possessed bottom marks. Three to five grain orientation measurements were made in each sample using a method developed by Shell Development Company (Fig. 13). There is close agreement (7 degrees) between the average orientation of fabrics and sole marks. Preferred orientation of plant and fossil fragments in sandstone beds was also measured and found to be essentially east–west. Colton (1967) found a similar relationship for plant fragments in siltstone and sandstone beds in western New York. However, Colton noted that specimens of *Fucoides graphica* Vanuxem and plant fragments in shale and mudrock are oriented northeast–southwest. He concluded that this direction represented the orientation of *normal* marine currents flowing parallel to the axis of an arm of the Devonian sea.

In small or poor exposures it was often difficult to distinguish "Chemung" from "Portage" lithofacies throughout much of the area. For this reason paleocurrent directions could not be satisfactorily mapped separately. Consequently the current data for all marine sandstones were grouped. Tests of the current variation for the two facies, both vertically and laterally, do not show a significant difference at the 95 percent confidence level between the "Portage" and the "Chemung" facies.

The average current directions based on approximately 2,400 readings (one average direction per bed) at 297 exposures of marine sandstones and siltstones in the Virginias, Maryland, Pennsylvania, and New York are shown in Figure 14. The regional average current directions are indicated by rose diagrams. The directions are characterized by pronounced consistency, both locally and regionally. The mean dispersion

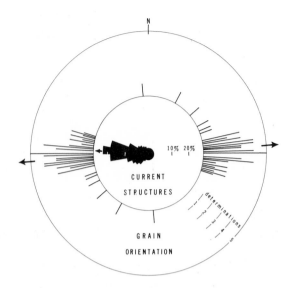

FIGURE 13. Preferred grain orientation compared with associated current structure azimuths for 57 "Portage" sandstones.

(difference between the mean and an individual reading) for all locations at which 10 or more readings were taken is only 13.4 degrees. Similar orientations and consistency were noted in New York by Sutton (1959) and Colton (1967).

Sandstone Genesis and Paleogeography

Deposition of the "Portage" sandstones in relatively deep water (below wave base) by turbidity currents is based on the following lines of evidence:

(1) regular alternation of even, continuous interbeds of shale, siltstones, and sandstones;

(2) graded upper contacts and sharp lower contacts of the sandstone beds;

(3) abundant, well-developed, and well-preserved bottom marks of current origin;

(4) regional and local consistency of current directions;

TABLE 1. Frequency Distribution of Bedding Types and Thicknesses for 400 "Portage" Beds Measured at Woodmont, Md. (Column 1) and 400 "Portage" Beds Measured at Sunbury, Penna. (Column 2)

			Bedding thickness (inches)					
	Frequency (%)		Mean		Maximum		Minimum	
Type of bedding	1	2	1	2	1	2	1	2
Laminated	53.4	45.0	2.9	3.3	27.0	14.0	0.5	0.5
Laminated and cross-laminated	26.8	21.1	3.8	4.1	13.0	25.0	0.5	1.0
Cross-laminated	10.6	12.1	1.7	1.8	3.5	7.5	0.5	0.5
Structureless	2.8	7.3	2.1	2.9	4.0	11.0	1.0	1.5
Composite	6.5	14.9	6.1	4.9	17.0	33.0	1.0	1.0
All beds			3.02	3.49	27.0	33.0	0.5	0.5

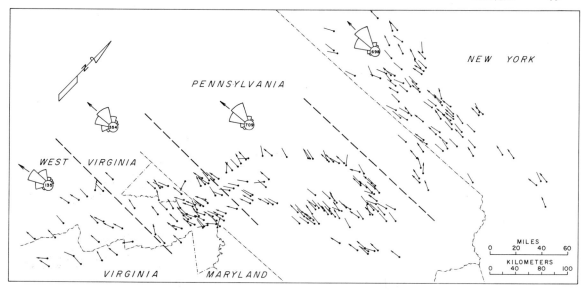

FIGURE 14. Average paleocurrent directions of Upper Devonian marine sandstones. Current directions are mainly based on "Portage" facies bottom markings except east of the Susquehanna River Valley and east of the central Finger Lakes where the Chemung facies predominates. Directions are summarized in rose diagrams for four subregions, the number of readings and average current direction are indicated for each.

(5) restricted, thin-shelled marine fauna compared to the laterally equivalent "Chemung" fauna;

(6) absence of medium- or large-scale cross-bedding.

Similar interpretation was made by Kuenen (1956), Sutton (1959), and Colton (1967), for these rocks in New York, and by Frakes (1967) in Pennsylvania.

Maximum depth of the Upper Devonian basin was probably less than 4,000 ft (1,220 m). This conclusion is based on the maximum accumulated thickness of the "Genesee" and "Portage" rocks plus approximately 600 ft (183 m) for a neritic interval. We assume the basin was not tectonically uplifted during deposition because there are no known marine unconformities. Prior to shale compaction the deposits would be somewhat thicker at the expense of the overlying water body. It is reasonable to assume some subsidence during deposition which would further reduce the maximum bathymetry. The presence of very fine to fine sands only, without pebbles or shale clasts, suggests relatively slow currents on gentle slopes. The alternating "Portage" and "Chemung"-like rocks in western New York certainly suggest a shallow sea in that area and the occurrence of periodic subsidence. It is likely that maximum paleobathymetry did not exceed 2,000 ft (610 m, upper bathyal depth).

Comparison of Martinsburg and Portage

Both sedimentary sequences are underlain by dark gray to black shales and argillaceous limestones and are overlain by shallow marine and/or continental sandstones. Pettijohn (1957, p. 64) noted these common sequences of "flysch" and "molasse" which make up two geosynclinal or tectonic cycles. However, from the description presented here it is apparent that these cycles are different.

The "Portage" facies intertongues laterally and grades vertically into shallow marine and continental regressive sediments. The Martinsburg, however, is unconformably overlain at its eastern margin by sands of shallow-water origin. The fossiliferous graywackes of the Shochary Ridge may be somewhat shallower than the Martinsburg but are still considered deep-water deposits, perhaps near the basin edge but not shelf deposits such as the "Chemung." Basinward the "Portage" passes into a widespread black-shale facies whereas the distal equivalent (and in part younger) facies of the Martinsburg is the silty Reedsville. It is possible that the real facies difference between these sequences of rocks is in the state of preservation. Inasmuch as we do not see the eastern shelf or continental equivalents of the Martinsburg we cannot be sure that deltas did not develop there. The Ordovician Queenston red and green shales and sandstones may be a deltaic facies but these rocks are considerably younger (Upper Cincinnatian) and are recognized only in western New York.

The Martinsburg sandstones are coarser and more poorly sorted than those of the "Portage." Mineralogically and texturally the Martinsburg graywackes are less mature. Conglomerate and shale clasts make up a minor but significant component of the Martinsburg turbidites but they are absent in the "Portage." The two units are similar in their general lack of fossil material in the sands. However, the Martinsburg shales contain no counterpart to the "Naples" megafauna.

Perhaps as a consequence of available grain size, the development of graded bedding is not the same in these sequences. Excellent graded bedding is common in the Martinsburg sandstones whereas only the upper contacts of the "Portage" turbidites are distinctly graded. Except for differences in bedding thickness, other primary fabrics and structures are equally well developed.

The most significant difference is in the paleocurrent patterns because they reflect differences in basin geometries and tectonics. The "Portage" paleocurrents define a broad, uniform paleoslope dipping gently from continental to basin environments. The Martinsburg paleoslope is poorly defined by diverse currents. They seem to indicate a linear trough into which currents flowed longitudinally from both ends as well as transversely and obliquely to the trough axis.

Conclusions

The Martinsburg clearly qualifies on sedimentological features to be classified as "flysch" according to the modern definition of Potter and Pettijohn (1963, p. 241–242), and others. The "Portage" probably does not, and at best can only be called flysch-like. Compared to most "flysch" the Upper Devonian turbidites are unusually shaly, fine-grained, and thin. Furthermore, the water depth was probably no greater than the minimum estimates of paleobathymetry for most flysch basins (Potter and Pettijohn, 1963, p. 241). A similar assessment of these rocks was made by Dzulynski (personal communication). Seilacher (1967, p. 197) considered the "Portage" biologically atypical of flysch deposits.

The "Portage" is a distal, relatively deep-water facies of the Catskill deltaic complex. Gross thickness distribution and sand : shale ratio (Ayrton, 1963) indicate that the depocenters were in the continental to strand areas. The meager sand supply swept over the shelf edge was deposited on gentle prodelta slopes. The important fact is that most of the sand was trapped in the deltas and was not available for turbidity current deposition.

There is no preserved record of Martinsburg deltaic deposition. Nor is there any evidence for shallow water sediments intertonguing with Martinsburg sands. If an analogy can be drawn with Recent deep-sea fan deposits, the coexistence of a delta is not necessary. Many submarine fans, such as those on the southern California continental borderland or the Astoria fan off the Columbia River of the Pacific Northwest, are accompanied by little or no present delta. Furthermore, it has yet to be demonstrated that fans in front of modern deltas such as the Mississippi, Niger, or Ganges contain appreciable contemporaneous sand deposits. The conclusion reached is that the differences noted in Martinsburg and "Portage" sand distribution reflect fundamental differences in development of the sedimentary basins.

Seilacher (1967) and others have recognized that the flysch suite is significantly different from other facies and that intermediate types are rare. In other words, there is not a complete range from shallow-water types to the flysch facies. Why should this be so? The basic difference between a deltaic basin and a flysch basin must be the nature of basin tectonics. The depocenters of thick deltaic sequences are localized by "load tectonics"—subsidence in response to deposition of sediment. In most cases the uneven balance between basin subsidence and deltaic progradation leads to complex intertonguing of facies. Flysch basins, on the other hand, form with an early, major structural depression to produce a deep trough. Many such deep troughs of structural origin are seen in the modern oceans. Paleobathymetric evidence shows that some Tertiary basins of California underwent substantial subsidence (bathyal water depth) before sand deposition began (Natland and Rothwell, 1954). A similar structural development of a Martinsburg trough may have preceded sand deposition. In final conclusion the proposition is made that deltaic and flysch deposits will tend to be mutually exclusive and will not merge within a single basin.

REFERENCES

Ayrton, W. G., 1963, Isopach and lithofacies maps of the Upper Devonian of northeastern United States: Pa. Geol. Survey Rpt. G39, p. 3–6.

Chadwick, G. H., 1935, Chemung is Portage: Geol. Soc. America Bull., v. 46, p. 343–354.

Clarke, J. M., 1917, Strand and undertow markings of Upper Devonian time as indicators of the prevailing climate: N. Y. St. Mus. Bull., 196, p. 199–238.

Colton, G. W., 1967, Late Devonian current directions in western New York with special reference to Fucoides graphica: Jour. Geol., v. 75, p. 11–22.

de Witt, W., Jr., and Colton, G. W., 1959, Revised correlations of the Lower Upper Devonian rocks in western and central New York: Am. Assoc. Pet. Geol. Bull., v. 43, p. 2810–2828.

Frakes, L. A., 1967, Stratigraphy of the Devonian Trimmers Rock in eastern Pennsylvania: Penna. Geol. Surv. Bull. G51, 208 p.

Hall, James, 1843, Geology of New York, Pt. 4, Survey of the fourth district: Albany, Carrol and Cook, 683 p.

Hopson, C. A., 1964, The crystalline rocks of Howard and Montgomery Counties: The Geology of Howard and Montgomery Counties, Md. Geol. Survey, 336 p.

Krynine, P. D., 1940, Petrology and genesis of the Third Bradford Sand: The Penna. State Coll. Bull. 29, 126 p.

Kuenen, Ph. H., 1956, Problematic origin of the Naples rocks around Ithaca, New York: Geol. en Mijnb., v. 18, p. 277–283.

McBride, E. F., 1962, Flysch and associated beds of the Martinsburg Formation (Ordovician), Central Appalachians: Jour. Sed. Pet., v. 32, p. 39–91.

McIver, N. L., 1961, Upper Devonian Marine Sedimentation in the Central Appalachians: Ph.D. Thesis, The Johns Hopkins University, 530 p.

Moseley, J. R., 1948, Ordovician–Silurian contact in eastern Penna.: Ph.D. Thesis, Harvard Univ., 139 p.

Natland, M. L., and Rothwell, W. T. Jr., 1954, Fossil Foraminifera of the Los Angeles and Ventura regions, California: Calif. Div. of Mines Bull. 170, Chap. III, p. 33–42.

Pettijohn, F. J., 1957, Sedimentary rocks (2nd Ed.): New York, Harper and Brothers, 718 p.

_____, and Potter, P. E., 1964, Atlas and glossary of primary sedimentary structures: New York, Springer-Verlag, 370 p.

Potter, P. E., and Pettijohn, F. J., 1963, Paleocurrents and basin analysis: Berlin, New York, Springer-Verlag, 296 p.

Seilacher, A., 1967, Tektonischer, sedimentologischer oder biologischer Flysch: Geol. Rundsch, v. 56, p. 189–200.

Sutton, R. G., 1959, Use of flute casts in stratigraphic correlation: Am. Assoc. Pet. Geol. Bull., v. 43, p. 230–237.

_____, 1963, Correlation of Upper Devonian strata in south-central New York: Pa. Geol. Survey Rpt. G39, p. 87–101.

Woodward, H. P., 1943, Devonian system of West Virginia: West Va. Geol. Survey Bull., v. 15, 655 p.

_____, 1963, Upper Devonian stratigraphy and structure in northeastern Pennsylvania: Pa. Geol. Survey Rpt. G39, p. 279–301.

Chapter 5

Loyalhanna Limestone—Cross-bedding and Provenance

ROBERT W. ADAMS

INTRODUCTION

THE LOYALHANNA LIMESTONE (Mississippian) of southwestern Pennsylvania is a markedly cross-bedded arenaceous calcarenite associated with Paleozoic alluvial sandstones. The calcareous composition and large-scale cross-bedded nature of the unit have led previous investigators to conclude that the Loyalhanna is of marine or eolian origin. Documentation of a cross-bedded marine or eolian facies within this region will contribute to a broader understanding of the geologic history of the central Appalachians.

The Loyalhanna of Pennsylvania and its lateral equivalents in Maryland and West Virginia were examined in outcrops along the crest and flanks of several northeast–southwest trending anticlines within the Appalachian Plateau province of southwest Pennsylvania, western Maryland, and northern West Virginia (Fig. 1). Additional outcrops of presumed correlative formations in Pennsylvania and West Virginia were examined to clarify stratigraphic relations. Published and unpublished subsurface data aided in the determination of the geometry and regional extent of the Loyalhanna Limestone. An integrated study of stratigraphy, petrology, and paleocurrents was made in order to determine provenance, environment of deposition, and dispersal system.

ACKNOWLEDGMENTS

This paper is part of a dissertation submitted to The Johns Hopkins University in partial fulfillment of the requirements for a Doctor of Philosophy degree. The writer expresses his appreciation to Dr. Gordon Rittenhouse for suggesting the study and Dr. F. J. Pettijohn for supervising it. The assistance of the following people is gratefully acknowledged: N. K. Flint, University of Pittsburgh; W. S. Lytle and V. C. Shepps, Pennsylvania Geological Survey; Alan Donaldson and Dana Wells, University of West Virginia; and W. R. McCord, West Virginia Geological Survey. Financial assistance for the study came from an American Association of Petroleum Geologists grant-in-aid and two Shell Oil Company Foundation Fellowships.

METHODS OF STUDY

Deep weathering of the calcareous Loyalhanna and subsequent concealment by associated sandstones and shales precluded systematic sampling. Active and abandoned quarries provided the principal exposures and were supplemented by road cuts and natural exposures. Outcrops were examined for primary structures, lithologic characteristics, and relationships to formations above and below. Well-sample descriptions, published, unpublished, and by the writer, provided additional stratigraphic and lithologic control.

Petrographic studies for compositional and regional trend analyses were made on approximately 60 thin sections. Modal analyses based on 200 point counts per thin section were used to determine the general composition of the Loyalhanna. Estimates of terrigenous and carbonate detrital varieties were made by additional counts of 200 grains for each component. Heavy mineral analyses for 109 outcrops and wells were made by counting 200 grains per slide of the 0.063–0.125 mm nonopaque fraction. The round-

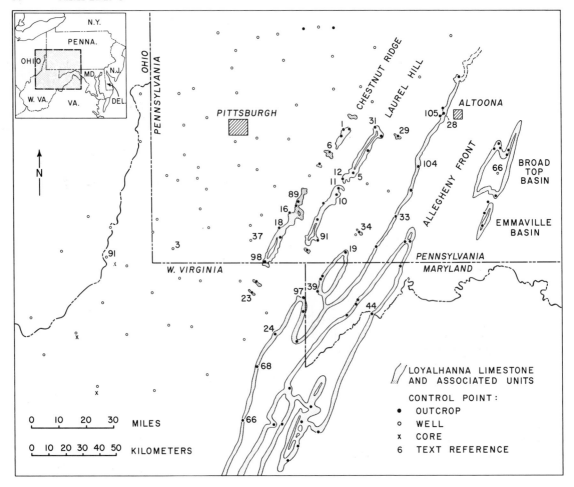

FIGURE 1. Outcrop and control map, Loyalhanna Limestone and associated units.

ness of tourmaline in the 0.125–0.250 mm size class was estimated for a minimum of 50 and maximum of 100 grains per slide by visual comparison with the roundness chart of Krumbein (1941, p. 68). Selected samples of the Loyalhanna were treated with acid in order to obtain the insoluble residue for size analyses.

All accessible cross-bedded sedimentation units in the Loyalhanna were measured for orientation of dip azimuth, angle of foreset dip, and maximum thickness. Correction for tectonic tilt was made wherever the true bedding was inclined more than 5 degrees. The mean current direction for each locality was determined by vector summation of cross-bedding dip azimuths (Curray, 1956). Vector summation and other statistical parameters including vector magnitude, consistency ratio, and variance were calculated using an IBM 7094 computer. A moving average of cross-bedding vector means was constructed by the graphic grid method (see Pelletier, 1958, Fig. 2) utilizing four adjacent 15-min quadrangles. A

contoured moving average of cross-bed unit thickness was constructed using the same technique.

STRATIGRAPHY

General Statement

The Loyalhanna Limestone was named by Charles Butts (1904, p. 5) from exposures along Loyalhanna Creek across Chestnut Ridge, Westmoreland County, Pennsylvania (Fig. 1, No. 6). The Loyalhanna occupies the stratigraphic interval above the Pocono Formation and below red beds of the Mauch Chunk Formation. Martin (1908, p. 4) correlated the Loyalhanna with the lower member of the Greenbrier Formation of western Maryland. Reger (1931, p. 323) suggested the Loyalhanna represented the Fredonia Limestone member of the Union Limestone, Greenbrier Series, in northern West Virginia and was equivalent to a portion of the Trough Creek Limestone member of the Mauch Chunk Formation in the Broad Top Basin of Pennsylvania

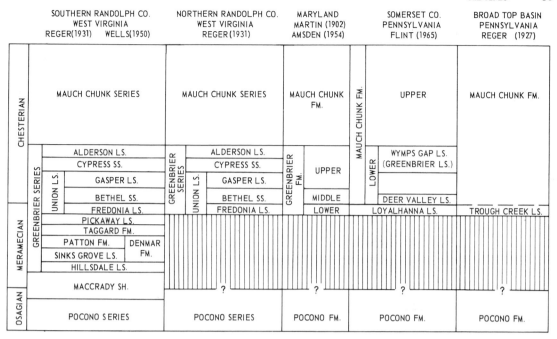

	SOUTHERN RANDOLPH CO. WEST VIRGINIA REGER(1931) WELLS(1950)	NORTHERN RANDOLPH CO. WEST VIRGINIA REGER(1931)	MARYLAND MARTIN (1902) AMSDEN (1954)	SOMERSET CO. PENNSYLVANIA FLINT (1965)	BROAD TOP BASIN PENNSYLVANIA REGER (1927)
CHESTERIAN	MAUCH CHUNK SERIES	MAUCH CHUNK SERIES	MAUCH CHUNK FM.	MAUCH CHUNK FM. · UPPER	MAUCH CHUNK FM.
	GREENBRIER SERIES — UNION L.S. : ALDERSON LS. / CYPRESS SS. / GASPER LS. / BETHEL SS. / FREDONIA L.S.	GREENBRIER SERIES — UNION L.S. : ALDERSON LS. / CYPRESS SS. / GASPER LS. / BETHEL SS. / FREDONIA LS.	GREENBRIER FM. : UPPER / MIDDLE / LOWER	LOWER : WYMPS GAP LS. (GREENBRIER LS.) / DEER VALLEY LS. / LOYALHANNA LS.	TROUGH CREEK LS.
MERAMECIAN	GREENBRIER SERIES — PICKAWAY L.S. / TAGGARD FM. / PATTON FM. DENMAR FM. / SINKS GROVE LS. / HILLSDALE LS.	?	?	?	?
OSAGIAN	MACCRADY SH. / POCONO SERIES	POCONO SERIES	POCONO FM.	POCONO FM.	POCONO FM.

FIGURE 2.   Correlation chart of the Mississippian System in the study area.

(1927, p. 408). Figure 2 summarizes the regional stratigraphic correlations.

South and southeast of the type locality a relatively pure limestone unit, the Deer Valley Limestone (Flint, 1965, p. 24), is present between the Loyalhanna and the overlying Mauch Chunk red beds. The Loyalhanna rests disconformably on the Pocono Formation throughout the area with two local exceptions, at Oglestown, Pennsylvania (Fig. 1, No. 104) and Westernport, Maryland (Fig. 1, No. 44), where thin, very slightly arenaceous limestones are present at the base of the Loyalhanna. It is not known whether the limestones are isolated units or northern representatives of pre-Loyalhanna limestones to the south. The regional physical stratigraphy is shown in Figure 3.

Age

The paucity of megafossils and the lack of a micropaleontological investigation precludes assignment of a specific age to the Loyalhanna. Correlation with the basal part of the Union Limestone (Fredonia member) by Reger (1931, p. 323) dated the Loyalhanna as upper Meramecian (Reger, 1926, p. 460–462). Wells (1950, p. 917) found that Meramecian carbonates in the Greenbrier Series are not present north of Randolph County, West Virginia and placed the entire Union Limestone in the lower part of the Chesterian Series of the type area. The Loyalhanna is upper Meramecian or lower Chesterian.

Lower Contact

The Pocono–Loyalhanna contact is a disconformity throughout the area of investigation. Brownish gray to gray, medium-grained, noncalcareous micaceous sandstones of the Pocono contrast with red to gray calcareous arenites of the Loyalhanna. The surface of the unconformity is remarkably planar as exhibited by extensive areas where the Loyalhanna has been stripped off in quarry operations. Pocono-type sandstone lithoclasts are absent in the basal Loyalhanna and carbonate grains do not occur in the uppermost Pocono. At many localities there are cobbles of slightly arenaceous calcarenite up to 6 in. (15 cm) in maximum diameter scattered on the surface of unconformity. The cobbles are not of normal Loyalhanna composition but resemble the local limestone units at the base of the Loyalhanna.

Gross Lithology and Aspect

The lithologic character of the Loyalhanna is dependent on the proportion of grain types, texture, internal structures, and color. The proportion of terrigenous and carbonate detrital grains varies between adjacent laminations, between sedimentation units and regionally in such a manner that one term cannot describe the general lithology. Basically the unit is a cross-bedded arenaceous calcarenite to calcareous sandstone composed of terrigenous clasts (quartz, feldspar, rock fragments and others), carbonate grains,

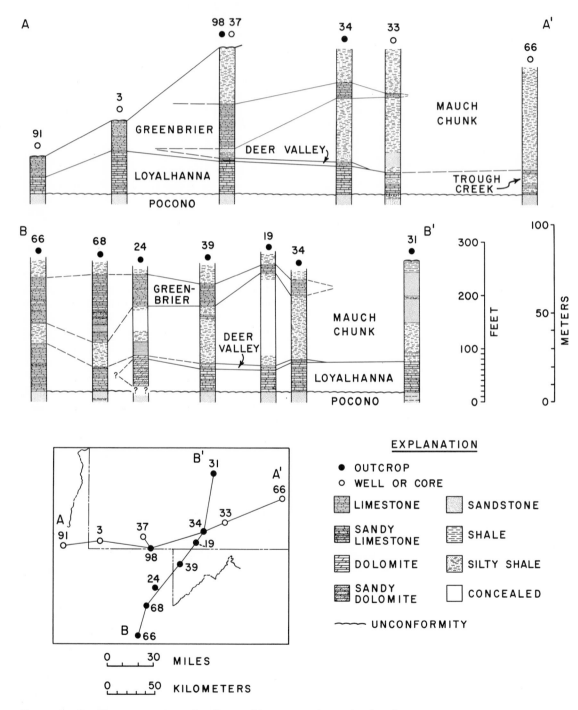

FIGURE 3. Graphic cross sections, Loyalhanna Limestone and associated units.

and carbonate cement. Composition and grain size variations between adjacent laminations result in the clear definition of internal structures in sedimentation units. These structures, principally cross-bedding, are particularly enhanced on weathered surfaces due to differential weathering between the carbonate and terrigenous constituents.

The general aspect of the formation is that of a vertical sequence of cross-bedded sedimentation units from several inches to several feet thick. Beds which are wavy-bedded, horizontally laminated, or massive occur within the general cross-stratified sequence, but generally represent a low proportion of the formation and fail to detract from the impression that the Loyalhanna is composed of cross-bedded units. Thin lenses of red micaceous siltstone occur locally in the eastern portion of the outcrop area.

Superimposed on the cross-stratified aspect of the Loyalhanna is a regional color zonation (Fig. 4). Bluish gray to greenish gray colors characterize the formation to the north and west while grayish red is typical of the southeastern outcrops. The red coloration is caused by an abundance of red hematitic material as grains, matrix in rock fragments, coatings on clasts, and as an impurity incorporated in micrite fillings of skeletal chambers and in oölith sheaths. Gray Loyalhanna lacks these red pigments and commonly contains pyrite as small nodules and disseminated crystals within carbonate clasts.

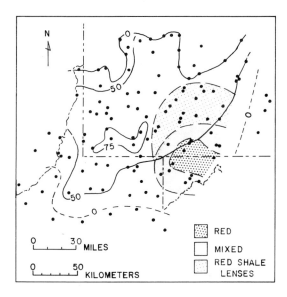

FIGURE 4. Isopach (in feet) and lithofacies map, Loyalhanna Linestone.

Distribution and Thickness

The distribution of the Loyalhanna is shown in Figure 4. The area encompassed by the zero isopach is approximately 17,000 sq mi (44,000 sq km). The maximum thickness is 103 ft (31.5 m) and the average is 60 ft (18.5 m). Assuming an average thickness of 60 ft, the ratio of lateral extent to thickness is about 12,000:1, indicative of a very thin sheet or blanket sand body. The Loyalhanna exhibits a tendency to thin to the southeast and northwest. A similar trend is apparent to the south before tracing of the Loyalhanna is terminated by a facies change or lateral pinchout. Post-Loyalhanna erosion removed the unit to the north, northeast, and west prior to Pennsylvanian deposition.

The present eastern limit of the Loyalhanna coincides with the Allegheny Front through Pennsylvania, Maryland, and West Virginia. Some insight to the lateral relationships to the east is provided by outliers of Mississippian strata in the Broad Top and Emmaville Basins of Pennsylvania, 30–40 mi (48.5–64.5 km) east of the Allegheny Front. Outcrops of marine limestones (Trough Creek Limestone) at or near the base of the Mauch Chunk Formation attest to marine conditions during early Mauch Chunk time. These limestones could not be directly correlated with the Loyalhanna but their stratigraphic position suggests they may represent contemporaneous carbonate deposits.

Lateral relationships between the Loyalhanna of Maryland and northern West Virginia and the Greenbrier limestones to the south are difficult to ascertain due to the paucity of outcrops, lateral facies changes, and the lack of detailed paleontological work. Distinct differences in petrology between the Loyalhanna and Greenbrier arenaceous calcarenites indicate the units are not lithologically correlative.

Upper Contact

The Loyalhanna is conformably overlain by Mauch Chunk red beds or the Deer Valley Limestone. The term Deer Valley is herein applied to relatively pure limestone (not quartzitic) which rests on the Loyalhanna Limestone. It is a light gray to reddish gray calcarenite containing marine fossils and fossil fragments. The limestone is conformably overlain by lower Mauch Chunk red clastics or thin beds of arenaceous calcarenite similar in composition to the Loyalhanna. The Loyalhanna–Deer Valley contact is commonly abrupt, changing from cross-bedded quartzitic arenite to fossiliferous calcarenite. At one locality narrow tubular structures filled with

Deer Valley calcarenite penetrate the uppermost Loyalhanna to a depth of 12 in. (30 cm). At the Deer Valley type locality 6–8 in. (15–20 cm) of shale and silty shale separate the two limestones (Fig. 1, No. 19).

Figure 5 shows the distribution and thickness of the Deer Valley Limestone; approximately the southern one-third of the Loyalhanna in outcrop is capped by this marine limestone. In the subsurface of western Pennsylvania the Deer Valley merges with younger limestones higher in the Mauch Chunk and cannot be identified as a distinct unit. Eastward tracing of the unit terminates at the Allegheny Front where the Mississippian has been removed by erosion.

In southwestern Maryland and northern West Virginia 20–40 ft (6–12 m) of dark gray to black argillaceous and fossiliferous lutite and calcarenite may represent a southern facies of the Deer Valley. These limestones grade into the underlying Loyalhanna with a decrease in argillaceous components and an increase in quartz sand, and are conformably overlain by red calcareous clastics of the Mauch Chunk Formation. The area of dark argillaceous Deer Valley is associated with the region of apparent pinchout of the Loyalhanna Limestone. South of this area the Loyalhanna is absent and its stratigraphic position is occupied by fossiliferous limestones of the Fredonia member of the Union Limestone in West Virginia. The Fredonia limestones may represent a southern facies of Loyalhanna Limestone, a southward extension of the Deer Valley Limestone, or a combination of both.

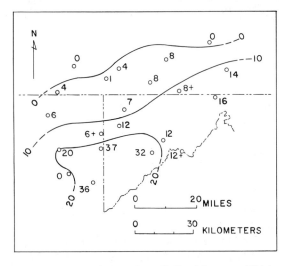

FIGURE 5. Isopach map, Deer Valley Limestone. Thickness and contour interval in feet.

PETROLOGY

General

The Loyalhanna is a greenish and bluish gray to grayish red (5G 6/1 and 5B 6/1 to 10R 4/2 on the Munsell Soil Color Chart) arenaceous calcarenite and calcareous orthoquartzite. Carbonate and terrigenous sand and silt form the framework with carbonate cement as an interstitial filling. The cement is finely crystalline clear calcite (spar) composed of crystals from a few microns to 60 μ in diameter which form a mosaic of anhedral crystals completely filling all original voids. Pressure-solution between carbonate and noncarbonate components commonly has resulted in the interpenetration of grains and an accompanying decrease of cement. Dolomite is present in northern West Virginia (Fig. 1, Nos. 23 and 24) where the cement is a mosaic of 5–10 μ dolomite crystals. Carbonate detritus has been replaced by a mosaic of fine to coarse (to 0.5 mm) dolomite.

Lenses of red siltstone and shale are interbedded with arenites in the southeastern area of outcrop. They occur scattered throughout the vertical extent of the Loyalhanna, range from 1 to 10 ft (0.3–3.0 m) thick and from 15 ft (4.5 m) to several tens of feet wide. In all exposures these units grade laterally into the Loyalhanna and thus appear as lenses in two dimensions. The three-dimensional geometry of the units was nowhere observed.

Arenites

Figures 6A and 6B illustrate the composition of the terrigenous component of the arenites which average 95.5 percent quartz with feldspar (3.8 percent), rock fragments and heavy minerals (0.7 percent) comprising the remainder. Monocrystalline quartz dominates the assemblage, averaging 81.5 percent, 80–90 percent of which is the common or igneous type exhibiting straight to slightly undulatory extinction and containing scattered vacuoles and microlites. Polycrystalline quartz, quartz composed of two or more crystals, averages 12.5 percent and microcrystalline quartz, chalcedony and chert, averages 1.5 percent of the terrigenous fraction. The ratio of monocrystalline to polycrystalline quartz (M:P) averaged 8.2 for 30 samples of gray Loyalhanna and 5.9 for 12 samples of red Loyalhanna. This indicates a relatively low amount of polycrystalline quartz in the Loyalhanna and a relatively higher proportion in the red facies of the unit toward the southeast.

The quartz ranges in size from silt to very coarse sand with few grains larger than 2.0 mm.

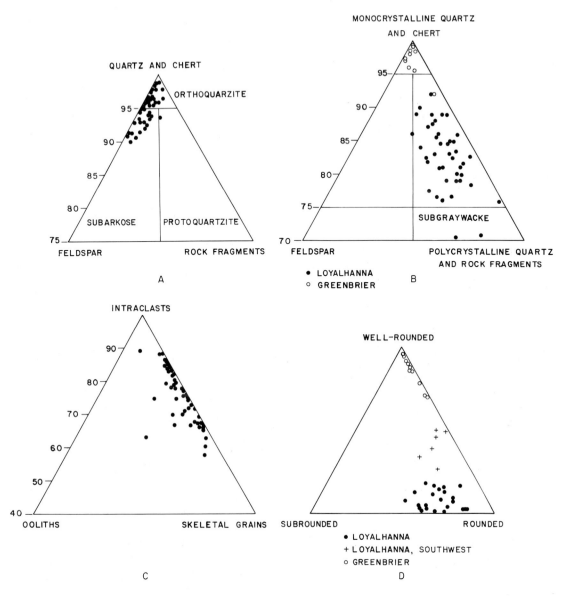

FIGURE 6. Compositional diagrams for the Loyalhanna Limestone and associated units. (A) Loyalhanna Limestone terrigenous composition. (B) Terrigenous composition comparison of the Loyalhanna Limestone and Greenbrier limestones. (C) Loyalhanna Limestone carbonate composition. (D) Proportions of tourmaline roundness types in the 0.125–0.250 mm size class for the Loyalhanna Limestone and Greenbrier limestones.

Coarse to very coarse grains fall in the well-rounded class of Pettijohn (1957, p. 59). Fine to medium grained quartz exhibits poorer rounding and would be classed as subrounded to rounded. Very fine sand and silt is angular to subangular.

Orthoclase, microcline, and plagioclase feldspar are present and commonly are partially replaced by calcite. Rock fragments are principally sericite–chlorite aggregates which contain variable amounts of silt-sized quartz. Fragments of silt- to very fine sand-sized quartz particles with

a hematitic matrix are present in all samples of the red facies of the Loyalhanna. Heavy minerals include tourmaline, zircon, staurolite, garnet, rutile, sphene, monazite, sillimanite, zoisite, kyanite, apatite, muscovite, biotite, ilmenite, magnetite, leucoxene, and hematite (red facies only).

A slight modification of the petrographic terminology of Pettijohn (1957, p. 291) is followed for the classification of the terrigenous fraction of the Loyalhanna. Carbonate clasts

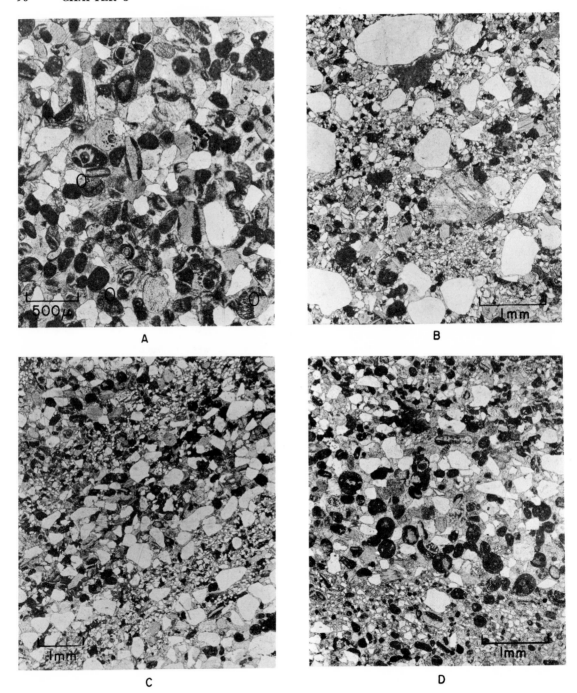

FIGURE 7. Photomicrographs of Loyalhanna Limestone. (A) Lime sand types and partial oöliths (0) (30 ×). (B) Poorly sorted arenite. Feldspar partially replaced by calcite near center (20 ×). (C) Coarse- and fine-grained laminations from a cross-bed (13.5 ×). (D) Fine- and very fine-grained laminations from a cross-bed (20 ×).

and cement are omitted and polycrystalline quartz is totalled with monocrystalline quartz and chert. The terrigenous fraction is an orthoquartzite. Sufficient amounts of feldspar are present in some samples to warrant the name feldspathic orthoquartzite. If polycrystalline quartz grains are included with rock fragments the term protoquartzite is applicable.

Carbonate detritus comprises from 6.5 to 96.5 percent of the clasts in the Loyalhanna. Carbonate grains tend to be concentrated in the very fine to medium grained size classes, there is a negligible amount of silt or coarse sand. Rounding of all carbonate grains is exceptionally good and they range from spherical to ovoid in shape. Figure 6C illustrates distribution of carbonate grain types in the Loyalhanna. The Loyalhanna carbonate fraction is lithic calcarenite (Pettijohn, 1957, p. 405) or intrasparite (Folk, 1959, p. 14).

Carbonate grain types include intraclasts, skeletal grains, and oöliths (Fig. 7A). The term intraclast refers to homogeneous microcrystalline (1–4 μ) calcite aggregates (micrite) ranging from very fine to medium sand size. There is a dominance of micrite grains in the Loyalhanna with no apparent compositional differences between grains of various sizes. Although most of the intraclasts are micrite, microspar or spar is present as small patches and included skeletal fragments are not uncommon. Quartz silt is very rare as an impurity in spite of its abundance in the Loyalhanna depositional environment. The local presence of thin beds and pebbles of micrite within the Loyalhanna and the fine to medium sand size of the micrite aggregates suggest that the clasts are principally fragments of penecontemporaneous carbonate sediments. Fecal pellets may be present but they could not be differentiated from intraclasts.

Identifiable skeletal grains are bryozoa fragments, echinoderm plates, and foraminiferal tests. Oöliths in the Loyalhanna have an intraclast or skeletal grain as a nucleus; the outer layers exhibit concentric and radial structure. Fragments of oöliths are present in most samples and consist of one-quarter or more of a normal oölith which has been broken and subsequently rounded. The abundant intraclasts, lack of terrigenous grains in intraclasts and oöliths, and the presence of partial oöliths impart an allochthonous aspect to the carbonate components of the Loyalhanna.

Petrologic classification of the Loyalhanna arenite is dependent upon the relative proportion of terrigenous and carbonate grains. Where carbonate grains dominate it should be termed an orthoquartzitic lithic calcarenite or orthoquartzitic intrasparite. Terrigenous sand-dominated phases are lithic calcarenaceous orthoquartzite or intrasparitic orthoquartzite.

A plot of the regional distribution of the percentage of terrigenous grains demonstrates a general decrease in their abundance from north to south (Fig. 8). The gray facies of the Loyalhanna grades from orthoquartzite into calcarenite to the south. The red facies of the Loyalhanna appears to be generally more arenaceous than the gray as is demonstrated by contrasts at individual localities and the presence of red orthoquartzites in the calcarenite region of the gray facies.

Siltstones and Shales

Lenses of red siltstone and shale are composed of quartz silt, muscovite, biotite, chlorite, hematitic matrix and calcite cement. In the gray facies of Loyalhanna these red beds grade laterally into gray arenites by a change in their coloration to green and then a loss of finer material and matrix accompanied by a gain of sand-sized detritus. Gradations at the base and top of the fine-grained units are similar to the lateral gradations; however, the upper contact is commonly a local unconformity with gray Loyalhanna arenites lying on red siltstone or shale.

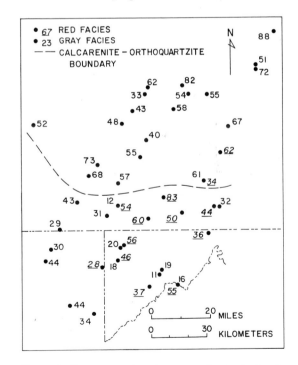

FIGURE 8. Percentage of terrigenous grains in the Loyalhanna Limestone. South of the calcarenite–orthoquartzite boundary the samples containing greater than 50 percent terrigenous detritus are from the red facies.

The aspect of the red lenses in the gray arenite is one of local, fine-grained oxidized material preserved in a generally reduced environment.

Heavy Minerals

The nonopaque heavy mineral composition of the Loyalhanna is mature, consisting of greater than 85 percent tourmaline and zircon. Garnet and rutile are present in most samples; staurolite is locally common (to 12 percent). The dominant color of tourmaline is yellow-brown (80–90 percent) with blue, green, pink, and varicolored types present in most samples. Rounding of the tourmaline averages between 0.4 and 0.6 on the roundness chart of Krumbein (1941, p. 68) and is generally uniform throughout Pennsylvania. In the subsurface of extreme southwest Pennsylvania and adjacent portions of West Virginia the proportion of well-rounded tourmaline increases and is intermediate between the Loyalhanna to the north and east, and the well-rounded tourmaline characteristic of the Greenbrier limestones to the south and southwest (Fig. 6D). The boundaries between the roundness groups is abrupt; occurring within a distance of 15 mi (24 km) in the subsurface and on the outcrops.

The higher proportions of well-rounded tourmaline to the south and southwest may indicate prolonged abrasion or an influx of multicycle rounded grains from a different source area. The magnitude of the difference requires appreciable abrasion to produce well-rounded grains from rounded or subrounded types, particularly in an arenite with a high proportion of relatively soft carbonate grains. An influx of a more highly rounded tourmaline is indicated and is supported by the terrigenous compositional contrasts between the Loyalhanna and arenaceous Greenbrier limestones (Fig. 6B). Both arenites are highly quartzose, however, when polycrystalline and monocrystalline quartz are differentiated the two units occupy distinct fields within a compositional triangle.

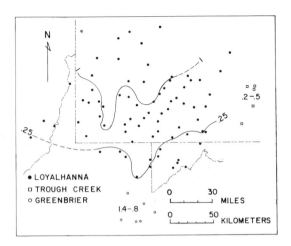

FIGURE 9. Tourmaline–zircon ratios in the Loyalhanna and associated limestones. The Trough creek limestone values are similar and on trend with the Loyalhanna values. Greenbrier limestone samples depart from the Loyalhanna trend.

Colorless, euhedral to rounded zircon is common in the very fine (0.063–0.125 mm) fraction of heavy minerals, colored varieties are rare. A plot of the relative abundance of tourmaline and zircon (T : Z ratio) in the very fine sand size illustrates three fields of relative abundance (Fig. 9): tourmaline exceeds zircon to the northwest, a general dominance of zircon occurs in the central portion, and tourmaline is subordinate in the southern region. The lack of any marked differences in tourmaline rounding within the major portion of the Loyalhanna minimizes the effect of differential abrasion on the apparent tourmaline loss. The influence of grain size on the T : Z ratio is significant (Table 1) and it is concluded that the trend of decreasing T : Z ratio reflects a general north to south decrease in the grain size of the terrigenous component of the Loyalhanna. T : Z ratios in the arenaceous limestones of the Greenbrier of West Virginia are larger than those of the southern Loyalhanna samples. As with tourmaline roundness and quartz-type contrasts,

TABLE 1.   Comparative Nonopaque Heavy Mineral Composition for Two Size Classes (200 grain count)

Sample and size interval (mm)	Percent composition*						
	T	Z	S	G	R	O	T:Z
W81b-1							
0.125–0.250	55.0	8.5	32.0	4.5	0	0	6.47
0.063–0.125	35.5	45.5	12.0	6.0	1.0	0	0.78
105B							
0.125–0.250	79.0	2.5	14.0	3.5	0.5	0.5	31.60
0.063–0.125	28.5	54.0	10.0	2.5	2.5	2.5	0.53

* T, tourmaline; Z, zircon; S, staurolite; G, garnet; R, rutile; O, other.

the T : Z ratios suggest different sources for the terrigenous component of the units.

The mature composition of the Loyalhanna heavy mineral suite indicates a long period of physical and chemical action or a multicycle derivation. Numerous color varieties of tourmaline and the range of rounding of tourmaline and zircon suggest a source terrain principally composed of sedimentary rocks. The abundance of staurolite and garnet in some samples, and trace amounts of kyanite and sillimanite indicate a contribution from some metamorphic terrain.

TEXTURAL ANALYSIS

Examination of the Loyalhanna on the outcrop and in thin section (Figs. 7B and 7C) discloses an abnormally high amount of fines (very fine sand and silt) for a deposit which has been considered eolian or a dune–beach complex. Size analyses of the noncarbonate fraction of the arenites show a high proportion of fine material (9.7–29 percent less than sand size) and a bimodal character for some size distributions (Fig. 10).

Comparison of the Loyalhanna size distribution parameters (Table 2) with known beach and eolian sands* show the Loyalhanna to be markedly different. Beach and eolian sands are strongly unimodal whereas bimodal samples characterize the Loyalhanna. A minor proportion of beach and eolian sands have very fine sand as the modal class, the common mode for the Loyalhanna, and comparisons of the measures of central tendency

* The following workers provide size analyses or statistical parameters for beach and eolian sands: Wentworth, 1931; Twenhofel, 1946; Shotten, 1937; Inman, 1953; Inman and Chamberlain, 1955; Shepard and Young, 1961; Friedman, 1961; Schlee et al., 1964; Giles and Pilkey, 1965; and Inman et al., 1966.

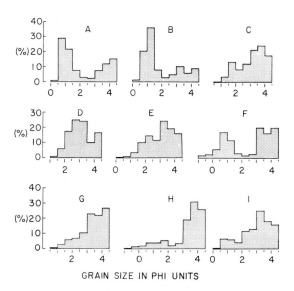

FIGURE 10. Histograms of the size distribution of Loyalhanna Limestone terrigenous grains. (A to E) cross-bedded units. (F) horizontally laminated unit. (G) wavy-bedded unit. (H and I) massive units.

(median and mean) show the Loyalhanna to be generally finer grained. In comparison with Gulf Coast eolian sands (Inman and Chamberlain, 1955 and Mason and Folk, 1958) the medians and means of the Loyalhanna are similar; however, the sorting of the Loyalhanna is markedly poorer.

The size analyses of the Loyalhanna demonstrate the general lack of characteristics of beach or eolian sand size distributions. A fluviodeltaic origin for the Loyalhanna is rejected because of the high carbonate content, large-scale cross-bedding, and lack of attributes of a fluvial or deltaic deposit. A shallow-water marine shelf is the most likely depositional environment which

TABLE 2. Size Parameters for Loyalhanna Terrigenous Component

Sample	Median		Phi mean,* M_ϕ	16th Percentile		84th Percentile		Phi deviation,† σ_ϕ	Fines,‡ wt %
	mm	phi		mm	phi	mm	phi		
A	0.365	1.45	2.41	0.560	0.83	0.063	3.98	1.08	15.5
B	0.415	1.27	2.20	0.530	0.92	0.091	3.47	1.28	9.7
C	0.098	3.35	3.00	0.260	1.94	0.060	4.06	1.06	17.9
D	0.124	3.01	3.18	0.203	2.30	0.060	4.06	0.93	16.6
E	0.107	3.21	3.03	0.238	2.07	0.063	3.98	0.96	15.9
F	0.108	3.20	2.46	0.600	0.73	0.055	4.18	1.73	19.9
G	0.090	3.47	3.47	0.179	2.47	0.045	4.47	1.00	27.3
H	0.081	3.64	3.05	0.280	1.83	0.052	4.27	1.22	26.6
I	0.110	3.18	2.98	0.254	1.97	0.063	3.98	1.01	15.9

* After Inman (1952) $M_\phi = (16\phi + 84\phi)/2$, 16th percentile and 84th percentile from cumulative curve plot.
† After Inman (1952) $\sigma_\phi = (16\phi - 84\phi)/2$.
‡ All material less than 0.063 mm.

could provide a distribution of material as is found in the Loyalhanna. Shepard and Cohee (1936, p. 445), Shepard and Moore (1955, Fig. 26), Inman and Chamberlain (1955, p. 117 and Fig. 8), Curray (1960, Figs. 8 and 10B), and Inman (1953, Appendix 1) report off-shore sands with sorting and size characteristics similar to those of the Loyalhanna.

CROSS–BEDDING

Complete exposures of the Loyalhanna Limestone consist of a high proportion of cross-bedded sedimentation units (Potter and Pettijohn, 1963, p. 69). Horizontal massive to wavy-bedded units, generally 1–3 ft (0.3–1.0 m) thick, occur scattered throughout the Loyalhanna and constitute less than 10 percent of most sections. Figure 11

A

B

C

FIGURE 11. Outcrops of Loyalhanna Limestone. (A) Large-scale cross-bedded units. Outcrop 64 ft. (19.5 m) high, lowermost cross-bedded unit 9 ft (2.7 m) thick, circled hammer for scale. (B) Medium-scale cross-bedded units. (C) Small-scale cross-bedded units.

illustrates the outcrop aspect and cross-bed characteristics, Table 3 summarizes properties of the cross-bedding.

The large size and limited exposure of the cross-bedded units precludes observation of individual unit geometry. Partial bedding plane or *ab* sections (Potter and Pettijohn, 1963, Fig. 4-2) exhibit laminations slightly concave downdip (maximum observed width was 15 ft or 4.5 m). Longitudinal *ac* sections show lower and upper contacts which are planar and erosional. Generally the contacts appear horizontal, but slight concave-up curvature is not uncommon. Traces of laminations in *ac* sections are characteristically straight in the upper part of the unit and invariably become tangent at the base. Foreset laminations in transverse *bc* sections are concave-up and parallel to a gently curved concave base which commonly truncates the internal structure of underlying unit(s). The aspect of a transverse section is a shallow trough or festoon (Knight, 1929, p. 56). In partially concealed transverse exposures the foresets appear to be horizontal or slightly inclined to normal bedding.

Figure 12 shows the distribution of Loyalhanna cross-bedding thickness. Large-scale cross-bedding is impressive (Fig. 11A) and has prompted some workers to assume an eolian origin for the formation. Eighty percent of the cross-bedding is less than 4 ft (1.2 m) thick and the modal class is 1–2 ft (0.3–0.6 m). A regional moving average plot of cross-bedding thickness (See Fig. 15) illustrates a trend of decreasing values to the southeast.

The general shape of the cross-bedded unit is a very thin lens with a large lateral extent relative to thickness. This is a form of trough cross-bedding (McKee and Weir, 1953) which approaches a tabular planar shape. It is representative of a gradation between tabular and trough cross-bedding of Potter and Pettijohn (1963, p. 71 and Fig. 4-2) and the Omikron- and Pi-cross-stratification of Allen (1963, p. 108–111).

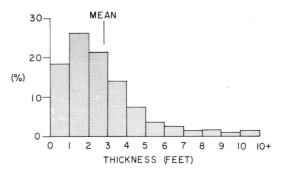

FIGURE 12. Histogram of Loyalhanna cross-bedding thickness. The mean thickness is 2.78 ft (0.85 m).

Observation of cross-bedding in modern and ancient sediments has shown that the azimuth of maximum foreset dip is parallel or subparallel to the direction of the average local current vector (Potter and Pettijohn, 1963, p. 81). Current rose diagrams for localities with 15 or more observations illustrate the distribution and variability of Loyalhanna cross-bedding (Fig. 13). Bimodal diagrams with one or two 30 degree classes separating the modes may have been caused by the trough-like form of the cross-beds as has been documented by Meckel (1967a, p. 84–85). Bimodal cross-bedding distributions with widely spaced modes cannot be explained as functions of cross-bed shape. Such cross-bedding azimuth patterns have been reported in ancient marine sands (Selley, 1967, Fig. 4; Sedimenta-

TABLE 3.   Statistics of Loyalhanna Cross-bedding

Azimuth	Mean direction: 072
	Modality: unimodal and bimodal
	Variance (standard deviation): 5776 (76°)
Scale (feet, meters)	Mean: 2.78, 0.85
	Mode: 1–2, 0.3–0.6
	Range: 0.3–16, 0.1–4.9
Inclination (degrees)	Mean: 20.0
	Mode: 20–25
	Range: 5–40

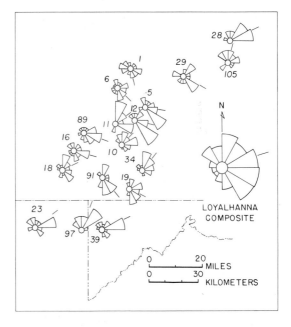

FIGURE 13. Rose diagrams with vector means for Loyalhanna cross-bedding dip azimuths. Localities with 15 or more observations, numbers indicate localities, 651 readings for composite diagram.

tion Seminar, 1966, Fig. 6; Tanner, 1955, Fig. 7 and 1963, Fig. 3C), inferred in recent marine barrier bar migrating channels (Hoyt and Henry, 1967, p. 83 and Fig. 6), reported in recent marine carbonate sand belts and belts of tidal bars (Ball, 1967, p. 559 and Fig. 9, p. 569 and Fig. 19) and recorded from tidal sand waves in estuaries and tidal inlets (see Klein, 1967, p. 368–370). Reversing tidal currents, littoral currents, seasonal currents (Selley, 1967, p. 221), offshore and onshore winds (Selley, 1967, p. 22), intermittent high-energy events (Ball, 1967, p. 560), and changing current conditions in barrier island channels (Hoyt and Henry, 1967, p. 83) have been cited as possible causes for marine bimodal cross-bedding distributions. Bimodal distributions in the Loyalhanna may have resulted from any one or a combination of such factors. Variations of current direction with time may also be quite important. At two localities, 89 and 97, strong unimodal orientation of azimuthal data is present over a 16-ft (5 m) interval. Measurements over larger vertical ranges may reflect changes in current patterns through time.

A general easterly direction of sediment transport can be visualized for most localities in the rose diagrams of Figure 13 and is reflected in the composite rose diagram for all Loyalhanna cross-bed azimuths. A plot of outcrop cross-bedding vector means (Fig. 14) demonstrates the same trend and a moving average plot of

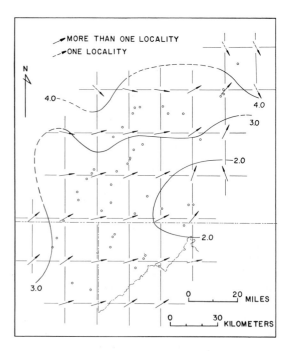

FIGURE 15. Moving average of cross-bedding vector means and contoured moving average of cross-bedding thickness (in feet). Loyalhanna Limestone.

vector means (Fig. 15) illustrates an easterly to northeasterly trend.

ORIGIN

Previous Interpretations

The Loyalhanna Limestone has been considered to be eolian (Butts, 1924; Hickok and Moyer, 1940) or a nearshore marine complex of beach, dune, bar, and sheet sands (Campbell, 1902; Rittenhouse, 1949; Flowers, 1955; and Flint, 1965). Large-scale cross-bedding, frosting of some quartz grains, and general lack of fossils have been cited as evidence for an eolian derivation. A marine origin was inferred from tabular form of the sand body, presence of fossils, minor proportion of frosted quartz, and correlation of the Loyalhanna with marine limestones of northern West Virginia.

Present Interpretation

A high proportion of carbonate clasts in the arenites, including skeletal debris and oöliths, demonstrates an affinity to marine clastic carbonates. The presence of some marine megafossils, an overlying normal marine limestone through much of the area, and the local occurrence of an underlying marine limestone supports the marine interpretation.

Gross attributes of Loyalhanna structure, texture, and composition are best explained by a

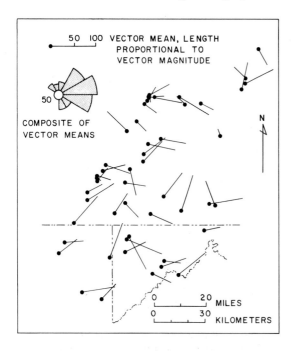

FIGURE 14. Locality vector means of Loyalhanna cross-bedding dip azimuths.

marine environment. Reducing conditions which characterize a large proportion of the Loyalhanna (pyritic greenish gray facies) are more typical of a subaqueous environment than a subaerial one. The range of grain sizes in the Loyalhanna sedimentation units is not typical of eolian or beach deposits but is found in nearshore marine sediments. Loyalhanna cross-bed thickness approaches that commonly attributed to eolian deposition; however, only 12.5 percent of the Loyalhanna cross-beds are in excess of 5 ft (1.5 m) thick. Eolian dunes similar in scale may be characteristic of the barrier island environment (McBride and Hayes, 1962; Land, 1964), but attributes of texture, geometry, and associated deposits negate such an origin for the Loyalhanna. Eolian limesands of Bermuda and Bahamas described by Mackenzie (1964) and Ball (1967) commonly contain slightly convex upward foresets which terminate sharply against underlying surfaces. In addition, brecciation of the bedding occurs within or between units. Convex foresets and internal brecciation are absent in the Loyalhanna.

In recent years increased attention has been given to submarine sand waves found on shallow marine shelves (see Off, 1963; Klein, 1967, p. 373; Ball, 1967). The sand waves attain an amplitude up to 40 ft (12 m) (Off, 1963, Fig. 20), are found to be oriented approximately normal to the paths of the strongest tidal currents and are generally parallel to sea-floor contours (Klein, 1967, p. 374). Medium-scale ripples with average amplitudes of 2 ft (0.6 m) characterize marine carbonate sand belts described by Ball (1967, p. 559 and 564). Orientation of these ripples is dependent upon tidal currents, storm currents, and local topography. A sand body similar to the Loyalhanna could be constructed by migration of such marine sand waves and sand belts coupled with an adequate supply of sediment.

A second marine environment exhibiting submarine sand waves has been discussed by Hoyt and Henry (1967). Sand waves with amplitudes up to 12 ft (3.5 m) and megaripples with amplitudes to 3 ft (1 m) occur in migrating channels associated with barrier islands. Layers of silt and clay are found in the inlet and in the adjacent nearshore–neritic environment. Stratification in the migrating inlet environment is described as having three dip directions, one dipping in the direction of migration and two dipping at right angles to the regional shore trend, either toward or away from the ocean. Under suitable conditions of sediment supply, lateral migration, and

preservation this type of environment could build a unit similar to the Loyalhanna.

BASIN ANALYSIS

Basin Geometry

Trends in the zircon–tourmaline ratio and increasing carbonate content suggest a subtle north to south regional paleoslope. The thickening sequence of Greenbrier limestones toward the south and the presence of the Deer Valley Limestone in the southern area of Loyalhanna outcrop supports this interpretation. Lithofacies to the west and southwest demonstrate a more marine aspect due to the apparent loss of red Mauch Chunk clastics between the Loyalhanna and Greenbrier Limestone (of Pennsylvania). Increasing proportions of shale within the Loyalhanna to the northeast (Fig. 3, Nos. 31 and 33) as well as the thin interval of Trough Creek Limestone suggest proximity to a more clastic facies, perhaps deltaic. The blanketlike geometry of the Loyalhanna, lithofacies gradients within it, and characteristics of associated formations indicate deposition on a shallow marine shelf or platform. The shelf was open to the sea on the south and southwest, shoaling to the north and northeast where there was a general influx of non-carbonate detritus.

Provenance and Source

The general north to south decrease in the terrigenous component of the Loyalhanna Limestone has been documented by earlier workers in portions of the region (Rittenhouse, 1949, Fig. 2; Flint, 1965, p. 38) and is now verified for the entire area. This distribution of quartz sand requires a general input of terrigenous materials at the northern edge of the basin (Fig. 16). The high quartz content, restricted heavy mineral suite, and low feldspar content of the terrigenous fraction indicates a source dominated by older Paleozoic sediments. Possible contribution by a metamorphic terrain is suggested by the presence of garnet and staurolite.

Additional sources of terrigenous material were present on the southeastern and southwestern flanks of the Loyalhanna shelf. To the east the red lithofacies of the Loyalhanna with lenses of red siltstone and shale represents the influx of iron-bearing solutions and hematitic debris. These sediments have a slightly higher proportion of polycrystalline quartz than the gray Loyalhanna and generally contain a higher proportion of non-carbonate grains. The possible stratigraphic equivalent to the east, the Trough Creek Limestone, contains abundant epidote and garnet

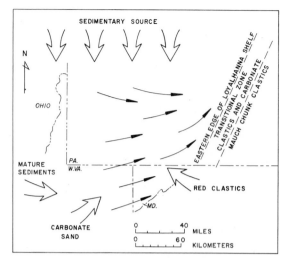

FIGURE 16. Loyalhanna Limestone depositional framework and dispersal system.

(Adams, 1964), reflecting a contribution from low-rank metamorphic terrain to the east or southeast. At this eastern flank of the shelf the basal portion of the Mauch Chunk probably represents an interfingering of red deltaic sediment with marine limestones and red beds along the eastern edge of the Loyalhanna shelf.

Toward the southwest the degree of rounding of tourmaline grains of the Loyalhanna becomes higher than normal. Characteristics of the non-carbonate fraction of arenaceous Greenbrier limestones to the south include highly rounded tourmaline as well as negligible polycrystalline quartz. These facts suggest an influx of very mature terrigenous grains from the west into the Greenbrier basin and possibly the Loyalhanna shelf. The composition of Devonian, Ordovician, and Cambrian sandstones in the upper Mississippi Valley and adjacent areas are characterized by very small amounts of polycrystalline quartz (0.3, 0.6, and 2.6 respectively) and a high proportion of well-rounded tourmaline (Potter and Pryor, 1961, Table 3). These sediments provide a plausible source for the Greenbrier and some Loyalhanna terrigenous detritus.

The general direction of sediment transport from the major terrigenous source areas on the north and east flanks of the shelf toward the south and west is not documented by cross-bedding orientation. Tidal reworking and perhaps the action of periodic storms nullified the effect of paleoslope control on cross-bedding orientation.

Carbonate grains in the Loyalhanna sedimentation units had a multiple derivation. Local limestone units and cobbles within the Loyalhanna and red staining of oölite sheaths and red micrite filling skeletal cavities attest to limestone deposition on the shelf. The general paucity of quartz silt or sand as components of oöliths or intraclasts and the common presence of partial oöliths suggests initial formation in a quartz-poor environment and subsequent mixing with terrigenous material. Significant areas of carbonate deposition occur on the south and west flanks of the Loyalhanna shelf (Greenbrier of West Virginia). In this general region the thickening limestones above and below the Loyalhanna attest to a significant site of carbonate deposition which may have contributed carbonate detritus to the Loyalhanna shelf.

Depositional Pattern

Carbonate units were deposited in West Virginia and locally in Maryland and Pennsylvania during the marine transgression which produced the Pocono–Loyalhanna disconformity. Terrigenous material from sedimentary and minor metamorphic sources was introduced along the northern flank of the basin and mixed with indigenous and introduced carbonate in a barrier–island shoreline complex or across a shallow marine shelf. Tidal and longshore currents associated with the nearshore complex or tidal currents on the shelf transported material in sand waves parallel to the shoreline (parallel to the strike of the paleoslope) and oblique to it (across the paleoslope).

To the southeast the influence of Mauch Chunk deposition was expressed through the addition of red terrigenous material into the basin. Local conditions along the eastern shore prevented the migration of the northerly derived, easterly transported terrigenous component into the Trough Creek depositional area. Incursions of oxidized terrigenous material from the east spread westward into the marine environment. The input of this material is reflected in the cross-bedding characteristics as well as facies distribution. The thinner cross-beds in this region and the orientation of the vector means (see Fig. 15) suggests shoaling conditions and northerly currents moving parallel to a general northeast–southwest oriented shoreline. The lack of coarse clastics in this eastern area attests to the lessened importance of the eastern tectonic source areas in Loyalhanna and Lower Mauch Chunk time as compared to Pocono (Pelletier, 1958) and Pottsville (Meckel, 1967b) time. Mature Paleozoic sediments of the Craton (Cincinnati Arch) made a minor contribution to the Loyalhanna but were important only in the Greenbrier Series further south.

At the close of Loyalhanna deposition marine limestone (Deer Valley) was deposited in northern West Virginia, Maryland, and southwestern Pennsylvania. Prograding Mauch Chunk sediments subsequently advanced across the basin from the east.

Mississippian Paleoslopes

The direction of sediment transport, as interpreted from cross-bedding dip azimuths, shows an apparent regional reversal during the Loyalhanna Limestone phase of Mississippian sedimentation in this portion of the Appalachians. Pelletier (1958) showed that the Pocono paleoslope was inclined to the west and northwest and Hoque (1966) has documented a westward inclined paleoslope for the Mauch Chunk. Loyalhanna deposition took place in a shallow marine shelf where paleoslope control on sediment dispersal was subordinate to local energy conditions. Cross-bedding orientation records tidal and possibly storm reworking of material and not sediment transport down the paleoslope from upslope source areas. As a result, the orientation of Loyalhanna paleocurrents is in marked contrast to all post-Cambrian paleocurrent trends within this portion of the central Appalachians.

REFERENCES

Adams, R. W., 1964, Loyalhanna Limestone, cross-bedding and provenance. Ph.D. thesis, Johns Hopkins Univ., Baltimore.

Allen, J. R. L., 1963, The classification of cross-stratified units, with notes on their origin: Sedimentology, v. 2, p. 93–114.

Amsden, T. W., 1954, Geology of Garrett County, in Geology and water resources of Garrett County: Maryland Geol. Survey Bull. 13, p. 1–95.

Ball, M. M., 1967, Carbonate sand bodies of Florida and the Bahamas: Jour. Sed. Petrology, v. 37, p. 556–591.

Butts, Charles, 1904, Description of the Kittanning quadrangle: U. S. Geol. Survey, Geol. Atlas of U. S., folio 115, 15 p.

————, 1924, The Loyalhanna Limestone of southwestern Pennsylvania—especially with regard to its age and correlation: Am. Jour. Sci., 5th ser., v. 8, p. 249–257.

Campbell, M. R., 1902, Description of the Masontown-Uniontown quadrangles: U. S. Geol. Survey, Geol. Atlas of U. S., folio 82, 21 p.

Curray, J. R., 1956, Analysis of two-dimensional orientation data: Jour. Geology, v. 64, p. 117–131.

————, 1960, Sediments and history of Holocene transgression, continental shelf, northwest Gulf of Mexico, in F. P. Shepard and others, eds., Recent sediments, northwest Gulf of Mexico: Am. Assoc. Petroleum Geologists Pub., p. 221–266.

Flint, N. K., 1965, Geology and mineral resources of southern Somerset County, Pennsylvania: Pa. Geol. Survey County Rept. C56A, 267 p.

Flowers, R. R., 1956, A subsurface study of the Greenbrier Limestone in West Virginia: West Virginia Geol. and Econ. Survey, Report of Inv. no. 15, 17 p.

Folk, R. L., 1959, Practical petrographic classification of limestones: Am. Assoc. Petroleum Geologists Bull., v. 43, p. 1–38.

Friedman, G. M., 1961, Distinction between dune, beach and river sands from their textural characteristics: Jour. Sed. Petrology, v. 31, p. 514–529.

Giles, R. T., and Pilkey, O. H., 1965, Atlantic beach and dune sediments of the southern United States: Jour. Sed. Petrology, v. 35, p. 900–910.

Hickok, W. O., IV, and Moyer, F. T., 1940, Geology and mineral resources of Fayette county, Pennsylvania: Pa. Geol. Survey 4th ser., Bull. C26, 530 p.

Hoque, Mominul, 1966, Mauch Chunk sediments—transport pattern and sediment sources (abs.): Program, 1966 Ann. Mtg., Northeastern Section, Geol. Soc. America, p. 24.

Hoyt, J. H., and Henry, V. J., Jr., 1967, Influence of island migration on barrier-island sedimentation: Geol. Soc. America Bull., v. 78, p. 77–86.

Inman, D. L., 1952, Measures for describing the size distribution of sediments: Jour. Sed. Petrology, v. 22, no. 3, p. 125–145.

————, 1953, Areal and seasonal variations in beach and nearshore sediments at La Jolla, California: Beach Erosion Board, Tech. Memo. 39, 82 p.

————, and Chamberlain, T. K., 1955, Particle-size distribution in nearshore sediments, in Finding Ancient Shorelines: Soc. Econ. Paleontologists and Mineralogists Spec. Pub. 3, p. 106–129.

————, Ewing, G. C., and Corliss, J. B., 1966, Coastal sand dunes of Guerrero Negro, Baja California, Mexico: Geol. Soc. America Bull., v. 77, p. 787–802.

Klein, G. deV., 1967, Paleocurrent analysis in relation to modern marine sediment dispersal patterns: Am. Assoc. Petroleum Geologists Bull., v. 51, p. 366–382.

Knight, S. H., 1929, The Fountain and the Casper Formation of the Laramie Basin: Univ. Wyoming Publ. Sci., Geology 1, no. 1, p. 1–82.

Krumbein, W. C., 1941, Measurement and geologic significance of shape and roundness of sedimentary particles: Jour. Sed Petrology, v. 11, p. 64–72.

Land, L. S., 1964, Eolian cross-bedding in the beach dune environment, Sapelo Island, Georgia: Jour. Sed. Petrology, v. 34, p. 389–394.

McBride, E. F., and Hayes, M. O., 1962, Dune cross-bedding on Mustang Island, Texas: Am. Assoc. Petroleum Geologists Bull., v. 46, p. 546–552.

Mackenzie, F. T., 1964, Geometry of Bermuda calcareous dune cross-bedding: Science, v. 144, no. 3625, p. 1449–1450.

Martin, G. C., 1902, The geology of Garrett County: in Garrett County, Maryland Geol. Survey, p. 55–182.

————, 1908, Description of the Accident and Grantsville quadrangles: U. S. Geol. Survey, Geol. Atlas of U. S., folio 160, 14 p.

Mason, C. C., and Folk, R. L., 1958, Differentiation of beach, dune, and aeolian flat environments by size analysis, Mustang Island, Texas: Jour. Sed. Petrology, v. 28, p. 211–226.

McKee, E. D., and Weir, G. W., 1953, Terminology of stratification and cross-stratification: Geol. Soc. America Bull., v. 64, p. 381–390.

Meckel, L. D., 1967a, Tabular and trough cross-bedding; comparison of dip azimuth variability: Jour. Sed. Petrology, v, 37, p. 80–86.

————, 1967b, Origin of Pottsville conglomerates (Pennsylvanian) in the Central Appalachians: Geol. Soc. America Bull., v. 78, p. 223–258.

Off, Theodore, 1963, Rhythmic linear sand bodies caused by tidal currents: Am. Assoc. Petroleum Geologists Bull., v. 47, p. 324–341.

Pelletier, B. R., 1958, Pocono paleocurrents in Pennsylvania and Maryland: Geol. Soc. America Bull., v. 69, p. 1033–1064.

Pettijohn, F. J., 1957, Sedimentary rocks (2d ed.): New York, Harper and Brothers, 718 p.

————, 1962, Paleocurrents and paleogeography: Am. Assoc. Petroleum Geologists Bull., v. 46, p. 1468–1493.

Potter, P. E., and Pettijohn, F. J., 1963, Paleocurrents and basin analysis: Berlin–Göttingen–Heidelberg, Springer–Verlag, 296 p.

————, and Pryor, W. L., 1961, Dispersal centers of Paleozoic and later clastics in the Upper Mississippi Valley and adjacent areas: Geol. Soc. America Bull., v. 72, p. 1195–1250.

Reger, D. B., 1926, Mercer, Monroe and Summers Counties: West Virginia Geol. Survey, 963 p.

————, 1927, Pocono stratigraphy in the Broadtop Basin of Pennsylvania: Geol. Soc. America Bull., v. 38, p. 397–410.

————, 1931, Randolph County: West Virginia Geol. Survey, 989 p.

Rittenhouse, Gordon, 1949, Petrology and paleogeography of Greenbrier Formation: Am. Assoc. Petroleum Geologists, v. 33, p. 1704–1730.

Schlee, John, and others, 1964, Statistical parameters of Cape Cod beach and eolian sands: U. S. Geol. Survey Prof. Paper 501-D, p. D118–D122.

Sedimentation Seminar, Indiana University, 1966, Cross-bedding in the Salem Limestone of central Indiana: Sedimentology, v. 6, p. 95–114.

Selley, R. C., 1967, Paleocurrents and sediment transport in nearshore sediments of the Sirte Basin, Libya: Jour. Geology, v. 75, p. 215–223.

Shepard, F. P., and Cohee, G. V., 1936, Continental shelf sediments off the Mid-Atlantic States: Geol. Soc. America Bull., v. 47, p. 441–457.

————, and Moore, D. G., 1955, Central Texas coast sedimentation: characteristics of sedimentary environment, recent history and diagenesis: Am. Assoc. Petroleum Geologists Bull., v. 39, p. 1463–1593.

————, and Young, Ruth, 1961, Distinguishing between beach and dune sands: Jour. Sed. Petrology, v. 31, p. 196–214.

Shotton, F. W., 1937, The Lower Bunter sandstone of North Worchestershire and East Shropshire: Geol. Mag., v. 74, p. 534–553.

Tanner, W. F., 1955, Paleogeographic reconstructions from cross-bedding studies: Am. Assoc. Petroleum Geologists Bull., v. 39, p. 2471–2483.

————, 1963, Permian shoreline of central New Mexico: Am. Assoc. Petroleum Geologists Bull., v. 47, p. 1604–1610.

Twenhofel, W. H., 1946, Mineralogical and physical composition of the sands of the Oregon coast from Coos Bay to the mouth of the Columbia River: Oregon State Dept. of Geology and Mineral Industries, Bull. 30, 64 p.

Wells, Dana, 1950, Lower Middle Mississippian of southeastern West Virginia: Am. Assoc. Petroleum Geologists Bull., v. 34, p. 902–917.

Wentworth, C. K., 1931, The mechanical composition of sediments in graphic form: Univ. of Iowa Studies in Natural History, v. XIV, no. 3, 126 p.

Chapter 6

Facies of Iron Sedimentation in the Clinton Group

RALPH E. HUNTER

INTRODUCTION

THE PROBLEM OF iron sedimentation has received considerable attention in recent years. Experimental and theoretical work summarized by Garrels and Christ (1965) and James (1966) has suggested that small changes in chemical parameters cause changes in iron mineralogy and that the iron minerals in sedimentary rocks may therefore be sensitive indicators of the chemical environments of deposition. In an application of this concept, the Precambrian iron formations of the Lake Superior region have been grouped into sedimentary "facies" by their iron mineralogy and other characteristics (James, 1954).

The Middle Silurian Clinton Group in the Central Appalachian Basin is especially favorable for the study of iron sedimentation. Unlike most Precambrian rocks, outcrops and deep wells provide widespread stratigraphic control (Fig. 1), precise faunal zonation is available (Fig. 2), metamorphism and extreme structural complexities are absent, and interpretation of the iron-rich rocks is facilitated by their close association with more normal sedimentary rocks.

In this study, the depositional environments of the various rock types are interpreted as far as possible from textures, sedimentary structures, fauna, stratigraphic relations, and content of nonferriferous minerals. The modes of occurrence of iron in the various rock types are then considered, and, finally, the environmental implications of the iron mineralogy and petrography are compared with the interpretations of environment made by other means.

GENERAL CHARACTER OF THE CLINTON GROUP

The Clinton Group forms the lower half of the Niagaran Series in the Central Appalachian Basin. It is predominantly shale, but sandstone is common in the eastern part of the basin and carbonate rock in the western part (Figs. 3 and 4). In much of the area sandstones of the Albion Series underlie the Clinton Group and carbonate rocks of the Lockport Group of the Niagaran Series overlie it. In the westernmost part of the basin the underlying sandstones grade into shales and carbonate rocks. The overlying carbonate rocks grade into shales and sandstones in the easternmost part of the basin.

Taken as a whole, the Albion–Niagaran sequence represents an eastward transgression over the land mass formed by the Taconic orogeny. The influx of coarse clastics decreased, with fluctuations, during this period. The Clinton Group was deposited long enough after the climax of the orogeny for most of the sandstones to be relatively mature protoquartzites and orthoquartzites, according to the terminology of Pettijohn (1957). The petrography of the Clinton terrigenous sand has been described in detail by Folk (1960).

The Clinton Group in the Central Appalachian Basin can be divided into two main tectonic facies. A geosynclinal belt trending northeast–southwest forms the southeastern facies. Along the geosynclinal axis, subsidence was relatively rapid and continuous, and thick strata accumulated (Fig. 3). To the southeast is a transitional

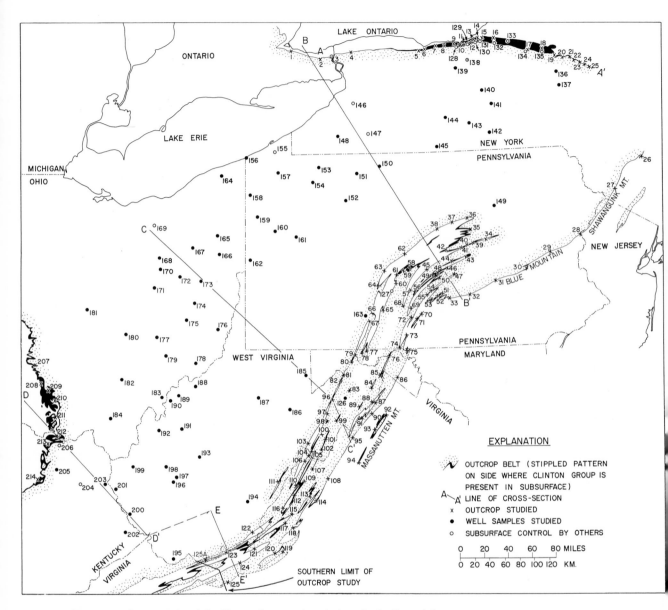

FIGURE 1. Outcrop belts of the Clinton Group and equivalents in the Central Appalachian Basin and locations of outcrop and subsurface control points. In the text, letters appended to locality numbers (for example, locality 22a) refer to one of several closely spaced control points that are assigned a single locality number and plotted at a single point on the maps. The locations of the control points are described by Hunter (1960, p. 244–395).

belt of thin strata between the geosynclinal axis and the positive area beyond the basin. To the northwest the geosynclinal facies grades into a shelf facies of thin strata with numerous disconformities. Thin strata undoubtedly extended far beyond the present limits of the Clinton Group and may have covered much of the Canadian Shield, which furnished little, if any, clastic material.

The ironstones for which the Clinton Group is noted are mainly oölitic hematites, but oölitic chamosites also are common. X-ray diffraction powder patterns show that the chamosite oölites are largely iron-rich septechlorite, a mineral of chloritic composition with 7 Å basal spacing (Hunter, 1960; Schoen, 1964). Among the more complete petrographic descriptions of the ironstones are those by Smyth (1918), Alling (1947),

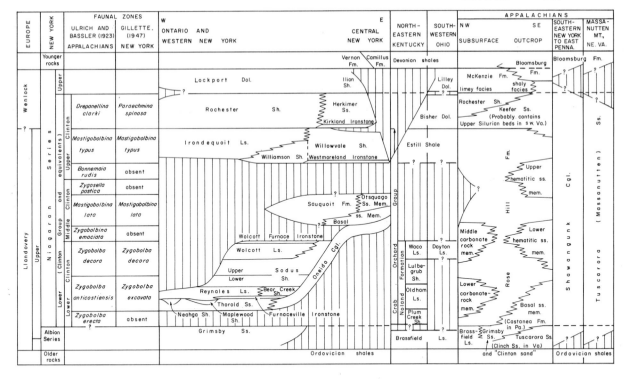

FIGURE 2. Correlation chart of the Clinton Group and equivalents in the Central Appalachian Basin. The stratigraphic nomenclature and correlations are based largely on the work of Bolton (1964), Fisher (1960), and Gillette (1947) in the New York and Ontario outcrop belt; Butts (1940), C. K. Swartz (1923), Swartz and Swartz (1931), F. M. Swartz (1934; 1935), and Woodward (1941) in the Appalachian outcrop belts; Foerste (1931) and Rexroad and others (1965) in the Kentucky and Ohio outcrop belt; Cate (1961), Fettke (1933; 1941; 1961), Freeman (1951), Martens (1939; 1945), and Rittenhouse (1949) in the subsurface; and C. K. Swartz and others (1942) and Ulrich and Bassler (1923) throughout the basin.

and Schoen (1964). The evidence that the oölites formed by precipitation of iron compounds on grains while they were being transported across the sea floor is considered conclusive (Hunter, 1960). The mineralogical forms in which the iron was originally precipitated, however, have not been conclusively determined. Hunter (1960) and Sheldon (1965) concluded that much of the hematite was formed by sea-floor oxidation of chamosite, but Schoen (1965) concluded that much of it was precipitated directly. For the purpose of this paper, the essential point is that no evidence of major postdepositional replacement was found. Minor changes in composition, such as the possible transformation of a hydrated iron oxide to hematite, and changes in crystal structure are of minor importance to the purpose of this paper.

FACIES OF THE CLINTON GROUP

The Clinton Group may be divided into several consanguineous associations of rock types. These associations will be referred to as "facies" and will be designated by their dominant rock type.

Silica–Cemented Sandstone Facies

The easternmost facies of the Clinton Group is composed predominantly of sandstone, mainly light gray in color, which is cemented by quartz overgrowths on the grains. Quartz–pebble conglomerate is common in parts of this facies, and shales ranging from light gray to black to greenish gray in color are minor constituents of the facies. Stratigraphic units forming this facies are (1) the eastern part of the Herkimer Sandstone, termed the Jordanville Member by Zenger (1966), (2) the southeastern part of the Keefer Sandstone, (3) parts of the undifferentiated Clinton Group along Blue Mountain, Pennsylvania, (4) the Oneida Conglomerate, (5) the Shawangunk Conglomerate, which is partly Albion and partly Clinton in age (Swartz and Swartz, 1931), and (6) the Tuscarora Sandstone, which is mainly of Albion age but which contains beds of

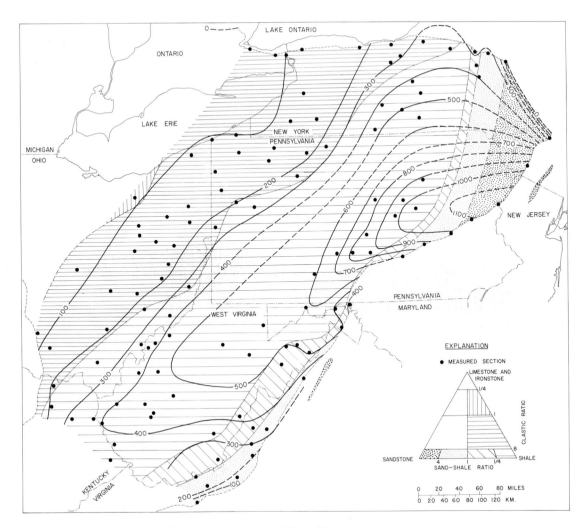

FIGURE 3. Isopach and general lithofacies map of the Clinton Group and equivalents on the Central Appalachian Basin. This and other stratigraphic maps in this paper are drafted on the present geographic base rather than on a palinspastic base.

probable Clinton age along Massanutten Mountain, Virginia (Bevan et al., 1938, p. 25).

Most beds in the silica-cemented sandstone facies are unfossiliferous. *Arthrophycus* and *Scolithus*, trace fossils of distinctive character, are the most common organic structures. They have been discussed in detail by Yeakel (1962). Fossils that are unquestionably marine are very rare in this facies and apparently are restricted to those parts of the silica-cemented sandstone units that are vertically or laterally transitional into marine beds. Unquestionably marine fossils have not been reported in beds with *Arthrophycus* or *Scolithus*.

Cross-bedding is fairly common to abundant in the silica-cemented sandstones. Yeakel found that the Tuscarora and Shawangunk cross-bedding

dips predominantly to the northwest. Cross-bedding in the silica-cemented sandstones of Clinton age also dips predominantly northwestward (Fig. 5).

The depositional environments of the silica-cemented sandstone facies and the *Arthrophycus–Scolithus* fauna are debated. The sandstones and the unusual fauna occur east of beds with undoubtedly marine faunas. Swartz (1946, 1948) and Amsden (1955) concluded that the Tuscarora Sandstone is at least partly a nearshore marine or littoral deposit but that the Shawangunk Conglomerate is probably nonmarine. Probably the best evidence for a littoral environment is the variation in grain roundness from bed to bed that Folk (1960) found in parts of the Tuscarora and Keefer Sandstones at locality 76. This varia-

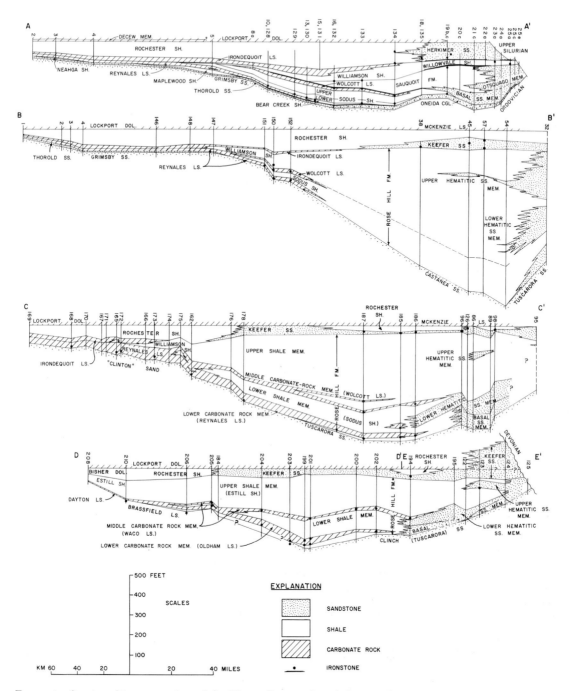

FIGURE 4. Stratigraphic cross sections of the Clinton Group and equivalents in the Central Appalachian Basin. Locations of the lines of cross section are shown on figure 1. Sections that are not located exactly on a line of cross section were plotted by projecting the sections perpendicularly onto the cross section.

tion implies strong abrasion at the depositional site.

Yeakel (1962), on the other hand, concluded that the Tuscarora Sandstone was deposited as fluvial channel deposits on an alluvial plain sloping to the northwest. In support of Yeakel's con-

clusion, it will be shown in the following sections that several facies that are laterally and vertically transitional between the silica-cemented sandstone facies and unquestionably marine beds are themselves probably in part littoral deposits. However, no available evidence rules out the

		STRIKES OF SYMMETRIC RIPPLE-MARKS (IN MARINE BEDS)	DIP AZIMUTHS OF CROSS-BEDS AND STEEP SIDES OF ASYMMETRIC RIPPLE-MARKS				
			MISCELLANEOUS SANDSTONES	CARBONATE-CEMENTED SANDSTONE FACIES	CHLORITE-CEMENTED SANDSTONE FACIES	HEMATITE-CEMENTED SANDSTONE FACIES	SILICA-CEMENTED SANDSTONE FACIES
UPPER PART OF UPPER CLINTON	NEW YORK	A. 23 HERKIMER SS. (LOCS. 19-21)	B. 21 HERKIMER SS. (LOCS. 19-21)				C. 15 HERKIMER SS. (LOCS. 22-24)
UPPER PART OF UPPER CLINTON	APPALACHIAN AREA	D. 11 KEEFER SS.	E. 24 KEEFER SS. (LOC. 48a)	F. 50 KEEFER SS.			G. 62 KEEFER SS.
CLINTON (EXCEPT UPPERMOST PART)	NEW YORK	H. 24 SAUQUOIT FM. (LOCS. 19-22)			I. 67 BASAL SAUQUOIT (LOCS. 19-25)	J. 42 SAUQUOIT FM. (LOCS. 19-25)	K. 36 ONEIDA FM. (LOCS. 18-23)
CLINTON (EXCEPT UPPERMOST PART)	APPALACHIAN AREA	L. 94 ROSE HILL FM.	M. 25 CRESAPTOWN HEM. SS. (LOCS. 80-82)		N. 16 BASAL ROSE HILL (LOC. 105)	O. 75 ROSE HILL FM.	P. 44 CLINTON GROUP (LOCS. 30-32)

Key to rose diagrams. Data are grouped in 30° classes and plotted on the basis of number frequency percent.

25 number of readings

FIGURE 5. Rose diagrams of cross-bedding and ripple-mark data from parts of the Clinton Group. The measurements were made and corrected for folding by the methods outlined by Yeakel (1962). The data in diagrams A and D are from the carbonate-cemented sandstone facies. Diagram E represents a locality at which the carbonate-cemented facies of the Keefer Sandstone is unusually well cross-bedded. The data in diagrams H and M are from the hematite-cemented sandstone facies and from thin sandstone beds in the green shale facies. The Cresaptown Hematitic Sandstone, data from which are shown in diagram N, is a tongue of the lower hematitic sandstone member of the Rose Hill Formation (Fig. 7).

possibility that marine waters occasionally invaded the area of silica-cemented sandstone deposition.

Very little iron sedimentation took place in the environment in which the silica-cemented sandstones were deposited. Roughly spherical spots of siderite and ankeritic dolomite cement occur in a few sandstone beds and traces of pyrite are commonly present in the sandstone.

Red Argillaceous Sandstone Facies

Many parts of the silica-cemented sandstone facies grade vertically or laterally to the west or northwest into a facies containing much grayish red argillaceous sandstone and siltstone. Other common and locally predominant rock types in this facies are greenish gray and gray argilla-

ceous sandstones and siltstones. Light gray silica-cemented sandstone and shales ranging in color from red to greenish gray to gray are present in small amounts. Examples of this facies are (1) most of the Castanea Formation, the upper part of which is probably of Clinton age, (2) parts of the basal sandstone member of the Rose Hill Formation in the outcrop belt of Virginia and West Virginia (Fig. 6), (3) parts of the Rose Hill Formation above its basal part along the eastern edge of the Appalachian outcrop belt (Figs. 7 and 8), and (4) the Grimsby Sandstone, of Albion age. Parts of the Thorold Sandstone are an entirely greenish gray subtype of this facies.

The fauna of the red argillaceous sandstone facies consists of *Arthrophycus* and *Scolithus*.

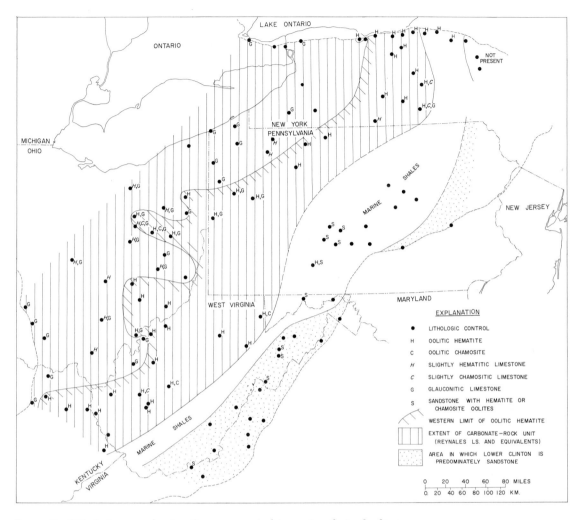

FIGURE 6. Lithofacies map showing the occurrence of iron minerals in the lower and middle parts of the Lower Clinton (*Zygobolba erecta*, *Z. anticostiensis*, and *Z. excavata* zones). Most of the ironstones are in or at the contacts of the Reynales Limestone of New York and its approximate equivalents elsewhere. The iron-rich beds in the Appalachian outcrop belts are in the basal sandstone member of the Rose Hill Formation, which forms the sandstone facies in Virginia and West Virginia, and in the upper part of the Castanea Formation in Pennsylvania.

Phosphatic brachiopod shell fragments and other marine fossils occur in some beds of the Castanea Formation and the basal sandstone member of the Rose Hill Formation. However, all of these unquestionably marine fossils occur in rocks referable to the hematite-cemented or chlorite-cemented sandstone facies and are not associated with *Arthrophycus* or *Scolithus*. A marine fauna occurs in the western part of the Grimsby Sandstone, but *Arthrophycus* is found without marine fossils in an eastern facies (Fisher, 1954, p. 1992).

The argillaceous sandstones of this facies contain as much as 30 percent argillaceous matrix. The clay is patchily distributed, and the clay mineral flakes tend to be randomly oriented. Both color mottling and textural mottling are

common. The bedding is even, and lamination within beds is poorly defined or lacking. Alling (1946) has described similar rocks in the Grimsby Sandstone. The common disruption of bedding and internal lamination and formation of textural mottling by *Scolithus* and by more irregularly shaped burrows in this facies strongly suggest that the argillaceous sandstones owe their high clay contents and lack of internal lamination to a continuation of burrowing to the point that originally interbedded sand and mud became thoroughly intermixed.

The red argillaceous sandstone facies, like the silica-cemented sandstone facies, occurs east of beds with unquestionably marine faunas and contains distinctive trace fossils not found in beds with definitely marine fossils. A nonmarine en-

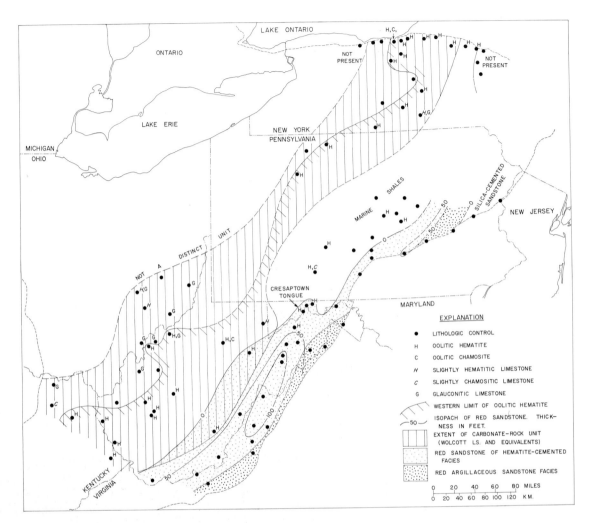

FIGURE 7. Lithofacies map showing the occurrence of iron minerals in the upper part of the Lower Clinton and the lower part of the Middle Clinton (*Zygobolba decora* and *Zygobolbina emaciata* zones). Most of the ironstones occur in or at the contacts of the Wolcott Limestone of New York and its approximate equivalents elsewhere. The lower hematitic sandstone member of the Rose Hill Formation, which forms the hematite-cemented sandstone facies shown in this map, occurs at or near the break between the Lower and Middle Clinton where it is thin, but occupies a broader interval where it is thick.

vironment therefore seems probable. Fisher (1954) reached a similar conclusion for the part of the Grimsby Sandstone that contains *Arthrophycus*. In contrast to the silica-cemented sandstones, which were most likely deposited in fluvial channels, the red argillaceous sandstones were probably deposited on floodplains on the low, deltaic parts of alluvial plains. However, intense burrowing such as took place in this facies is characteristic of parts of some present-day tidal flats (Evans, 1965; van Straaten, 1961), and such an environment cannot be ruled out.

Iron occurs in red argillaceous sandstones mainly in the form of very finely crystalline hematite associated with argillaceous matrix. In the greenish argillaceous sandstones, the main iron-bearing mineral is chlorite, which forms part of the argillaceous matrix. Small amounts of ankeritic dolomite, siderite, and pyrite occur in some of the green beds. The strongest red and green colors, and presumably the highest iron contents, occur in the most argillaceous rocks. The red argillaceous sandstones of the Clinton Group are similar in color, general lithology, and

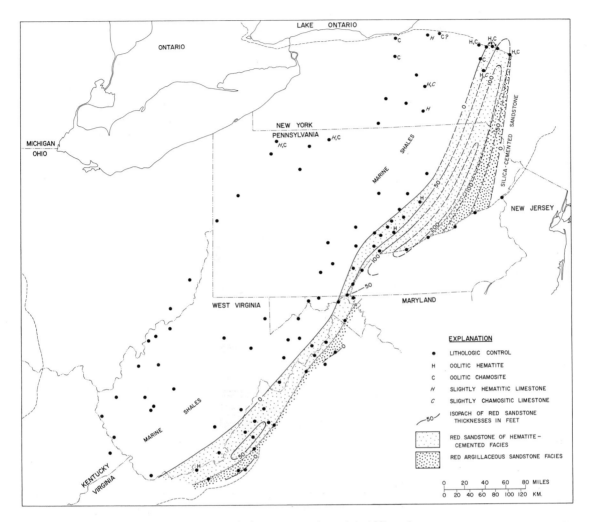

FIGURE 8. Lithofacies map showing occurrence of iron minerals in the middle and upper parts of the Middle Clinton and lower and middle parts of the Upper Clinton (*Mastigobolbina lata*, *Zygosella postica*, *Bonnemaia rudis*, and *M. typus* zones). The upper hematitic sandstone member of the Rose Hill Formation, which forms the hematite-cemented sandstone facies shown in the Appalachian outcrop belt on this map, occurs at or near the break between the Middle and Upper Clinton where it is thin but occupies a broader interval where it is thick. The best example of ironstone in this interval is the Westmoreland Ironstone at the base of the Willow-vale shale in New York.

probably in iron content to most other red beds in the Appalachian region.

The origin of the hematite pigment in the red argillaceous sandstones is uncertain. Precipitation of the hematite from solutions seems unlikely in view of its close association with detrital matrix. The separation of many parts of the red argillaceous sandstone facies from the source area by a facies of light gray, silica-cemented sandstone would seem to rule out derivation of the iron oxide from the source area, especially when it is considered that even shale partings between sandstone beds and rock fragments within sandstone beds of the silica-cemented sandstone facies are rarely red. However, Friend (1966) has explained the lack of red color in channel sandstones in the Catskill red beds as the result of postdepositional reduction of iron oxide by organic material in sediments that were below the water table most of the time. A similar explanation for the lack of red color in the silica-cemented sandstone facies of the Clinton Group cannot be ruled out.

If the hematite in the red argillaceous sandstones is formed by postdepositional oxidation, the only apparent source of the iron is iron-bearing chlorite, such as occurs in the matrix of the greenish argillaceous sandstone. Friend (1966) cites the iron contents of red portions of the Catskill beds, which are higher than those of associated green portions, as evidence against formation of the red beds by postdepositional oxidation of green beds. Although some green mottles in the Clinton red argillaceous sandstones probably formed by postdepositional reduction, a prior oxidation of the entire bed followed by partial reduction is possible.

Although it is uncertain whether the red argillaceous sandstones of the Clinton Group owe their hematite content to the postdepositional oxidation of iron-bearing chlorite or to the preservation of fine-grained iron oxide derived from soils of the source area, it can be stated definitely that they are indicative of an oxidizing environment of deposition. This is implied by the evident ease with which any existing detrital iron oxide was reduced in the environment of greenish argillaceous sandstone deposition. It can also be stated, from the evidence of thin red beds interbedded with thin green beds, that postdepositional changes were caused by chemical environments existing within a few feet of the sediment surface. Very likely postdepositional oxidation and reduction both were common processes in this facies.

Hematite–Cemented Sandstone Facies

Many parts of the red argillaceous sandstone facies grade westward into a facies dominated by grayish red hematite-cemented sandstone (Figs. 7 and 8). Minor rock types in the hematite-cemented sandstone facies are red and greenish gray shales and oölitic hematite. This facies makes up the lower and upper hematitic sandstone members, named the Cabin Hill and Centre Members in Pennsylvania (Miller, 1961), of the Rose Hill Formation, the Otsquago Member of the Sauquoit Formation, and some beds in the upper part of the Castanea Formation.

Trails and castings are common in the hematite-cemented sandstones. However, none of these trace fossils resembles *Arthrophycus* or *Scolithus*. The only other fossils found in most outcrops are thin, arcuate fossil fragments composed of collophone, and these are rare. Their similarity in thin section to *Lingula*, a common fossil in parts of the Clinton Group, suggests that they are *Lingula* fragments. Other marine fossils are found in a few beds, especially in the western parts of the units.

Intraformational shale chips are very common in this facies. Chamosite and collophane pellets are common in the more iron-rich beds.

Ripple marks and cross-bedding are common sedimentary structures in these sandstones. Most of the cross-laminated beds are wavy or lenticular. Much of the unevenness of bedding appears depositional rather than erosional, and most of the cross-bedding appears to be the internal structure of large-scale current ripple marks averaging about 6 in. (15 cm) in ripple height and more than 3 ft (0.9 m) in wavelength. Small-scale ripple marks of oscillation, current, and interference types are common on thinner beds. The cross-bedding and current ripple marks in most parts of the hematitic sandstone facies indicate currents mainly to the west and northwest, as does the cross-bedding in the silica-cemented sandstones (Fig. 5). Oscillation ripple marks strike predominantly normal to the current direction (Fig. 5, diagrams H and L). In the Cresaptown Member of the Rose Hill Formation, a tongue that extends northward from the main body of the lower hematitic sandstone member (Fig. 7), the cross-bedding is more complexly oriented (Fig. 5, diagram M).

The sparse fauna of the hematite-cemented sandstones is marine, but its restriction at most localities to probable *Lingula* suggests a near-shore, shallow marine environment (Craig, 1952). The occurrence of current ripple marks averaging about 6 in. in ripple height suggests depths averaging about 5 ft (1.6 m) (Allen, 1963a, Fig. 6), but the lack of abundant data on sedimentary structures on present sea floors makes the relationship of ripple height to water depth questionable.

In at least one outcrop (Loc. 125A, on Route 52, 4 miles east of Bluefield, West Virginia, lat. 37°16', long. 81°08'), however, an intertidal origin seems certain. A large bedding plane exposure at this point reveals current ripple marks about 1 ft (0.3 m) in ripple height and more than 10 ft (3.2 m) in wavelength with small-scale ripple marks located in the trough between two ripple crests and striking at right angles to the trend of the trough. Such a feature has been found in modern intertidal sediments, where the small-scale ripple marks form in the submerged troughs of larger ripple marks whose crests are above water (Land and Hoyt, 1966, p. 196).

Although parts of the hematite-cemented sandstone facies are probably tidal flat deposits, other parts were probably deposited in shallow marine waters below low tide line. The predominantly basinward sediment transport indicated by cross-bedding was probably caused by ebb tidal currents, perhaps assisted by reflux from on-shore waves and influx of river water. The tonguelike shape of the Cresaptown Member suggests a bar, and its polymodal cross-bedding may have been formed by wave-produced currents as well as tidal currents washing over the bar crest.

The hematite-cemented sandstones are, next to the ironstones, the most iron-rich rocks in the Clinton Group. The sandstones are cemented principally by quartz overgrowths and hematite with smaller amounts of iron-rich chlorite and traces of carbonate. Hematite and iron-rich chlorite commonly form about 30 percent of the cement, or 10 percent of the rock. The hematite and chlorite occur as small platy crystals partially rimming the sand grains and projecting into the quartz cement.

Although the delicate hematite and chlorite crystals must have grown in open pore spaces after deposition, as noted by Folk (1960, p. 35), they are probably in large part diagenetic reorganization products of iron oxide and iron-rich silicate that were present at the time of deposition as thin oölitic coatings on the sand grains. This is suggested by (1) the occurrence, in the more iron-rich beds, of some grains with relatively thick hematite coatings that can be seen in reflected light to be concentrically laminated, and (2) the occurrence of thin but sharply defined, concentrically laminated, oölitic coatings of hematite and chamosite on sand grains in patches of carbonate cement, which occur in a few beds. The absence of hematite cement at many grain contacts, taken by Folk (1960, p. 35) as evidence of postdepositional precipitation, may alternatively be explained by intrastratal solution at points of grain contact, for the quartz grains themselves are commonly slightly sutured at those points.

The contradictory evidence of pre- and postdepositional origin of the hematite and iron-rich chlorite cement is best explained, I feel, by the precipitation of most of the iron compounds as oölitic coatings on sand grains before their final deposition, followed by crystallization, recrystallization, or solution and reprecipitation after burial, except where the pores were cemented at an early date by carbonate. However, some of the hematite may have been precipitated after deposition of the sand but very near the sediment–water interface. Hematite remained stable after burial evidently because the Eh of the interstitial solutions was kept high by the large amount of ferric oxide within the sediment from the time of deposition.

Chlorite–Cemented Sandstone Facies

In parts of the Clinton Group a facies of chlorite-cemented sandstone with subordinate greenish gray shale takes the place of the red argillaceous sandstone facies and hematite-cemented sandstone facies in positions between silica-cemented sandstones and green marine shales. The basal sandstone member of the Sauquoit Formation and parts of the basal sandstone member of the Rose Hill Formation belong to this facies.

Collophane fossil fragments of the type found in the hematite-cemented sandstones are the only fossils that were found in the chlorite-cemented sandstones. Their identification as *Lingula* is strengthened by the findings of a few whole shells of *Lingula* in the basal sandstone member of the Sauquoit Formation at locality 23b. Collophane and chamosite pellets are common in the more iron-rich sandstones.

Small-scale ripple marks are much less common in the chlorite-cemented sandstones than in the hematite-cemented sandstones, but cross-bedding is more common and on a larger scale. The cross-laminated beds are as much as 7 ft (2.1 m) thick and average about 3 ft (0.6 m) thick. Most of the cross-laminated beds are approximately tabular. Bedding surfaces are well exposed in present-day stream beds at localities 19 to 25, where it can be seen that the cross-bedding strikes are nearly straight to slightly concave downcurrent. The relatively straight strikes and tabular bedding suggest that the cross-bedding was formed by the migration of large-scale current ripples with relatively straight crests (Allen, 1963b).

In contrast to the basinward orientation of cross-bedding dip directions in the hematite-cemented and silica-cemented sandstones, the distribution of cross-bedding dip directions in the chlorite-cemented sandstones is bimodal, even in single outcrops, with one mode having a northwesterly or basinward direction and the other mode having a southeasterly or shoreward direction (Fig. 5). Diametrically opposed modes in cross-bedding distributions have been found in modern sediments of the North Sea, where they occur in large-scale ripples near and below low tide line (Hulsemann, 1955; Reineck, 1963). They have also been found in some ancient deposits, such as shallow marine calcarenites (Sedimentation Seminar, 1966). In both modern and ancient deposits the bimodal distributions have been attributed to ebb and flow tidal currents.

If the cross-bedding in the chlorite-cemented sandstones is the internal structure of large-scale current ripples, the heights of the ripples in this facies averaged about four times as great as the current ripples in the hematite-cemented sandstones, suggesting that the chlorite-cemented sandstones formed in deeper water than the hematite-cemented sandstones (Allen, 1963a).

Petrographically, the chlorite-cemented sandstones are very similar to the hematite-cemented sandstones except that the chlorite takes the place of most or all of the hematite. X-ray diffraction powder patterns and indices of refraction indicate that the chlorite is an iron-rich 14 Å type. Patches of ankeritic dolomite and siderite cement are common in a few beds. The evidence advanced for the origin of the hematite in the hematite-cemented sandstones as oölitic grain coatings also holds true for the chlorite in the chlorite-cemented sandstones. The oölitic nature of the grain coatings in carbonate-cemented patches is especially obvious in this facies.

Green Shale Facies

The hematite-cemented and chlorite-cemented sandstones intertongue westward with greenish gray and generally subordinate grayish red shales with thin beds of greenish gray sandstone and siltstone and thin beds of carbonate rock. These shales make up the bulk of the Rose Hill and Sauquoit Formations and the Maplewood, Neahga, Bear Creek, Sodus, Williamson, Willowvale, Plum Creek, Lulbegrud, and Estill Shales. The dark gray to black shales in the basal part of the Williamson Shale are included in the green shale facies.

Much of the green shale facies is only slightly fossiliferous, but some beds contain abundant marine fossils such as brachiopods, ostracodes, trilobites, *Tentaculites*, gastropods, pelecypods, and the problematical *Buthotrephis*. Fossils are more common in parts of the green shale facies that grade westward or vertically into the carbonate-rock facies or carbonate-cemented sandstone facies.

The clay minerals identified in X-ray diffraction powder patterns of six samples of Rose Hill green and red shales were illite and iron-rich chlorite. Orientation of the clay mineral flakes parallel to bedding results in a normal shaly fissility. The carbonate contents of typical shales in this facies are negligible.

The greenish gray sandstone and siltstone beds in this facies average about 2 in. (5 cm) thick, and some beds can be seen to pinch-out when traced over distances more than 10 times their maximum thicknesses. Small-scale ripple marks of oscillation, current, and interference types are

fairly common. The beds are thinly laminated and have sharp upper contacts. As pointed out by Folk (1960, p. 41), the characteristics of these sandstones do not suggest deposition by turbidity currents.

Most of the carbonate rocks in this facies are coquinoid limestones, or their dolomitized equivalents, composed of poorly sorted, commonly unbroken fossils. Large amounts of argillaceous or fine-grained carbonate matrix are commonly present. Carbonate rocks are common only in parts of the green shale facies that grade westward or vertically into the carbonate-rock facies or carbonate-cemented sandstone facies, both of which are described in subsequent sections.

Among the environmental factors that localized the green shale facies between the several sandstone facies to the east and the carbonate-rock facies to the west, water depth was probably of prime importance. The westward decrease in ratio of sand to shale (Fig. 3), an accompanying decrease in average grain size of the sandstones, and the dominantly westward and northwestward sediment transport indicated by cross-bedding (Fig. 5) leave little doubt that the shales were deposited in deeper water than the sandstones. A comparison of textures of the carbonate rocks in the green shale facies with those in the carbonate-rock facies suggests that the shale facies also formed in deeper water than the carbonate-rock facies. The carbonate-rock facies contains calcarenites that are texturally more mature, by the criteria proposed by Folk (1962), than the coquinoid limestones of the shale facies. These more mature rocks must have formed in more agitated waters than those in which the coquinoid limestones of the green shale facies formed, and water depths were presumably less.

Red shale is the predominant rock type in the Rose Hill Formation of the western half of West Virginia, eastern Kentucky, and southern Ohio. This is the same area that, during latest Clinton time, was the site of shallow marine Keefer Sandstone deposition bounded on the east, north, and west by deeper marine Rochester Shale deposition (Fig. 9). The similarity in areal distribution of the red shale and the overlying Keefer Sandstone suggests that the red shale is a relatively shallow water deposit compared to the green shale to the east, north, and west.

Most of the iron in the green shales probably occurs in the chlorite. However, in some of the green shales a significant portion of the iron may occur in green illite or glauconite. In the red shales, hematite occurs in addition to iron-bearing silicates. The sandstones in this facies contain

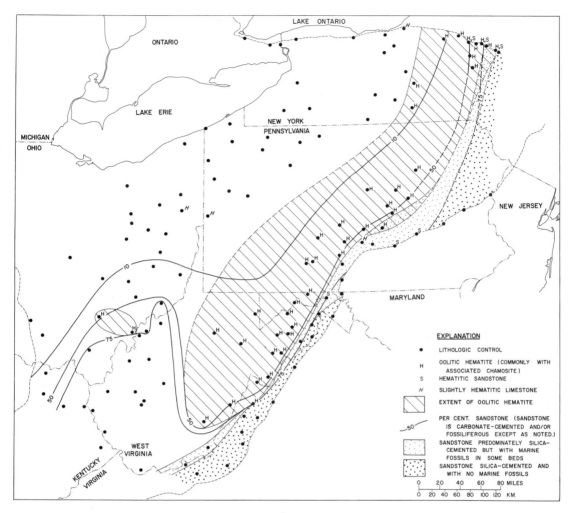

FIGURE 9. Lithofacies map showing the occurrence of iron minerals in the upper part of the Upper Clinton (*Drepanellina clarki* and *Paraechmina spinosa* zones). Most of the ironstones occur in or at the contacts of the Keefer and Herkimer Sandstones.

small amounts of iron-rich chlorite cement and detrital matrix but are cemented mainly by carbonate. They also commonly contain detrital chlorite flakes, pale green muscovite flakes that are probably iron-bearing, and, in the western parts of the shale units, glauconite pellets.

A chemical analysis of a typical greenish gray shale from locality 52b showed the Fe_2O_3 content to be 2.36 percent and the FeO content to be 4.08 percent. A typical grayish red shale from locality 56 was found to contain 4.16 percent Fe_2O_3 and 3.68 percent FeO. Such iron contents are not unusually high for shales (Pettijohn, 1957, p. 344).

Carbonate–Rock Facies

West of the green shale facies of the Clinton Group, and intertonguing with it, is a facies consisting of limestones, dolomites, and subordinate greenish gray to gray calcareous shales. The Reynales, Wolcott, and Irondequoit Limestones of New York, the Oldham, Waco, and Dayton Limestones of Kentucky and Ohio, and the carbonate-rock members of the Rose Hill Formation in the subsurface are examples of this facies.

The most common types of carbonate rock in this facies are biocalcarenites and coquinoid limestones and their dolomitized equivalents. The most common fossils are crinoid fragments, brachiopods, and bryozoans. Most of the fossils in the calcarenites are broken or disarticulated, but whole brachiopod valves are fairly common. Most of the calcarenites are fairly well sorted, but the fossils have been only moderately abraded. No carbonate oölites were found in the carbonate rocks. The interstices of the calcarenites are filled largely by sparry carbonate cement, but partial filling or flooring of interstices by fine-grained detrital carbonate is fairly common. The calcarenites are texturally fairly, but not extremely, mature, according to the criteria of Folk (1962).

Calcilutites are rare in the carbonate-rock facies, but dololutites are fairly common in parts of the Reynales Limestone. Some of these dololutites were considered by Weber (1964) to be of primary origin.

The carbonate-rock facies was deposited far from sources of terrigenous material. At times, however, the water was sufficiently shallow and turbulent for calcarenites to form.

Very little iron sedimentation took place in the carbonate-rock facies except in tongues extending eastward into the green shale facies, where oölitic ironstones are associated with carbonate rocks (Figs. 6 and 7). Only traces of iron minerals, mainly pyrite, hematite, and glauconite, occur in typical rocks of this facies. Glauconite, although not very important quantitatively, was found in every carbonate-rock unit of this facies. It occurs principally as pellets. Several randomly oriented powder samples of the pellets were identified as a 1M mica polymorph by the X-ray diffraction criteria described by Burst (1958a). Glauconite occurs west of the area in which chamosite occurs (Figs. 6 and 7).

Carbonate–Cemented Sandstone Facies

The red argillaceous sandstone facies, hematite-cemented sandstone facies, and chlorite-cemented sandstone facies are not found in the upper part of the Upper Clinton. Instead, the silica-cemented facies of the Keefer and Herkimer Sandstones grade westward into a facies dominated by light gray carbonate-cemented sandstone. Subordinate rock types in the carbonate-cemented sandstone facies are dark gray shales, carbonate rocks, and ironstones. The carbonate-cemented facies of the Herkimer Sandstone has been named the Joslin Hill Member (Zenger, 1966), whereas the carbonate-cemented facies of the Keefer Sandstone has received no formal name.

Many beds in the carbonate-cemented sandstone facies contain abundant marine fossils. Crinoid columnals are a major constituent of many sandstone beds. Brachiopods and bryozoans are also common.

The bedding of the carbonate-cemented sandstones is typically gently wavy. Cross-bedding is much less common in these sandstones than in the silica-cemented sandstones. Oscillation ripple marks ranging from 3 in. to 2 ft (0.1–0.6 m) in wavelength are fairly common.

In contrast to the sandstones, the calcarenites and oölitic hematites associated with the sandstones are commonly gently cross-bedded. The tops of the cross-laminated beds are commonly gently wavy, and much of the cross-bedding is obviously the internal structure of current ripple marks having wavelengths of from 6 in. to 3 ft (0.15–0.9 m).

Cross-bed dip directions in the carbonate-cemented sandstone facies have a more nearly random distribution than those of the silica-cemented facies of the same formations, but the principal modes are approximately basinward, to the northwest or north (Fig. 5). The cross-bedding in the Keefer Sandstone at locality 48a is plotted separately (Fig. 5, diagram E) from the other cross-bedding in the carbonate-cemented facies of the Keefer Sandstone because this is the only locality at which cross-bedding is common in the sandstones themselves. Oscillation ripple marks in this facies strike predominantly at right angles to the principal cross-bedding modes (Fig. 5, diagrams A and D).

Disruption of bedding by burrows is common in the western parts of the carbonate-cemented sandstone facies where interbedded shale is common. However, the bedding is rarely so completely obliterated as it is in the red argillaceous sandstone facies. The burrows in this facies are more irregular in form than *Arthrophycus* and *Scolithus*. Similar texturally mottled beds produced by burrowing have been described in recent marine sediments by Moore and Scruton (1957).

The predominance of well-sorted, fossiliferous sandstone with oscillation ripple marks in this facies indicates a shallow marine shelf environment. Nearness to shoreline is suggested by the location of this facies immediately west of the silica-cemented sandstone facies. In none of the underlying Clinton beds are the rocks immediately west of the silica-cemented sandstone facies so abundantly fossiliferous.

The total amount of iron sedimentation in the carbonate-cemented sandstone facies is quite small compared to that in nearshore sand facies in underlying parts of the Clinton Group. However, owing to a relatively efficient segregation of iron sedimentation from clastic sedimentation, thin beds of ironstone continued to form. The ironstones are closely associated with calcarenites. Sandstones cemented by hematite or iron-rich chlorite are rare and occur mainly in the transitional zone between the silica-cemented and carbonate-cemented facies (Fig. 9). The distinct change in the amount of iron sedimentation and the change in fossil content in the uppermost part of the Clinton Group compared to underlying parts are evidence for a striking change in chemical environment of deposition.

Gray Shale Facies

The Rochester Shale, which is the western equivalent of the Keefer and Herkimer Sandstones, differs considerably from the underlying Clinton green shale facies, just as the carbonate-cemented facies of the Keefer and Herkimer Sandstones differ considerably from any of the older Clinton sandstone facies. The Rochester Shale forms a facies consisting mainly of medium to very dark gray, silty, calcareous and/or dolomitic shale. Subordinate rock types are carbonate-cemented siltstone, silty or argillaceous, finely crystalline carbonate rock, and coquinoid limestone. Argillaceous, fine-grained dolomite is the predominant rock type in the western part of this facies and forms the Bisher Dolomite, the Rochester correlative in the Ohio outcrop belt.

The gray shale facies ranges from slightly to very fossiliferous. Fossils are much more common in the gray shale facies than in the green shale facies. Brachiopods, bryozoans, and crinoid fragments are the most common fossils.

Illite and chlorite are the main clay minerals in the Rochester Shale (Folk, 1962, p. 541), as they are in the green shale facies. Rochester Shale differs from typical green and red shales of the Clinton Group in its higher content of carbonate and carbonaceous matter.

The westward decrease in ratio of sand to shale (Fig. 9) and the westward to northward sediment transport indicated by cross-bedding (Fig. 5) indicate that the gray shale facies formed farther offshore and in deeper water than the carbonate-cemented sandstone facies. Folk (1962) interprets the Rochester Shale at locality 76 as a lagoonal deposit behind a Keefer Sandstone offshore bar. Although the very dark colors of the Rochester Shale in the Appalachian Basin south of Pennsylvania may be related to restricted circulation in a semilagoonal area east of the tongue of the Keefer Sandstone in western West Virginia and southern Ohio (Fig. 9), the facies relations do not suggest a lagoonal origin for the gray shale facies as a whole.

The gray shale facies has a low content of iron minerals. Iron occurs mainly in the chlorite and in pyrite, which is a minor constituent of the rocks in this facies. A dark gray shale in the western part of the Keefer Sandstone (loc. 50d) was found to contain 1.05 percent Fe_2O_3 and 3.78 percent FeO, and typical Rochester Shale probably contains similar amounts of iron.

Facies Relations of the Ironstones

Ironstones occur in the hematite-cemented sandstone facies, green shale facies, carbonate-rock facies, and carbonate-cemented sandstone facies of the Clinton Group. Most of the ironstones are oölitic hematites. Most oölite nuclei in ironstones associated with the hematite-cemented sandstone facies are quartz sand grains. Most oölite nuclei in the other ironstones are carbonate fossil grains. The fossil grains are typically partially replaced by iron minerals, and their internal cavities are filled by iron minerals. All of the oölitic ironstones are cemented mainly by carbonate, largely calcite and dolomite.

Thin beds of oölitic chamosite are commonly associated with oölitic hematite in the Clinton Group (Figs. 6, 7, 8, and 9). The chamosite occurs mainly as transitional beds between oölitic hematite and overlying or underlying shale. In a few places oölitic chamosite is interbedded with shale and is not associated with oölitic hematite (Fig. 8). These relations suggest that oölitic chamosite formed in deeper water than that in which oölitic hematite formed.

The ironstones in the hematite-cemented sandstone facies occur mainly in the thin, western parts of the stratigraphic units forming the facies (Figs. 7 and 8). These ironstones occur at or near the contacts of the Lower and Middle Clinton and the Middle and Upper Clinton. The Westmoreland Ironstone at the base of the Willowvale Shale in New York is the best example of an ironstone associated with the hematite-cemented sandstone facies.

The ironstones in the carbonate-rock facies occur mainly in the eastern parts of the stratigraphic units forming the facies (Figs. 6 and 7). The ironstones in this facies occur near the base of the Clinton Group and at or near the contact of the Lower and Middle Clinton. The best examples of ironstones associated with the carbonate-rock facies are the Furnaceville Ironstone at the base of the Reynales Limestone in New

York and the Wolcott Furnace Ironstone at the top of the Wolcott Limestone in New York.

Most or all of the ironstones in the green shale facies are probably correlative with ironstones in the carbonate-rock and hematite-cemented sandstone facies, and some of the ironstones in the green shale facies are lithologically continuous with ironstones in those facies. The Westmoreland Ironstone, for example, extends westward beyond the limit of associated hematite-cemented sandstones, and the Furnaceville Ironstone extends eastward beyond the limit of associated carbonate rocks.

Most of the ironstones associated with the carbonate-cemented sandstone facies occur at or near the base of that facies, which marks the contact of the middle and upper parts of the Upper Clinton. The best example is the Kirkland Ironstone at the base of the Herkimer Sandstone in New York.

As noted by Berman (1963), the horizons at which most of the ironstones occur are major lithologic and faunal breaks in the Clinton Group, and at least some of these horizons are represented by unconformities or diastems in parts of the basin (Fig. 2). At these horizons the shale facies was greatly restricted in area by the extension of tongues of carbonate rock and sandstone into the area generally occupied by shale deposition (Fig. 4). These horizons represent times of shallowing, evidently caused by decreased subsidence or actual uplift of the basin. The fact that carbonate rocks as well as sandstone are more extensive at these horizons than elsewhere suggests that the clastic influx was no greater than normal. It may actually have been less if the decreased subsidence in the basin was accompanied by decreased uplift or actual subsidence in the source area, such as might have been produced by rotation around a hinge line at the edge of the basin.

GEOCHEMISTRY OF IRON SEDIMENTATION IN THE CLINTON GROUP

Source and Transportation of the Iron

The high iron contents of facies deposited in shallow marine water near the eastern shoreline of the Central Appalachian Basin during Clinton time, together with the low iron contents of the shallow marine carbonate-rock facies in the western part of the basin, are considered here to be conclusive evidence that the iron was derived from east of the basin, and presumably from the same land area that supplied clastic material to the basin. In view of the mineralogical maturity of the clastic material supplied to the Appalachian Basin during Clinton time (Folk, 1960), the iron most likely originated from the weathering of iron-bearing rocks in the source area.

Folk (1960, p. 56) concluded that the abundance of iron in the Clinton Group was caused by the weathering of plutonic igneous rocks that were not widely uncovered by erosion until Clinton time. Although granitic to granodioritic igneous rocks are no higher in iron content than the average shale (Nockolds, 1954; Pettijohn, 1957, p. 344), it is conceivable that iron is more easily released from igneous rocks than from sedimentary and low-grade metamorphic rocks. As evidence of a change in source rock, Folk cites stratigraphically upward increases in the ratio of quartz to metamorphic rock fragments, chert fragments, and metaquartzite fragments, even in poorly rounded sands. The absence of an accompanying increase in feldspar, according to Folk, was caused by intense weathering.

It is possible, however, that aspects of the petrographic composition apart from feldspar content are controlled more by weathering than by source rock. The relative instability of chert and polycrystalline quartz fragments demonstrated by Blatt and Christie (1963) may be due more to weathering than to abrasion. If so, the upward decrease in these components even in poorly rounded sands does not necessarily indicate a change in source rock but may indicate an increase in weathering intensity. Given the fact that the intensity of weathering would be expected to increase as relief in the source area decreased after the climax of the Taconic orogeny, the change in detrital mineralogy and iron content can be explained as the result of more intense weathering, without an extensive change in source rock type.

The decrease in iron content in the uppermost part of the Clinton Group and the lack of iron-rich rocks in strata of Lockport age are probably related to the gradual marine transgression over the source area that supplied both the clastic material and iron to the Appalachian Basin. Another possible cause is a decrease in rainfall, resulting in less transportation of iron to the seas. A decrease in rainfall would be expected under the geologic conditions of gradually decreasing relief in the source area and the consequently decreasing ability of the uplands to force air currents to rise, cool, and precipitate their moisture.

The manner in which the iron was transported from the source area to the seas in which it was deposited is uncertain. Some workers (Huber and Garrels, 1953) have suggested transport of ferrous iron in solution in rivers of low Eh and pH. Recently, James (1966, p. W48) and Schoen (1965) have suggested transportation in ground water of low Eh and pH. Unfortunately, there is little evidence in the nonmarine facies of the manner in which the iron was transported through these areas.

Deposition of the Iron

The low iron content of the silica-cemented sandstone facies indicates that this was a realm of transportation of iron, a conclusion that is in accord with the nonmarine environment deduced for this facies. The reduced state of the small amount of iron that is present suggests that either the rivers were of low Eh or that sufficient organic matter was deposited with the sediment to produce ground water of low Eh and to eliminate any oxidized iron compounds.

Although the red argillaceous sandstone facies has a higher iron content than the silica-cemented sandstone facies, it is probably not unusually high for rocks with such amounts of clay, and probably little if any iron was precipitated in this facies. Rather, most of the iron was transported through this facies, probably in channels or channel sediment represented by the subordinate silica-cemented sandstones in this facies. The lack of iron-rich rocks is consistent with the interpretation of nonmarine environment favored for this facies. The predominantly oxidized state of the iron in this facies is consistent with the periodic drying that is likely to occur either on a floodplain or on the upper part of a tidal flat.

Whether the iron was carried by rivers or by ground water that was discharged subaqueously near the shoreline, precipitation took place soon after the iron-bearing water reached the marine environment. Iron contents are relatively high in the nearshore marine hematite-cemented and chlorite-cemented sandstone facies, indicating that much of the iron was precipitated before currents could distribute it uniformly throughout the basin. In the upper part of the Upper Clinton also, the rocks richest in iron occur in a nearshore marine facies, the carbonate-cemented sandstone facies. Such rapid precipitation of iron is to be expected when water having the low pH and moderately low Eh necessary for high solubility of iron is introduced into a body of water, such as a shallow marine basin, that has a higher pH and Eh (Castaño and Garrels, 1950; Huber and Garrels, 1953).

The iron contents of the shales of the Clinton Group are not unusually high, doubtless because much of the iron was precipitated before it reached the area of shale deposition. Much of the iron that is present in this facies is in the chlorite of the shales. It is uncertain whether the relatively iron-rich chlorite is inherited from the source rocks or soils of the source area, was produced by fixation into detrital chlorite of iron from river or seawater during transportation or on the sea floor, or was formed by the recrystallization of predecessor clays after deposition.

Very little iron reached the westernmost part of the Central Appalachian Basin during Clinton time.

The iron contents of rocks of the Clinton Group are controlled only partly by the rate of iron sedimentation. Other types of sedimentation tended to mask iron sedimentation. The iron contents were lowered by sand deposition in the eastern part of the basin, mud deposition in the central part, and carbonate deposition in the western part.

The segregation of precipitated iron compounds in the southeastern part of the basin from calcium and magnesium carbonates in the northwestern part of the basin is a problem in chemical sedimentation to which no final answer can be given here. However, it is suggested that a salinity gradient may have been a major cause of the segregation. High salinity is an important factor favoring calcium carbonate precipitation (Trask, 1938; Cloud, 1962). High salinity also favors iron precipitation by the accompanying increase in pH, but the Eh–pH diagrams of Huber and Garrels (1953) suggest that a rather moderate rise in pH above that of the transporting solution would probably result in precipitation of the iron without accompanying carbonate precipitation, especially if the rise in pH was accompanied by a rise in Eh. A salinity gradient ranging from relatively low in the southeastern part of the basin to relatively high in the northwestern part of the basin would therefore explain the segregation of iron and carbonate. The magnitude of the gradient need not have been large. Trask (1938) found that where salinities are less than 34 ppt, the carbonate content of recent marine sediments is generally less than 5 percent, whereas in regions where the salinity exceeds 36 ppt, carbonate generally forms more than 50 percent of the sediment.

A salinity gradient across the Central Appalachian Basin during Clinton time might be expected. Water depths were probably shallow enough, relative to the width of the basin, for circulation to be somewhat restricted and for intrabasinal variation in water chemistry to be maintained. The influx of fresh water into the southeastern part of the basin, which is implied by the high iron contents of the nearshore marine rocks, could then have produced a nearshore zone of relatively low salinity.

Some facts tend to support northwestwardly increasing salinity across the Appalachian Basin during Clinton time. The low fossil content and small number of species in much of the Rose Hill Formation of the Appalachian outcrop belt have been cited as suggesting subnormal salinity in this part of the basin (Swartz, 1946; Folk, 1960, p. 41). Fine-grained dolomites in the Decew Dolomite, which immediately overlies the Rochester Shale in Ontario and western New York, and in the Reynales Limestone in the Ontario outcrop belt have been interpreted as primary in origin from their petrologic character, and chemical evidence has been given favoring a high-salinity environment for such rocks (Weber, 1964). Also, gypsum nodules occur in the Rochester Shale at locality 1 in Ontario. If the gypsum was not introduced into the sediment after deposition, a highly saline environment in the northwestern part of the basin during Rochester time is indicated.

During the deposition of the upper part of the Upper Clinton, the salinity in the southeastern and central parts of the marine basin was probably nearly normal, as suggested by the high carbonate content and abundant fossils in the carbonate-cemented sandstone facies and gray shale facies. This increase in salinity during latest Clinton time is in accord with the decrease in fresh-water influx inferred from the low iron contents of the latest Clinton rocks.

Iron–Mineral Facies

The principal iron-mineral facies of the Clinton Group are a hematitic facies and a chloritic facies. A glauconitic facies may also be distinguished, although none of the glauconitic rocks are rich in iron. Rocks in which siderite or pyrite is the main iron mineral are rare in the Clinton Group, unless the silica-cemented sandstones, which contain very little iron in any form, be considered examples of such rocks.

The rock types in which hematite is the main iron mineral are hematite-cemented sandstone, red argillaceous sandstone, red shale, and oölitic hematite. Chlorite-cemented sandstone, green argillaceous sandstone, green shale, and oölitic chamosite are the rock types in which iron-rich chlorite or septechlorite is the main iron mineral.

With the exception of the red and green argillaceous sandstones, which are probably nonmarine, petrologic evidence has been presented that indicates the chloritic rocks formed in deeper water than their hematitic analogs. Moreover, most of the hematitic rocks were deposited southeastward, or shoreward, of most of the chloritic rocks, again suggesting that hematite tended to form in shallower water than iron-rich chlorite. This evidence is in agreement with theoretical data suggesting that ferrous silicate should form under conditions of lower Eh than those under which ferric oxide forms (Garrels and Christ, 1965).

The facies relation between glauconite and chamosite in the Clinton Group, with glauconite on the northwest and chamosite on the southeast, may have been caused by the northwestwardly increasing salinity suggested in the preceding section. High salinity would presumably favor the formation of glauconite rather than chamosite because of the accompanying high concentration of potassium. Unfortunately, no experimental data on the stability relations between glauconite and chamosite are available. However, some observational data may be cited that tend to support salinity as a factor controlling glauconite–chamosite relations. It is generally agreed that glauconite forms only in marine waters and possibly in waters of no less than normal marine salinity (Cloud, 1955). Chamosite, on the other hand, has been found forming from basalt in soils (Carroll, 1958, p. 3), which suggests that it can form in water of negligible salinity.

Glauconite–chamosite facies relations have been reported by other workers. Berg (1944, p. 63) interpreted glauconite in European iron-bearing strata as having formed farther offshore than chamosite, and probably in water of higher salinity and lower iron concentration. Burst (1958b, p. 320) reported that some glauconite from Tertiary inner neritic and marsh environments contains 7 Å chamosite, whereas mid-neritic glauconite contains little or no chamosite. Although this is attributed by Burst to more strongly reducing conditions in the nearshore environments, lower salinity in these environments should be considered as a possible cause for the variation.

CONCLUSIONS

Most of the facies of the Clinton Group are not unusually rich in iron and resemble rocks in other parts of the Appalachian stratigraphic section. The hematite-cemented and chlorite-cemented sandstone facies, on the other hand, are rich in iron, and these rock types have not been found in other parts of the Appalachian section. These facies show that iron sedimentation was proceeding at unusually rapid rates in nearshore water through much of Clinton time. At times when clastic sedimentation was relatively slow, ironstones formed.

Several factors were probably important in producing relatively high rates of iron sedimentation during Clinton time. The land area from which the iron was derived evidently had low enough relief for erosion to be relatively slow and for weathering to be relatively thorough. It is doubtful that the source area contained unusually iron-rich rocks; more probably it was large enough for the weathering of common rock types to supply the iron found in the Clinton Group.

The circulation and physical chemistry of the water in the marine basin were such that the iron was deposited before it became diluted throughout the basin and was efficiently segregated from carbonate. Probably the shallowness of the marine body of water and the resultant restricted lateral circulation were the main factors in preventing dilution. The relatively slow circulation also probably allowed the influx of fresh water to produce a nearshore zone of relatively low salinity having a pH high enough to cause precipitation of iron but low enough to suppress carbonate deposition.

In the Birmingham, Alabama, area an offshore barrier of sand separated a lagoon from open marine water during Clinton time (Sheldon, 1965). This basinal geometry was evidently very effective in segregating and preventing dilution of the iron and permitted thick ironstone beds to be deposited in the lagoon. The absence of such an effective barrier in the Central Appalachian Basin is probably the main reason for the ironstones in this area being thinner than those in the Birmingham area.

The most basic reason for iron-rich rocks being common in the Clinton Group and absent or rare elsewhere in the Appalachian stratigraphic section is probably the tectonic setting of depositional basin and source area. Iron-rich rocks are absent or rare in thick sequences of terrigenous clastic rocks deposited during periods of orogeny because (1) rapid erosion in the land area prevented complete weathering of the source rocks, and (2) the high rate of clastic sedimentation masked any iron sedimentation that did take place. In the Appalachian region, examples of such sequences are the shale–graywacke or "flysch" assemblage of the Martinsburg Forma-

tion (Ordovician) and its equivalents (Pettijohn, 1957, p. 641). At the other extreme of tectonic activity, orthoquartzite–carbonate rock assemblages deposited on stable shelves rarely contain iron-rich rocks because little if any source area was above sea level and undergoing weathering. The Oriskany Sandstone and associated Lower Devonian carbonate rocks make up an assemblage of this type in the Appalachian area.

The Clinton Group was deposited in a tectonic setting transitional between orogeny and extreme quiescence. It postdates the climax of the Taconic orogeny, when the Juniata Formation was being deposited, and predates the early Devonian period of extreme quiescence. Clinton time apparently represented the optimum conditions of a source area having both sufficiently large area and sufficiently low relief to furnish maximum amounts of iron together with relatively small amounts of clastic material.

The Lake Superior Huronian iron formations were deposited during another type of transitional interval between orogeny and stability. Their deposition followed a period of orthoquartzite–carbonate shelf sedimentation and preceded geosynclinal shale–graywacke–volcanic deposition (James, 1954). According to James, it was at this time that an offshore buckle had risen enough to form a restricted basin between the welt and its foreland but was still too low for rapid erosion. Ironstones are found in a similar tectonic setting in the Appalachian region between the Lower Devonian orthoquartzite–carbonate sequence and overlying black shales that represent the beginning of another orogeny (Sheppard and Hunter, 1960; Pettijohn, 1957, p. 641).

Probably most iron-rich rocks form, as did those of the Clinton Group and of the Lake Superior Huronian, between times of orogeny and extreme quiescence. To the extent that sedimentary sequences illustrate the orogenic cycle defined by Pettijohn (1957, p. 636–644) therefore, ironstones should be looked for in the intervals between orthoquartzite–carbonate rock sedimentation on stable shelves and the succeeding geosynclinal flysch sedimentation and in the intervals between molasse sedimentation and the succeeding orthoquartzite–carbonate rock sedimentation.

REFERENCES

Allen, J. R. L., 1963a, Asymmetrical ripple marks and the origin of water-laid cosets of cross-strata: Liverpool Manchester Geol. Jour., v. 3, p. 187–236.

————, 1963b, The classification of cross-stratified units, with notes on their origin: Sedimentology, v. 2, p. 93–114.

Alling, H. L., 1946, Quantitative petrology of the Genesee Gorge sediments: Rochester Acad. Sci. Proc., v. 9, p. 5–63.

————, 1947, Diagenesis of the Clinton hematite ores in New York: Geol. Soc. America Bull., v. 58, p. 991–1018.

Amsden, T. W., 1955, Lithofacies map of lower Silurian deposits in central and eastern United States and Canada: Amer. Assoc. Petroleum Geologists Bull., v. 39, p. 60–74.

Berg, G., 1944, Vergleichende Petrographie oolithischer Eisenerze: Arch. Lägerstättenforschung, v. 76, 128 p.

Berman, B. L., 1963, Hematite facies in the Silurian of the Appalachian basin, (abs.): Geol. Soc. America Spec. Paper 73, p. 114–115.

Bevan, A. C., and others, 1938, Guidebook, field conference of Pennsylvania Geologists, Virginia: Virginia Geol. Survey, 44 p.

Blatt, Harvey, and Christie, J. M., 1963, Undulatory extinction in quartz of igneous and metamorphic rocks and its significance in provenance studies of sedimentary rocks: Jour. Sed. Petrology, v. 33, p. 559–579.

Bolton, E. E., 1964, Pre-Guelph, Silurian formations of the Niagara Peninsula, Ontario, in Geology of central Ontario: Am. Assoc. Petroleum Geologists–Soc. Econ. Paleontologists Mineralogists Guidebook, p. 57–80.

Burst, J. F., 1958a, Mineral heterogeneity in "glauconite" pellets: Amer. Mineralogist, v. 43, p. 481–497.

————, 1958b, "Glauconite" pellets: their mineral nature and applications to stratigraphic interpretations: Amer. Assoc. Petroleum Geologists Bull., v. 42, p. 310–327.

Butts, C., 1940, Geology of the Appalachian Valley in Virginia: Virginia Geol. Survey Bull. 52, 568 p.

Carroll, Dorothy, 1958, Role of clay minerals in the transportation of iron: Geochim. Cosmochimica Acta, v. 14, p. 1–27.

Castaño, J. R., and Garrels, R. M., 1950, Experiments on the deposition of iron with special reference to the Clinton iron ore deposits: Econ. Geology, v. 45, p. 755–770.

Cate, A. S., 1961, Stratigraphic studies of the Silurian rocks of Pennsylvania—Pt. 1, Stratigraphic cross sections of Lower Devonian and Silurian rocks in western Pennsylvania and adjacent areas: Pennsylvania Geol. Survey, 4th Ser., Spec. Bull. 10, 3 p.

Cloud, P. E., Jr., 1955, Physical limits of glauconite formation: Amer. Assoc. Petroleum Geologists Bull., v. 39, p. 484–492.

————, 1962, Environment of calcium carbonate deposition west of Andros Island, Bahamas: U. S. Geol. Survey Prof. Paper 350, 138 p.

Craig, G. Y., 1952, A comparative study of the ecology and paleoecology of Lingula: Trans. Edinburg Geol. Soc., v. 15, p. 110–120.

Evans, Graham, 1965, Intertidal flat sediments and their environments of deposition in the Wash: Quart. Jour. Geol. Soc. London, v. 121, p. 209–245.

Fettke, C. R., 1933, Subsurface Devonian and Silurian sections across northern Pennsylvania and southern New York: Geol. Soc. America Bull., v. 44, p. 601–660.

————, 1941, Subsurface sections across western Pennsylvania: Pennsylvania Geol. Survey, 4th Ser., Progress Rept. 127, 51 p.

————, 1961, Well-sample descriptions in northwestern Pennsylvania and adjacent states: Pennsylvania Geol. Survey, 4th Ser., Bull. M40, 691 p.

Fisher, D. W., 1954, Stratigraphy of Medinan group, New York and Ontario: Amer. Assoc. Petroleum Geologists Bull., v. 38, p. 1979–1996.

————, 1960, Correlation of the Silurian rocks in New York State: New York State Mus. and Sci. Service Geol. Survey Map and Chart Ser., no. 1.

Foerste, A. F., 1931, The Silurian fauna, in The paleontology of Kentucky: Kentucky Geol. Survey, series 6, v. 36, p. 167–214.

Folk, R. L., 1960, Petrography and origin of the Tuscarora, Rose Hill, and Keefer formations, Lower and Middle Silurian of eastern West Virginia: Jour. Sed. Petrology, v. 30, p. 1–58.

————, 1962, Petrography and origin of the Silurian Rochester and McKenzie shales, Morgan County, West Virginia: Jour. Sed. Petrology, v. 32, p. 539–578.

————, 1962, Spectral subdivision of limestone types, in Classification of carbonate rocks—a symposium: Amer. Assoc. Petroleum Geologists Mem. 1, p. 62–84.

Freeman, L. B., 1951, Regional aspects of Silurian and Devonian stratigraphy in Kentucky: Kentucky Geol. Survey, series 9, Bull. 6, 565 p.

Friend, P. F., 1966, Clay fractions and colours of some Devonian red beds in the Catskill Mountains, U. S. A.: Geol. Soc. London Quart. Jour., v. 122, p. 273–292.

Garrels, R. M., and Christ, C. L., 1965, Solutions, minerals, and equilibria: Harper and Row, New York, 445 p.

Gillette, Tracy, 1947, The Clinton of western and central New York: New York State Museum Bull. 341, 191 p.

Huber, N. K., and Garrels, R. M., 1953, Relation of pH and oxidation potential to sedimentary iron mineral formation: Econ. Geology, v. 48, p. 337–357.

Hulsemann, Jobst, 1955, Large ripples and incline-bedded structures in the tidal marshes of the North Sea and in the molasse: Senkenbergiana Lethaea, v. 36, p. 359–388.

Hunter, R. E., 1960, Iron sedimentation in the Clinton Group of the Central Appalachian basin: Ph.D. dissertation, The Johns Hopkins University, 416 p.

James, H. L., 1954, Sedimentary facies of the Lake Superior iron formations: Econ. Geology, v. 49, p. 235–293.

————, 1966, Chemistry of the iron-rich sedimentary rocks, in Data of geochemistry, 6th edition: U. S. Geol. Survey Prof. Paper 440-W, p. W1–W61.

Land, L. S., and Hoyt, J. H., 1966, Sedimentation in a meandering estuary: Sedimentology, v. 6, p. 191–207.

Martens, J. H. C., 1939, Petrography and correlation of deep-well sections in West Virginia and adjacent states: West Virginia Geol. Survey, v. 11, 255 p.

————, 1945, Well-Sample records: West Virginia Geol. Survey, v. 17, 889 p.

Miller, J. T., 1961, Geology and mineral resources of the Loysville quadrangle: Pennsylvania Geol. Survey, 4th Ser., Atlas 127, 47 p.

Moore, D. G., and Scruton, P. C., 1957, Minor internal structures of some recent unconsolidated sediments: Amer. Assoc. Petroleum Geologists Bull. v. 41, p. 2723–2751.

Nockolds, S. R., 1954, Average chemical compositions of some igneous rocks: Bull. Geol. Soc. America, v. 65, p. 1007–1032.

Pettijohn, F. J., 1957, Sedimentary rocks: 2nd ed., Harper, New York, 718 p.

Reineck, H. E., 1963, Sedimentgefüge im Bereich der südlichen Nordsee: Abhand. Senckenberg. Naturforsch. Gesellschaft, no. 505, 90 p.

Rexroad, C. B., Branson, E. R., Smith, M. O., Summerson, Charles, and Boucot, A. J., 1965, The Silurian formations of east-central Kentucky and adjacent Ohio: Kentucky Geol. Survey, Ser. X, Bull. 2, 34, p.

Rittenhouse, G., 1949, Early Silurian rocks of the northern Appalachian basin: U. S. Geol. Survey Oil and Gas Invest. Prelim. Map 100.

Schoen, Robert, 1964, Clay minerals of the Silurian Clinton ironstones, New York State: Jour. Sed. Petrology, v. 34, p. 855–863.

————, 1965, Origin of ironstones in the Clinton group (abs.): Geol. Soc. of America, Spec. Paper 82, p. 177.

Sedimentation Seminar (Bloomington, Indiana, U. S. A.), 1966, Cross-bedding in the Salem Limestone of central Indiana: Sedimentology, v. 6, p. 95–114.

Sheldon, R. P., 1965, Barrier island and lagoonal iron sedimentation in the Silurian of Alabama (abs.): Geol. Soc. of America Spec. Paper 82, p. 182.

Sheppard, R. A., and Hunter, R. E., 1960, Chamosite oolites in the Devonian of Pennsylvania: Jour. Sed. Petrology, v. 30, p. 585–588.

Smyth, C. H., Jr. 1918, On the genetic significance of ferrous silicate associated with the Clinton iron ore: New York State Museum Bull. 208, p. 175–198.

Swartz, C. K., 1923, Stratigraphic and paleontologic relations of the Silurian strata of Maryland: Maryland Geol. Survey, Silurian Volume, p. 25–51.

————, and Swartz, F. M., 1931, Early Silurian formations of southeastern Pennsylvania, Geol. Soc. America Bull., v. 42, p. 621–662.

————, and others, 1942, Correlation of the Silurian formations of North America: Geol. Soc. America Bull., v. 53, p. 533–538.

Swartz, F. M., 1934, Silurian sections near Mount Union, central Pennsylvania: Geol. Soc. America Bull., v. 45, p. 81–134.

————, 1935, Relations of the Silurian Rochester and McKenzie formations near Cumberland, Maryland, and Lakemont, Pennsylvania: Geol. Soc. America Bull., v. 46, p. 1165–1194.

————, 1946, Faunal development, conditions of deposition, and paleogeography of some Appalachian mid-Paleozoic sediments: Anais de Segundo Congresso Panamericano de Engenharia, v. 3, p. 9–33.

————, 1948, Late Ordovician and Silurian facies, conditions of deposition, and paleogeography in north-central Appalachians (abs.): Amer. Assoc. Petroleum Geologists Bull. 32, p. 2160.

Trask, P. D., 1938, Relation of salinity to the calcium carbonate content of marine sediments: U. S. Geol. Survey Prof. Paper 186N, p. 273, 299.

Ulrich, E. O., and Bassler, R. S., 1923, Paleozoic ostracoda: their morphology, classification, and occurrence: Maryland Geol. Survey, Silurian volume, p. 271–391.

Van Straaten, L. M. J. U., 1961, Sedimentation in tidal flat areas: Jour. Alberta Soc. Petroleum Geologists, v. 9, p. 203–226.

Weber, J. N., 1964, Trace element composition of dolostones and dolomites and its bearing on the dolomite problem: Geochim. Cosmochimica Acta, v. 28, p. 1817–1868.

Woodward, H. P., 1941, Silurian system of West Virginia: West Virginia Geol. Survey, v. 14, 326 p.

Yeakel, L. S., Jr., 1962, Tuscarora, Juniata, and Bald Eagle paleocurrents and paleogeography in the Central Appalachians: Geol. Soc. America Bull., v. 73, p. 1515–1540.

Zenger, D. H., 1966, Redefinition of the Herkimer Sandstone (Middle Silurian), New York: Geol. Soc. America Bull., v. 77, p. 1159–1166.

THE VALLEY AND RIDGE AND APPALACHIAN PLATEAU—STRUCTURE AND TECTONICS

Chapter 7

Introduction

K. N. WEAVER

DOES THE BASEMENT participate in the folding of the supracrustal rocks in the Valley and Ridge Province? The various attempts to answer this question of "thin-skinned" vs. "thick-skinned" tectonics (Rodgers, 1949) form the major thrust of this section. The reader should not, however, expect to find any pat answers to this question in the papers which follow. Although the Valley and Ridge Province is a classical area of structural geology, we have literally merely scratched the surface in our knowledge of its three-dimensional tectonics. Because of the dearth of subsurface information structural interpretations rely on surface exposures, supplemented at widely spaced intervals with drilling data.

Geophysical data have increased significantly over the past 8 to 10 years, but these are not a substitute for the more tangible drilling information. All of the papers in this section emphasize in one way or another the importance of deep drilling in the interpretation of regional and local tectonics.

The Valley and Ridge province parallels and lies north and west of the Reading Prong and the Blue Ridge belt from eastern Pennsylvania to Alabama. Its western boundary is defined by the Appalachian structural front which coincides largely with the topographic features known as the Catskill Escarpment, the Allegheny Front and the Cumberland Escarpment. The width of the belt varies from a minimum of about 20 mi (32 km) in extreme eastern Pennsylvania to a maximum of about 70 mi (113 km) in central Pennsylvania. However, throughout most of its length, the width of the Valley and Ridge belt varies from 30 to 50 mi (48–81 km).

Structurally, the southeastern boundary of the Valley and Ridge approximately coincides with the "tectonite front" where rocks to the south-

east have been subjected to penetrative deformation, and those to the northwest have been folded and faulted but have not been penetratively deformed (Fellows, 1943). The Appalachian structural front, which defines the boundary between the Valley and Ridge and Appalachian Plateaus, generally marks a sharp transition between very gently folded rocks on the northwest and intensely folded and faulted rocks on the southeast.

The single most striking feature of the Valley and Ridge Province is the apparent change in style of deformation between the northeastern and southwestern portions; folding predominates northeast of Roanoke, Va. (lat. 37°15′ long. 79°55′) and thrust faulting predominates southwest. This change in style of deformation coincides with a change in Blue Ridge structure from an overturned anticlinorium northeast of Roanoke to a complexly thrust-faulted structure southwest. However, the change in deformation plan in the Valley and Ridge may be more apparent than real. In this section of the volume Gwinn describes one drill hole in the Pennsylvania Valley and Ridge which penetrates two thrust faults in an area where only very small displacement thrusts are detectable at the surface; one of the subsurface thrusts is of major magnitude with Lower Ordovician Beekmantown Dolomite resting on an Upper Silurian to Middle Ordovician sequence. Similar repetitions are found in several other deep wells in central Pennsylvania. Wood and Bergin also report intensely faulted strata in the Anthracite region of Pennsylvania. They conclude that the intensity and style of faulting is dependent on the position of the strata in relation to the stress direction and to the competency of the strata.

Harris points out in his contribution that thrust faults in the Valley and Ridge of the southern

Appalachians do not maintain the same angle of dip on a regional basis. Instead, the thrusts are initially developed as bedding plane thrusts, but may break upward into higher stratigraphic units along moderate- to high-angle ramps. He demonstrates these features by numerous examples from the Pine Mountain fault system of Kentucky, Tennessee, and Virginia. Harris sees no major break in style of deformation from one physiographic province to another in the southern Appalachians. The same types of structures exist in the Appalachian Plateau and the Valley and Ridge, although there is a gradual decrease in intensity from east to west.

Rodgers examines the Pulaski fault system which extends for some 350 mi (564 km) in Virginia and Tennessee. He suggests that a spectacular breccia—the Max Meadows breccia—may be similar to the salt-bearing "Haselgebirge" breccias of the Austria Alps. He suggests that the Max Meadows breccia may be derived from a tectonically deformed Cambrian evaporite sequence and cites several examples where evaporites have been identified in the Rome and Elbrook Formations by deep drilling. He further postulates that the gypsum and anhydrite beds, along with impervious shale layers, extended further north and south and that they helped to provide the major décollement zone that controls the structure of the Valley and Ridge province.

Cooper, in his reply to Rodgers' paper, sees no need to attribute the features in the Max Meadows breccia to the presence of evaporites. The reason for generation of the large amounts of breccia along the Pulaski thrust is the unique abundance of extremely incompetent thin-bedded shaly dolomites. He can find no evidence for the presence of salt in the fresh, unleached crush conglomerates, and he points out that Haselgebirge-type salt breccias show unmetamorphosed rock clasts, whereas the Max Meadows breccia contains clear evidence of rock-on-rock calaclasis that metamorphosed red shale fragments into chlorite phyllites.

Pettijohn (Chapter 1 this volume) poses the question: "What was the nature of the eastern limit of the huge (early Paleozoic) carbonate section?" Unfortunately, structural evidence does not provide an unequivocal answer. Gwinn concludes from available drilling, seismic, gravity, and aeromagnetic data that the Valley and Ridge in Pennsylvania is underlain by a gently southeast-dipping essentially undeformed basement and that the Paleozoic strata yielded in accordance with thin-skinned tectonics. He estimates that the Paleozoic sequence was shortened 50 mi (80 km) and postulates that the Paleozoic carbonates were originally deposited southeast of the present position of the Martic Line in Pennsylvania and Maryland, an area now occupied by the upper units of the Glenarm Series. If the upper Glenarm is Precambrian, Gwinn calls upon gravity sliding-off of the Baltimore anticlinorium as the mechanism for transporting the Paleozoic cover to its present position. If the upper Glenarm is Paleozoic, it has been thrust northwestward into the position formerly occupied by the Paleozoic carbonates.

Whatever the actual mechanisms involved, whether by gravity sliding-off of a highly uplifted central metamorphic–plutonic core (the Piedmont), or by gravitationally flattening of the uplifts in the central core creating a tangential force which crowded the sedimentary rocks of the Valley and Ridge Province northwestward (King, 1964, p. 27), the central Piedmont zone certainly played a primary role in the deformation of the Appalachians.

Thus, the basic question—do surface structures continue downward with little modification into the basement, or are they rootless?—remains elusive. We get fascinating and fleeting glimpses of these structures in widely scattered deep wells, and geophysical data are gradually accumulating, but the same data lead different observers to varying and sometimes diametrically opposed conclusions. Until much more subsurface information is available, the precise nature of the deep Valley and Ridge structures will remain in question.

REFERENCES

Fellows, R. E., 1943, Recrystallization and flowage in Appalachian quartzite: Geol. Soc. America Bull., v. 54, p. 1399–1431.

King, P. B., 1964, Further thoughts on tectonic framework of southeastern United States; in Tectonics of the southern Appalachians, Va. Polytech. Inst., Dept. of Geol. Sci. Memoir 1, p. 5–31.

Rodgers, J., 1949, Evolution of thought on structure of the middle and southern Appalachians: Amer. Assn. Petrol. Geol. Bull., v. 33, p. 1643–1654.

Chapter 8

Kinematic Patterns and Estimates of Lateral Shortening, Valley and Ridge and Great Valley Provinces, Central Appalachians, South-Central Pennsylvania

VINTON E. GWINN

INTRODUCTION

GEOLOGIC AND GEOPHYSICAL data to late 1963 from the Plateau and northwestern Valley and Ridge Provinces of the Central Appalachians indicated that Precambrian basement crystalline rocks and the lowest portions of the sedimentary sequence did not participate concordantly in Paleozoic deformations observed in surface rocks in the outer part of the deformed geosyncline. Throughout this outermost region of Appalachian folding, skin-deep concentric folds and imbricate thrust faults appear to terminate downward against master thrusts or sole thrusts which accommodated radial outward (westward, northwestward, and northward) translation of the upper part of the sedimentary sequence (Woodward, 1959, 1964; Rodgers, 1964a; Gwinn, 1964).

New surface and subsurface data from central and southern Pennsylvania raise the possibility that the thin-skinned dermal detachment pattern typical of the outer folded Appalachian region may persist southeastward through the inner Valley and Ridge into the northwestern margin of the Great Valley. New geophysical data permit one to speculate, further, that essentially undeformed basement may descend gently still farther to the southeast beneath the Great Valley to the foot of or beneath the South Mountain Anticlinorium, the northern tip of the Blue Ridge physiographic–structural province.

This paper analyzes data drawn from an almost-pie-shaped sector at the center of the Central Appalachian fold arc, bounded roughly on the east by the Susquehanna River in Pennsylvania, on the south–southwest by the Maryland–Pennsylvania border, and on the southeast by downwarped basins of Triassic redbeds (Fig. 1).

HYPOTHESES TO BE TESTED

Three major hypotheses will be tested by geological and geophysical data from the south-central Pennsylvania region and are illustrated schematically in Figure 2. They are the following:

(1) That the basement of the Great Valley is intimately involved in the surface deformation and is linked kinematically with the basement deformation of a rooted South Mountain Anticlinorium (Fig. 2A) (see Espenshade, Chapter 14 this volume, Fig. 1);

(2) South Mountain Anticlinorium is a rooted basement uplift reflecting primary upper crustal shortening, but the Great Valley surface deformation is fundamentally discordant to a very deep, perhaps undeformed basement; major shortening of the Paleozoic sequence occurs above a now folded zone of detachment which crops out in the northwestern flank of South Mountain Anticlinorium (Fig. 2B);

(3) Great Valley surface structures, discordant to deep basement, reflect shortening of the Paleo-

127

zoic rocks above a detachment zone which persists southeast beneath a rootless South Mountain anticlinorial allochton (Fig. 2C).

A brief, nonexhaustive review of the surface geology across south-central Pennsylvania based on previous work of other geologists and the writer precedes presentation of new geophysical data and their integration with results of prior geophysical investigations. For more complete treatment of the regional geology of the South Mountain–northern Blue Ridge Province, the reader is referred to Espenshade's paper in this volume (Chapter 14).

ACKNOWLEDGMENTS

I would like to express my gratitude to Arthur A. Socolow, State Geologist of Pennsylvania, and MacKenzie L. Keith, Head, Mineral Conservation Section, Pennsylvania State University, for continuing personal encouragement and financial support during the preparation of the Tectonic Map. Services extended by McMaster, Pennsylvania State, and Louisiana State Universities in the period 1964–1967 are gratefully acknowledged. Mrs. Mary Johnson and Mrs. Barbara Cornay typed the manuscript, Mr. P. Larimore and staff drafted the illustrations.

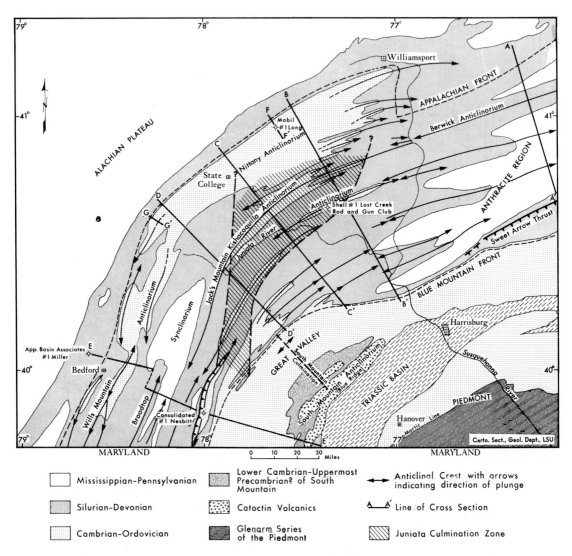

FIGURE 1. Index map of the central Appalachians in south-central Pennsylvania, locating critical wells, major anticlinal crests, directions of fold plunge, and lines of cross sections in following figures. McConnellsburg Cove Anticline and the Little Scrub Ridge Thrust on its northwestern flank are immediately east of the label, "Consolidated #1 Nesbitt," in the lower left-middle of the map. The Juniata Culmination is marked by the diagonally lined pattern.

Especially large debts of gratitude must be expressed to Samuel I. Root, David B. MacLachlan, John Clark, and W. R. Wagner (Penn. Geol. Survey), and Gordon H. Wood, Jr., and Randolph Bromery (U.S. Geol. Survey), Jim Griffiths, and John Rodgers for mutual considerations of Appalachian tectonic problems and access to then unpublished data. Helpful criticism of drafts of this paper were provided by K. N. Weaver and G. W. Fisher. I would not, however, wish them to be held accountable for mistakes of fact or speculative interpretation presented, for which only I am responsible.

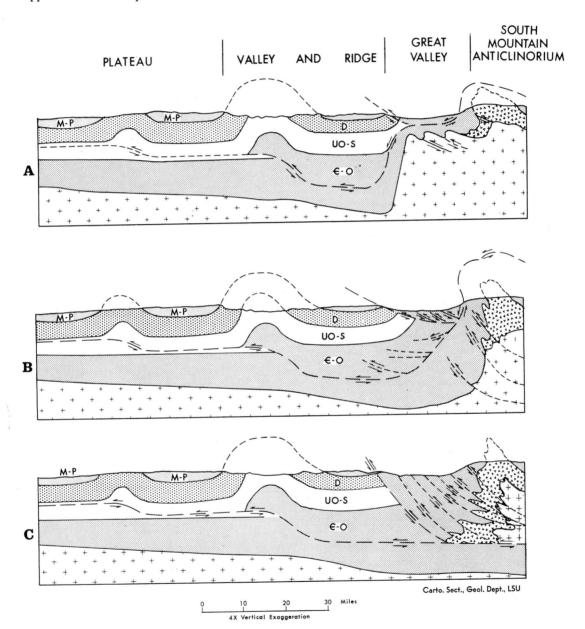

FIGURE 2. Hypothetical configuration of basement and master detachment (sole) thrusts in southern Pennsylvania as along section E-E' of Figure 1: (A) intimate basement involvement in Great Valley and South Mountain, minimum shortening; (B) basement involvement only in South Mountain, with the shortened Paleozoic sequence of the southeastern Great Valley transported into its present position along major detachment zone which now rises above South Mountain Anticlinorium; (C) no basement involvement. (See Fig. 3 for explanation of lithologic units.)

GEOLOGY

South Mountain and Great Valley

Structural style and deformation features change consistently along any radial traverse (westward, northwestward, or northward) across the Central Appalachians of Pennsylvania. In the South Mountain uplift on the southeastern side of the region, penetrative deformation accompanied by the development of uniformly southeastward dipping cleavage, smear limeations, and elongated oöids and clasts predominate. Both limbs of an anticlinorium are present to the southwest in Maryland, but Triassic normal faulting has apparently resulted in the depression of the southeastern flank in Pennsylvania; if once present, it is now covered by Triassic sedimentary rocks of the Gettysburg Basin. The South Mountain uplift rises more than two miles above the Great Valley Cambrian sequence at its local culmination in northwestern Adams County. From there, the anticlinorium plunges northeastward toward the Susquehanna River and southwestward toward the Maryland border. The uplift is oversteepened on the northwest and broken by longitudinal thrust faults and high-angle faults breaking the distended steep limb (Stose and Bascom, 1929; Cloos, 1947, 1951; Nickelsen, 1956; Fauth, in press).

Basement gneisses appear at the surface in the core of the Blue Ridge Anticlinorium southwestward in southernmost Maryland and the Virginias; basement is not exposed northeastward where stratified Precambrian and Cambrian rocks form the core of the uplift. In the core of the uplift in Maryland, mild metamorphism of greenstones and overlying terrigenous sedimentary rocks to the lower greenschist facies is reported (Cloos, 1951; Nickelsen, 1956).

To the northwest, in the structurally lower Great Valley terrain, deformation style changes with the appearance of flexural-slip parallel folds in Upper Cambrian and Ordovician rocks a few miles northwest of similar shear folds in the Middle Cambrian outcrop belt in the northwestern limb of the anticlinorium (Fig. 3, which

is section E-E of Fig. 1) (Root, in MacLachlan and Root, 1966). The transition in deformation style is recorded also in rapid northwestward reductions in the amount of elongation of oöids by penetrative shear and flow (Cloos, 1947). Root's work demonstrates that flexural-slip folding, with shortening accommodated primarily by interlayer slip, becomes the prevalent mode of deformation in the Great Valley far southeast of the Valley and Ridge.

Southeast-dipping high-angle thrust faults are revealed in all large-scale maps of the Great Valley. Root (in MacLachlan and Root, 1966) has illustrated that one of the thrusts turns eastward in southern Franklin County as a dextral tear fault which continues into the trans-South Mountain Carbaugh–Marsh Creek shear fault (Stose and Bascom, 1929). This fault system has apparently accommodated relative westward concomitant motion of the southern end of the Pennsylvania portion of South Mountain and the southeastern Great Valley.

Valley and Ridge—Plateau

The central Appalachian folded belt is displayed in its most classic form in south-central Pennsylvania. Gently curving, doubly plunging anticlines and synclines tens of miles long alternate across a 120–150 mi (96–121 km) wide terrain including the Valley and Ridge and Plateau Provinces. In an overall sense, intensity of deformation, magnitude of displacement on thrust faults, amount of appression of fold limbs on one another, and fold amplitude decreases northwestward, as noted by Rogers and Rogers (1843), Willis (1892), and others.

The writer has elsewhere described (Gwinn, 1964, 1967, 1968) surface and subsurface evidence indicating the presence of sole faults subparallel to bedding in tectonically weak stratigraphic units beneath the concentrically folded northwestern Valley and Ridge and the Plateau region. The sole zones of detachment or "ungluing" have accommodated relative centrifugal outward transport of the upper parts of the sedimentary sequence. During translation, the thrust

Explanation of abbreviations of lithologic units in cross sections of Figures 3–5: Miss. = Mississippian; Dmu = Middle and Upper Devonian; Dls (Fig. 3) or Dor-Scl (Figs. 4 and 5) = Lower Devonian and Post-Tuscarora–Upper Ordovician Juniata and Bald Eagle Fms.; Omb = Upper Ordovician Martinsburg Fm. (Fig. 3) and Or = Upper Ordovician Reedsville Fm. (Figs. 4 and 5); Oml = Middle and Lower Ordovician carbonates; C = Conococheague and Elbrook Fms. (Fig. 3) and Mines, Gatesburg, Warrior, and Pleasant Hill Fms. (Figs. 4 and 5) except in Fig. 5, in which Cm = separated Mines Formation; Cw = Waynesboro Fm.; C-PC? = basal Cambrian–Precambrian? Chilhowee Group and Tomstown Fm. (Fig. 3) and Chilhowee Group and Tomstown Fm. analogs (Figs. 4 and 5); PCcv = Precambrian Catoctin greenstones.

sheets buckled and shortened above the detachment thrusts by concentric folding. Concentric thrust faults (Price, 1965; Gwinn, 1964, 1967), rising out of bedding slip zones in the fold flanks, and imbricate thrusts, which branched from the sole-thrust systems at depth, aided in the shortening of the thrust mass(es). Deep wells in the region (Figs. 3–5) reveal marked downward reduction in and local disappearance of evidence of deformation below the sole faults. Anticlines terminate downward against the thrust zones in the Cambrian Waynesboro, Upper Ordovician Martinsburg–Reedsville shales, and the shales and evaporites of the Upper Silurian sequence.

New data from recent drilling at three critical sites in the Valley and Ridge and one site in the northwestern Great Valley are presented to further document the universality of tectonic thickening of the normal stratigraphic section by imbricate thrusts and concentric thrusts above the deep décollement zone.

SUBSURFACE GEOLOGIC DATA

Nittany Arch

The Birmingham Window on the crest of the Nittany Anticlinorium in Blair County, Pennsylvania, has been interpreted as a major overthrust sheet breached by erosion (Moebs and Hoy, 1959). In 1964, I speculated (Fig. 4), that the thrust was broadly folded, having been elevated from depth by folding of the hanging wall and its parent detachment thrust. Folding of the thrusts was thought to have been induced by reinitiation of detachment slip in the subsurface at points west of the original site of upward shearing of the first "stepped" imbricate thrust from the master sole thrust (Gwinn, 1964, p. 880–881).

The Mobil No. 1 Long well, drilled to a depth of 15,662 ft (4,774 m) on the crest of the anticlinorium about 40 mi (64 km) northeastward of Birmingham along strike in eastern Centre County, spudded in the sandy dolomites of the

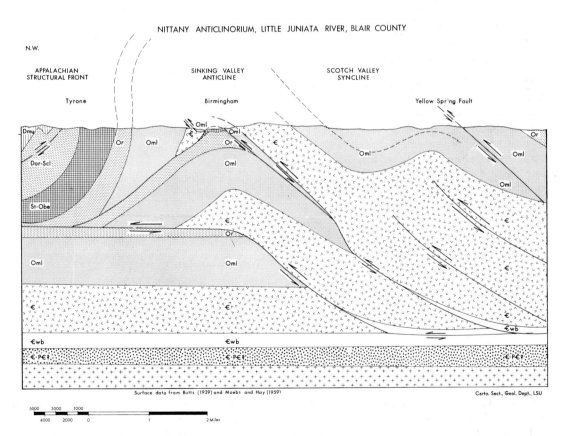

FIGURE 4. Section G-G' of Figure 1. Thin-skinned interpretation of southwestern Nittany Anticlinorium based on shallow drill holes in the vicinity of the Birmingham window (Moebs and Hoy, 1959), presence of low-angle thrusts in deep wells to the southwest and northeast along strike, and aeromagnetic, gravity, and seismic data. See p. 130 for explanation of abbreviations of lithologic units. (From Gwinn, in press.)

Upper Cambrian Gatesburg Formation. The bore penetrated a thrust fault at 1,100 ft (336 m) depth and entered tectonically thickened Lower Ordovician Beekmantown dolomites to a drilling depth of 8,650 ft (2,636 m) (Fig. 5). At that depth, the well bore crossed a major thrust, below which was penetrated an apparently normal, upright Upper Silurian to Middle Ordovician sequence. Dipmeter surveys revealed regional southeastward low dips and local intervals of low westward and northward inclined strata. Downward extrapolation to estimate depth to basement using an estimate of 8,200 ft (2,499 m) of older Ordovician and Cambrian strata (assumed also to be essentially undisturbed on the basis of dipmeter and seismic data) yields a geological estimate of basement at 23,800 ft (7,254 m), a figure in excellent accord with the seismic estimate of 25,000 ft ± 3,000 ft (7,620 m ± 914 m) to be discussed later. Only very small displacement thrusts have been detected at the surface. The major thrusts are assumed to continue westward beneath the Plateau as the sole thrust for the folding observed in that region (Gwinn, 1964, p. 883–884, Figs. 16, 19, and 20).

Shade Mountain Anticlinorium

Shell Oil Co.'s No. 1 Lost Creek Rod and Gun Club well on Shade Mountain Anticlinorium in easternmost Mifflin County (Fig. 1) encountered tectonic thickening by faulting of the Ordovician Martinsburg–Beekmantown interval. Structural relief below the total depth of the well is more than 3,000 ft (914 m) less than that observed at the surface because of the fault duplication of the section high in the well. Because none of these faults were detected in the thin Silurian formations on the flank of the folds by 1:24,000 scale mapping (Conlin and Haskins, 1962), the faults must flatten into bedding slip zones in the northwestern flank of the fold in the pattern apparently typical of the folded Central Appalachians.

Northwestern Great Valley Margin, Fulton County

The McConnellsburg Cove Anticline on the boundary between the Valley and Ridge and Great Valley provinces in Eastern Fulton County is broken on its western flank by the Little Scrub Ridge thrust fault. Separation on the Little Scrub Ridge fault reaches a maximum at the surface of at least 9,500 ft (2,895 m) immediately opposite the anticlinal culmination. The fault dies out in both directions along its strike in the steep western flank of the doubly plunging anticline.

Consolidated Gas Corporation drilled a deep exploratory well on the crest of McConnelsburg Cove Anticline (Figs. 1 and 3) late in 1965. Robert E. Bayles, Chief Geologist, communicated that "the well . . . commenced in . . . the Cambrian Conococheague. At a (sample) depth of 4,900 ft (1,493 m) a fault was encountered, the well bore passing from Middle Cambrian (Elbrook) dolomite into lower Middle Ordovician limestones. At . . . 5,460 ft (1,664 m) a second fault was encountered (where the well passed) into (Upper Ordovician Martinsburg) shale . . ., the well remaining in Martinsburg to its total depth of 8,648 ft (3,281 m)." (Words in parentheses are the writer's). An average dip of 40° is indicated on the fault zone from the surface east–southeastward to the well (Fig. 3).

This low-angle thrust fault with large displacement thus dips southeastward beneath the northwestern margin of the Great Valley. Duplication of lower Paleozoic strata by overthrusting accounts for at least 8,648 ft (3,281 m) of the local structural relief on the anticline.

Southern Pennsylvania Valley and Ridge

The western two-thirds of Figure 3 is a detailed section constructed west–northwestward across the Valley and Ridge in Southern Pennsylvania from McConnellsburg Cove to the Schellsburg Dome on the eastern side of the Appalachian Plateau. Two deep wells on the Schellsburg uplift again reveal the pattern of overthrust fault-induced tectonic thickening high in the sequence. Evidence that the thrusts encountered in the wells cropout is lacking, which suggests that these thrusts are folded and disappear into west-dipping beds parallel to slip zones west of the wells.

Juniata Culmination

All major anticlinal and anticlinorial trends of the Valley and Ridge in Pennsylvania reach their highest points in a zone of culmination trending obliquely across the strike of the fold belt (Fig. 1, diagonally lined zone). The culmination zone is narrow on the south near the Great Valley in east-central Franklin County and widens gradually near its northern end to its widest point in eastern Centre, Mifflin, and northern Juniata Counties. Nickelsen (1963) referred to the culmination as the Pennsylvania Culmination; Rodgers (1964b) independently named it the Susquehanna Culmination. The writer has suggested elsewhere (Gwinn, in Gwinn, Wise, and Wood, 1968) that it should be renamed again as the Juniata Culmination. Rejec-

NORTHWESTERN FLANK OF NITTANY ANTICLINORIUM

CENTRE COUNTY, PENNSYLVANIA

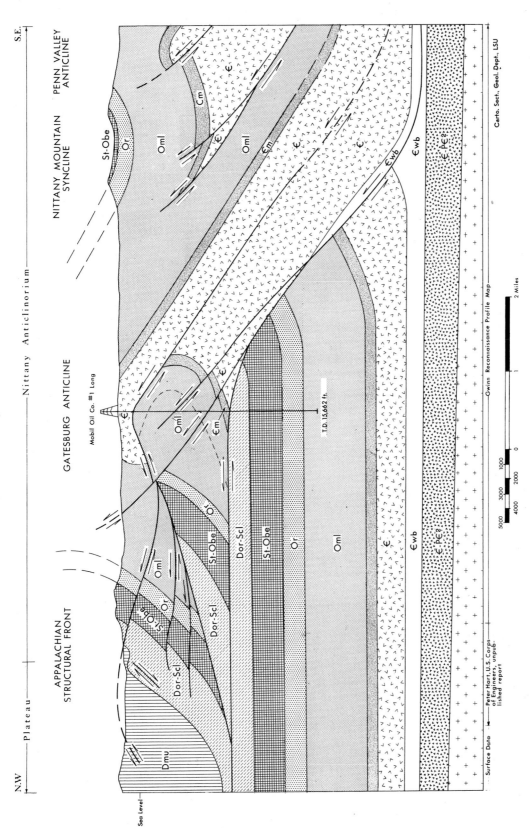

FIGURE 5. Section F-F' of Figure 1. Gatesburg Anticline on the northwestern flank of the Nittany Anticlinorium; data on northwestern flank contributed by Peter Hart, U.S. Corps of Engineers; Mobil No. 1 Long well calibrates the seismic profiles referred to in the text, providing a basis for the indicated relatively undisturbed pre-Upper Silurian sequence. See p. 130 for explanation of abbreviations of lithologic units. (From Gwinn, in press.)

tion of the first term, Pennsylvania Culmination, is suggested because of the presence nearby of two other culminations crossing the Valley and Ridge in Pennsylvania. Because one of these is directly adjacent to and parallel to the Susquehanna River, it is suggested that we utilize Rodgers' term, Susquehanna Culmination, for that feature and eliminate its application to the larger culmination to the west.

Culmination points on the anticlines are arranged in a left lateral en echelon pattern northward in the Juniata Culmination zone. The culmination points seldom occur opposite high or low points in adjacent synclines, but rather opposite plunging segments of the adjacent downwarps. The plunges of the synclines vary; some synclines plunge northeastward through the culmination zone and some plunge southwestward.

It is important to note that both Nickelsen and D. U. Wise (*in* Nickelsen, 1963) and Rodgers (1964b) have speculated that this culmination zone is a product of post-Appalachian, Triassic basement uplift transverse to trends of terminal Paleozoic folding.

Subsurface drilling data and geophysical information from the Juniata Culmination play a critical role in evaluating the degree of possible basement involvement in the vaguely defined zone.

GEOPHYSICAL DATA

Gravity, magnetic, and seismic data available in 1964 gave no indication that the basement of the outer (northwestern) portions of the Valley and Ridge and Plateau was involved in the deformation observed in the sedimentary cover (Gwinn, 1964, p. 887–889).

Since 1964, while compiling a 1:250,000 tectonic map of the Great Valley and Valley and Ridge for the Pennsylvania Geological Survey (1968), the writer has been allowed to view and analyze seismic profiles in four widely scattered, but strategically located regions of the Valley and Ridge and Great Valley in Pennsylvania.* Moreover, recently reported preliminary analyses of regional aeromagnetic and gravity investigations of a transcontinental crustal strip including the southern 50 mi (80 km) of Pennsylvania (Zietz and others, 1966) have a crucial bearing on the problem.

* At the request of the donors of the information, the areas of seismic investigation can be located in only a very general fashion.

Reflection Seismograph Studies

Seismic data from the four following localities, only generally located as explained in the footnote on this page were analyzed; the Nittany Anticlinorium, Centre–Blair Counties; Berwick Anticlinorium, less than 40 mi (64 km) east of the Susquehanna River; Shade Mountain Anticlinorium less than 40 mi (64 km) northeast of the Juniata River; and the Cambrian–Ordovician terrain on the west–northwestern margin of the Great Valley in Franklin County and Fulton County. Previously cited drilling data is available near all of the locales except the Berwick area.

Nittany Anticlinorium. Seismic records reveal a series of sharp, clear subhorizontal velocity interfaces at considerable depth in longitudinal and transverse profiles of the high central portion of the anticlinorium. The series of reflections can be correlated in number and sequence with similar velocity contrasts in the Cambrian system and at the top of the basement in deep wells drilled northwestward in the Plateau. The deepest interface, here judged to be the contact between basement and basal Cambrian sandstones, was estimated to be at a depth of 25,000 ft $\pm$ 3,000 ft (7,620 m $\pm$ 814 m) (error due to inadequate knowledge of velocities of wave transmission in the deep Cambrian and Ordovician stratigraphic units in the deeper part of the Appalachian Basin).

This inference on the depth to basement and the presence of undeformed deep sedimentary rocks was confirmed and calibrated by the Mobil Oil Co. No. 1 Long well in northeastern Centre County (Fig. 5, described in section on Geology).

Reflections in shallow portions of the profiles were observed to have steep northwestward dips. The deformation appeared to terminate abruptly downward, where the inclined reflections were succeeded by subhorizontal deep reflections.

Easily correlatable deep reflecting interfaces appeared to persist southeastward subhorizontally 3–6 mi (5–10 km) beneath the apex of the anticlinorium across a surface zone where surface structural relief exceeds 2 mi (3 km).

Berwick Anticlinorium. Nearly horizontal to slightly north-dipping reflections in the lower portion of profiles across the Berwick Anticlinorium; correlatable to the Nittany profiles, were estimated to be at depths of 17,000–25,000 ft $\pm$ 3,500 ft (5,181–7,620 m $\pm$ 1,067 m). At the surface, the Berwick uplift rises 9,500–15,000 ft) 2,895–4,575 m) above the Northcumberland Synclinorium on the south (Wood and Carter, *in* Gwinn, Wood, and Wise, 1968).

Shade Mountain Anticlinorium in the Juniata Culmination Zone. Subhorizontal deep seismic reflections were recognized at depths estimated at 18,000–25,000 ft ± 3,500 ft (5,486–7,620 m ± 1,067 m) in Shade Mountain Anticlinorium. Steep dips were observed in the upper 15,000 ft (4,573 m) of the profiles. Thrust-thickened sections beneath this anticline, penetrated in the aforementioned deep well, is another example of creation of anticlinal relief by imbricate thrust faulting of rocks at shallow depths in the cores of anticlines.

The 30 mi-long (48 km), almost nonplunging crestal segment of Shade Mountain Anticline falls within the northeastern portion of the Juniata Culmination. The deep subhorizontal reflecting interfaces here and in the above-mentioned Berwick area a short distance off the Culmination are at comparable seismic depths in the Valley and Ridge. The surface structural relief measured longitudinally across the zone at the surface ranges up to 15,000 ft (4,572 m). Basement involvement beneath the Juniata Culmination is not reflected in the scanty data.

Great Valley Margin, Fulton and Franklin Counties, Penna. As in the other locales, subhorizontal deep reflecting interfaces stand out clearly in longitudinal and transverse seismic profiles in the carbonate terrain on the northwestern margin of the Great Valley in Fulton and Franklin Counties. The Cambrian and Ordovician rocks of the Great Valley stand 1.0–2.8 mi (1.6–4.8 km) higher than Devonian and Mississippian strata immediately to the west in the south-southeastern margin of the Broadtop Basin. Depths to the subhorizontal reflectors in a transverse profile across the margin of the Great Valley are estimated at depths of 23,000–30,000 ft ± 4,000 ft (7,010–9,144 m ± 1,219 m). The previously cited deep well on McConnellsburg Cove Anticline encountered thrust faults causing tectonic duplication of more than 9,000 ft (2,743 m) of Cambrian–Ordovician stratigraphic section. Most of the fold relief can, thus, be attributed to thrust duplication in the upper part of the sedimentary column, rather than to alleged basement uplift, undetectable in the seismic profiles.

Additional Aeromagnetic and Gravity Data

Study of aeromagnetic and gravity patterns in a 200-mi wide (320 km) transcontinental strip crossing the Appalachian Mountains of southern Pennsylvania and states immediately southward by Zietz and others (1966) reveals continuously east–southeastwardly descending gradients across the entire Valley and Ridge and Great Valley Provinces (Fig. 6). Moving east–southeastward from the Pittsburgh–western Plateau area, zones of steeper gradient appear to widen and the gradients within the zones to decrease. This must be an indication that the depths to the more highly magnetized bodies, presumably within the basement and likely to have their tops at the surface of the basement (a profound unconformity), are increasing east–southeastwardly (Fig. 6). The strikes of the minor anomalies, both nosings and closures, are oblique to the strikes of the surface structural features. Also the distances between the gravity and magnetic anomalies are greater than the distances between the crests of the surface features, resulting in poor coincidence between the surface features and anomalies. All the aspects of poor correlation suggest a complete independence of the causes of the geophysical anomalies from those of the surface structures—a good hint of a décollement condition. Removal of the earth's magnetic field from the total intensity map reduces the magnetic gradient across the region, but serves also to make the residual magnetic pattern even more regular. Zietz and others (1966, p. 1433) attributed the smooth southwestward decrease in magnetization to a southeastward sloping, deeply downwarped basement surface.

Both the gravity and magnetic maps (Fig. 6) illustrate that narrower and more intense anomalies, possibly but not necessarily indicative of shallower basement, occur generally within 5–10 mi (8–16 km) of the Catoctin–Chilhowee contact on the northwestern flank of South Mountain Anticlinorium. The change is not coincident with the surface exposure of the highly magnetic Catoctin greenstones in the core of the anticlinorium, but instead is locally as much as 5 mi (8 km) northwest of it and 10 mi (16 km) southeast of it (Fig. 6).

Also note that no gravity or magnetic anomaly is associated with the northwestern margin of the Great Valley, where the surface rocks rise abruptly 1.0–2.8 mi (1.6–4.8 km) in structural elevation.

Analysis of magnetic data revealed the disappearance at depth in the upper crust of the reversal and high magnetic anomaly associated with the South Mountain area (Zietz and others, 1966, p. 1442). At the deeper levels, a highly magnetic area appears 40–80 mi (64–128 km) southwestward in the Piedmont northwest of Baltimore.

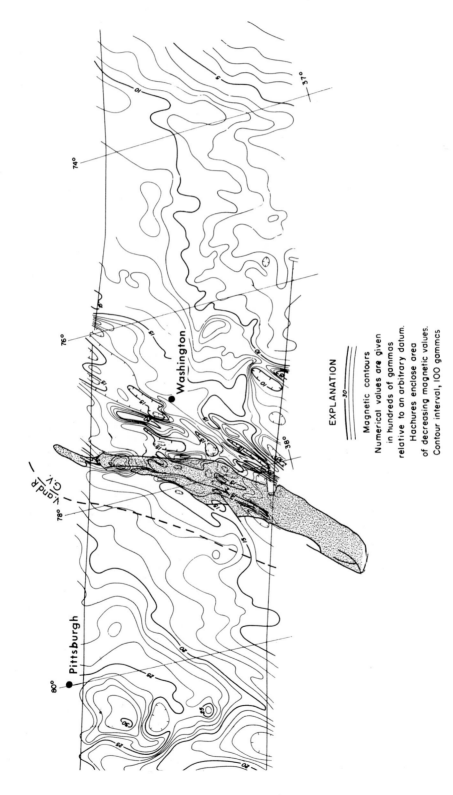

FIGURE 6. Geophysical data and basement map in a 200-mi wide strip crossing southern Pennsylvania (at top of figure) (modified slightly after Zietz and others (1966) with the permission of Ed., Bull. Geological Society of America). Surface distribution of Precambrian Catoctin Volcanics and other Precambrian units of the Blue Ridge–South Mountain mass indicated by light stipple.

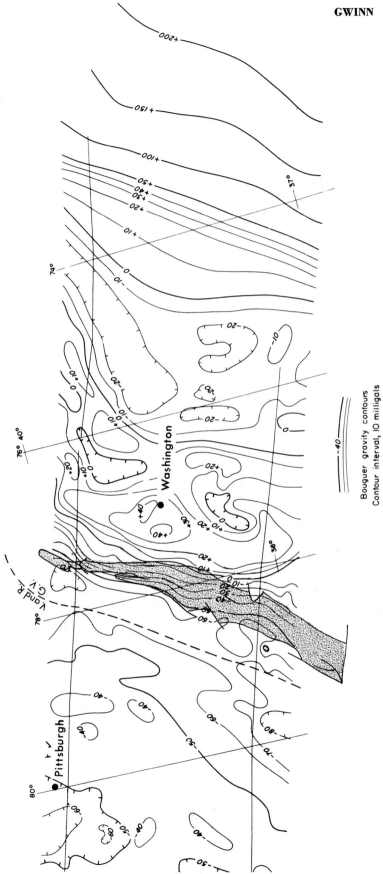

Bouguer gravity contours
Contour interval, 10 milligals

Figure 6. *Continued*

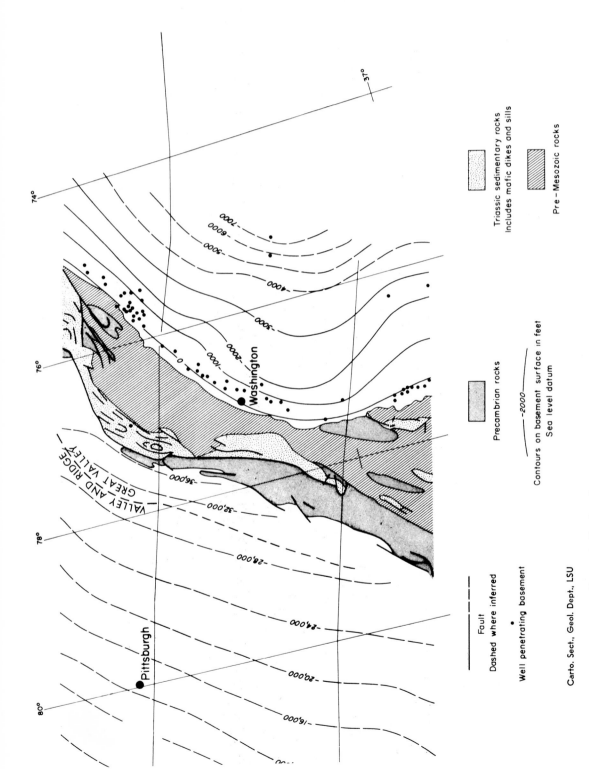

Carto. Sect., Geol. Dept., LSU

Figure 6. *Continued.*

INTERPRETATION

Folding and Faulting in the Valley and Ridge

Geological and geophysical data available to the writer indicate that deformation northwest of the Great Valley in the Valley and Ridge and Plateau Provinces is epidermal and do not indicate that the crystalline basement is appreciably, if at all, involved in the major lateral shortening observed in the anistropic sedimentary sequence at the surface. All major highly uplifted subprovinces of the Valley and Ridge yield surface, subsurface, and seismic data which indicate abrupt downward terminations in deformation and major reductions in and/or disappearance of structural relief on folds at depth.

The three most highly uplifted subprovinces in the Valley and Ridge are:
(1) The highly upwarped southeastern common boundary zone with the Great Valley;
(2) The transverse Juniata Culmination zone;
(3) The Nittany Anticlinorium.

If basement were deformed concordantly or semiconcordantly with surface folding in the Valley and Ridge area, one or all of these areas of maximum uplift might have been expected to yield at least some direct or indirect geologic or geophysical evidence of basement deformation. No such evidence is available.

On the Nittany Anticlinorium, addition of a normal pre-Upper Ordovician stratigraphic section beneath the total depth of the thrust thickened sequence in the Mobil No. 1 Long well yields a depth to basement of approximately 23,500 (7,162 m), nearly the same depth determined by seismic methods.

Evidence of basement involvement beneath the broad Juniata Culmination zone trending obliquely transverse to the Valley and Ridge is lacking (Fig. 1). Both the Shade Mountain well and the No. 1 Nesbitt well near McConnellsburg fall on the Juniata Culmination zone; the Nesbitt well is, moreover, in the highly uplifted boundary zone between the Great Valley and the Valley and Ridge. Seismically determined depths to basement on and off the culmination zone in the Valley and Ridge are approximately 25,000 ft (7,620 m). Concordant basement uplifts beneath the uplifted subareas would have been within 13,000 ft (3,962 m) or less of the surface. The seismic profiles, extending as much as 8 mi (12.9 km) along and across the strike, equaling the local distance from synclinal trough to synclinal trough at the surface, reveal essentially planar basement surfaces.

Aeromagnetic and gravity anomalies which could be correlated with the surface structures in a one-to-one fashion in amplitude, wavelength, or trend, as predicated in a hypothetical concordant thick-skinned deformation are not noted.

In conclusion, available data suggest that a nearly planar basement descends southeastward toward the Great Valley at a rate between 130–200 ft/mi (25–38 m/km) from a position about 23,500–25,000 ft (7,162 m) below the surface beneath the Nittany Arch on the northwest. An apparently thin-skinned pattern of dermal shortening above subhorizontal sole thrusts persists southeastward to the Great Valley (Fig. 7).

Deformation in the Great Valley

Available geologic data are inconclusive as to whether the deformation observed at the surface in the Great Valley persists downward into rooted concordant basement structural features. However, seismic, aeromagnetic, and gravity data singly and collectively fail to reveal anomalies which could be linked to major structural relief concordant with the deformed surface rocks. Instead, on balance, the available evidence—the absence of basement geophysical anomalies, a single deep well revealing lower structural relief at depth, and the apparent prevalence of a thin-skinned detachment thrusting pattern in the Valley and Ridge Province on the northwest—may indicate a nonconcordant, deeply buried, perhaps undeformed basement surface beneath perhaps all but the southeasternmost Great Valley.

Traditionally, it has been assumed that the highly folded and thrust-faulted Cambrian and Ordovician rocks of the Great Valley at the foot of the Central Appalachian Blue Ridge overlay an intensely deformed concordant basement linked to the rooted, deep-seated, autochthonous uplift and shortening of the South Mountain–Blue Ridge Anticlinorium (King, 1951, p. 129–130; Eardley, 1951, p. 105–106; Rodgers, 1964b, p. 79; Espenshade, Chapter 14 this volume, Fig. 1). Structural evidence (Bryant and Reed, Chapter 15 this volume) and geophysical data (King, 1964, p. 18–19) from the Blue Ridge Province of eastern Tennessee and western North Carolina strongly suggest, however, that the crystalline rocks and flanking Precambrian and early Cambrian strata on the southern Blue Ridge may indeed be allochthonous and overlie low-angle thrust faults serving as detachment zones in the Great Valley on the northwest.

The readiest support for a thick-skinned interpretation of the Great Valley deep subsurface structure is found in the appearance, proceeding

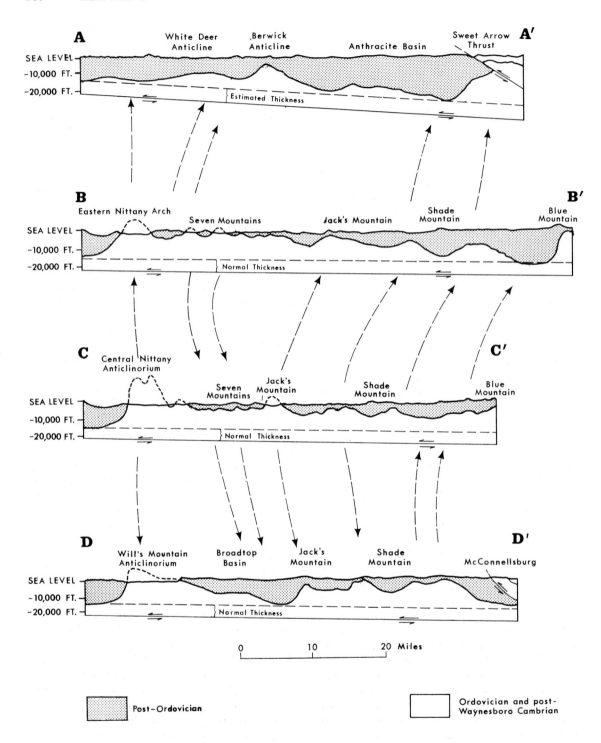

FIGURE 7. Schematic cross sections of the Valley and Ridge illustrating decreases of dermal shortening along strike ENE and SSW of the approximate axis of the Juniata Culmination, modified after the original sections by H. B. Rogers (in Willis, 1892, Pl. LVI). Lines of sections are indicated on Figure 1. Note plunge of anticlines correlated with curvature. (From Gwinn, in press.)

southeastward across the province, of similar folds, pervasive gently southeastward dipping flow cleavage, mild metamorphism in the core of South Mountain Anticlinorium, and finally, in the appearance of basement at the surface in the core of the uplift southwestward along strike in Maryland and the Virginias. Though suggestive of Paleozoic deformation of the core of the South Mountain Anticlinorium at formerly slightly higher temperatures and pressures, these features do not constitute direct, *ipso facto* proof of rooted, concordant basement deformation in the Great Valley or, for that matter, in South Mountain. The presence of basement crystalline rocks at the surface in the Blue Ridge of Maryland indicates involvement of basement but does not necessarily indicate that the basement is rooted and concordant with structure at deeper levels.

In support of the speculation entertained here, that the thin-skinned detachment thrusting of the Valley and Ridge persists southeastward, attention is focused on the subsurface data from the Nesbitt well near McConnellsburg, the seismic data from Franklin and Fulton counties, and the absence of aeromagnetic and gravity anomalies in the vicinities of anticlines and fault zones in the Great Valley. All of these examples indicate the absence of basement deformation.

Aeromagnetic and Gravity Data. The absence of either aeromagnetic or gravity anomalies along the 1.0–2.8 mi (1.6–4.8 km) structural rise from the Valley and Ridge southeastward into the Great Valley in no way suggests the involvement of basement in the surface folding and faulting (Fig. 6).

Utilizing the maximum stratigraphic thicknesses reported from the northwestern flank of South Mountain and the Great Valley in Franklin County (Stose, 1909; Stose and Bascom, 1929; Root, in MacLachlan and Root, 1966; Fauth, in press), the thickness of rocks between the Upper Ordovician Martinsburg Formation and the top of the highly magnetic Precambrian Catoctin greenstones is only about 10,000 ft (3,048 m). Assuming a 50 percent underestimation of true thickness, the revised thickness would be only 15,000 ft (4,572 m). If the basement were anywhere near concordant to the surface sequence, it thus would be within 10,000–15,000 ft (3,048–4,572 m) of the surface under the numerous anticlines exposing Cambrian rocks in the Great Valley. No aeromagnetic and gravity anomalies have been defined which would indicate concordant folding or locally correlatable faulting of the Catoctin and underlying basement.

Can the airborne magnetometer and the gravimeter resolve structure at these considerable depths or do their powers of resolution decrease rapidly downward, so as to leave unanswered the basic questions? W. J. Dempsey (personal correspondence, *in* Wood and Carter, *in* Gwinn, Wood, and Wise, 1968), addressing this question directly, contends that only below 25,000 ft (7,620 m) does the airborne magnetometer have limited ability to discern short wavelength anomalies such as would correspond to the Catoctin or basement uplifts beneath surface anticlines and thrusts. Zietz and others (1966) describe the ability of the magnetometer to discern longer wavelength anomalies at considerably greater depth.

Dempsey (see preceding paragraph) proposed three interpretations of the absence of the magnetic anomalies in the Valley and Ridge and the Great Valley:

(1) The basement and Catoctin greenstones are uplifted in the cores of anticlines and concordant to the highly deformed surface strata, but the crystalline rocks are virtually nonmagnetic in these provinces, while being locally highly magnetic in adjacent surface areas of South Mountain and the Piedmont and in subsurface localities beneath the Plateau Province;

(2) The basement contains highly magnetic rocks locally and is deformed, but is everywhere more than 25,000 ft (7,620 m) beneath the surface;

(3) A gently southeastward dipping nearly horizontal nonconcordant basement at greater than 25,000 ft depth (7,620 m) beneath the Great Valley and Valley and Ridge Provinces is locally highly magnetic along zones wider than and at high angles to the surface structures.

The first model seems untenable in light of the presence in nearby Appalachian provinces of basement rocks which are highly magnetic in local areas and everywhere more magnetic than the Paleozoic sedimentary rocks (Joesting and others, 1949; Zietz and others, 1966).

Hypothesis 2 cannot be disproved. However, if a deformed basement were at depths greater than 25,000 ft (7,620 m) over the entire area and the post-Catoctin sequence is only 10,000–15,000 ft (3,048–4,572 m) thick, lateral shortening in excess of 100 percent and thin-skinned dermal deformations are indicated. As only 15 mi (24.0 km) of shortening can be observed at the surface (Fig. 3), extensive unobserved subsurface shortening is implied. With basement at a depth two or more times the thickness of the normal stratigraphic sequence in the area, in which concentric folds prevail, dermal shortening of surface rocks above a detachment thrust zone strongly discordant to basement is assured.

The third model, a centripetally inward (southeastward)-dipping basement surface sloping gently toward the foot of South Mountain, is quite consistent with the seismic and aeromagnetic data. This interpretation of the Central Appalachians was presented on the basement map of the United States (Amer. Assoc. Petroleum Geol. and U. S. Geol. Survey, 1967).

The writer thus suggests further analysis of hypotheses 2 and 3 which embody a deep deformed discordant or deep undeformed nonconcordant basement underlying a highly shortened surface sequence and detachment thrust faults in the Great Valley. The conclusion rests on the geophysical and geological data cited and three principal assumptions derived from the data:

(1) that the basement could hardly be nonmagnetic over the entire Great Valley–Valley and Ridge region;

(2) that the airborne magnetometer would be able to discern shallow concordant or semiconcordant uplifts of basement rocks if they were present; and

(3) that the normal, tectonically undisturbed Great Valley stratigraphic sequence is less than 15,000 ft (4,572 m) thick, probably closer to 10,000 ft (3,048 m) in aggregate.

The top panel of Figure 3 presents data available from drilling, surface mapping, and seismic exploration of the Valley and Ridge, Great Valley, and Blue Ridge across southernmost Pennsylvania. The basic assumptions defended above have been utilized in the construction of the figure.

In the absence of geophysical evidence of basement involvement, one must utilize a thin-skinned tectonic model and tectonic processes to maintain a deep nearly undeformed basement. Figure 3 reveals no magic solution to the manner in which the 10,000 ft (3,048 m) stratigraphic column fills the much thicker zone between the deep basement and the surface. Inasmuch as a considerable number of hypothetical sections like Figure 3 could be drawn across the region illustrating different thickening mechanisms, with no deep subsurface data to justify one over another, it was decided to illustrate only very generally the surface folds and thrusts assumed to overlie a deep hypothetical detachment zone. The great shortening postulated here was probably accomplished by repeated low-angle thrust detachment and superposition of section with extensive folding of thrusts. The presence of major nappe overfolds in the Great Valley, such as those described by Drake (Chapter 19 this volume) in the eastern Pennsylvania Great Valley, is not apparent in the surface geology.

Estimates of Lateral Shortening

The possibility of a deep nonconcordant Great Valley basement has great implications in determining the magnitude of lateral shortening of the lower Paleozoic rocks and their original sites of deposition.

If one assumes thick-skinned deformation in the Great Valley and thin-skinned deformation in the Valley and Ridge (Fig. 2A), lateral shortening along a radial traverse from the Blue Ridge to the Plateau is minimal (19.3 mi; 31 km) calculated by the sinuous bed method (Fig. 3, lower panel). The existing Valley and Ridge sole-thrust system would have to either crop out presently somewhere in the Great Valley, to have climbed up section southeastward in the northwestern Great Valley into now-eroded younger rocks, or presently crop out on the flank of South Mountain.

Considerably more shortening is documented by drilling in the Valley and Ridge–McConnellsburg Cove segment of the cross section alone than the 19.3 mi indicated by sinuous bed reconstruction. The proven presence of noncropping thrust faults in the Central Appalachians, which introduce shortening not readily apparent at the surface, makes it imperative to utilize the equal area method of calculation of shortening (Bucher, 1955, p. 357).

The lower panel and the underlying inset tables of Figure 3 illustrate the assumed rock dispositions, give the data on normal stratigraphic intervals and present a series of calculations of shortening by the equal area method. Minimum estimates of shortening are calculated from the lower panel of Figure 3 assuming that South Mountain was allochthonous and overlay a deep sole thrust, as in Figure 2C. Approximately equal shortening would be present if the sole thrust either has been folded in the subsequent rise of a rooted South Mountain Anticlinorium or has been employed in thrusting of the Paleozoic sediments up and over an already extant South Mountain Anticlinorium. Shortening would be greater in any of these situations if South Mountain has an overturned northwestern limb at depth. The figure of 50 mi (80 km) shortening assumes an upright limb at depth.

The outermost portion of the Plateau is thought to have slipped and shortened 1–2 mi (1.6–3.2

km); strata in the southeasternmost Valley and Ridge appear to have slipped northwestward another 25.3 mi (40.7 km); additional shortening and translation postulated in the Great Valley gives an estimated net northwestward motion of more than 50 mi (80 km) for the Cambrian rocks at the foot of South Mountain.

Alternate lower estimates of shortening would result for some segments of Figure 3 if the actual normal stratigraphic columns are thicker than the "best estimate" thickness used in the initial calculations. The absence of indications of major local thickness variations induced on fold limbs by flow and shear and the cross-check provided by multiple sources of stratigraphic data make it seem improbable that the estimated Great Valley stratigraphic thicknesses would be greatly in error.

South Mountain: Rooted or Unrooted?

If the estimations of shortening presented are valid, a detachment thrust of the first magnitude must pass either over or under the South Mountain Anticlinorium (Fig. 2B-C). As Rodgers (1964a) has argued, a given amount of shortening in the sedimentary cover, in this instance 50 mi (80 km), demands equal shortening in all rock units through which the thrust passes. On the one hand, if deformation involves the basement, net relative translation of an equal magnitude must affect the crystalline rocks somewhere in the system. On the other hand, if the detachment–sole fault system of the outer folded Appalachians remains entirely within the layered anisotropic sequence, as in a detached thrust plate, basement need not be involved at all, and the shortening observed in the stratified sequence need not be present in the crystallines.

No evidence for shortening of basement equivalent in magnitude to shortening in the layered stratigraphic sequence can be found either below the Great Valley or in the surface of South Mountain Anticlinorium. Chilhowee strata in the anticlinorium appear to have been shortened only 10–15 mi. Thus rooted shortening of basement in South Mountain in excess of 10–15 mi is precluded on quantitative considerations. An additional 35–40 mi of shortening and/or translation observed in the Paleozoics to the west of South Mountain must have occurred below (or above the present erosion surface). This is a minimum figure for the unobserved shortening in the vicinity of South Mountain, because the shortening has been determined utilizing a conservative in-place rooted sub-Chilhowee South Mountain with no overturning to the west at depth. Westward overturning and an

allochthonous South Mountain (as in Fig. 2C) would increase the estimate of shortening greatly. In my desire to demonstrate the minimum, however, it seems advisable to portray South Mountain conservatively.

Thus, the writer concludes that basement shortening kinematically linked to that possibly occurring beneath South Mountain has not occurred beneath the southeastern Great Valley. This conclusion demands that detachment faults with very large displacement (ca. 40 mi + (64 km+)) must pass over or under the South Mountain Anticlinorium.

On the one hand, no data are available to adequately evaluate the possibility that South Mountain overlies the very deep regional sole thrust system. Drilling programs to probe for over-ridden sedimentary rocks beneath the possibly allochthonous crystalline rocks are faced with the high probability that such a thrust would lie at depths so great that only a deep, presently unjustifiable petroleum wildcat well could present any reasonable test. The presence of gneissic crystalline rocks beneath the Catoctin at modest depths would in no way preclude the crystalline rocks being allochthonous, in the fashion of the crystalline core of the Blue Ridge farther southwest in the southern Appalachians (see Bryant and Reed's (Chapter 15) and Espenshade's (Chapter 14) accounts in this volume).

On the other hand, it seems probable that shortening of the cover on a detachment thrust now passing over South Mountain would have had to have occurred prior to the major uplift of a rooted South Mountain Anticlinorium. Either early or late intraformational detachment thrusts could now be highly folded, lying subparallel to bedding in tectonically weak zones on the northwestern flank and plunging noses of the South Mountain folds. However, late shortening and thrust transport of an allochthonous plate up and over a *preexisting* uplift must contend with objections to the immense energy required for such movement and the inadequacy of rock materials to transmit the immense stresses and unbalanced forces needed to initiate such motion.

Whereas the Precambrian crystallines of the Reading Prong in easternmost Pennsylvania lie over the regional gravity minimum extending along strike in the Appalachians, as do the gneisses of the Blue Ridge southwestward in the Carolinas and east Tennessee (King, 1964), South Mountain and the northern tip of the Blue Ridge lie to the southeast of the longitudinal gravity minimum in the intervening area. Avery Drake (personal communication, 1967) and the writer (Gwinn, 1968) have speculated that the gravity data may be used to infer a rooted South Mountain uplift. This would also demand that the inferred sole pass over South Mountain as a folded zone of intraformational detachment, with detachment presumably occurring prior to uplift of South Mountain, as reasoned above. Cloos and Hietanen (1941) and Wise (this volume) present data from the Martic zone immediately east of the Susquehanna supporting a similar sequence of early low-angle thrust detachment preceding later shear folding and metamorphism. It will be recalled that Cloos (1951) and Nicklesen (1956) have reported a temporal sequence of flexural slip folding followed by shear folding and metamorphism in South Mountain.

In further support of detachment prior to South Mountain uplift, a major zone of thrust discordance along a contact between the late overriding allochthon and an already highly deformed footwall is not recognized.

Root Zones of the Allochthon. If the Great Valley basement is uninvolved in the Appalachian deformations, as concluded herein, the sites of derivation of the highly shortened Paleozoic rocks of the southeastern Great Valley are at least 40–50 mi (64–80 km) southeastward and east–southeastward. The probable root zone is in the Piedmont southeast of the Martic Line (see Wise, Chapter 22 this volume) in southern York County, Penna., and adjacent Frederick County, Maryland. This Piedmont terrain comprises rocks of the Glenarm Series, a thick sequence dominated by pelitic metamorphic rocks, but including minor metamorphosed carbonates and volcanics of Catoctin aspect.

Hopson (1964) proposed that the Glenarm sequence is Precambrian in age on the base of radiogenic age dates from the base of the sequence on the crest of the Baltimore–Washington Anticlinorium. He correlated the volcanics of the upper Glenarm in the Piedmont with the Catoctin greenstone sequence of South Mountain. Alternately, speculation that the upper Glenarm is a metamorphosed southeastern terrigenous facies of the Cambrian–Ordovician carbonates to the northwest appeals strongly to many workers (Wise, Chapter 22 this volume).

If Hopson is correct, the Piedmont core of the Appalachians in Maryland and Pennsylvania comprises the oldest rocks in the orogen and the area of maximum structural uplift, rather than a depressed geosynclinal axis containing early Paleozoic rocks. In the former, the grossly homoclinal, northwestward younging of rocks from the Piedmont into the outer Appalachians would re-

flect maximal uplift along the Baltimore–Washington Anticlinorium accompanied by a northwestward shift of the now-shortened stratified cover by tectonic and/or gravity thrusting processes. In a general sense, the outward translation of the Paleozoic cover toward the folded Appalachians would simply be a secondary manifestation of primary crustal shortening and vertical uplift in the Piedmont core of the Appalachian orogen.

Proof that the Glenarm sequence is Lower Paleozoic in age, on the other hand, would not invalidate the conclusions based on shortening considerations that the zone of derivation of the Great Valley Paleozoic sequence lay in the northwestern part of the Piedmont. In this event, the Glenarm Series rocks would make up a far-traveled allochthon, thrust northwestward over the zone previously vacated by the southeastern Great Valley carbonate rocks.

Determination of whether the crystalline rocks in the core of South Mountain Anticlinorium came into their present site atop the regional sole-fault system (subsequently shear folded and mildly metamorphosed) or whether they comprise a postdetachment autochthonous massif in the Alpine sense must await further investigation. Detailed reflection seismology supplemented by additional aeromagnetic and gravity surveys of the South Mountain southeastern Great Valley region constitute the brightest hope for a resolution of the problem of the deep structural configuration along the northwestern flank of the Blue Ridge in southern Pennsylvania.

CONCLUSIONS

(1) Seismic and magnetic data in the Valley and Ridge indicate a gently southeast-dipping essentially undeformed basement. Data do not support the contention that the obliquely transverse Juniata plunge culmination crossing the central and southeastern Valley and Ridge is a basement arch of Triassic age.

(2) Geological data from recent deep wells on the Nittany and Shade Mountain Anticlinoria and from the borehole on the McConnellsburg Cove Anticline at the Great Valley's outer edge in Franklin County support the geophysical conclusion of "no-basement" tectonics.

(3) Magnetic and gravity data, the deep McConnellsburg well, and seismic profiles on the northwest margin combine to imply that the Great Valley basement may not be involved in the extensive shortening observed at the surface of the Great Valley.

(4) Calculations of shortening assuming a deep, subplanar basement across the Valley and

Ridge and Great Valley to the foot of South Mountain, yield an estimate of 50 mi (80 km) of shortening in the Paleozoic sequence. Even greater shortening has occurred along northwest–southeast profiles along the axis of the Pennsylvania Salient from central South Mountain to the Nittany Anticlinorium.

(5) The weight of current geological opinion holds that South Mountain is a rooted uplift in south-central Pennsylvania, but, in the absence of proof to this effect, the speculation that it overlies a Valley and Ridge–Great Valley detachment zone seems worthy of further testing by integrated seismic, magnetic, and gravity investigation. Considerations of the shortening, however, demand that the major sole fault system either pass beneath South Mountain or rise over it as an intraformational detachment thrust zone.

(6) The probable sites of deposition of the Paleozoic carbonates of the southeastern Great Valley are southeast of the Martic Line in the northwestern Piedmont in zones occupied by rocks of the upper part of the Glenarm Series. If the Glenarm Series is Precambrian, the Appalachian Piedmont is highly uplifted and an ideal terrain from which the now-shortened and detached Paleozoic cover could have slipped northwestward. If the Glenarm is Paleozoic, it has been thrust northwestward into positions formerly occupied by Paleozoic carbonate rocks.

REFERENCES

Amer. Assoc. Petroleum Geologists and U. S. Geol. Survey, 1967, Basement map of North America: A.A.P.G., Tulsa, Okla.

Bucher, W., 1955, Deformation in orogenic belts: Geol. Soc. America Spec. Paper 62, p. 343–368.

Cloos, E., 1947, Oolite deformation in the South Mountain fold, Maryland: Geol. Soc. America Bull., v. 58, p. 843–918.

————, 1951, Washington Co., State of Maryland: Dept. of Geology, Mines, and Water Resources (report), p. 1–333.

————, and Hietanen, Anna, 1941, Geology of the "Martic overthrust" and the Glenarm Series in Pennsylvania and Maryland: Geol. Soc. America Special Paper 35.

Conlin, R. R., and Hoskins, D. M., 1962, Geology and mineral resources of the Mifflintown quadrangle: Penna. Geol. Survey, 4th Series, Atlas 126, 46 p.

Eardley, A. J., 1951, Structural Geology of North America: Harpers, New York, 624 p.

Fauth, J. L., in press, Geology of the Caledonia Park Quadrangle area, South Mountain, Pennsylvania: Penna. Geol. Survey 4th Series, Atlas 129a.

Gwinn, V. E., 1964, Thin-skinned tectonics in the Plateau and northwestern Valley and Ridge Provinces of the Central Appalachians: Geol. Soc. America Bull., v. 75, p. 863–900.

————, 1967, Lateral shortening of layered rock sequences in the foothills regions of major mountain systems: Mineral Industries, v. 36, no. 5, p. 1–7.

————, in press. Tectonics of the Plateau, Valley and Ridge, Great Valley, South Mountain, and the Reading Prong: in Gwinn V. E., Wood, G. H., Jr., and Wise, D. U., Tectonics of Pennsylvania: Penna. Geol. Survey, 4th Series.

Hopson, C. A., 1964, The Crystalline rocks of Howard and Montgomery Counties, in The Geology of Howard and Montgomery Counties, p. 27–215, Maryland Geological Survey.

Joesting, H. R., Keller, R., and King, E., 1949, Geologic implications of aeromagnetic survey of Clearfield–Phillipsburg area, Pennsylvania: Amer. Assoc. Petroleum Geologists Bull., v. 33, p. 1747–1766.

King, P. B., 1951, Tectonics of Middle North America: Princeton Univ. Press, Princeton, N.J. 203 p.

————, 1964, Further thoughts on tectonic framework of the southeastern United States, in Lowry, W. D., Ed., Tectonics of the Southern Appalachians: Va. Polytech. Inst., Department of Geol. Sci., Mem. 1, p. 5–31.

Knowles, R. R., 1966, Geology of a portion of the Everett 15-minute quadrangle, Bedford County, Pennsylvania: Penna. Geol. Survey, 4th Series, Prog. Report, Pr 170, 90 p.

MacLachlan, D. B., and Root, S. I., 1966, Comparative tectonics and stratigraphy of the Cumberland and Lebanon Valleys (Penna.): Guidebook for the 31st Annual Field Conference of Pennsylvania Geologists, Penna. Geol. Survey, 90 p.

Moebs, N. N., and Hoy, R. B., 1959, Thrust faulting in Sinking Valley, Blair and Huntingdon counties, Pennsylvania: Geol. Soc. America Bull., v. 70, p. 1079–1088.

Nickelsen, R. P., 1956, Geology of the Blue Ridge near Harpers Ferry, W. Va.: Geol. Soc. America Bull., v. 67, p. 239–270.

————, 1963, Fold patterns and continuous deformation mechanisms of the Central Pennsylvania Appalachians: Guidebook to Field Conference, Tectonics and Cambrian–Ordovician Stratigraphy in the Central Appalachians of Pennsylvania: Pittsburgh Geological Society and Appalachian Geological Society, p. 13–29.

Price, R. A., 1965, Flathead map area British Columbia and Alberta: Geol. Survey of Canada, Mem. 336, 221 p.

Rodgers, J. 1964a, Mechanics of Appalachian foreland folding in Pennsylvania and West Virginia: Am. Assoc. Petroleum Geologists Bull., v. 47, p. 1527–1536.

————, 1964b, Basement and no-basement hypotheses in the Jura and the Appalachian Valley and Ridge, in Lowry, W. D., ed., Tectonics of the Southern Appalachians: Va. Polytech. Inst., Dept. of Geol. Sci., Mem. 1, p. 71–80.

Rogers, W. B., and Rogers, H. D., 1843, On the physical structure of the Appalachian Chain, as exemplifying the laws which regulated the elevation of great mountain chains generally: Assoc. American Geologists, Rept. (Trans.), p. 474–531; reprinted in The Geology of the Virginias, by W. B. Rogers, New York, 1884.

Stose, A. J., and Stose, G. W., 1946, Geology of Carroll and Frederick Counties, in The Physical Features of Carroll County and Frederick County: Maryland Dept. of Geology, Mines, and Water Resources, p. 11–131.

Stose, G. W., 1909, Description of Mercersburg–Chambersburg district, Penna.: U. S. Geol. Survey Atlas of the U. S., Folio 170, 19 p.

————, and Bascom, F., 1929, Description of the Fairfield and Gettysburg (Adams, Franklin and Cumberland Counties) quadrangles: U. S. Geol. Survey, Geol. Atlas of U. S. Folio 225, 22 p.

Willis, B., 1893, Mechanics of Appalachian structure: U. S. Geol. Survey Ann. Rept. 13, Part 2, 221–281.

Wood, G. H., Jr., and Carter, M. D., in press, Tectonics of the Anthracite Region: *in* Gwinn, V. E., Wood, G. H., Jr., and Wise, D. U., Tectonics of Pennsylvania: Penna. Geol. Survey, 4th Series.

Woodward, H. P., 1959, Structural interpretations of the Burning Springs Anticline, *in* Woodward, H. P., Editor, A Symposium on the Sandhill deep well, Wood County, West Virginia: Geol. Survey, Rept., Inv. 18, p. 159–168.

————, 1964, Central Appalachian tectonics and the deep basin: Amer. Assoc. Petroleum Geologists, v. 48, p. 338–356.

Zietz, I., King, E. R., Geddes, W., and Lidiak, E. G., 1966, Crustal study of a continental strip from the Atlantic Ocean to the Rocky Mountains: Geol. Soc. America Bull., v. 77, p. 1427–1448.

Chapter 9

Structural Controls of the Anthracite Region, Pennsylvania*

GORDON H. WOOD, JR. AND M. J. BERGIN

INTRODUCTION

FOR MORE THAN a century geologists have recognized a northwestward decrease in the intensity of folding in the Pennsylvania Anthracite region. About twenty years ago a similar decrease in faulting was recognized and was believed to be largely confined to the coal fields (Arndt and Wood, 1960). Recent geologic mapping has shown that not only does the complexity of faulting decrease northwestward, but also that the complexity of deformation varies stratigraphically, some rock sequences being more deformed or less deformed than adjacent sequences.

Detailed geologic mapping and mining data clearly show that the local structural complexity and geometry were governed by the relative competence of both individual beds and thin to thick sequences of beds. These data also indicate that the regional complexity and geometry were governed by the relative competence of five region-wide rock sequences or lithotectonic units whose average thicknesses range from 2,000 to 9,000 ft (610–2,740 m) (Table 1). Each lithotectonic unit contains characteristic assemblages of structural features that maintain their individuality throughout the region. The complexity of each assemblage gradually decreases northwestward in conformance with the regional northwestward geographic decrease.

The purpose of this paper is to identify and discuss these lithotectonic units and to demonstrate that development of the tectonic framework

* Publication authorized by the Director of the United States Geological Survey.

was governed by differences in their abilities (relative competence) to transmit stress.

LOCATION

The Anthracite region lies mostly in the Valley and Ridge province and partly in the Allegheny and Pocono Plateaus (Fig. 1). It is the principal area containing coal of anthracite and semi-anthracite rank in the Central Appalachian Mountains. The coal-bearing areas are the Northern, Eastern Middle, Western Middle, and Southern anthracite fields (Fig. 2).

STRUCTURAL SETTING

The Anthracite region is in a structural depression between highly deformed rocks on the Pennsylvania culmination to the west and gently deformed rocks of the Pocono Plateau to the east. Its southern boundary is sharply defined by the Blue Mountain structural front (Rodgers, 1953); the northern border is poorly defined, generally paralleling the Appalachian structural front which separates the Valley and Ridge and Appalachian Plateaus provinces (Fig. 1).

The trough of the structural depression trends sinuously N. 20° E. transverse to the east–northeast structural grain of the Valley and Ridge province. A series of narrow, linear to arcuate, complexly deformed anticlinoria and synclinoria plunge northeastward off the Pennsylvania culmination into the depression (Fig. 2). Several of these irregularly reverse plunge at the trough, and rise toward the Pocono Plateau where they die out. The others maintain their northeastward plunge to their terminations on the plateau. Most

TABLE 1. Lithotectonic Units of the Anthracite Region

Lithotectonic unit	Rocks included (in ascending order)	Lithology	Thickness: maximum, minimum, average	Competence
Unit 5	Upper member Mauch Chunk Formation, Pottsville and Llewellyn Formations of Late Mississippian to Late Pennsylvanian age.	Sandstone and conglomerate with lesser amounts of shale, siltstone, and coal.	$5,500+$ ft or $1,680+$ m in south; $2,400+$ ft or $730+$ m in north; $4,000\pm$ ft or $1,220\pm$ m average.	Rocks of upper member and Pottsville are competent, greatly faulted, and moderately folded. Rocks of Llewellyn range from competent to incompetent, are greatly faulted and folded.
Unit 4	Middle member of Mauch Chunk Formation of Late Mississippian age.	Shale, siltstone, and fine-grained sandstone.	$6,500\pm$ ft or $1,980\pm$ m in south; absent in northeast; $2,000\pm$ ft or $610\pm$ m average.	Incompetent, greatly deformed; bounded at many places by décollements; functioned as tectonic buffer separating lithotectonic units 3 and 5.
Unit 3	Trimmers Rock Sandstone, Catskill and Pocono Formations, and lower member of Mauch Chunk Formation of Late Devonian to Late Mississippian age.	Sandstone and siltstone with lesser amounts of shale and conglomerate.	$13,200\pm$ ft or $4,020\pm$ m in south; $8,000\pm$ ft or $2,440\pm$ m in north; $9,000\pm$ ft or $2,740\pm$ m average.	Competent, areas underlain by unit are characterized by large open folds and relatively few faults.
Unit 2	Marcellus Shale, Mahantango Formation, and where present, Tully Limestone and Harrell and Brallier Shales of Middle and early Late Devonian age.	Shale with siltstone and sandstone interbeds. Locally contains thick medial sandstone.	$3,000\pm$ ft or $915\pm$ m central and east; $1,100\pm$ ft or $335\pm$ m southwest and west; $2,000\pm$ ft or $610\pm$ m average.	Incompetent; basal unit, the Marcellus Shale, commonly is greatly deformed and is site of décollements.
Unit 1	Tuscarora Sandstone, Clinton Formation, Bloomsburg Red Beds, Wills Creek Shale, Tonoloway and Keyser Limestones, Helderberg Formation, Oriskany Group, Needmore Shale, Selinsgrove Limestone, and lateral equivalents of Early Silurian to early Middle Devonian age.	Sandstone and conglomerate with lesser amounts of siltstone and shale. Upper part above Bloomsburg Red Beds consists of limestone, dolomite, and shale with lesser amounts of sandstone, siltstone, chert, and evaporites. Upper part absent in southwest part of region.	$5,200\pm$ ft or $1,580\pm$ m in north; $2,800\pm$ ft or $850\pm$ m in southwest; $4,400\pm$ ft or $1,340\pm$ m average.	Moderately competent; areas underlain by unit are characterized by long, open folds. Upper part of unit is less competent and could be considered a sublithotectonic unit locally.
Total thickness			$31,600\pm$ ft or $9,630\pm$ m maximum, in south;* $17,600\pm$ ft or $5,360\pm$ m minimum, in northeast;* $21,400\pm$ ft or $6,520\pm$ m average.	

* Because of the nondefinitive geographic location of maximum and minimum thickness of Units 1 and 2, the average thickness of these units was used in computing total maximum and minimum thickness.

of these fold systems strike N. 60° to 85° E., but in the northeastern part of the depression the strike of the two northernmost systems swings to N. 10° E. (Fig. 2).

From southeast to northwest the subsidiary folds of the anticlinoria and synclinoria grade from high-amplitude, concentric to disharmonic, asymmetric to overturned, tight anticlines and synclines to low-amplitude, concentric, symmetric, open anticlines and synclines. Concomitant with this northwestward decrease in fold complexity is a comparable decrease in fault complexity. In the southeastern part of the region hundreds of thrust, reverse, tear, and bedding faults and several décollements are present whereas in the northern part faults are sparse, generally reverse, and of small displacement.

VERTICAL DIVERSITY OF STRUCTURAL FEATURES

In addition to the northwestward decrease in structural complexity there is a vertical diversity of structural features controlled by the differential abilities (relative competence) of rocks to transmit stress. For example, where thin relatively incompetent rocks are interbedded with thicker relatively more competent rocks, the former commonly are folded disharmonically and the latter concentrically. Wherever this thickness relationship is reversed, folds in both the competent and incompetent rocks commonly are harmonic and the bedding is parallel to subparallel.

The amplitude, wavelength, internal harmony, and symmetry of many folds change abruptly at sharp contrasts in rock competence, and in some

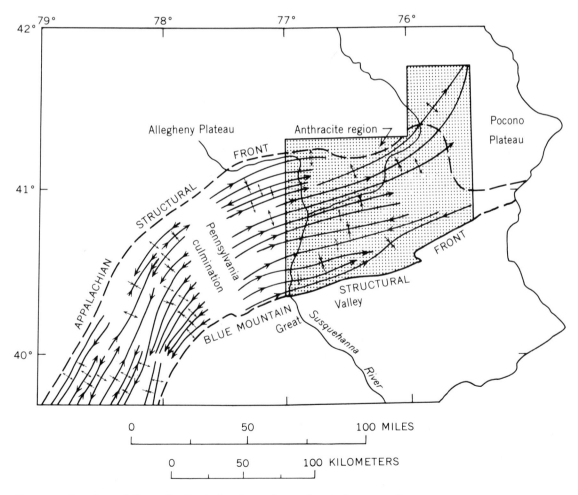

FIGURE 1. Location of Pennsylvania Anthracite region and principal tectonic elements in central and eastern Pennsylvania.

places individual folds or suites of folds disappear.

Most high-angle reverse faults and higher angle thrust faults are in relatively competent rocks whereas décollements and low-angle thrust and bedding faults are generally in incompetent rocks. Although low-angle faults occur throughout the region, they are most abundant in the southern part.

Many thrust faults shear upward through competent rocks and utilize incompetent rocks as glide zones. Small thrust faults that step stratigraphically from shale to shale by shearing across intervening sandstone or conglomerate are exposed in numerous road cuts and strip pits. Anticlines subsidiary to the major fold systems commonly have formed in upper plates where thrusts cease gliding and shear upward through com-

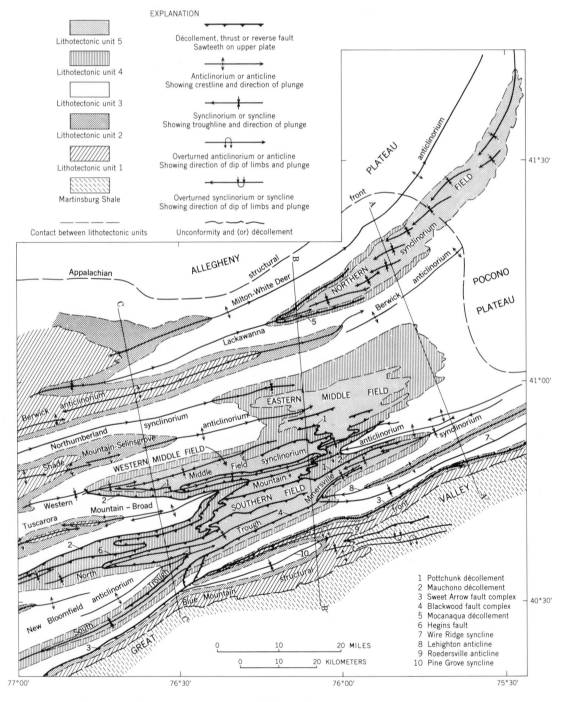

FIGURE 2. Tectonic map of the Anthracite region.

petent rocks (Gwinn, 1964, p. 883, Fig. 14). Similarly, subsidiary synclines commonly have developed in upper plates where thrusts cease shearing and commence gliding within incompetent rocks.

Evidence of displacement on many large thrust and reverse faults commonly disappears near the contacts between thick sandstone or conglomerate sequences and underlying thin to thick shale or siltstone sequences. Detailed geologic mapping and mining data in some localities prove that these faults are imbricate fractures rising from larger displacement low-angle thrust faults or décollements which developed near contacts between these rock types. At other localities low-angle thrusts and décollements have been interpreted where evidence of displacement on higher angle thrust and reverse faults, presumably imbricates, disappears near the contact of competent and incompetent rocks. The known low-angle thrusts and décollements have been difficult to identify and map because of approximate parallelism to bedding and extensive surficial cover. Undoubtedly, other similar faults, as yet unidentified, are present in the region.

PRINCIPAL TECTONIC ELEMENTS OF THE ANTHRACITE REGION

The principal tectonic elements of the Anthracite region are located on Figure 2 and interpreted in cross section on Figure 3. Their characteristics are discussed briefly here, listed in Table 2, and amplified in the section on lithotectonic units.

There are nine anticlinoria and synclinoria in the Anthracite region. Several deep wells between the crest of the Pennsylvania culmination and the trough of the Anthracite region depression indicate that these fold systems affect many thousands of feet of Paleozoic rocks in the subsurface which do not crop out (Fig. 3).

The principal faults are low-angle glide thrusts and décollements whose upper plates moved northwestward. Some broke during the early stages of the Appalachian orogeny and continued developing until so closely folded that further northwestward movement was impossible (Table 2). A few broke during folding and were intensely warped. Those that formed late in the orogeny generally were not warped. The low-angle thrust faults and décollements that developed early in the orogeny are as complexly folded as the enclosing rocks and commonly crop out across several fold systems; whereas, those that formed at any later time generally are confined to a single fold system or cut across several fold systems without regard to the fold pattern.

A major décollement or a series of décollements apparently exists within the Marcellus Shale of lithotectonic unit 2 (Table 1) throughout all but the southwest and northern parts of the region. This décollement has not been traced at the surface because of cover and the scarcity of Marcellus outcrops, but its presence is shown by disrupted outcrops, by the differing styles, amplitudes, and wavelengths of folds above and below the Marcellus, and by wells that penetrated either duplicated sections or greatly thickened sections. It has been identified at the surface to the east of the region by J. B. Epstein (oral communication, 1966), in wells in the subsurface of the region (Wood and Carter, in press), and from differing fold styles to the west of the region by V. E. Gwinn (in press).

Another décollement may be responsible in part for the discordant relations between the Tuscarora Sandstone at the base of lithotectonic unit 1 (Table 1) and the Martinsburg Shale in the Blue Mountain structural front. The contact between these formations has been long considered to be either the Taconic unconformity (Fig. 2) or a fault (Fig. 3). Wood and Carter (in press) and Gwinn (in press) believe that fracturing occurred at or near the contact and that the Taconic unconformity has been amplified by a décollement.

The décollements within the Marcellus and at the Tuscarora–Martinsburg contact are not shown on Figure 2 because they have not been proven by detailed geologic mapping. The authors, however, believe that both décollements exist as shown on Figure 3 and that they developed where the greatest contrasts in relative competence existed between thick sequences of competent and incompetent rocks.

LITHOTECTONIC UNITS OF THE ANTHRACITE REGION

Wood and Carter (in press), in describing the lithologic control of structural features in the Anthracite region, subdivided the Silurian through Pennsylvanian stratigraphic succession into five rock sequences or lithotectonic units, each of which contains a characteristic suite of structural features. They postulated that these five lithologic units governed the development of the tectonic framework of the region because of inherent differences in competence. This thesis is amplified by description in Table 1 and discussion in subsequent paragraphs.

Structural Features of Lithotectonic Unit 1

Rocks of unit 1 (Table 1) crop out on the Pennsylvania culmination. Within the region

TABLE 2. Principal Tectonic Elements of the Anthracite Region

A. Fold Elements

Name	Approximate length		Average wavelength	Amplitude: Maximum, average, minimum	Plunge	Subordinate folds (principal characteristics in italics)	Faults
	In region	Total					
Milton–White Deer anticlinorium (Wood and Carter, in press)	80 mi (130 km)	80 mi (130 km)	8 mi (13 km)	8,300± ft (2,530 ± m), 5,000–5,500 ft (1,520–1,680 m), 0 at northeast termination.	<3° NE.	*Low amplitude, short, concentric*, en echelon, *symmetric* to asymmetric, locally disharmonic, *open*.	Few small reverse faults.
Lackawanna synclinorium (Wood and Carter, in press)	120 mi (190 km)	120 mi (190 km)	8–9 mi (13–14 km)	9,000± ft (2,740 ± m), 4,000–5,000 ft (1,220–1,520 m), 0 at northeast termination.	Doubly plunging: <3° NE to west of lat. 41°12′N.; long. 75°57.5′ W.; <6° SE to east of lat. 41°27.5′ N.; long. 75°37.5′ W.	Short, left lateral, *en echelon, symmetric* to asymmetric, concentric, locally disharmonic, *low* to moderate amplitude, open.	Small to moderate reverse and thrust faults; probable décollements in lithotectonic units 2 and 4 (See Table 1).
Berwick anticlinorium (Wood and Carter, in press)	70 mi (110 km)	170 mi (270 km)	10 mi (16 km)	15,000± ft (4,570± m), 8,500–9,000 ft (2,590–2,740 m), 0 at northeast termination.	Variable west of lat. 41°59′ N.; long. 76°32.5′ W.; 1°–2° NE to east.	*Symmetric* to asymmetric, *concentric* to disharmonic, open, *low* to moderate amplitude.	Small to moderate reverse and thrust faults; probable décollement in unit 2.
Northumberland synclinorium (Wood and Carter, in press)	32+ mi (50+ km)	130 mi (210 km)	9 mi (14 km)	12,500± ft (3,800± m), 6,000± ft (1,830± m), 0 at northeast termination.	<3° NE.	Short to long, *symmetric* to asymmetric, left lateral en echelon, open, *concentric* to disharmonic, *low* to moderate amplitude.	Small to moderate reverse and thrust faults; probable décollement in unit 2.
Shade Mountain–Selinsgrove anticlinorium (Wood and Carter, in press)	35+ mi (55+ km)	130 mi (210 km)	10 mi (16 km)	13,000± ft (3,960± m), 6,000± ft (1,830± m), 0 at northeast termination.	4° E.	*Short, en echelon, symmetric* to asymmetric, open, *concentric* to disharmonic, low to moderate amplitude.	Small to moderate reverse and thrust faults; probable décollements in unit 2.
Western Middle field synclinorium (Wood and Carter, in press)	65+ mi (105+ km)	75 mi (120 km)	8 mi (13 km)	16,000± ft (4,880± m), 6,500± ft (1,980± m), 0 at east termination.	Doubly plunging.	Short to *long, symmetric* to asymmetric, as much as 4,000-ft (1,220-m) amplitude, en ech-	Numerous small to large thrust (as much as 2,000-ft [610-m] displacement) and reverse

	Length		Width	Structural relief	Plunge	Folds	Faults
						elon, concentric to disharmonic, open to closed.	faults (as much as 1,000-ft [305-m] displacement); two décollements in unit 4.
Tuscarora Mountain– Broad Mountain anticlinorium (Wood and others, 1958)	70 mi (110 km)	150 mi (240 km)	8 mi (13 km)	18,000± ft (5,490± m), 6,000± ft (1,830± m), 0 at east termination.	Doubly plunging; 5° NE to west of lat. 40°47′ N.; long. 76°14′ W.; 2° SE to east of lat. 40°49′ N.; long. 76°09′ W.	As much as 55 mi (88 km) long, amplitudes as much as 10,000 ft (3,050 m), en echelon, *symmetric* to asymmetric, *concentric* to disharmonic, open, anticlines generally broad, synclines generally V-shaped in midparts, U-shaped near terminations.	Numerous small to large thrust and reverse faults. Décollements in units 2 and 4. Many of thrust and reverse faults are imbricates from décollements.
Minersville synclinorium (consists of two troughs to west) (Wood and others, 1958)	90 mi (145 km)	160 mi (255 km)	9 mi (14 km) Main Trough, 5 mi (8 km) South Trough, 10 mi (16 km) North Trough.	25,000+ ft (7,620+ m) South and Main Trough, 18,000± ft (5,490± m) North Trough, 15,000 ft (4,570 m) all, 0 at east termination.	Doubly plunging: <2° NE to west of lat. 40°41.5′ N.; long. 76°15.5′ W.; <3° SW to east.	As much as 64 mi (100 km) long, amplitudes as much as 20,000 ft (6,100 m), en echelon, symmetric, to *asymmetric*, concentric to *disharmonic*, many overturned, locally isoclinal.	Numerous small to large thrust and reverse faults. Décollements in units 2 and 4, possibly in unit 1 to east; many thrust faults are imbricates from décollements; high-angle reverse faults commonly cut out axial areas of isoclinal folds.
New Bloomfield anticlinorium (Miller, 1961)	40 mi (65 km)	85 mi (135 km)	11 mi (18 km)	12,000 ft (3,660 m), 8,000 ft (2,440 m), 0 near Minersville.	6° NE.	As much as 26 mi (42 km) long, amplitudes to 12,000 ft (3,660 m), symmetric to *asymmetric*, *concentric* to disharmonic, open, synclines U-shaped at termination, V-shaped in midparts.	Few thrust and reverse faults. Décollements in unit 4.

(continued)

TABLE 2.—Continued

B. Fault Elements

Name	Type	Relation to beds	Maximum displacement	Stratigraphic displacement	Time of development	Imbricate faults	Folded or nonfolded
Pottchunk (Wood and Carter, in press)	Décollement.	Generally parallel to overlying beds, truncates underlying beds.	Unknown, probably in miles (km).	5,000± ft (1,520± m)	Early in Appalachian orogeny.	Numerous imbricate thrusts in Southern and Western Middle anthracite fields.	Folded.
Mauchono (Wood and Carter, in press)	Décollement.	Parallels underlying beds, truncates overlying beds.	Unknown, probably in thousands of feet (meters) to one or more miles (km).	5,000± ft (1,520± m)	During folding of Appalachian orogeny.	One known south of Broad Mountain anticlinorium.	Folded.
Sweet Arrow complex (Wood and Kehn, 1961)	Low-angle thrusts spooning to north.	Truncates overlying and underlying beds.	3.5–4 mi (5.5–6.5 km)	2,000± ft (610± m)	After folding of Appalachian orogeny was nearly over.	Locally consists of basal fault and several imbricates.	Nonfolded.
Blackwood complex (Wood and Carter, in press)	Low-angle thrusts spooning to north.	Truncates overlying and underlying beds.	5,000± ft (1,520± m)	2,000–3,000 ft (610–910 m)	After folding of Appalachian orogeny was nearly over.	Locally consists of basal fault and several imbricate thrusts, Sweet Arrow complex.	Nonfolded.
Mocanaqua (Named herein)	Décollement.	Parallels overlying and underlying beds.	Unknown.	Unknown.	Early in Appalachian orogeny.	Numerous imbricate thrusts in west part of Northern anthracite field.	Folded.
Hegins (Wood and Carter, in press)	Low-angle thrust.	Truncates overlying and underlying beds.	1 mi (1.6 km)	Several thousand feet (meters).	Early in Appalachian orogeny.	None known.	Folded.

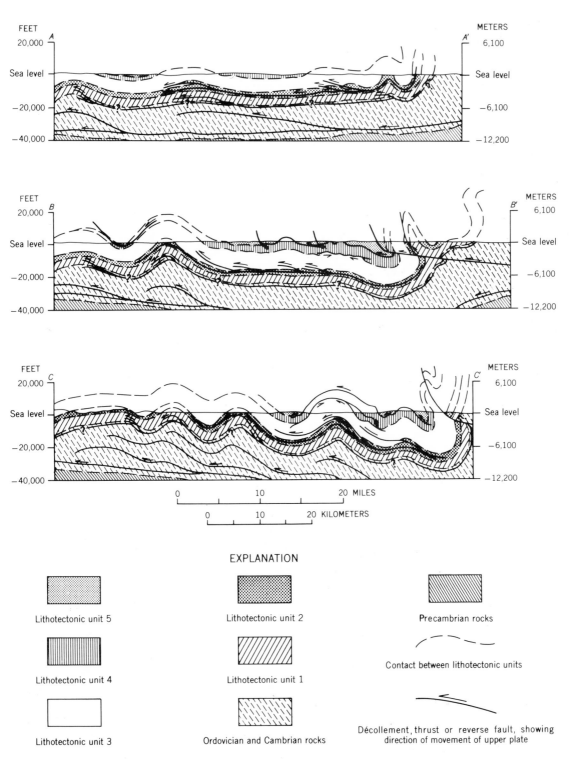

FIGURE 3. Interpretive cross sections of the Anthracite region.

EXPLANATION

Lithotectonic unit 5

Lithotectonic unit 4

Lithotectonic unit 3

Lithotectonic unit 2

Lithotectonic unit 1

Ordovician and Cambrian rocks

Precambrian rocks

Contact between lithotectonic units

Décollement, thrust or reverse fault, showing
direction of movement of upper plate

they are exposed at four localities: the combined Berwick, Lackawanna, and Milton–White Deer fold systems; the Shade Mountain–Selinsgrove anticlinorium; the Tuscarora Mountain–Broad Mountain anticlinorium; and the Blue Mountain structural front (Fig. 2).

East of the culmination and at the first three localities listed, the unit is deformed into long, narrow, symmetric to asymmetric, open, concen-

tric to locally disharmonic, en echelon folds (Fig. 4). Regionally, these folds plunge northeastward so that younger rocks crop out successively in that direction (Gray and others, 1960). The fold style in unit 1 west of the region contrasts markedly with the generally broader, shorter, less numerous, more open and concentric folds in unit 3 within the region. The change in fold style takes place in the Marcellus Shale (unit 2)

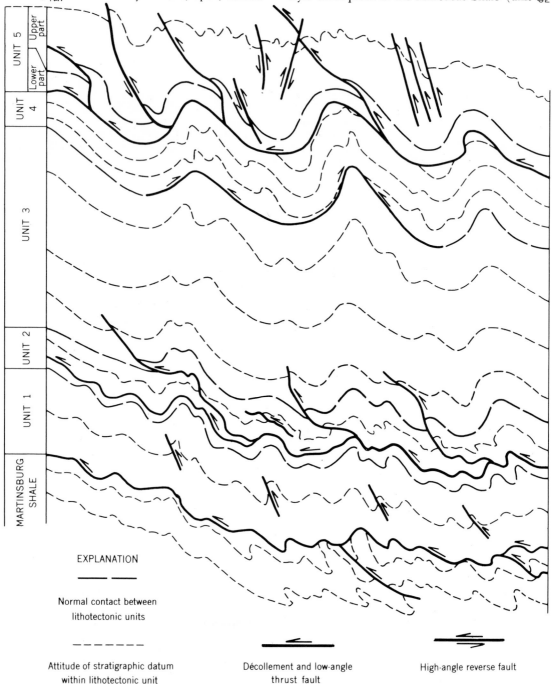

FIGURE 4. Diagrammatic section showing structural features characteristic of lithotectonic units in the Anthracite region.

which appears to have functioned during deformation as a ductile sedimentary blanket (Fig. 4). The stress that formed many large anticlines and synclines in units 1 and 3 appears to have been absorbed within the ductile rocks of unit 2. Wood and Carter (in press) and Gwinn (in press) attributed this absorption to flexual slip, disharmonic folding, and glide thrusting or décollement faulting.

The Wills Creek Shale in the upper part of unit 1 appears to have been the site of a décollement west of the region (Gwinn, in press). The existence of this décollement is supported by: (1) well data and (2) the fact that folds in rocks younger than the Wills Creek are more closely spaced and have smaller amplitudes and wavelengths than folds in rocks older than the Wills Creek (Gray and others, 1960, see outcrop patterns of Helderberg Formation and Oriskany Group in Perry, Juniata, Mifflin, Snyder, and Union Counties). The authors believe that this décollement dies out before entering the Anthracite region.

Because of décollements, flexural slip, and disharmonic folding in the Wills Creek Shale and the Marcellus Shale, only the largest folds present in unit 1 west of the region extend northeastward through unit 2 into unit 3 within the region.

In the western part of the Blue Mountain structural front the rocks of unit 1 generally are overturned northwestward and contain few distinguishable folds (Fig. 2). In contrast, in the central part of the front they are deformed into open, symmetric, concentric, en echelon folds fractured by northwest-dipping reverse faults. These folds generally resemble those in unit 1 west of the region.

The rocks of unit 1 have been intricately deformed in the eastern part of the structural front into several series of short, southwest-plunging, locally overturned to recumbent, low amplitude, en echelon anticlines and synclines (J. B. Epstein, oral communications, 1966 and 1967). These folds are discordant with those in unit 3 and in the Martinsburg Shale. Epstein (oral communication, 1966) believes that the discordance with the Martinsburg is due to a décollement superimposed upon the Taconic unconformity and that the discordance with unit 3 is due to a décollement in unit 2.

The characteristics of folds in unit 1 in the subsurface of the region are unknown, but are probably similar to those of folds exposed to the west. That is: long, narrow, concentric, open, en echelon folds broken by reverse, thrust, and bedding faults. Also, unit 1 is probably separated

from overlying and underlying units at many places by décollements (Fig. 4).

Structural Features of Lithotectonic Unit 2

Rocks in unit 2 (Table 1) crop out in six belts in the Anthracite region. Four are in the west part on the combined Berwick, Lackawanna, and Milton–White Deer fold systems, and on the Shade Mountain–Selinsgrove, Tuscarora Mountain–Broad Mountain, and New Bloomfield anticlinoria. The fifth belt lies south of the Sweet Arrow fault complex and the sixth is on the Lehighton anticline (Fig. 2).

Little is known about unit 2 on the Berwick, Lackawanna, and Milton–White Deer fold systems because of soil and forest cover and a lack of modern geologic mapping. A section of rocks of unit 2, believed to be much thickened by faulting and disharmonic folding, was penetrated by the P. Good No. 1 well (lat. 41°06.5′ N.; long. 75°58′ W.) on the crest of Berwick anticlinorium.

Only the upper part of unit 2 crops out on the New Bloomfield anticlinorium. West of the region on this anticlinorium, unit 2 is highly faulted and appears to separate closely spaced lower amplitude folds in unit 1 from widely spaced higher amplitude folds in unit 3.

The upper shale member of the Mahantango Formation and the Marcellus Shale, which comprise the bulk of the upper and lower parts of unit 2, underlie valleys on the Shade Mountain–Selinsgrove and Tuscarora Mountain–Broad Mountain anticlinoria. The medial part, the Montebello Sandstone Member of the Mahantango, underlies an intervening ridge. The Marcellus generally is more deformed than the overlying rocks of unit 2.

The most severely deformed exposed rocks of unit 2 are at the crest of the Shade Mountain–Selinsgrove anticlinorium on the east bank of the Susquehanna River (lat. 40°48.5′ N.; long. 76°50.5′ W.). There, shale and siltstone of the Marcellus and Mahantango are greatly thickened by flowage, fracture cleavage, low-angle thrusts, bedding faults, and disharmonic folds. Evidence of bedding has been nearly destroyed by fracture cleavage and flowage. The differing fold styles of units 1 and 3 indicate that a décollement is in unit 2 and may be the underlying cause of the severe deformation.

The R. Fox No. 1 (lat. 40°52.5′ N.; long. 76°39.5′ W.), G. Krick No. 1 (lat. 40°52′ N.; long. 76°35′ W.), and P. Knorr No. 1 (lat. 40°53′ N.; long. 76°24′ W.) wells penetrated greatly thickened sections of the Marcellus on the Shade Mountain–Selinsgrove anticlinorium. These sections and the distinctly different fold styles of rocks in units 1 and 3 indicate a décollement in the Marcellus.

Rocks of the Marcellus also are deformed intensively on the crest of the Tuscarora Mountain–Broad Mountain anticlinorium (lat. 40°38′ N. to 40°40′ N.; long. 75°54′ W. to 76°56′ W.). Two large en echelon anticlines and an intervening syncline in unit 1 (Gray and others, 1960) disappear at or near the contact with unit 2 and are overlain eastward by a single anticline in the medial and upper parts of unit 2. This abrupt change in fold style also indicates a décollement in the Marcellus.

Rocks of unit 2 south of the Sweet Arrow fault complex and west of long. 76°30′ W. are largely overturned to the northwest and deformed similarly to adjacent rocks in units 1 and 3. Between long. 76°30′ W. and long. 75°45′ W. and north of the Roedersville anticline, faults of the complex have fractured and attenuated the rocks of unit 2. South of the anticline the unit is deformed comparably with units 1 and 3.

The upper part of the Mahantango Formation crops out on the crest of Lehighton anticline. There, the H. Smith No. 1 and Grover No. 1 wells penetrated greater thicknesses of the Mahantango and the Marcellus Shale than expected from nearby outcrops. This suggests that the décollement identified in unit 2 by J. B. Epstein (oral communication, 1966) to the south of the Wire Ridge syncline extends northward in the subsurface to the Lehighton anticline. The authors believe that this décollement may be the same as that in the Marcellus on the Shade Mountain–Selinsgrove and Tuscarora Mountain–Broad Mountain anticlinoria.

Structural Features of Lithotectonic Unit 3

Rocks of unit 3 (Table 1) crop out over more of the Anthracite region than those of any other unit (Fig. 2). Although unit 3 thins northward, it is the most competent lithotectonic unit as shown by its relatively simple folds and faults.

Generally, anticlines and synclines in unit 3 are long, broad, open, concentric, and symmetric; they have wavelengths measurable in miles (kilometers) and amplitudes measurable in thousands of feet (meters). Locally, however, rocks of unit 3 are folded into tight asymmetric anticlines and synclines on the limbs of the Tuscarora Mountain–Broad Mountain and the New Bloomfield anticlinoria.

Compared to folds in other units, those in unit 3 are larger, simpler, less acute and disharmonic, more symmetric and concentric, less faulted, and less commonly en echelon (Fig. 4). The contrast between the long, broad, symmetric, and concentric folds of unit 3 in the region and the more closely spaced, less symmetric, and concentric folds in unit 1 west of the region is impressively displayed on the Geologic Map of Pennsylvania (Gray and others, 1960). The compressive forces that caused such differing fold styles must have been divided into two stratigraphically controlled stress fields separated by the incompetent rocks of unit 2. During deformation unit 3 apparently functioned as an extremely competent strut, unit 1 as a somewhat less competent strut, and unit 2 as a ductile or buffer zone. Early in deformation units 1 and 3 were capable of transmitting stress for considerable distances and moved differentially on décollements which formed in unit 2. Later, when folding became more intensive, units 1 and 3 flexed into discordant fold styles and unit 2 functioned as an intervening adjustment zone. During intensive folding the décollements in unit 2 were locally rejuvenated and developed overlying imbricate faults and rocks adjacent to the décollements adjusted by flexural slip, fracture cleaving, and disharmonic folding.

Structural Features of Lithotectonic Unit 4

Lithotectonic unit 4 (Table 1) underlies the central part of the Anthracite region (Fig. 2). It thins northward and is composed largely of incompetent ductile rocks. Folds in the unit are diverse, being short to long, open to tight, symmetric to asymmetric, concentric to disharmonic, and en echelon.

Many anticlines and synclines in unit 4 are discordant with folds in units 3 and 5 (Fig. 4). Discordance is most pronounced on the margins of the Southern and Western Middle anthracite fields where only the largest amplitude folds are present in all three units. Many lesser folds, too small to show on Figure 2, are confined to outcrops of one or two units.

The Mauchono and Pottchunk décollements underlie and overlie unit 4 on the margins of much of the Southern, Western Middle, and Eastern Middle anthracite fields. Displacements on these décollements are unknown, but probably are measurable in miles (kilometers). Mocanaqua décollement disrupts unit 4 at the southwest end of the Northern anthracite field. It has not yet been mapped in detail and may underlie much of the west part of the field.

Unit 4 is thickened by low-angle thrust faulting and rock flowage in the troughs of the Lackawanna, Minersville, and Western Middle field synclinoria and on the crest of the Tuscarora Mountain–Broad Mountain anticlinorium. It is also thickened on the limb between the Minersville synclinorium and the Tuscarora Mountain–Broad Mountain anticlinorium. Elsewhere on the limbs of the other fold systems, unit 4 generally has been thinned by flowage toward crests and troughs.

Unit 4 served as a buffer between the more competent rocks of units 3 and 5 (Fig. 4). Prior to intensive folding units 3 and 5 tore loose from unit 4 in the southern part of the region and moved differentially northwestward on décollements. When intensive folding commenced, the more competent rocks moved differentially above and below unit 4 by flexural slippage toward anticlinorial crests and synclinorial troughs. At the same time, unit 4 flowed differentially towards the fold axes, was broken by low-angle thrusts, and developed variable fold geometries.

Structural Features of Lithotectonic Unit 5

Unit 5 (Table 1), which thins northward, underlies the coal fields of the Anthracite region (Fig. 2). Rocks in the lower part are relatively competent and commonly are deformed into long, concentric, symmetric to slightly asymmetric, open, en echelon folds broken by low-angle thrust and bedding faults and fewer reverse faults (Fig. 4). In contrast, rocks in the upper part range from incompetent to competent and are folded more tightly into numerous shorter, narrower, lower amplitude, commonly disharmonic, en echelon anticlines and synclines broken by reverse faults and fewer low-angle thrust and bedding faults.

The intensity of deformation in unit 5 decreases northward from tight, high-amplitude, disharmonic, overturned, highly faulted folds in the south part of the Southern anthracite field to gentle, low-amplitude, concentric, unfaulted folds in the north part of the Northern anthracite field. The transition from much-faulted folds to unfaulted folds is sharp and occurs in the Northern anthracite field in a narrow zone paralleling the stratigraphic wedgeout of unit 4 shown on Figure 2. Similarly, the transition from generally disharmonic, overturned folds to generally concentric folds is sharp and takes place in the trough of Minersville synclinorium.

Folds in unit 5 commonly decrease in amplitude stratigraphically downward. Some fade out within the lower part of the unit; others disappear abruptly as they intersect the Pottchunk décollement and are discordant with folds in unit 4.

Many low-angle thrusts break through competent rocks in the lower part of unit 5 and are gradually absorbed by folding in the less competent rocks of the upper part (Fig. 4). In the Southern, Western Middle, and Eastern Middle anthracite fields most low-angle thrusts are imbricates from the Pottchunk décollement, and in the west half of the Northern anthracite field many may be imbricates from Mocanaqua décollement.

The most intensively deformed outcropping rocks in the region are in unit 5 in the Southern anthracite field. There, the Pottsville and Llewellyn Formations have been folded into a multitude of low- to high-amplitude, symmetric to overturned, locally to areally disharmonic anticlines and synclines that are broken by an intricate maze of low-angle thrust, reverse, bedding, tear, and décollement faults. Some folds in the central part of the field are isoclinal and are broken by high-angle reverse faults which truncate troughs and crests. On the southern margin of the field the rocks are overturned at many places and broken by low-angle thrusts.

During the early stages of deformation the rocks of unit 5 tore loose from underlying rocks of units 4 along the Pottchunk and Mocanaqua décollements and possibly along related unidentified décollements. Rocks above these décollements moved northwestward and were slightly to moderately folded. Low-angle thrust faults commonly imbricated from the décollements wherever displacement was impeded by friction, by stratigraphic irregularities, or by warping of the planes of the décollements. During the intensive folding of the region the rocks of unit 5 were warped into a multitude of anticlines and synclines; the décollements and their imbricate thrust faults were folded; numerous high-angle reverse faults developed; and all the structural features of unit 5 came into being. These features, including the décollements, were controlled in their geometric development by the moderate contrast in competence between the lower and upper parts of unit 5 and by the strong contrast in competence between units 4 and 5.

CONCLUSIONS

Northwestward directed stresses of the late Paleozoic Appalachian orogeny were largely responsible for the development of the tectonic framework of the Anthracite region and the remainder of the Valley and Ridge province in Pennsylvania. The framework of the south part of the region may have been outlined during the Late Ordovician Taconic orogeny and extended slightly during the Late Devonian Acadian orogeny, but evidence supporting these earlier events is scanty.

Vectoral resolution of stresses in the Silurian to Pennsylvania rocks was controlled by five lithotectonic units. This control resulted in each unit developing characteristic suites or assemblages of structural features. Each suite maintains a distinctive individuality throughout the region, even though its constituent structural features decrease in complexity northwestward. This decrease conforms with an overall "away from source" decrease in that direction.

The rocks of the lowermost, medial, and uppermost units acted as blanketlike struts which transmitted the deforming stresses northwestward across the region with a relative competence that ranged from moderate to great. The rocks of the two intervening units functioned as incompetent ductile blankets where differential movements between the more competent struts were localized. As a result, the rocks of the more competent units characteristically folded into generally concentric, symmetric to asymmetric anticlines and synclines broken variably by faults. The rocks of the less competent units developed disharmonic folds broken by décollements, low-angle thrust, and bedding faults and commonly separate discordant folds in the more competent rocks.

REFERENCES

Arndt, H. H., and Wood, G. H., Jr., 1960, Late Paleozoic orogeny in eastern Pennsylvania consists of five progressive stages: U. S. Geol. Survey Prof. Paper 400-B, p. B182–B184.

Gray, C., and others, 1960, Geologic map of Pennsylvania: Pennsylvania Geol. Survey.

Gwinn, V. E., 1964, Thin-skinned tectonics in the Plateau and northwestern Valley and Ridge provinces of the Central Appalachians: Geol. Soc. America Bull., v. 75, no. 9, p. 863–900.

——, [in press] Tectonics of the Valley and Ridge province, in Tectonic map of Pennsylvania: Pennsylvania Geol. Survey, 4th ser.

Miller, J. T., 1961, Geology and mineral resources of the Loysville quadrangle, Pennsylvania: Pennsylvania Geol. Survey, 4th ser., Geol.; Atlas 127, 47 p.

Rodgers, J., 1953, The folds and faults of the Appalachian Valley and Ridge province, in McGrain, P., ed., Southeastern Mineral Symposium Proc., 1950: Kentucky Geol. Survey, ser. 9, Spec. Pub. 1, p. 150–166.

Wood, G. H., Jr., and Carter, M. D., [in press] Tectonics of the Anthracite region, in Tectonic map of Pennsylvania: Pennsylvania Geol. Survey, 4th ser.

——, and Kehn, T. M., 1961, Sweet Arrow fault, east-central Pennsylvania: Am. Assoc. Petroleum Geologists Bull., v. 45, no. 2, p. 256–263.

——, Trexler, J. P., Yelenosky, A., and Soren, J., 1958, Geology of the northern half of the Minersville quadrangle and a part of the northern half of Tremont quadrangle, Schuylkill County, Pennsylvania: U. S. Geol. Survey Coal Inv. Map C-43.

Chapter 10

Details of Thin-Skinned Tectonics in Parts of Valley and Ridge and Cumberland Plateau Provinces of the Southern Applachians*

LEONARD D. HARRIS

INTRODUCTION

STRUCTURE OF THE Valley and Ridge province in the Southern Appalachians is dominated by a series of thrust faults. Northeastward these faults die out as surface features and are gradually replaced by folds until in the Central Appalachians folds are the dominant features. This apparent contrast in structural style has contributed in the past to the development of two distinctly different concepts concerning mechanics of deformation in the Appalachians. Rodgers (1949) suggested that these concepts might be termed "thick-skinned" and "thin-skinned." The thick-skinned school of thought, which is the more traditional concept, reasons that all folds and faults extend into basement and their existence depends on support from basement. It postulates that major deformation during the Appalachian orogeny occurred mainly in the basement and the sediments simply mimic those structures.

The thin-skinned school of thought, which was largely developed by geologists concerned with the Southern Appalachians, reasons that the Valley and Ridge structures are features marginal to the main area of deformation and were produced by tangential forces acting from the southeast only upon the sedimentary prism. These forces produced huge bedding plane thrust plates with miles of displacement without involvement of the basement. Movements of these sheets

toward the northwest produced a series of imbricate thrust faults and rootless folds.

Both of these concepts developed largely out of interpretation and extrapolation of surface features into the subsurface. Only in recent years have enough subsurface data been accumulated from oil and gas tests to furnish additional insight into the mechanics of deformation in the Valley and Ridge and Appalachian Plateaus (Miller and Fuller, 1954; Miller and Brosgé, 1954; Young, 1957; Wilson and Stearns, 1958; Gwinn, 1964; and Harris, 1967). This additional information has tended to reaffirm the validity of thin-skinned tectonics in the Cumberland Plateau and Southern Appalachians and to confirm the suggestion by Rodgers (1953 and 1963) that thin-skinned tectonics have played a major role in the development of structures in the Central Appalachians and adjacent Allegheny Plateau.

Although the gross mechanics of thin-skinned tectonics is well established (Rich, 1934; King, 1960) only recently has attention been directed toward a clearer understanding of the details (Gwinn, 1964). As more detailed information becomes available and the ideas concerning mechanics of thrusting begin to be better focused, it becomes increasingly clear that certain concepts and descriptive terms need to be abandoned or redefined and new terms proposed. In the past, the low-angle nature of a thrust fault has been emphasized by the use of the term "overthrust." Regional studies of the habits of thrust faults in the Valley and Ridge have pointed up

---

* Publication authorized by the Director of the United States Geological Survey.

the fact that such faults do not maintain the same low-angle of dip regionally. Rather, they develop initially as bedding plane thrusts in particular stratigraphic positions but may break upward from one stratigraphic position to another along moderate to high-angle ramps. Thus, there is no clear-cut distinction between high-angle and low-angle thrust faults, for the same fault may assume either attitude. The geologist, in mapping a thrust-faulted area, is faced with the vagaries of erosion. If the bedding plane thrust part of the "overthrust" has been removed leaving only the high-angle ramp fault exposed, one is tempted to assign a different origin to this high-angle thrust when, in fact, it is but a segment of an "overthrust." For that reason, throughout this paper the more general terms—bedding plane thrust, thrust fault, thrust plate, allochthonous, and autochthonous—are used in place of the terms overthrust and overthrust block.

THIN-SKINNED MECHANICS OF THE PINE MOUNTAIN FAULT SYSTEM

The fundamental concepts of thin-skinned tectonics in the southern Valley and Ridge province were largely developed from field studies by Butts (1927) of the Pine Mountain thrust plate (formerly called the Cumberland overthrust block) and the subsequent analysis of the mechanics of thrusting by Rich (1934). This key area (first described by Wentworth, 1921) is a quadrilateral thrust plate about 125 mi (201 km) long and 25 mi (40 km) wide occurring in parts of Kentucky, Tennessee, and Virginia (Fig. 1). It is bounded on the northwest by the Pine Mountain thrust fault, on the southwest by the Jacksboro fault, on the northeast by the Russell Fork fault, and limited on the southeast by the Clinchport thrust fault (Harris, 1965a). The plate can be divided into two major features—the Middlesboro syncline on the northwest, and the Powell Valley anticline to the southeast. Heretofore, most authors (Butts, 1927; Rich, 1934; Miller and Fuller, 1954; and Harris and Zietz, 1962) have shown these folds in regional index maps as simple structures that could be defined by a single axis. However, as more data become available it is obvious that both the Middlesboro syncline and the Powell Valley anticline are extremely complex structures and neither is defined by a single axis (Fig. 1). As a matter of fact, Englund and Roen (1963) have suggested that the circular fault system at Middlesboro, Kentucky, is unrelated to thin-skinned tectonics and may be a meteor impact scar (Fig. 1).

The discovery in 1923 and 1926 by Butts (1927) of fensters in the core of the Powell Valley anticline established the true nature of the Pine Mountain thrust. In a series of sections (Butts, 1927, Pl. 2; Butts in Butts and others, 1932, Pl. 23) he clearly recognized the Pine Mountain fault as a bedding plane sole thrust with the Wallen Valley and Hunter Valley thrusts branching off from the sole. Butts further suggested that the sole thrust developed in the incompetent Cambrian shales, ramped upward in the fenster area to another incompetent shale zone (Devonian and Mississippian), continued beneath the Middlesboro syncline, and finally came to the surface along Pine Mountain. Rich (1934) accepted Butts' inspired interpretations and with equal perception presented an interpretation of the mechanics of low-angle thrusting.

Rocks in the Pine Mountain thrust plate were considered by Rich (1934) to have been initially nearly flat-lying. The Pine Mountain thrust, taking advantage of contrasts in rock competence, developed as a bedding plane thrust in an incompetent shale zone, ramped upward along a diagonal shear plane through competent beds, and continued in another higher incompetent shale zone as a bedding plane thrust. Movement of the thrust plate to the northwest developed an anticline simply by duplication of beds (Fig. 2). Folds of this type are rootless structures confined to the thrust plate and rocks below the plate are left relatively undisturbed. Size and shape of such an anticline are controlled by the amount of movement. Thus, a rounded, crested anticline is produced by small movement and a broad, flat-topped anticline by large movement.

Warping of the Pine Mountain Thrust Fault

Regional detailed mapping and subsurface study of the Powell Valley anticline by Miller and Fuller (1954) and Miller and Brosgé (1954) tended to substantiate the main views of Rich (1934). However, these authors differ substantially in detail from Rich in their interpretation of the development of some of the structure within the area. Rich (1934), p. 1589, noting that Butts (1927, Pl. 2) had shown the Pine Mountain fault as being arched beneath the Powell Valley anticline concluded that the fault was only apparently warped and that the present structure developed from tilting the block northward. Data from detailed geologic surface and subsurface studies by Miller and Fuller (1954) showed conclusively that the Pine Mountain thrust is arched on the order of 5,000 ft (1,524

m) beneath the Powell Valley anticline. They suggested that warping of the fault into an arch occurred in the final stage of deformation and that folding was largely responsible for the Powell Valley anticline. Harris and Zietz (1962) modified their theory by suggesting that with over 5,000 ft (1,524 m) of vertical warping, basement could be involved both before and after thrusting. Cooper (1964, p. 107) has suggested a similar conclusion for the Blacksburg–Pulaski and Draper Mountain areas in southwest Virginia.

The deep Bales well, drilled by the Shell Oil Company near Ewing, Virginia, in the Powell Valley anticline (Fig. 1) in 1965 has revealed an alternative mechanism for arching of the Pine Mountain thrust fault that is contrary to all previous interpretations (Harris, 1967). Arching of the Pine Mountain thrust is due to the massive subsurface duplication of 5,650 ft (1,722 m) of beds by a nonoutcropping thrust fault (Fig. 3) called the Bales thrust. Apparently this large slice, originally part of the autochthonous plate, was emplaced after the initial development of the Powell Valley anticline. This massive duplication of beds modified the configuration of the Pine Mountain thrust beneath the initial Powell Valley anticline by changing its attitude from a nearly flat-lying to an arched surface. Heretofore, displacement along the Pine Mountain thrust has been measured as the distance between broken edges of the same formation above and below the Pine Mountain thrust—a distance on the order of 6 mi (9.6 km). Data from the Bales well suggest that this is only part of the total movement and that additional displacement has occurred along the nonoutcropping Bales–Pine Mountain thrust fault. An increment estimated to be as much as 3 mi (4.8 km) should be added to original figure of 6 mi (9.6 km).

Drilling has not completely outlined the subsurface areal extent of the nonoutcropping Bales thrust; however, some inferences can be made based on regional geologic relationships. Enough detailed mapping has been done in the area of the Powell Valley anticline to show that a series of fensters or near-fensters is exposed along its axial region from the Virginia–Tennessee boundary to near Big Stone Gap, Virginia—a distance of about 36 mi (57.9 km) (Miller and Fuller, 1954; Miller and Brosgé, 1954; and Miller, 1962). In contrast, southwest of the Virignia–Tennessee boundary the Pine Mountain thrust plate is not greatly warped and no fentsers have been found (Fig. 1). Indeed, information from mapping and sparse drilling (wells 1 and 2, Fig.

1) suggests that southwest of the fensters the Pine Mountain fault may be nearly flat. Thus, it would seem that areal distribution of the fensters probably very nearly outlines areas where in the subsurface massive duplication has warped the Pine Mountain thrust.

Transverse Faults

Coincidental with the change in the attitude of the Pine Mountain thrust fault near the Tennessee and Virginia boundary from warped to the northeast to apparently unwarped to the southwest, there is a change along strike in the subsurface of the stratigraphic position occupied by the thrust. Near the west edge of the last fenster, the Pine Mountain thrust has been found from drilling (well No. 5, Fig. 1) to have developed 2,200 ft (670.6 m) lower stratigraphically (Rome Formation) than the position of the fault in the fenster area (Maynardville Formation, Miller and Fuller, 1954, p. 258). A restored longitudinal section suggests that this change in stratigraphic position occurred along a steeply dipping subsurface transverse fault (Fig. 4C). These relationships seem to confirm the inference of Wilson and Stearns (1958, p. 1292, Fig. 7), that transverse faults may play a dual role in the mechanics of thin-skinned deformation. In their diagram, transverse faults that come to the surface limit the lateral extent of the Cumberland Plateau thrust plate; whereas, nonoutcropping subsurface transverse faults confined to particular stratigraphic intervals do not necessarily limit bedding plane thrust faults, but act as connecting links that enable bedding thrusts to develop along strike at different stratigraphic positions. The restored longitudinal section through the Pine Mountain thrust plate suggests that the Pine Mountain thrust developed in conjunction with subsurface transverse faults as a bedding plane thrust at different stratigraphic levels (Fig. 4). Later northwest movement of the Pine Mountain plate rotated the steeply dipping transverse faults to a nearly horizontal plane. The position of rotation is reflected in the Powell Valley anticline by a series of step plunges to the northeast. Other examples of the influence exerted by transverse faults on synclines and anticlines include the deflection of the axis of the Middlesboro syncline by incipient and outcropping transverse faults (Fig. 1), the abrupt ending of the Powell Mountain anticline (Fig. 5), and the ending of small scale synclines and anticlines against a transverse feature (Fig. 6). Thus, it is evident that transverse faults not only play a dual role in the development of a thrust plate, they also influence

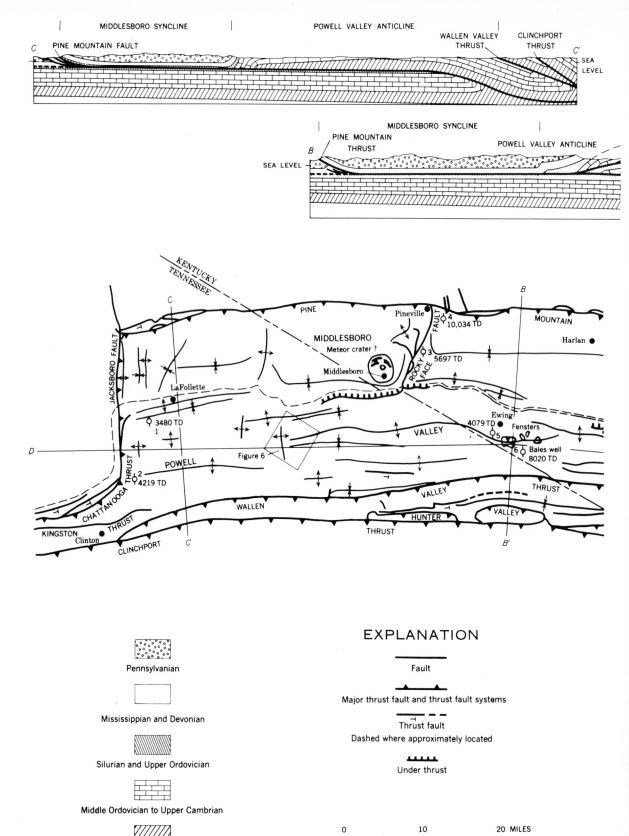

FIGURE 1. Structure map of the Pine Mountain thrust plate, illustrating the complexities of the Middlesboro syncline and the Powell Valley anticline. Compiled from Ashley and Glenn (1906), Bates (1939), Brent (1963), Brokaw and others (1966), Butts (1914, 1933), Crider (1916), Eby (1923), Englund (1957, 1958, 1964), Englund and Roen (1963), Englund and others (1961, 1963, 1964),

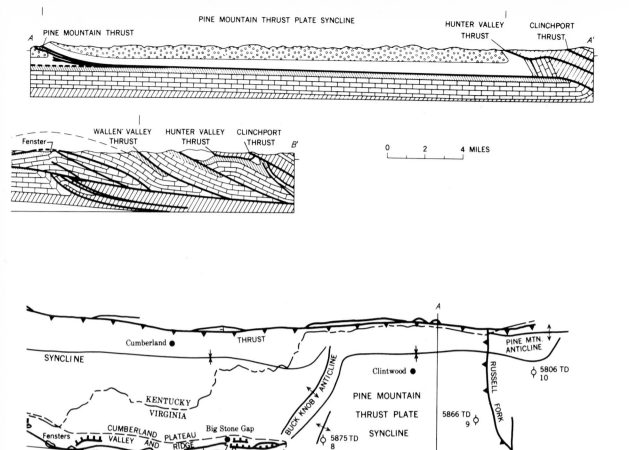

PINE MOUNTAIN THRUST PLATE SYNCLINE

PINE MOUNTAIN THRUST   HUNTER VALLEY THRUST   CLINCHPORT THRUST

WALLEN VALLEY THRUST   HUNTER VALLEY THRUST   CLINCHPORT THRUST

Fenster

0   2   4 MILES

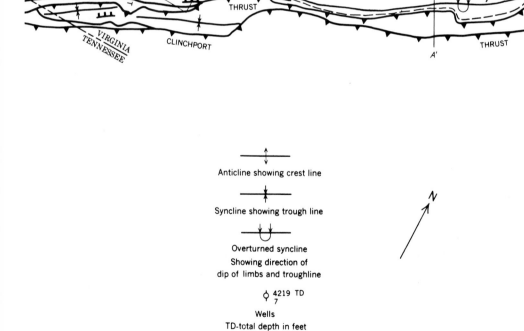

A

Cumberland   THRUST   PINE MTN. ANTICLINE

SYNCLINE   Clintwood   RUSSELL FORK   5806 TD 10

KENTUCKY   BUCK KNOB ANTICLINE   PINE MOUNTAIN THRUST PLATE SYNCLINE

VIRGINIA

Fensters   CUMBERLAND VALLEY AND PLATEAU RIDGE   Big Stone Gap   5866 TD 9

ANTICLINE   5875 TD 8   D'

4861 TD   FAULT

Figure 5

POWELL MTN. ANTICLINE   THRUST

VIRGINIA   CLINCHPORT

TENNESSEE   THRUST

A'

Anticline showing crest line

Syncline showing trough line

Overturned syncline
Showing direction of
dip of limbs and troughline

4219 TD
7

Wells
TD-total depth in feet

N

Giles (1921), Hardeman and others (1966), Harris (1965a,b,c), Harris and
Miller (1958, 1963), Harris and others (1962), Hinds (1918), Hodge (1912),
Miller (1962, 1965), Miller and Brosgé (1954), Miller and Fuller (1954),
Swingle (1960a,b), Wentworth (1922, 1927), and Wilson and others (1956).

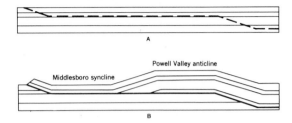

FIGURE 2. Rich's (1934) interpretation of the development of the Pine Mountain thrust plate. (A) Area before faulting, showing initial development of Pine Mountain thrust fault. (B) After movement, which produced the Powell Valley anticline and Middlesboro syncline by duplication of beds.

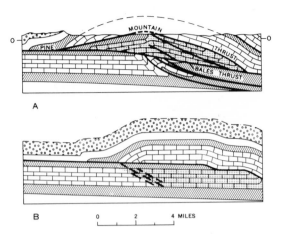

FIGURE 3. Arching of the Pine Mountain thrust fault by subsurface duplication of beds (section is part of section B-B', Fig. 1). (A) Cross section interpreted from deep drilling; stippled area outlines beds duplicated by the Bales thrust. (B) Restored section; stippled area outlines the incipient development of the Bales thrust plate. See Figure 1 for explanation of symbols.

the size and shape of anticlines and synclines within the plate. That this is not an isolated structural mechanism is illustrated by the inference of similar features in the subsurface of part of the Central Appalachian Plateau by Rodgers (1963) and Gwinn (1964).

Russell Fork and Jacksboro Transverse Faults

In the past the Pine Mountain thrust plate has been considered to be limited at either end by transverse faults—Russell Fork on the northeast and Jacksboro on the southwest. This concept does not explain the enigma of the Jacksboro fault changing along strike to the southwest from a transverse to a thrust fault (Chattanooga) without cutting completely across the southwest "corner" of the Pine Mountain thrust plate. The sequence of rock from Lower Cambrian to Silurian on the southeast limb of the Powell Valley anticline forms an uninterrupted strike sequence with the same rocks considered to be part of the Chattanooga thrust plate (Fig. 7). In contrast, rocks that are part of the northern limb of the Powell Valley anticline and the Middlesboro syncline do not form an uninterrupted sequence with rocks across the Jacksboro fault. Geologic mapping by Englund (1957) clearly shows the Pine Mountain fault ending against the Jacksboro and a projection of the Jacksboro with little or no strike-slip displacement continuing several miles northwestward beyond the Pine Mountain plate. This confinement of the Jacksboro as a transverse fault to only part of the Pine Mountain plate suggests that it developed only in the northern half of the plate after the Pine Mountain fault had ramped up from the Cambrian to a younger stratigraphic level. Southwest of the Jacksboro fault, the Pine Mountain–Chattanooga fault instead of ramping to the same younger stratigraphic level continued upward, probably to the surface (Fig. 8). Regional stratigraphic analysis

suggests that the deflection of the Pine Mountain fault to higher levels along the Jacksboro fault coincides with a marked decrease in the thickness of the Chattanooga Shale (Devonian and Mississippian) from about 1,000 ft (304.8 m) on the northeast end of the plate to about 50 ft (15 m) on the southwest edge of the plate. Detailed geologic studies and sparse drilling (Englund, 1957, 1958) have shown that near the Jacksboro fault the Pine Mountain thrust beneath the Middlesboro syncline and Powell Valley anticline actually developed below the thin Chattanooga near the contact of Silurian (Rockwood Formation) and Upper Ordovician (Sequatchie Formation) rocks.

Displacement along the Jacksboro fault is probably nearly equal to the distance the upturned edges of Ordovician to Pennsylvanian rocks in the Pine Mountain plate are offset from the same rocks in front of the Chattanooga thrust (Fig. 8). Drag at the junction of the Jacksboro–Chattanooga faults may have warped the autochthonous plate somewhat, but a conservative estimate of displacement would be on the order of 10–12 mi (16–19 km).

Relations of the Russell Fork fault on the northeast end of the Pine Mountain plate are not as clear; however, Bates (1936, Pl. 18) and Woodward (1938, Pl. 1) indicate that from northwest to southeast the fault changes in character from a strike slip to a thrust fault. Regional mapping (Harnsberger, 1919, Pl. 2) suggests that the Russell Fork fault joins a zone of thrust faults that die

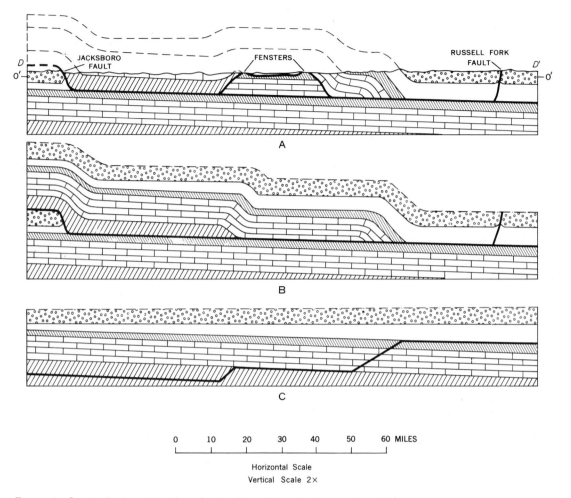

FIGURE 4. Longitudinal section through the Pine Mountain thrust plate. (A) Present structure. (B) Restored section with the Bales fault slice removed, illustrating the step plunge nature of the Powell Valley anticline. (C) The Pine Mountain thrust plate resolved to the horizontal, showing the initial position of the fault before movement of the thrust plate to section line D-D′ (Fig. 1). See Figure 1 for explanation of symbols.

out on the surface toward the northeast in the Abbs Valley anticline. These relationships do not seem to differ materially from the structural relations of the Jacksboro–Chattanooga faults on the southwest end of the plate (Fig. 1).

The greatest difference between the northeast and southwest ends of the plate is the manner in which the Pine Mountain thrust seems to end at the Jacksboro fault but does not end at the Russell Fork fault, where its trace has been mapped for about 10 mi (16.1 km) beyond the Russell Fork fault (Hinds, 1918; Giles, 1921). Drill holes west of the Russell Fork fault in the Pine Mountain thrust plate have cut a consistent shear zone (the Pine Mountain thrust) near the base of the Devonian shale (Young, 1957). Sudden high pressure gas "blowouts" occur upon penetration

of the shear zone. East of the Russell Fork fault, drilling has intersected a shear "blowout" zone at the same stratigraphic position; however, the intensity of shearing is not as great as that in the Pine Mountain plate (Young, 1957, p. 2573). These relationships suggested to Young that the Pine Mountain fault developed at the base of the Devonian on both sides of the Russell Fork fault, but that movement was greatest in the Pine Mountain plate. Wentworth (1921, p. 65) has estimated a displacement on the order of 2 mi (3.2 km) for the plate. Accordingly the Russell Fork fault would be a late development, limiting major thrusting to the Pine Mountain plate. What caused the development of the Russell Fork fault? Perhaps as suggested by King (1960, p. 122) the strength of thrust sheets are limited by the materials of which they are composed; if this strength is exceeded they cannot move as a single mass.

The realization that thrust faults and transverse faults form an integrated structural system that can aid, as well as limit, development of thrust plates suggests that the Pine Mountain thrust may be only a part of a more extensive system, which includes the Russell Fork, Pine Mountain, Jacksboro, and Chattanooga faults. The surface trace of this system extends from Virginia to Georgia, a distance of 300 mi (482.7 km). Drilling east of the Russell Fork fault has established the presence of the Pine Mountain thrust in the subsurface; however, not enough deep wells have been drilled to determine the subsurface extension. The possibility does exist that the Pine Mountain system may link with the extensive subsurface

thrust system described by Gwinn (1964) in West Virginia, Virginia, Maryland, and Pennsylvania.

Imbricate and Underthrust Faults

Geologic mapping by Englund (1957, 1958) along the Pine Mountain thrust has revealed a series of complex slices lying in front of the main fault trace. Drilling (well No. 4, Fig. 1) just south of the outcrop of the Pine Mountain thrust at Pineville, Kentucky, indicates that in the subsurface a thousand feet of duplication occurred just above the fault (Thomas, 1960); however, below the fault beds must be nearly horizontal, as penetrated thicknesses closely match nearby surface sections. Both surface mapping and subsurface information suggests that imbricate

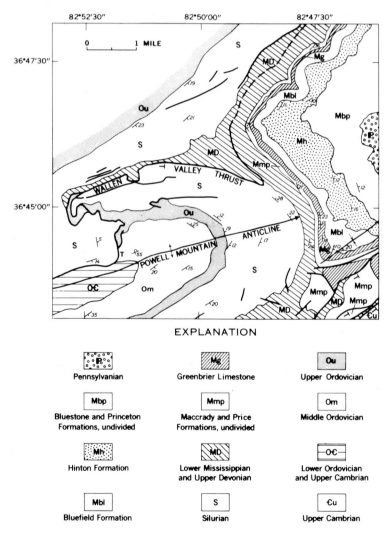

FIGURE 5. Abrupt plunge out of a large scale anticline along a transverse feature (see Fig. 1 for explanation of structure symbols and map location). Modified from Harris and Miller (1958) and Miller (1965).

thrusting may be largely responsible for deflecting the Pine Mountain thrust upward to higher stratigraphic levels or to the surface (Fig. 1). These relations are similar to those described by Gwinn (1964) in the subsurface of the central Appalachian Plateau. Gwinn (p. 891–894) also points out that underthrusting or high-angle reverse faults are commonly associated with these structures and may be related to the concentric folding process whereby tightening of the core zone of an anticline is resolved by symmetrical thrusting. Underthrusts, though not extensive, have been mapped at several localities within the Pine Mountain plate (Miller and Fuller, 1954; Miller and Brosgé, 1954; Harris and Miller, 1958; and Englund, 1964) and are thought to be associated with Chattanooga thrust (Swingle, 1964; Swingle and Luther, 1964).

GENERALIZATIONS CONCERNING THIN-SKINNED DEFORMATION

Dissection of the Pine Mountain fault system into its component parts has served to point up the complexities of thin-skinned tectonics. Not only is there a bedding plane thrust that ramps from one incompetent zone to another, but also there is an integrated system whereby transverse, bedding plane, underthrust, and imbricate faults all play a part in regional development of a thrust plate. The interaction of the different parts of a fault system produces certain clearly definable structures from which the following generalizations can be made.

(1) Thrust faults develop initially along the plane of the bedding in deep-seated incompetent units that are nearly flat-lying, shear steeply upward across more competent units either to younger incompetent beds or to the surface. These diagonal shears develop both parallel (as ramps) and perpendicular (as transverse faults) to the strike of the thrust. Longitudinal and transverse structure sections of the thrust plates do not differ materially (Figs. 2 and 4).

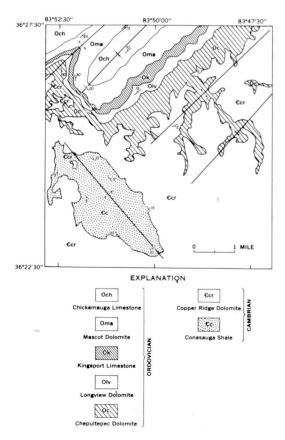

EXPLANATION

Och	Chickamauga Limestone
Oma	Mascot Dolomite
Ok	Kingsport Limestone
Olv	Longview Dolomite
Oc	Chepultepec Dolomite

ORDOVICIAN

| €cr | Copper Ridge Dolomite |
| €c | Conasauga Shale |

CAMBRIAN

FIGURE 6. Abrupt termination of small scale synclines and anticlines by a transverse structure. Modified from Brokaw and others (1966).

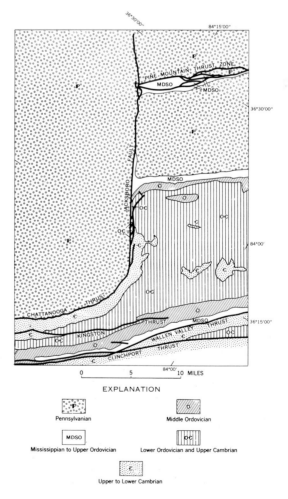

EXPLANATION

| P | Pennsylvanian |
| MDSO | Mississippian to Upper Ordovician |

| Middle Ordovician |
| OC | Lower Ordovician and Upper Cambrian |

| C | Upper to Lower Cambrian |

FIGURE 7. Geologic map of the southwest corner of the Pine Mountain thrust plate. Modified from Hardeman and others (1966).

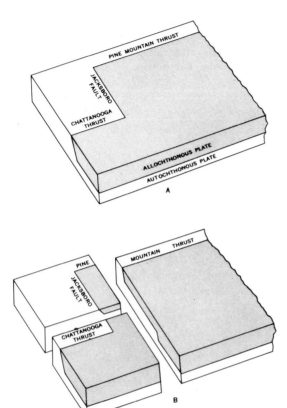

FIGURE 8. Block diagram illustrating the relationship of the Pine Mountain, Jacksboro, and Chattanooga faults.

(2) The Pine Mountain thrust plate is not unique, it simply illustrates the relation of transverse and thrust faults on a large scale. The same relationship can be seen even in small-scale structures (Fig. 6).

(3) Both transverse and ramp faults assume dual functions in the regional development of a thrust plate. Neither of these faults necessarily cuts the entire sedimentary prism; instead they may be confined to specific stratigraphic intervals. Those that extend to the surface, usually from shallow depth, serve to limit the present regional development of a thrust plate. However, subsurface nonoutcropping transverse and ramp faults confined to certain stratigraphic intervals do not limit regional development; instead they act as connecting links that enable bedding thrusts to develop at different stratigraphic intervals to form a coherent thrust plate.

(4) Rootless folds are the principal constructional features produced as a consequence of thin-skinned deformation. Anticlines result from either large scale duplication of beds or relatively small scale duplication by imbricate or splay thrusting. In contrast, synclines are more passive features that largely occur as a consequence of anticline development, or as the result of drag in a ramp zone. The manner in which these folds terminate is controlled either by decreasing movement along strike of a thrust or by the relationship of transverse faults to the thrust. The Powell Valley anticline is an example of a complex step-plunging fold involving several transverse faults at different stratigraphic levels (Figs. 1 and 4). Figure 9 illustrates other less complex but common anticlines resulting from variation in transverse and thrust-fault relationships.

(5) Restored sections of the Pine Mountain plate suggest that major anticlines play an important part in deflecting later developing thrust faults up to higher stratigraphic beds or possibly to the surface (Fig. 10). These later thrusts tend to situate themselves on the south limb of anticlines, parallel to the previous ramps. If the fold is relatively small the thrust may break completely across the core of an anticline (Fig. 10, section A-A'); however, if the anticline is large the fault tends to break across the lower competent beds and develop as a bedding plane thrust on the south limb (Fig. 10, section C-C').

(6) Deep drilling indicates that "folded" thrusts are not the result of the folding process operating within the sedimentary prism, rather arching has resulted from subsurface duplication of beds (Fig. 3). The closure of the arch is equal to the amount of duplication. The source of the duplicated beds lies within the autochthonous plate, which suggests that rocks of that plate are not passive elements, instead they may play an active role in thin-skinned tectonics.

(7) Major thrust plates must have limits beyond which they cannot develop and move as cohesive masses; however, this limit must be great for major thrust faults—the Pine Mountain-Chattanooga, Clinchport, Copper Creek, Saltville, and Pulaski—have surface traces that extend for several hundred miles. Even though the surface traces of these faults end as they enter the Central Appalachians, at least some of the faults may continue in the subsurface and link with the extensive subsurface thrust system described by Gwinn (1964).

(8) No sharp change in structural style coincides with physiographic boundaries. The same type of structures exist in both the Appalachian Plateau and the Valley and Ridge. The principal difference is a gradual decrease in intensity of deformation from east to west. Deep drilling has not as yet delineated the regional extent of thrusting in the Plateau.

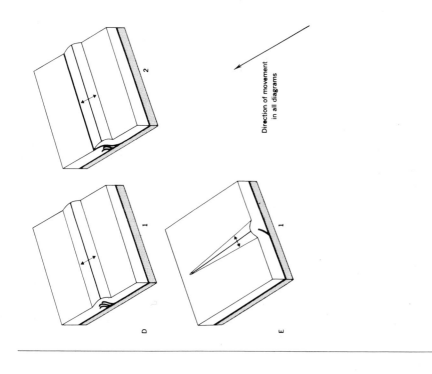

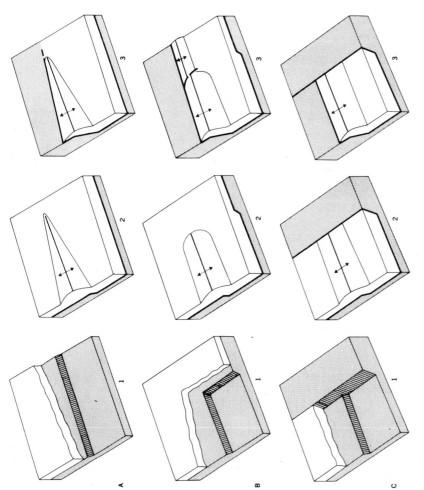

FIGURE 9. Block diagrams illustrating anticline development by varying the relationship of ramp, transverse, and thrust faults. Stage 1, subsurface relations (A1, B1, and C1). Stage 2, anticline resulting from movement (A2, B2, C2, D1, and E1). Stage 3, anticline resulting from intersection of thrust with surface (A3, B3, C3, and D2). (A) Sharp-nosed anticline produced by decreasing movement along strike. (B) Blunt-nosed anticline resulting from subsurface intersection of transverse fault with a thrust fault. (C) Chopped-off anticline resulting from the intersection of a transverse fault with the surface. (D) Anticline developed by subsurface imbricate thrusting. (E) Transverse anticline development, associated with an incipient transverse fault.

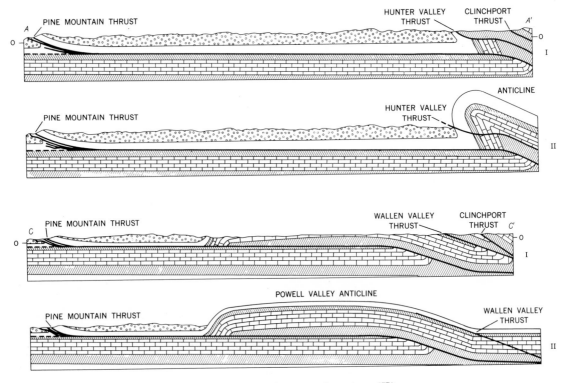

FIGURE 10. Present structure sections (I) and partially restored sections (II) illustrating the control exerted by anticlines in deflecting later developing thrust to higher stratigraphic levels. See Figure 1 for explanation of symbols and lines of section.

REFERENCES

Ashley, G. H., and Glenn, L. C., 1906, Geology and mineral resources of part of the Cumberland Gap coal field, Kentucky: U. S. Geol. Survey Prof. Paper 49, 239 p.

Bates, R. L., 1936, The Big A Mountain area, Virginia: Virginia Geol. Survey Bull. 46-M, p. 167–204.

————, 1939, Geology of Powell Valley in northeastern Lee County, Virginia: Virginia Geol. Survey Bull. 51-B, p. 31–104.

Brent, W. B., 1963, Geology of the Clinchport quadrangle, Virginia: Virginia Division of Mineral Resources, Rept. Inv. 5, 45 p.

Brokaw, A. L., Rodgers, John, Kent D. F., Laurence, R. A., and Behre, C. H., Jr., 1966, Geology and mineral deposits of the Powell River area, Claiborne and Union Counties, Tennessee: U. S. Geol. Survey Bull. 1222-C, 56 p.

Butts, Charles, 1914, The coal resources and general geology of the Pound quadrangle of Virginia and Kentucky: U. S. Geol. Survey Bull. 541-F, p. 165–221.

————, 1927, Fensters in the Cumberland overthrust block in southwestern Virginia: Virginia Geol. Survey Bull. 28, 12 p.

————, 1933, Geologic map of the Appalachian Valley of Virginia with explanatory text: Virginia Geol. Survey Bull. 42, 56 p.

————, Stose, G. W., and Jonas, A. I., 1932, Southern Appalachian region: Internat. Geol. Cong., 16th, Washington 1933, Guidebook 3, Excursion 3-A, 94 p.

Cooper, B. N., 1964, Relation of stratigraphy to structure in the southern Appalachians *in* Tectonics of the southern Appalachians: Virginia Polytech. Inst., Dept. Geol. Studies Mem. 1, p. 81–114.

Crider, A. F., 1916, The coals of Letcher County, Kentucky: Kentucky Geol. Survey ser. 4, v. 4, 234 p.

Eby, J. B., 1923, The geology and mineral resources of Wise County and the coal-bearing portion of Scott County, Virginia: Virginia Geol. Survey Bull. 24, 617 p.

Englund, K. J., 1957, Geology and coal resources of the Pioneer quadrangle, Scott and Campbell Counties, Tennessee: U. S. Geol. Survey Coal Inv. Map C 39.

————, 1958, Geology and coal resources of the Ivydell quadrangle, Campbell County, Tennessee: U. S. Geol. Survey Coal Inv. Map C 40.

————, 1964, Geology of the Middlesboro South quadrangle, Tennessee–Kentucky–Virginia: U. S. Geol. Survey Geol. Quad. Map GQ-301.

————, Landis, E. R., and Smith, H. L., 1963, Geology of the Varilla quadrangle, Kentucky-Virginia: U. S. Geol. Survey Geol. Quad. Map GQ-190.

————, and Roen, J. B., 1963, Origin of the Middlesboro basin, Kentucky: U. S. Geol. Survey Prof. Paper 450-E, p. 20-22.

————, Roen, J. B., and DeLaney, A. O., 1964, Geology of the Middlesboro North quadrangle, Kentucky: U. S. Geol. Survey Geol. Quad. Map GQ-300.

————, Smith, H. L., Harris, L. D., and Stephens, J. G., 1961, Geology of the Ewing quadrangle,

Kentucky and Virginia: U. S. Geol. Survey Geol. Quad. Map GQ-172.

Giles, A. W., 1921, The geology and coal resources of Dickenson County, Virginia: Virginia Geol. Survey Bull. 21, 224 p.

Gwinn, V. E., 1964, Thin-skinned tectonics in the Plateau and northwestern Valley and Ridge provinces of the Central Appalachians: Geol. Soc. America Bull., v. 75, no. 9, p. 863-900.

Hardeman, W. D., and others, 1966, Geologic map of Tennessee: Nashville, Tennessee Div. Geology, 4 sheets, scale 1:250,000.

Harnsberger, T. K., 1919, The geology and coal resources of the coal-bearing portion of Tazewell County, Virginia: Virginia Geol. Survey Bull. 19, 195 p.

Harris, L. D., 1965a, The Clinchport thrust fault—a major structural element of the southern Appalachian Mountains: U. S. Geol. Survey Prof. Paper 525-B, p. B49-B53.

————, 1965b, Geologic map of the Wheeler quadrangle, Claiborne County, Tennessee, and Lee County, Virginia: U. S. Geol. Survey Geol. Quad. Map GQ-435.

————, 1965c, Geologic map of the Tazewell quadrangle, Claiborne County, Tennessee: U. S. Geol. Survey Geol. Quad. Map GQ-465.

————, 1967, Geology of the L. S. Bales well, Lee County, Virginia—a Cambrian and Ordovician test: *in* Kentucky Geol. Survey, Proc. of the Tech. Sess. Kentucky Oil and Gas Assoc., 29th Ann. Mtg., June 3-4, 1965, Spec. Pub. 14, ser. 10.

————, and Miller, R. L., 1958, Geology of the Duffield quadrangle, Virginia: U. S. Geol. Survey Geol. Quad. Map GQ-111.

————, 1963, Geology of the Stickleyville quadrangle Virginia: U. S. Geol. Survey Geol. Quad. Map GQ-238.

————, Stephens, J. G., and Miller, R. L., 1962, Geology of the Coleman Gap quadrangle, Tennessee and Virginia: U. S. Geol. Survey Geol. Quad. Map GQ-188.

————, and Zietz, Isidore, 1962, Development of Cumberland overthrust block in vicinity of Chestnut Ridge fenster in southwest Virginia: Am. Assoc. Petroleum Geologists Bull., v. 46, no. 12, p. 2148-2160.

Hinds, Henry, 1918, The geology and coal resources of Buchanan County, Virginia: Virginia Geol. Survey Bull. 18, 278 p.

Hodge, J. M., 1912, Report on the upper Cumberland coal field, the region drained by Poor and Clover Forks in Harlan and Letcher Counties: Kentucky Geol. Survey Bull. 13, ser. 3, 223 p.

King, P. B., 1960, The anatomy and habitat of low-angle thrust faults: Am. Jour. Sci., v. 258-A (Bradley Volume), p. 115-125.

Miller, R. L., 1962, The Pine Mountain overthrust at the northeast end of the Powell Valley anticline, Virginia: U. S. Geol. Survey Prof. Paper 450-D, p. D69-D72.

————, 1965, Geologic map of the Big Stone Gap quardrangle, Virginia: U. S. Geol. Survey Geol. Quad. Map GQ-424.

————, and Brosgé, W. P., 1954, Geology and oil resources of the Jonesville district, Lee County, Virginia: U. S. Geol. Survey Bull. 990, 240 p.

————, and Fuller, J. O., 1954, Geology and oil resources of the Rose Hill district—the fenster County, Virginia: Virginia Geol. Survey Bull. 71, 383 p.

Rich, J. L., 1934, Mechanics of low-angle overthrust faulting as illustrated by Cumberland thrust block, Virginia, Kentucky, and Tennessee: Am. Assoc. Petroleum Geologists Bull., v. 18, no. 12, p. 1584-1596.

Rodgers, John, 1949, Evolution of thought on structure of middle and southern Appalachians: Am. Assoc. Petroleum Geologists Bull. v. 33, no. 10, p. 1643-1654.

————, 1953, The folds and faults of the Appalachian Valley and Ridge province, *in* McGrain, P., ed., Southeastern Mineral Symposium 1950: Kentucky Geol. Survey, ser. 9, Special Pub. no 1, p. 150-166.

————, 1963, Mechanics of Appalachian foreland folding in Pennsylvania and West Virginia: Am. Assoc. Petroleum Geologists Bull., 47, no. 8, p. 1527-1536.

Swingle, G. D., 1960a, Geologic map of the Lake City quadrangle, Anderson County, Tennessee: Tennessee Div. Geology Geol. Map GM 137-NW.

————, 1960b, Geologic map of the Jacksboro quadrangle, Campbell County, Tennessee: Tennessee Div. Geology Geol. Map GM 136-SW.

————, 1964, Geologic map of the Graysville quadrangle, Tennessee: Tennessee Div. Geology Geol. Map GM 111-NE.

————, and Luther, E. T., 1964, Geologic map of the Soddy quadrangle, Tennessee: Tennessee Div. Geology Geol. Map GM 111-SW.

Thomas, G. R., 1960, Geology of recent deep drilling in eastern Kentucky: Kentucky Geol. Survey, ser. 10, Spec. Pub. 3, p. 10-28.

Wentworth, C. K., 1921, Russell Fork fault *in* The geology and coal resources of Dickenson County, Virginia: Virginia Geol. Survey Bull. 21, p. 53-67.

————, 1922, The geology and coal resources of Russell County, Virginia: Virginia Geol. Survey Bull. 22, 179 p.

————, 1927, The geology and coal resources of the Middlesboro basin in Kentucky: Kentucky Geol. Survey, ser. 6, v. 29, p. 161-235.

Wilson, C. W., Jr., Jewell, J. W., and Luther, E. T., 1956, Pennsylvanian geology of the Cumberland Plateau: Nashville, Tennessee Div. Geology, 21 p.

————, and Stearns, R. G., 1958, Structure of the Cumberland Plateau, Tennessee: Geol. Soc. America Bull., v. 69, no. 10, p. 1283-1296.

Woodward, H. P., 1938, Outline of the geology and mineral resources of Russell County, Virginia: Virginia Geol. Survey Bull. 49, 91 p.

Young, D. M., 1957, Deep drilling through Cumberland overthrust block in southwestern Virginia: Am. Assoc. Petroleum Geologists Bull., v. 41, no. 11, p. 2567-2573.

Chapter 11

The Pulaski Fault, and the Extent of Cambrian Evaporites in the Central and Southern Appalachians

JOHN RODGERS

THE PULASKI FAULT, one of the major thrust faults of the Valley and Ridge province of the Appalachians, has a number of unusual, not to say unique, features, which have been carefully described by Byron Cooper in a series of publications (see References). It is actually a complex fault system, especially complex in the vicinity of Pulaski, Virginia, where it was first named by Campbell and Holden (Campbell and others, 1925, p. 43), and from there eastward to and beyond Roanoke, 50 mi (80 km) east–northeast of Pulaski (Fig. 1). In this region, some faults of the system are warped and cut by others, producing sinuous fault traces and a number of windows and semiwindows that demonstrate an across-strike displacement of not less than 9 mi (15 km) on the basal fault of the system (the Pulaski fault, *sensu stricto,* which brings Middle Cambrian over Lower Mississippian rocks), not to mention the displacement on the higher faults. In either direction along strike, the fault traces become somewhat more regular, though outlying klippen and semiklippen still attest considerable displacement; northeastward the Staunton fault of the system can be traced into the Shenandoah Valley of northern Virginia, 170 mi (275 km) from Pulaski, and southwestward the Seven Springs fault can be traced into East Tennessee, where it divides into several faults that finally disappear under the frontal thrust fault of the Blue Ridge province, 140 mi (225 km) from Pulaski. Thus the total length of the fault system is 310 mi (500 km). Moreover, the Pulaski fault or fault system is the

only structural feature in the Valley and Ridge province that can be traced entirely through the angular Roanoke recess, from the southern wing of the Pennsylvania or central Appalachian arcuate salient into the northern wing of the Tennessee or southern Appalachian salient. The other major structures of the two salients interfere and intersect (at an angle of about 25°) in the recess, whereas the Pulaski fault system reaches its maximum throw and complexity there and in the next 60 mi (100 km) to the southwest along strike.

But perhaps the most extraordinary feature of the Pulaski fault system is the constant association with it of a spectacular breccia, named the Max Meadows fault breccia or formation by Cooper, who has described it and shown its tectonic nature (earlier it had been considered a cave breccia or a sedimentary conglomerate). In the Pulaski region, where it is best developed, it is associated particularly with Cooper's Max Meadows fault, which generally lies a few hundred meters above the basal Pulaski fault for a strike distance of at least 30 mi (50 km). This breccia is no ordinary fault breccia, however; it is sharply set off from the rocks on either side, though they are fractured, and between them it forms a blanket whose thickness averages perhaps 33 ft (10 m) but ranges up to 115 ft (35 m) and from which dike-like bodies project irregularly into the walls for as much as 1640 ft (500 m). The hanging wall is mainly red and green shale of the Lower Cambrian Rome Formation, though including some thin-bedded dolo-

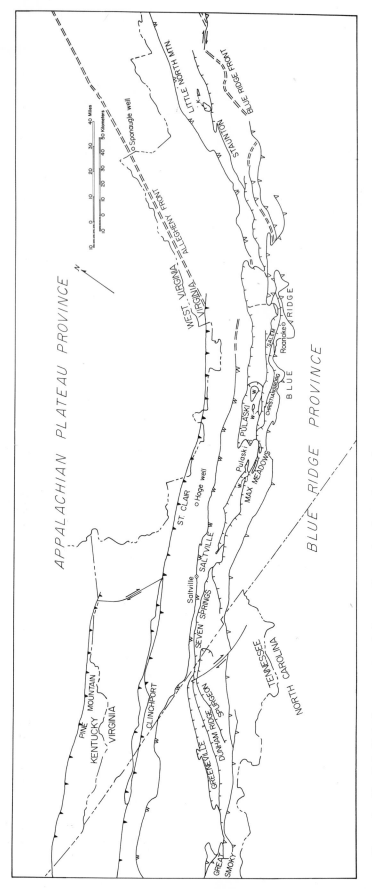

FIGURE 1. The Pulaski fault system, in relation to other major thrust faults of the Valley and Ridge province, Virginia and Tennessee.

Ticks and triangles are along overthrust side of faults; ticks indicate faults of Pulaski fault system; single open triangles, faults representing the Blue Ridge structural front, along border of Blue Ridge province; double open triangles, faults representing the mid-province structural front; solid triangles, faults representing the Allegheny structural front, along border of Appalachian Plateau province; double dashed lines, structural fronts unbroken by major faults. Names of faults are placed close to type localities. (W) window in a fault of the Pulaski system; (K) klippe of a fault of the Pulaski system.

stone and limestone; the footwall (which is the hanging wall of the basal Pulaski fault) is mainly thin-bedded dolostone of the Middle Cambrian Elbrook Formation, though including some gray shaly beds. Those parts of the breccia nearest the walls consist of fragments of all sizes of the adjacent wall-rock in a powdery gouge of similar material; the shale fragments are notably contorted, macerated, slickensided, and even phyllitized, and red color has been destroyed. Away from the walls, the maximum fragment size drops off to less than 6 in. (15 cm), the materials become more and more mixed, and the fragments more and more rounded; a few blocks not assignable to either wall are present, perhaps derived from older formations down the dip of the fault. The matrix is largely powdery gouge in fresh cuts but quickly hardens on exposure. Vein material (calcite and quartz) occurs but is minor. Cooper (1959) points out that the breccia is overlain by a distinctive soil type.

The Max Meadows breccia is particularly widespread in the Pulaski region, but it can be found in many places along the entire length of the Pulaski fault system. Yet no such breccia is known associated with any of the other major thrust faults in the central and southern Appalachians, although many of them are at least as long (for example, the Saltville fault) or have equal or even greater displacements (for example, the Great Smoky and other faults along the northwest side of the Blue Ridge province). The breccia clearly records some unique feature of the Pulaski faulting; its unusual character and especially its quasi-intrusive relations indicate that somehow it was plastic or fluidized during deformation.

To me, the appearance of much of the Max Meadows breccia is very reminiscent of the salt-bearing breccias—"Haselgebirge"—mined in the Austrian Alps, especially of surface outcrops of those breccias from which the salt has been leached. "Haselgebirge" forms where probably Permian salt-bearing strata have been involved in complex thrust-faulting or nappe formation (Schauberger, 1955; Mayrhofer, 1955—Schauberger has demonstrated a stratigraphy in the Austrian Haselgebirge and infers from it that the breccia is sedimentary, but Mayrhofer shows that a tectonic origin is more likely). On the other hand, some varieties of the Max Meadows breccia, especially where weathered, resemble the "Rauhwacke" or "cargneule" of the Alps, a rock formed by the deep weathering of mixtures of dolostone with anhydrite or gypsum, whether interbedded or intermixed tectonically (Brückner, 1941).

The presence of such rocks along a fault between Middle and Lower Cambrian strata suggests the possibility of their derivation from a tectonically deformed Cambrian evaporite sequence. Although evaporites are widespread in the Upper Silurian rocks of the central Appalachians and also occur in the Mississippian near Saltville, Virginia, 55 mi (90 km) west–southwest of Pulaski and elsewhere in southwest Virginia and vicinity, it is only recently that their presence has been recognized in Cambrian strata in the Appalachians, and even now the records are scanty. In the Sponaugle well (Perry, 1964), drilled in 1960 to a depth of 13,001 ft (about 4,000 m) in Pendleton County, West Virginia, 124 mi (200 km) northeast of Pulaski, traces of anhydrite were encountered in dolostone assigned to the Middle Cambrian Elbrook Formation, not far above a major thrust fault bringing the dolostone over Middle Ordovician limestone. Anhydrite was also encountered in the Hoge well, drilled in 1949 on the Burkes Garden dome in Tazewell County, Virginia, 30 mi (50 km) west of Pulaski (Cooper, 1961, p. 109–110), interbedded with dolostone through about 300 ft (100 m) of beds either beneath or in the lower part of the Lower Cambrian Rome Formation. Cooper considers the anhydrite-bearing rocks in this well very unusual for the Rome, but others who logged the well (see Virginia Minerals, v. 11, p. 41, November, 1965) considered them merely as more dolostone-bearing than normal (the Rome contains an especially high proportion of dolostone in parts of the southeastern belts of the Valley and Ridge province in East Tennessee and southwest Virginia). Beneath the anhydrite-bearing strata lie more than a hundred meters of cherty dolostone, some oölitic, classed as pre-Rome by Cooper but as part of the Upper Cambrian Copper Ridge Formation by James Griffith (personal communication, 1963) on the basis of the same well samples; if the latter assignment is correct, here again the evaporite-bearing strata overlie a thrust fault.

According to data given by Withington (1965), all but one of the 17 springs in southwest Virginia that show more than 11 ppm of sulfate, including two that have 1,000 ppm or better, occur in the Pulaski block where either the Rome or the Elbrook Formation is at the surface. Withington (1965, p. B32) infers indeed that the sulfate is coming up from the Mississippian Maccrady Formation beneath the Pulaski fault, but many of the springs, including the two strongest, lie either near parts of the fault where the underlying rocks are Devonian or far back from the fault trace where the fault is presumably fairly deep. A sulfate source within the upper plate of the Pulaski fault seems more probable.

That the Rome Formation represents a regressive phase in the thick marine Cambro-Ordovician carbonate section of the Valley and Ridge province was pointed out years ago by Grabau (1936, p. 13); in and near it is a logical place for evaporites. Accordingly I suggest that during this regression both salt and gypsum were deposited in a large evaporite basin along the eastern side of the great bank on which the Cambro-Ordovician carbonate sediments were being deposited, and that during later deformation the presence of salt localized a major low-angle thrust system, the Pulaski fault, whose position and extent thus record where there was salt in the original evaporite basin. Gypsum or anhydrite beds extended farther north and west and, along with impervious shale layers, helped to provide the major décollement zone that I believe controlled the structure of the Valley and Ridge province (for this debate, cf. Rodgers, 1964, with Cooper, 1964, and see the references by each there cited). Where the salt was thick, flowage of the salt-bearing layers produced the Max Meadows breccia, from the presently outcropping part of which the evaporites have been removed by solution during the present erosion cycle.

ACKNOWLEDGMENTS

Anyone dealing with the Pulaski fault is first of all indebted to Prof. Byron N. Cooper for his detailed work on its many facets; I am further grateful to him for demonstrating the Max Meadows breccia to me on field trips and arguing with me about its origin. Several geologists have supplied me with unpublished information on drill holes in the Appalachians. A paper by Jesús Nájera, prepared for my two-week course at the University of Texas in March 1966, helped me to bring the problem into focus. Finally, I thank my friends who went over the manuscript for me—Byron Cooper, James Griffith, Charles Withington; they are not, of course, to be held responsible for the "outrageous hypotheses" it presents.

REFERENCES

Brückner, W., 1941, Über die Entstehung der Rauhwacken und Zellendolomite: Eclogae geol. Helvetiae, v. 34, p. 117–134.

Campbell, M. R., and others, 1925, The Valley coal fields of Virginia: Virginia Geol. Survey Bull. 25, 322 p.

Cooper, B. N., 1938, Duality of the Pulaski fault in the type locality (abst.): Geol. Soc. America Proc., 1937, p. 74.

———, 1939, Geology of the Draper Mountain area, Virginia: Virginia Geol. Survey Bull. 55, 98 p.

———, 1946, Metamorphism along the "Pulaski" fault in the Appalachian Valley of Virginia: Am. Jour. Sci., v. 244, p. 95-104.

———, 1959, Max Meadows formation: (Virginia Polytech. Inst.) Mineral Industries Jour., v. 6, no. 4, p. 6, 8.

———, 1960, The geology of the region between Roanoke and Winchester in the Appalachian Valley of western Virginia: Johns Hopkins Univ. Studies in Geology: no. 18, Gdbk. 2 (Am. Assoc. Petroleum Geologists Ann. Mtg., Atlantic City 1960), 84 p.

———, 1961, Grand Appalachian field excursion: Virginia Polytech. Inst., Eng. Ext. Ser., Geol. Gdbk. 1 (Geol. Soc. America Ann. Mtg., Cincinnati 1961, Gdbk. 1) 187 p.

———, 1964, Relation of stratigraphy to structure in the southern Appalachians: Virginia Polytech. Inst., Dept. Geol. Sci., Mem. 1, p. 81-114.

———, and Haff, J. C., 1940, Max Meadows fault breccia: Jour. Geology, v. 48, p. 945-974.

Grabau, A. W., 1936, Paleozoic formations in the light of the pulsation theory, Vol. I. Lower and Middle Cambrian pulsations (2nd ed.): Univ. Press, Nat. Univ. Peking, 680 p.

Mayrhofer, H., 1955, Beiträge zur Kenntnis des alpinen Salzgebirges: Deutsche geol. Gesell. Zeitschr., v. 105, p. 752-775.

Perry, W. J., Jr., 1964, Geology of Ray Sponaugle well, Pendleton County, West Virginia: Am. Assoc. Petroleum Geologists Bull., v. 48, p. 659-669.

Rodgers, J., 1964, Basement and no-basement hypotheses in the Jura and the Appalachian Valley and Ridge: Virginia Polytech. Inst., Dept. Geol. Sci., Mem. 1, p. 71-80.

Schauberger, O., 1955, Zur Genese des alpinen Haselgebirges: Deutsche geol. Gesell. Zeitschr., v. 105, p. 736-751.

Withington, C. F., 1965, Suggestions for prospecting for evaporite deposits in southwestern Virginia: U. S. Geol. Survey Prof. Paper 525B, p. B29-B33.

Chapter 12

The Max Meadows Breccias: A Reply

BYRON N. COOPER

INTRODUCTION

RODGERS (1949, 1950, 1953a, 1953b, 1963, 1964) is a strong proponent of the décollement hypothesis for the origin of Appalachian structures and coined the name "thin-skinned" as a colloquial term referring to the supposed superficiality of fold and thrust structures in the Folded Appalachians. It was expectable, therefore, that Rodgers would consider the possible presence of evaporitic sequences to have localized horizontal or bedding-plane-oriented lateral displacements along one or more great sole faults below which the Appalachian Paleozoic beds are supposedly undeformed. It is only natural to consider this possibility, because the décollement or Abscherung hypothesis was proposed by Buxtorf (1916) to have functioned in the Jura because of the presence of Mid-Triassic evaporites. The germ of Rodger's thesis is that the Max Meadows breccias resemble the Austrian Haselgebirge. The purpose of the writer's reply is to indicate why such a hypothesis, despite any superficial similarities of the two breccias, is unlikely to have been a significant factor in the operating mechanism forming the Max Meadows breccias.

GENERAL NATURE AND EXTENT OF MAX MEADOWS BRECCIAS

As shown on Figure 1, breccias of so-called Max Meadows type occur along the Pulaski–Staunton fault from U. S. Route 33 east of Harrisonburg, Rockingham County, Virginia, southwestward to and beyond the Virginia–Tennessee line—over a strike distance of nearly 275 mi (442 km); and thence on southwestward for an unknown distance, possibly all the way to the disappearance of the Pulaski thrust under other thrust sheets—75 mi (121 km) southwest of the Virginia–Tennessee line. The entire length of the exposed trace of the fault is, therefore, about 350 mi (563 km).

The Pulaski fault as mapped by Butts (1933), is really two faults (Fig. 1). In Smyth County, the southwest end of the Pulaski fault is the Hungry Mother fault of Cooper (1936), and the Pulaski fault which picks up as an en echelon off-set on the southeast side of the Marion dome is the fault that extends into Tennessee. This fault was called the Seven Springs thrust by Cooper (1936). William W. Cashion, a graduate student at Virginia Polytechnic Institute, and the writer will shortly publish a revision of the Pulaski fault to show the relation between the Hungry Mother fault and the Seven Springs fault as previously mapped by the writer.

The breccias were first noticed and illustrated by Campbell and Holden (1925, p. 17–19, Pl. 8A) who ascribed their occurrence to formation of pebbly calcareous tufas at places along New River gorge where upland streams flowing over limestone–dolomite terranes transported limestone–dolomite clasts and dumped them into New River gorge. Butts (1940, Pl. 60) described the Max Meadows breccias as "cave fillings?."

The writer (Cooper, 1939; Cooper and Haff, 1940) encountered these breccias in 1936 while studying the geology of the Draper Mountain area of Pulaski and Wythe counties, Virginia, and he worked out the tectonic relations of three main facies of the breccias.

In 1944, while working with R. S. Edmundson, the writer became aware of the occurrence of breccias of Max Meadows type along the so-called Staunton fault. Butts (1933, map) had shown the

Staunton fault as an unrelated thrust whose trace overlapped the northeastern end of the Pulaski fault as mapped by him, which was delineated as ending near Greenville, Augusta County, Virginia. The Staunton thrust, as shown by Butts (1933, map), was mapped as extending northeastward to the vicinity of Endless Caverns, Rockingham County. Reconnaissance mapping by Edmundson and the writer (Cooper, 1946; Edmundson, 1945) showed that the Pulaski thrust was continuous with the Staunton fault and that they constituted one and the same thrust zone.

The subsequent discovery of occurrences of the breccias down to and beyond the Tennessee line was made by the writer between 1940 to 1953, but he made no attempt then or subsequently to establish the most southerly extent of the breccias, which is still unrecorded. He has alluded to the Virginia occurrences of Max Meadows breccias in subsequent papers (Cooper, 1961, 1963, 1964, 1967).

During the past 21 years, the writer has covered and recovered much of the areal geology of Pulaski and Montgomery counties and during that time has gained a fuller appreciation of the nature of occurrence of the breccia. Probably the best studied area of its occurrence is Montgomery County, Virginia, where it has been extensively mapped by the writer and many of his students. H. C. Porter, Virginia Polytechnic Institute agronomist, has mapped the soils of Montgomery County, and he (personal communication) concluded that approximately 30 percent of the limestone–dolomite terranes of the County have a special alkaline soil type named by him "Pedlar Soil," which is formed from Max Meadows breccias.

In Wythe and Pulaski counties, where the writer has also studied the breccias intensively, a threefold zonation of the breccia is well developed along the Max Meadows and Pulaski thrusts— which are for general purposes of description two faults in the same thrust zone. Characteristic zonation of the Max Meadows breccia was described in detail and illustrated by Cooper and Haff (1940).

The normal disposition of these zones is as shown in Figure 2. In the typical condition, as conceived in the initial studies (Cooper, 1939; Cooper and Haff, 1940), these three distinct, but transitional, zones commonly occur between the Rome and Elbrook Formations, or two of the three zones may occur where the Rome forms the bottom of the Pulaski thrust mass or where the Elbrook Formation overrides younger formations. Since that time a tremendous amount of breccia

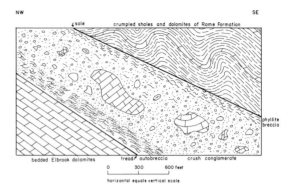

FIGURE 2.   Threefold zonation of Max Meadows breccias along "Pulaski" fault, where Rome beds have been thrust over Elbrook dolomites.

and crush conglomerate of Max Meadows type has been found to occur as irregular, ramifying masses widely distributed virtually over the entire outcrop width of the Elbrook Formation in certain belts. Careful study has shown that the breccias are not all the same, and, therefore, some brief additional description is necessary. These materials should be called "tectonic breccias" rather than "fault breccias," because much of the cataclastic material occurs away from faults.

The hanging wall zone of the Max Meadows Breccia is a breccia of special type, composed of disarranged slivers of green phyllite set in a matrix of secondary recrystalline dolomite or, where weathered, is composed of a loose aggregate of green phyllite slivers with rusty powdery interstices (Fig. 2). This breccia zone, composed almost wholly of sheared, crumpled fragments of Rome Shale, is made by the dismemberment and cataclastic rolling out of originally red shales along the fault during which process the hematite–red illite shales were mildly metamorphosed to green chlorite phyllites. The formation of chlorite indicates that this zone is the product of low-grade dynamic metamorphism that attended thrusting of the Pulaski–Max Meadows blocks. The relation of the red shale to the green phyllite is perhaps best shown along the N & W Railway in the bluffs just north of U.S. Route 11 where that highway comes very close to the river just beyond the east environs of Radford, Virginia. The completely decalcified green macerated phyllite breccia forms a lightweight soft powdery rock that can be broken in the hands. In the bluffs along the railroad just east of Radford there are large and very irregular vestigial masses of red Rome shale which fade out into green phyllite in all directions. Macerated green phyllite breccia never occurs except where the Rome Formation forms the base of the thrust mass and where this condition prevails it commonly constitutes a substantial part of the fault zone.

Where the Rome (right-side-up) lies in fault contact with the subjacent Elbrook, the Elbrook is intimately fractured and gash-veined (Fig. 3) and in places is a recrystalline autobreccia developed in thick-bedded black to blue-gray dolomites. Where the upper beds of the Elbrook are thin-bedded shaly dolomites, the Elbrook forming the tread of the overridden block is a curl or sliver breccia in which only incipient differential rotation of the cataclastic fragments has taken place.

These Elbrook autobreccias also characterize the top of the brecciated zone where the well-bedded Elbrook dolomites have been thrust over other younger formations ranging in age from the Upper Cambrian Conococheague Formation to Chesterian red shales of the Stroubles Formation. But neither Elbrook autobreccias nor macerated Rome phyllite breccias constitute the abundant and predominant rock-type characteristic of the Max Meadows tectonic breccias, which is crush conglomerate made up of cataclastically broken shaly dolomites rolled out into an indurated pebbly mass. All stages of crushing from intimate autobrecciation to rock flour are evident, but characteristic breakup of the bedded autobreccias was greatly promoted by prevalence of thin shaly dolomites which are not found in any other Appalachian formations. Thicknesses of crush conglomerate range widely; several hundred to a few thousand feet (60–1,000 m) are not uncommon between Roanoke and the Tennessee State Line.

Considerable breccia and crush conglomerate have been brought up along high-angle faults that cut the buried Pulaski–Max Meadows thrusts (Fig. 4), so that these peculiar rocks occur also along the branches of the Salem fault which cuts the Pulaski block and forms the fault traces

FIGURE 3. Abundant carbonate vein fillings in autobrecciated Elbrook, 150 ft south of gaging station on Middle Fork of Holston River, Seven Mile Ford, Smyth County, Virginia.

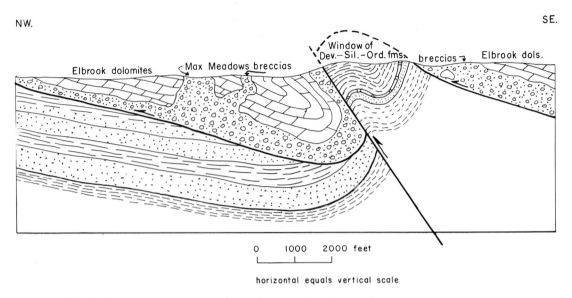

NW.   SE.

FIGURE 4. Max Meadows breccias and crush conglomerates brought up along a branch of the Salem Fault on north side of Berringer Mountain window, 3 mi west of Christiansburg, Virginia.

on the northwest sides of the Berringer, Ingles, East Radford, and Christiansburg windows (Cooper, 1963). This redistribution accounts for a substantial part of the excessive occurrence of Max Meadows breccia but it does not account for all the excess. The writer is convinced from re-study of many Montgomery County exposures of breccia that a great deal of the occurrences that he originally interpreted as intrusions of breccia were locally generated by interstratal shearing that crumpled, cracked, and rolled out the thin platy Elbrook dolomites very effectively.

In some places, large vestigial masses of carbonate rock (mostly blue-gray limestone) engulfed in the great thicknesses of crush conglomerate are exposed. Some of these irregular blocks show chaotic orientation of bedding so that they too must have been detached and somewhat crushed and rolled out, though not comparably reduced in size to the vast majority of cataclasts which range from 8 cm down to 0.10 mm. Limestone clasts up to 15 m have been measured in the crush conglomerate and in a number of places especially in road cuts, the incipient dismemberment of these larger clasts can be seen in various stages of separation. The finest place to view this characteristic phenomenon is in the cuts at the west approach to Montgomery Tunnel on the N & W Railway about 5 mi (8 km) east of Christiansburg. One very large fractured but essentially monolithic block shrouded in partially weathered fault breccia is exposed in the rock cut on the westbound lane of U. S. Route 11 about 1.5 mi west of Shawsville (Cooper, 1967).

Along the Den Creek Road (County Road 641) between the Montgomery underpass and the Salem fault about 2 mi (3.2 km) northwest of Montgomery underpass, the Elbrook beds including several blue-gray algal-matte magnesian limestones are very well exposed. Tectonic autobreccias in zones 5–100 ft (1.5–31 m) thick and zones of carbonate flour and crush conglomerate

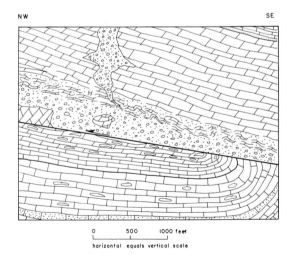

FIGURE 5. Elbrook bedded dolomites and associated Max Meadows breccias developed where the Elbrook has been thrust over younger Cambro-Ordovician dolomites.

occur in a 1,650 ft (503 m) thick exposure of Elbrook between the fault and the underpass. The writer counted 68 different zones of the latter sandwiched within the bedded dolomites occurring between the Montgomery underpass and the Salem thrust.

The great body of the Elbrook Formation in many places particularly in Montgomery County seems to have been dismembered in a chaotic manner and on a grand scale so that some of the breccias generated along the thrust near the base of the overriding Elbrook worked their way up along widening joints between enormous blocks of Elbrook to form dikes and "intrusions" of crush conglomerate (Cooper, 1939; this paper, Figs. 4 and 5). The relations of bedded Elbrook dolomite, commonly intensely brecciated, were formerly well shown along a lane leading south from Horseshoe Bend of Reed Creek and U. S. Route 11 (Cooper and Haff, 1940) but the exposures have been considerably altered since that time by quarrying and by construction of Interstate Route 81. However, the newer quarry exposures add certain features not seen in the 1930's. For example, there are large dismembered blocks of macerated Rome shale—portions of it still with vestiges of the original, characteristically red color—which are engulfed in crush conglomerate.

The irregular distribution of breccia and crush conglomerate is undoubtedly also the result of differential bedding–surface slippage or décollement in very shaly, thin-bedded dolomites where they overrode thicker-bedded limestones or black dolomites. Once the movement was initiated, the disharmonically folded and crumpled shaly rocks cracked and became partially disrupted into a disorganized mass of particles with a wide range in size (Fig. 6). Thus some brecciation in the Elbrook was generated neither along the Pulaski thrust nor along its related neighbor, the Max Meadows thrust, but within the Elbrook Formation as it overrode younger beds (Fig. 5).

The full extent and character of the brecciation in the Elbrook has never been adequately described in previous literature. The Elbrook, within the area of the Pulaski thrust contains zones of massive very fine-grained dolomite, some of which is very light gray and some of which is black and pyritic. These dolomites if closely examined megascopically are very intimately crisscrossed with minute hairline veinlets of dolomite and crushed aggregates of such dolomite consist of angular subcubical fragments that break along these myriad veinlets (Fig. 7). Unlike other Appalachian dolomites, those in the

Elbrook and equivalent Honaker Dolomite break down into fractured rubble.

In Virginia, there are hundreds of exposures that exhibit such dolomites, and it is a fact that the Virginia Department of Highways derives significant quantities of spalls just from the rapid disintegration of these autobrecciated rocks at the base of cuts in the dolomite.

In 1942 to 1945 the writer, in association with R. S. Edmundson, covered all the limestone and dolomite belts west of the Blue Ridge, from Winchester to Cumberland Gap and Bristol. Both of us noted many times that the Honaker dolomites or Honaker-like beds in belts west of Walker and Clinch Mountains invariably were intimately brecciated.

Perhaps the most spectacular development of intimate autobrecciation in Virginia is revealed in an agricultural limestone quarry (Mundy Quarry) worked first in World War II near the hanging wall of the North Mountain fault north of Cootes Store, in Rockingham County (Edmundson, 1945, p. 93, Pl. 9). In this quarry operated then by the late Norman Hottinger, the rock was quarried largely without use of explosives. The heel of the shovel was simply raked up and down over the exposed face to bring down tons of brecciated rock that was screened into fractions used for various purposes.

The same type of intimately fractured beds occurs along U. S. Route 11 just at the base of Christiansburg Mountain and on the Den Creek Road from south of Ellett to Montgomery. Beginning about 1.3 mi (2.1 km) northeast of Seven Mile Ford, Smyth County, along U. S. Routes 11 and 81 and continuing along the banks of the relocated Middle Fork of Holston River in the same county, the bedded Elbrook and interrelated Max Meadows breccia zones are well exposed.

Identical types of dolomite are exposed in the Honaker Formation between Honaker and Swords Creek in Russell County, where they occur near but not actually along the St. Clair thrust, along State Highway 80 and the Richlands–Wardell Road.

Intimate brecciation of the most massive dolomites in the Elbrook (and equivalent Honaker) is, therefore, the hallmark of these formations and in these two equivalent units the granulated, autobrecciated quality is the most reliable means for their ready identification. None of the other Appalachian dolomites are so brecciated. Much of these breccias is so finely comminuted that it is little more than a loose mass of mealy consistency.

FIGURE 7. Finely autobrecciated dolomites in Elbrook Formation, filled with vein carbonate in patterns typical of Honaker and massive Elbrook dolomite zones. Locality same as Figure 3.

FIGURE 6. Autobrecciation in Elbrook Dolomite along old Route 100 northwest of Stone House Dock on Claytor Lake, Pulaski County, Virginia.

However, it is not these granulated dolomites that led to generation of enormous volumes of crush conglomerate. It was, rather, the presence of thick zones of distinctly shaly dolomite, which abound variously near the base and in the middle of the Elbrook that apparently controlled crushing and rolling out with the position of the zones varying from place to place. The thin-bedded dolomites, some of which are millimeter laminated, are the source of the subangular pebbles in the typical crush conglomerate. During transposition of the Pulaski thrust mass, the Elbrook rode out over the younger beds contributing to the considerable breakup of the body of the formation. Bedding plane slippage was rendered easy and these conditions doubtless contributed to the development of the breccia. In literally hundreds of exposures of Elbrook and related Max Meadows breccias, the basic controlling factor in the origin of the breccias was the unique, shaly dolomite that lent itself to disharmonic crumpling, crushing, and rolling out. Another factor is the predominant subparallelism of the bedded Elbrook with the associated thrust surfaces. In most instances where fresh breccia or crush conglomerate is examined, the matrix is either recrystalline carbonate or a paste of finely milled dolomite microclasts (Fig. 8). The matrix and coarser fragments make a crush conglomerate commonly so compact that it is hard to break with a hammer and which is so firmly cemented that it breaks across the pebbles rather than around them. It is significant that there is absolutely no sign of brecciation in the southeastern belt of Elbrook in the Holston River syncline south of Seven Mile Ford, Smyth County, where the section was carefully studied and measured, but in the gently dipping northwest limb that extends to the Pulaski (Seven Springs) thrust, three major brecciation zones—two of which are hundreds of feet thick—are separated by bodies of bedded Elbrook in which limestone beds occur with well-behaved massive dolomites, both of which dip 25–35 degrees southeast (Fig. 10). The breccias along the banks of the Middle Fork of Holston River just downstream from the Seven Mile Ford Gaging Station are composed of especially coarse chunky to slabby cataclasts ranging up to 30 cm long (Fig. 9).

It is also significant that the crush conglomerates and breccias so profusely developed in the northwest limb of the Holston River syncline south of the Pulaski Fault at Seven Mile Ford are composed entirely of Elbrook clasts with no sign of green or red Rome shale clasts such as characterize the breccia and conglomerate in the

Pulaski–Montgomery counties. Because the Rome is not involved, the cataclastic material has a slightly different color and appearance, but the cataclastic textures are virtually the same as in other areas (Fig. 10).

The generation of cataclastic rock by interstratal shearing, crumpling, autobrecciation, crushing and rolling out along the directions essentially parallel to general bedding in the Elbrook is undoubtedly a major, though certainly not the only, cause for so much cataclastic rock within the Elbrook Formation. In Montgomery County, as much as 60 percent of the area underlain by Elbrook Formation is Max Meadows breccia occurring as multiple zones, "intrusions," and "islands," of crush conglomerates, crackle breccias, and autobrecciated rock.

The distinguishing feature of most of these intra-Elbrook cataclastics is that all the clasts are from the Elbrook itself. Crush conglomerates formed from only Elbrook beds lack the telltale greenish gray color of Pulaski–thrust-generated breccia.

Every exposure of the cataclastic deposits gives abundant evidence of their origin by *abrasion of particle upon particle* and by internal microbrecciation which leads to disruption of large dolomite clasts. The obvious characteristics of the Elbrook, some of whose dolomites were commonly microbrecciated to flour and others that are shaly and that were crumpled, autobrecciated, and finally rolled out with some cataclastic rounding to produce subangular clasts (Fig. 7), all indicate generation of breccias indirectly or directly by faulting.

It should also be mentioned that near the base of the actual Pulaski–thrust-generated breccia—that is, near the tread or footwall surface of the overridden block—the cataclastic rock invariably has incorporated significant quantities of the tread rocks which have been "dozed off" and worked into the breccia mass from place to place. These clasts were derived from rocks as young as the late Mississippian Stroubles Formation. For example, on Road 610, Botetourt County, about 1.6 mi (2.6 km) north of Rocky Point, fault breccia largely of macerated Rome phyllite material is loaded with cataclasts of the overridden Middle Ordovician Liberty Hall black shales and thin-bedded limestones. On the Radford Arsenal grounds along the road leading from the old Rad-Tech Barracks to Price Forks, horses of fossiliferous Martinsburg Shale are worked into the crush conglomerate exposed just up grade from the rail spur crossing southwest of the N & W underpass. In the same area, breccias back of

FIGURE 8. Typical crush conglomerate in Max Meadows Formation. Same locality as Figure 3.

FIGURE 9. Coarse phase of crush conglomerate in Max Meadows Formation. Same locality as Figure 3.

the U. S. Defense Department's Computing Center are loaded with red siltstone clasts out of the Mississippian Stroubles Formation (Cooper, 1961).

Along Interstate Route 81 in the east environs of Marion, Smyth County, Virginia, the Elbrook brecciated dolomites topped by macerated Rome shales have overridden the lower part of the Rich Valley Formation, a Middle Ordovician black shaly limestone, and the normal green-gray color of the breccia is obliterated by the local abundance of black limestone clasts.

On the north side of the Kent window in Wythe County, the wide frame of crush conglomerate contains clasts of Mosheim-type limestone, and oölitic chert and sandy dolomite—both suggestive of the Conococheague Formation which the Pulaski thrust mass overrode just to the south of the breccia zone. If the overriding thrust mass had been lubricated by a "well greased" décollement zone containing evaporites, the wholesale dismemberment of such great thicknesses of rock, especially of rocks forming the tread of the overridden block, would hardly be expectable.

It is more than coincidence that wherever the Pulaski fault zone has neither Rome nor Elbrook at the base of the thrust block, the brecciation practically disappears and all the characteristics of the Max Meadows breccia—macerated phyllite breccia, crush conglomerate, and gradational dolomite autobreccia—are lacking. None of the other Appalachian dolomites are so amenable to brecciation and rolling out as the Elbrook. The St. Clair and

Pulaski thrusts illustrate a typical contrast worthy of mention. Max Meadows breccias and crush conglomerates containing mainly Rome and Elbrook clasts but also fragments of virtually all of the formations overridden in the general vicinity of the Price Mountain window near Blacksburg occur along the Pulaski thrust in a wide, thick zone which drilling on the Radford Arsenal grounds has shown to be over 2,500 ft (763 m) thick. The breccia is commonly so thick that the overridden footwall rocks are relatively remote from the lowest well-bedded rocks of the overriding Pulaski thrust block. On the other hand, in the best-known exposure of the St. Clair fault in the Narrows of New River near Rich Creek, Giles County, Virginia (Cooper, 1961, Pl. 46a) the overthrust dolomites rest upon the overridden Devonian–Mississippian shales and sandstones of the Parrott and Price formations with absolutely no trace of hanging wall brecciation.

Although the writer's original interpretation that all the breccias were generated along the Pulaski–Max Meadows thrust has been somewhat revised after thirty years of continuing study of the Max Meadows breccias, he has never seen any evidence whatever favoring spatial redistribution of these most unusual of all Appalachian rocks by salt or gypsum flowage that encouraged dismemberment of the overthrust carbonate rocks.

Since 1946, the writer has examined in detail many hundreds of feet of core from the Rome Shale and has never seen a trace of salt or gypsum in it, even at depths down to 1,000 ft (305

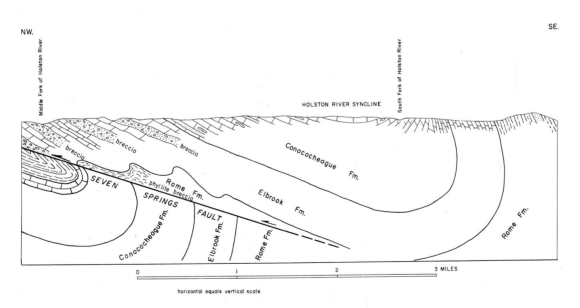

FIGURE 10. Distribution of Max Meadows breccias in the Elbrook Formation in the Holston River syncline, south of Seven Mile Ford, Smyth County, Va.

m) below the surface. Wells and core holes sunk in the Elbrook in the Blacksburg area have yielded fresh brecciated Elbrook rock but none has ever disclosed any sign of salt or gypsum that could possibly account for formation of the Max Meadows breccias at depth where they occur beyond reach of percolating waters. In so many places the breccia is quite as hard, compact, and impervious as any Appalachian carbonate rock (Fig. 8). In many places the breccia is not leached or oxidized to any significant depth and is characteristically fresh, pyritic, and green-gray in color. Only a few months of exposure is needed to change the color from normal faint olive to green-gray color of unleached, unoxidized breccia to rusty brown which characterizes the oxidized rock.

If, as Rodgers has suggested, the evaporites are now largely leached out of the Max Meadows breccias, the breccias should be loose and full of voids where the matrix of salt and/or gypsum, purportedly, once existed. Furthermore, any significant movement of circulating water through the breccias would have readily oxidized them to yellow-brown. Core data on the Elbrook and Max Meadows breccias indicate leaching action is entirely superficial and that it could not have removed any salt or gypsum if originally present as a paste or matrix in the breccia.

The Merrimac Tunnel under U. S. Route 460 between Blacksburg and Christiansburg offers a good place to test the hypothesis outlined by Withington (1965) and supported by Rodgers. The west portal of this tunnel is overridden rocks of the Stroubles Formation—the youngest sedimentary rock in the local area. The east portal is in fault breccia and autobrecciated Elbrook dolomites. Inside the tunnel where the dolomites and breccias are in contact with the impermeable Stroubles red shales and sandstones there is a strong egress of water which is ditched to issue out the east portal. A sample of this water tested in 1948 showed over 300 ppm of bicarbonate hardness, but no chlorides and only about 15 ppm of sulfates. Fresh unoxidized, unleached breccias are present above the tunnel so that evaporite matrix if present should show up as sulfate- and chloride-rich spring water in the tunnel, but the chlorides and sulfates do not occur.

The Max Meadows breccias are, except where weathered and leached, tight cataclastic rocks which are in many places so indurated that they make bold cliffs. Where they are hard, compact and dense, the fresh breccia or crush conglomerate has negligible porosity and permeability

and shows no sign of containing or having contained salt or gypsum or anhydrite. The exposure pictured by Campbell and Holden (1925, Pl. 8A) is superficially leached but has been subsequently case-hardened to produce an essentially monolithic mass.

The writer (Cooper, 1966, p. 11–34) has described true salt breccias occurring in the Maccrady Formation in the Saltville district of Smyth and Washington counties, Virginia. These breccias consist of angular clasts composed of red and green shale, red siltstone, or gray anhydrite and gray dolomite. The salt constitutes probably 75 percent of the volume of the breccias and is itself stained pink by contaminating red iron oxide derived from the Maccrady shales. Detailed study of cores shows that the salt breccias were formed by mobilization of predominating salt zones in sequences containing intercalated red shales and sandstones, dolomites, and anhydrite. The latter were disrupted by the mobilized salt into angular pieces up to possibly a foot or more in diameter but none of the tectonically disrupted material has been subjected to the particle-on-particle cataclasis evident in the Max Meadows breccia. The obvious particle disruption shown by virtually every exposure of Max Meadows breccia is one that would not have been promoted by evaporite lubricants acting as a mobile matrix. In the hundreds of feet of salt breccia studied in cores from the Saltville–Maccrady beds, the writer has never seen anything to suggest any similarities with the Max Meadows breccias along the Pulaski fault.

However, Max Meadows-type breccias do occur locally in a few localities along the Saltville thrust—the next major fault northwest of the Pulaski thrust—one of which is in the first railroad cut east of Plasterco on the N & W spur to Saltville. In that exposure a few feet of crush conglomerate and autobreccia have been formed along the base of the Saltville thrust block which from North Holston southwestward to Plasterco is made by dolomites of the Honaker Formation. Thus, the general statement made by Rodgers concerning the lack of development of breccias anywhere except along the Pulaski thrust is not entirely correct.

THE CASE FOR EVAPORITES IN THE ROME AND ELBROOK FORMATIONS

Rodgers alludes to two main points of evidence as possible indications of once extensive beds of evaporites in either the Rome or Elbrook, which could have encouraged décollement in those formations and development of Haselgebirge-type salt breccias by flowage of evaporites that dismember shales and dolomites to form each conglomerate. Rodgers cites Withington's (1965) ideas on the occurrence of soluble sulfates in spring waters in the Pulaski–Montgomery County area where the Pulaski thrust is so well known. Withington intimated that the high sulfate content indicated the presence of evaporite deposits in the Mississippian red beds of the Pulaski–Blacksburg district, and did not at all suggest saline zones in the Elbrook or Rome. The writer (Cooper, 1966) has objected to Withington's suggestions and surmises about sulfate waters being indicative of gypsum beds and possibly other associated evaporites.

As every Appalachian field geologist well knows, there are limonite deposits banked along the Pulaski and other Appalachian thrusts almost wherever carbonate rocks have been thrust over black shales that are characteristically loaded with pyrite. The fractured overthrust dolomites and limestones overriding black, pyrite-bearing shales probably have piped oxygenating waters to contact with shales whose pyrite has been converted to sulfates and hydrous iron oxides and the sulfates moved to egress as spring waters. The millions of tons of brown iron ore occurring along Appalachian thrusts owe their emplacement largely to interaction of carbonate rocks and black shales—but none of this suggests that the gypsum was primary in either the Rome or Elbrook formations. Rodger's curious attraction to Withington's (1965, p. B29–B33) suggestions, which the writer does not believe are at all well founded, should not be regarded too seriously because even Withington was not arguing for or suggesting the presence of Rome or Elbrook evaporites or that such evaporites were involved in the mechanics of generation of tectonic breccias.

Secondly, Rodgers cites the occurrence of anhydrite encountered in the United Producing Company's No. 1 J. M. Hoge well drilled in Burkes Garden in 1948. The writer and one of his former graduate students, Carl A. Meyertons, discovered that some of the red beds below the Honaker Dolomite, occurring in a 260-ft (80-m) Rome interval beginning probably at about 5,125 ft (1,563 m) depth and continuing down to about 5,385 ft (1,642 m), were anhydrite. Some gray anhydrite also occurs in the cuttings, but this occurrence is by no means indicative of a widespread or thick evaporite sequence in the Rome and/or the Elbrook formations. Rodgers secondhandedly interpreted the black and light

gray dolomites below the Rome gray and black dolomites, red beds, and intercalated anhydrites as "Copper Ridge Dolomite."

Beds below the anhydrite-bearing succession identified by the writer and others as "Rome," beginning at about 5,385 ft (1,563 m) and continuing down to the bottom of the well hole (total depth 5,632 ft), contain dolomite lithologies identical to those found in the transition zone between the Rome and immediately underlying Shady Dolomite as developed in the Speedwell–Cripple Creek–Ivanhoe area of Wythe County, Virginia. In Smyth, Wythe, and Pulaski counties, the lowermost Rome is a dolomitic succession that contains several prominently cherty zones including one that yields large masses of chalcedonic chert ranging from white to black in color. The very cherty zone is an excellent marker zone. Slightly higher in the succession near the base of the Rome in Wythe County are black and gray laminated chert containing oölites and abundant dolomolds. Chert of identical character occurs in the lower 250 ft (76 m) of the Burkes Garden well. The lithologic succession in the lowermost 400 ft (122 m) of the Burkes Garden well is so similar to the Rome–Shady transition zone of Wythe County that a number of the outstanding similarities should be emphasized, including (1) identical colors and textures of dolomites, (2) similar oölitic chert with dolomolds, (3) drusy quartz associated with chalcedony, and (4) black chalcedonic chert.

Rodgers' willingness and readiness to call these rocks Copper Ridge (Upper Cambrian) is hard to understand, but he seems committed to the currently popular idea that there is a master sole fault under the Appalachian region, which he believes to occur in or below the Rome. Therefore, beds younger than Rome would be expected by him below that formation in the Burkes Garden well. However, cherts, oölites, and dolomolds are hardly grounds for identifying the beds in question as "Copper Ridge," which thereby signifies that there *is* a fault between the Rome and the bottom beds in the Burkes Garden well. Perhaps his preoccupation with the décollement hypothesis has led Rodgers to make premature judgments about the dolomites in the lower part of the Burkes Garden well. Formations should never be identified on the basis of structural assumptions, and, of course, such formation identifications cannot be fed back as supporting the existence of a postulated sole fault in or at the base of the Rome.

The California Company's No. 1 Strader test near Bane, Giles County, throws light upon the identity of the beds in the lower part of the Burkes Garden well. In the vicinity of the Strader residence near Bane, there are curious black sharpstone chalcedonic cherts in gray to black dolomites like those in the upper part of the transition zone between the Rome and underlying Shady Dolomite in Wythe County, Virginia. Laminated chert resembling that which occurs in the Wythe County transition zone is also present in the Bane area. The dolomites themselves show the same range in color and texture as those in the transition zones between the Rome shales and the Shady dolomites in Wythe County. Furthermore, the Strader well cut hundreds of feet of creamy gray dolomite that is lithologically identical to the so-called "saccharoidal dolomite" of the Shady Formation in Wythe County, Virginia. The occurrence of beds identical to these distinctive Shady dolomites in the Strader well and the occurrence of overlying darker cherty dolomites like those in the Rome–Shady transition zone lead to the logical conclusion that the beds penetrated by the Strader well are Shady dolomite which occurs in normal stratigraphic position below the transition beds which in turn are so strongly linked with a similar zone between the Rome and Shady in Wythe County, Virginia. Because the beds in the lower portion of the Hoge well in Burkes Garden resemble the transition zone that occurs between the Rome and Shady in more southeasterly belts in Wythe County, it is logical to conclude that the Burkes Garden well bottomed in the transition zone and would have—if deepened—cut normal Shady "saccharoidal dolomite" just as was found below the transition beds in the Bane well. Thus the weight of the available evidence indicates that no sole fault exists in or at the bottom of the Rome rocks encountered in the Burkes Garden well. Furthermore, the occurrences of anhydrite in the Rome below Burkes Garden have no established tie with an evaporite succession that has controlled the major fault structures under that anticline or in any other part of Virginia.

This lengthy explanation has been given to put the Burkes Garden Rome anhydrites in their proper setting. They are not indicative of a thick evaporite sequence even locally. The writer believes that the data presented are strong evidence that there is no décollement at the stratigraphic position of the Rome in the Bane and Burkes Garden anticlines. Anhydrite in the Rome or subjacent transition beds is not related to the Max Meadows breccias.

The Max Meadows breccias, so Rodgers believes, were leached of their original saline matrix subsequent to faulting, but this is a necessary hypothesis that he must employ because no salt beds are present now. Contrary to what Rodgers implies, the Max Meadows crush conglomerates and breccias are absolutely fresh, unoxidized, and unleached. One can be very positive about this generalization, because the fresh breccias abound in finely disseminated sulfides, particularly pyrite. If percolating water sufficient to remove a salt matrix from these breccias had ever moved through those pyritiferous rocks, wholesale oxidation would have resulted. The breccias and conglomerates are so compact that there simply was no significant space in which a salt matrix could have been located. The breccias needed no salt lubricant to emplace them in the various ways in which they are found.

CONCLUSION

Transitions from normal, bedded Elbrook to crumpled and well-brecciated rock and to comminuted material enclosing subrounded pebbles of dolomites that are themselves in various states of progressive dismemberment, finally to what would be called gouge or carbonate flour is evident in so many places that the pebble-on-pebble attrition process for generation of the crush conglomerates of the Max Meadows Formation must be accepted as indicative of the cataclastic origin of these unusual rocks.

The real reason for generation of so much breccia along the Pulaski thrust is the unique abundance of thin-bedded to shaly dolomites that are extremely incompetent—especially when their dip approximates the low angle of the major thrust that lies below them in the Pulaski block. The writer agrees with Rodgers that this profound brecciation is rather unique, but it seems far more logical and reasonable to attribute the breccia formation to so much thin-bedded dolomite which can be seen in all stages of disharmonic folding and cataclastic dismemberment. To attribute the breccias to salt zones that are not now there—even where the breccias are fresh and unweathered—appears to the writer to be "grabbing at straws" to support the décollement interpretation of regional Appalachian structure. The Haselgebirge hypothesis simply falls for want of any field or valid subsurface evidence. All Elbrook–Honaker dolomites have a strong tendency to microbrecciate regardless of what fault they may occur along or near the Virginia Appalachians. Where only medium-bedded dolomites occur, as in the Honaker Formation, plenty of finely brecciated and ground-up dolomite has been generated along or close to several thrusts, including the St. Clair and Saltville faults. However, thin-bedded dolomites do not occur in the Honaker but do occur farther southeast in equivalent Elbrook beds along the Pulaski and related faults. Haselgebirge-type salt breccias contain unmetamorphosed rock clasts, as exemplified by the salt-matrix tectonic breccias in the Maccrady Formation at Saltville, Virginia. The Max Meadows breccias contain clear-cut evidence of rock-on-rock cataclasis that metamorphosed red illite shales into chlorite phyllites. Deductions of a salt breccia to be meaningful require a tangible "corpus delecti" of evaporites—something Professor Rodgers is admittedly unable to supply in support of his "outrageous" hypothesis.

REFERENCES

Butts, C., 1933, Geologic map of the Appalachian Valley of Virginia with explanatory text: Va. Geol. Survey Bull. 42, map.

———, 1940, Geology of the Appalachian Valley in Virginia (Part 1): Va. Geol. Survey Bull. 52, 568 p.

Buxtorf, A., 1916, Prognosen und Befunde beim Hauensteinbasis- und Grenchenberg-tunnel und die Bedeutung der letzeren fur die Geologie des Juragebirges: Naturf. Ges. Basel, Verh., v. 27, p. 184-205.

Campbell, M. R., and Holden, R. J., 1925, Valley coal fields of Virginia: Va. Geol. Survey Bull. 25, p. 17-19, pl. 8A.

Cooper, B. N., 1936, Geology of the Marion area, Virginia: Va. Geol. Survey Bull. 46-L, p. 125–170.

———, 1939, Geology and mineral resources of the Draper Mountain area, Virginia: Va. Geol. Survey Bull. 55, 98 p.

———, 1946, Metamorphism along the "Pulaski" fault in the Appalachian Valley of Virginia: Am. Jour. Sci., v. 244, pp. 95-104.

———, 1960, The geology of the region between Roanoke and Winchester in the Appalachian Valley of Western Virginia: Johns Hopkins Univ., Studies in Geology, No. 18, 84 p.

———, 1961, Grand Appalachian Excursion: Va. Polytechnic Inst., Engr. Exper. Sta. Exten. Ser., Geol. Guidebook 1, p. 1-129, 134-169, 182-187.

———, 1963, Blacksburg synclinorium and Pulaski overthrust; in Geological Excursions in southwestern Virginia: Va. Polytechnic Inst. Engr. Exper. Sta., Exten. Ser., Geological Guidebook 2, 19-47.

———, 1964, Relation of stratigraphy to structure in the southern Appalachians: V.P.I. Dept. Geol. Sci. Mem. 1, p. 81-114.

———, 1966, Geology of the evaporite deposits of the Saltville district, Smyth and Washington counties, Va.: Northern Ohio Geol. Soc., 2nd Salt Symposium, v. 1, p. 11-34.

———, 1967, Profile of the folded Appalachians of western Virginia: [Univ. Missouri (Rolla), McNutt Colloquium Volume]. UMR Jour., 1, 1, p. 27–63.

———, and Haff, J. C., 1940, Max Meadows fault breccia: Jour. Geol., V. 48, p. 945-974.

Edmundson, R. S., 1945, Industrial Limestones and Dolomites in Virginia; Northern and central Shenandoah Valley district: Va. Geol. Survey Bull. 65, 195 p.

Rodgers, J., 1949, Evolution of thought on structure of the middle and southern Appalachians: Amer. Assn. Petrol. Geol. Bull., v. 33, p. 1643-1654.

———, 1950, Mechanics of folding as illustrated by the Sequatchie anticline, Tennessee and Alabama: Amer. Assn. Petrol. Geol. Bull., v. 34, p. 672-680.

———, 1953a, The folds and faults of the Appalachian Valley and Ridge province; in Southeastern Min. Symposium of 1950: Ky. Geol. Surv. Spec. Publ. No. 1, Sec. 9, p. 150-166.

———, 1953b, Geologic map of East Tennessee with explanatory text: Tenn. Div. Geol. Bull. 58, 158 p. maps.

———, 1963, Mechanics of foreland folding in Pennsylvania and West Virginia: Am. Assn. Petrol. Geol. Bull., v. 47, p. 1527-1536.

———, 1964, Basement and No-Basement hypotheses in the Jura and the Appalachian Valley and Ridge; in Tectonics of the Southern Appalachians: V.P.I. Dept. Geol. Sci. Mem. 1, p. 71-80.

Withington, C. F., 1965, Suggestions for prospecting for evaporite deposits in southwestern Virginia: U. S. Geol. Survey Prof. Paper 525-B, p. B29-B33.

THE BLUE RIDGE
AND THE READING PRONG

Chapter 13

Introduction*

JOHN C. REED, JR.

FROM SOUTHERN PENNSYLVANIA to northwestern Alabama the unmetamorphosed Paleozoic rocks of the Valley and Ridge province are flanked on the southeast by a continuous belt of variously metamorphosed Precambrian rocks known as the Blue Ridge belt. The northwestern edge of the Blue Ridge coincides approximately with a change in the style of deformation between penetratively deformed rocks on the southeast and rocks to the northwest that have been folded and faulted but that have not undergone penetrative deformation. This boundary has been termed the "tectonite front" (Fellows, 1943; Cloos, 1953, 1957). The Blue Ridge belt is more than 700 mi (1,100 km) long and ranges in width from less than 5 mi (8 km) in Maryland and Pennsylvania to as much as 70 mi (110 km) in North Carolina and Tennessee. It is thus one of the largest outcrop areas of Precambrian rocks in North America outside the Canadian Shield.

The Blue Ridge geologic belt is roughly coincident with the Blue Ridge physiographic province, a chain of ridges and highlands that forms the southeastern edge of the Appalachian Mountain system. The Blue Ridge highlands die out in northwestern Georgia, however, whereas the Blue Ridge geologic belt (commonly referred to as the Ashland–Wedowee belt in Alabama and Georgia) continues southwestward across the Piedmont physiographic province to the edge of the Coastal Plain.

Throughout much of its length, the northwestern boundary of the Blue Ridge belt is sharp, but in several segments, narrow belts of somewhat different character lie between the Blue Ridge and the Valley and Ridge. These include the Unaka belt in northeastern Tennessee and

* Publication authorized by the Director, U. S. Geological Survey.

southwestern Virginia, composed largely of a series of thrust sheets of Lower Cambrian clastic rocks, and the Talladega belt in Alabama and Georgia, composed of low-grade metamorphic rocks thought to be largely of Paleozoic age.

In Pennsylvania, Maryland, and northern Virginia, the Blue Ridge belt is bounded on the southeast by downfaulted basins filled with unmetamorphosed sedimentary rocks of Triassic age. In central and southwestern Virginia, the southeastern edge of the belt is the James River synclinorium, which contains low- and medium-grade metamorphic rocks of Paleozoic age. In most of North Carolina and in South Carolina, Georgia, and Alabama, the southeastern boundary is generally considered to be the Brevard fault zone.

The Paleozoic rocks northwest of the Blue Ridge are almost all of miogeosynclinal facies, whereas the metamorphic rocks of the Piedmont southeast of the Blue Ridge are largely eugeosynclinal deposits—graywackes, shales, and associated volcanic rocks. At least some of these rocks must be equivalent in age to parts of the miogeosynclinal rocks northwest of the Blue Ridge, but in only a few places can stratigraphic units on one side of the Blue Ridge be recognized on the other. Establishing the correlation between the well-known miogeosynclinal sequence of the Valley and Ridge belt and the very poorly known eugeosynclinal rocks of the Piedmont has thus become one of the most vexing problems in Appalachian geology.

The pronounced change in facies across the Blue Ridge has been interpreted as indicating that the Blue Ridge occupies the site of a welt or "medial geanticline" that persisted throughout much of the Paleozoic (Bucher, 1933; Kay,

1951). Alternatively, it has been suggested that the Blue Ridge is entirely a superimposed feature whose position is controlled by changes in mechanical properties of the basement produced by temperature gradients or some other unknown factors (Rodgers, 1964). Whatever the explanation, the great length and geologic continuity of Blue Ridge belt is one of its most striking and enigmatic features.

The Blue Ridge belt ends at the northern end of South Mountain (lat. 40°05′; long. 77°10′) in southern Pennsylvania, and Precambrian rocks are absent north of the Gettysburg Triassic basin for a distance of about 70 mi (110 km) to the east, except for a small isolated outcrop area in Little South Mountain (lat. 40°20′; long. 76°10′). Near Reading, Penna. (lat. 40°20′; long. 75°55′), Precambrian crystalline rocks again appear at the surface in the Reading Prong, a discontinuous belt that extends northeastward across the Hudson River. The Prong is similar in tectonic position to the Blue Ridge and forms the connecting link between the Blue Ridge belt of the central and southern Appalachians and its analog in the northern Appalachians—the Green Mountain anticlinorium in Vermont.

The oldest rocks in the Blue Ridge are granitic gneisses, granites, and associated metamorphic rocks which yield radiometric dates of 1,000 to 1,100 million years. Locally, the basal clastic deposits of the Paleozoic miogeosynclinal sequence rest directly on these basement rocks, but throughout most of the Blue Ridge, the basement complex is unconformably overlain by thick sequences of upper Precambrian sedimentary and volcanic rocks that lack the imprint of the 1,000–1,100 m.y. episode of metamorphism and plutonism. Along the northwestern edge of the Blue Ridge belt, the upper Precambrian rocks are little metamorphosed and are readily distinguished from the basement rocks on which they lie. Farther southeast, the grade of Paleozoic metamorphism rises, and it becomes increasingly difficult to differentiate between rocks of late Precambrian age and the earlier Precambrian basement. Recent investigations have shown that large areas of rocks previously considered to be part of the basement, are, in fact, metamorphosed upper Precambrian rocks. Chapter 17 by Hadley and Chapter 16 by Rankin in this section discuss the possible correlations of some of the principal sequences of upper Precambrian rocks and the problem of their recognition at higher grades of metamorphism.

From its northern terminus in southern Pennsylvania southward to the vicinity of Roanoke, Va. (lat. 37°15′; long. 79°55′), the Blue Ridge is apparently an autochthonous or at least parautochthonous anticlinorium overturned toward the northwest and complicated by local faulting. The present knowledge of this segment of the belt, to which Ernst Cloos has contributed so much, is summarized by Espenshade (Chapter 14).

South of Roanoke, the structure of the Blue Ridge changes abruptly. The steep flexures that mark the northwestern boundary of the northern part of the belt give way to large low-angle thrust faults along which Precambrian rocks have moved northwestward across unmetamorphosed Paleozoic rocks of the Valley and Ridge belt. Geophysical data and the presence of windows exposing overridden rocks in several places in North Carolina and Tennessee suggest that the northwestward transport of the Blue Ridge rocks there can be measured in tens of miles and may rival that of the larger overthrust masses in the Alpine orogenic belt. The structural and metamorphic history of this segment of the Blue Ridge is described by Bryant and Reed (Chapter 15). The tectonic significance of the Brevard fault zone and its possible relation to northwestward thrusting of the Blue Ridge rocks is discussed by Reed, Bryant, and Myers (Chapter 18).

The structure of the Reading Prong has been a subject of controversey for several decades. Originally it was believed that the Precambrian rocks there were exposed in the cores of anticlinal folds and that the structure of the belt was analogous to that of the northern segment of the Blue Ridge. Later it was suggested that the Precambrian rocks were part of a thin, far-traveled thrust sheet, similar to those in the southern Blue Ridge. The history of this controversy is reviewed by Drake in Chapter 19. He concludes that the Precambrian rocks of the Prong are the core of an enormous nappe which is rooted somewhere to the southeast, perhaps beneath the Newark Triassic basin.

Although the geology of large parts of the Blue Ridge still remains poorly known, a general understanding of the broad regional relations has emerged from nearly seven decades of geologic study. These investigations began with the pioneer reconnaissance work of Arthur Keith between 1897 and 1912 and were continued by A. I. Jonas (later Mrs. Anna J. Stose) between 1914 and 1940 and later by many other geologists of the U. S. Geological Survey, the various State geological surveys, and several universities. Preeminent among the later geologists, of course, has been Ernst Cloos, who is perhaps best known for his investigations of the structure and tectonics of the Blue Ridge in Maryland and adjacent states. His studies began as early as 1930 and are continuing with scarcely diminished vigor even as this is being written.

The papers in this chapter are not intended to present a comprehensive review of the geology of the Blue Ridge and Reading Prong. They do, however, provide some idea of the present state of knowledge on some of the principal unresolved problems, and the directions in which research is currently proceeding. Clearly the Blue Ridge and Reading Prong will be fruitful fields for geologic study for many generations to come!

REFERENCES

Bucher, W. H., 1933, The deformation of the earth's crust: Princeton, N.J., Princeton Univ. Press, 518 p.

Cloos, E., 1953, Appalachenprofil in Maryland: Geol. Rundschau, Bd. 41, p. 145–160.

_____, 1957, Blue Ridge tectonics between Harrisburg, Pennsylvania, and Asheville, North Carolina: Natl. Acad. Sci. Proc., v. 43, p. 834–839.

Fellows, R. E., 1943, Recrystallization and flowage in Appalachian quartzite: Geol. Soc. America Bull., v. 54, p. 1399–1431.

Kay, M., 1951, North American geosynclines: Geol Soc. America Mem. 48, 143 p.

Rodgers, J., 1964, Basement and no-basement hypotheses in the Jura and the Appalachian Valley and Ridge, in Lowry, W. D., ed., Tectonics of the southern Appalachians: Virginia Polytech. Inst. Dept. Geol. Sci. Mem. 1, p. 71–80.

Chapter 14

Geology of the Northern Part
of the Blue Ridge Anticlinorium*

GILBERT H. ESPENSHADE

NATURE OF THE ANTICLINORIUM

THE BLUE RIDGE anticlinorium is a prominent uplift extending southwest from southern Pennsylvania for several hundred miles across Maryland and Virginia (see tectonic map at end of this volume). The part lying north of Roanoke, Virginia, is described in this chapter. Older Precambrian rocks—various types of plutonic gneisses—are exposed in the core of the anticlinorium and are flanked on the limbs and nose of the fold by younger Precambrian rocks of sedimentary and volcanic origin. Paleozoic sedimentary rocks of the miogeosynclinal facies occur in the tightly folded belt of the Valley and Ridge province to the west. Metamorphosed rocks, mainly eugeosynclinal and probably both late Precambrian and early Paleozoic in age, as well as Triassic sedimentary and igneous rocks in downfaulted basins, are present in the Piedmont province east of the anticlinorium. The Blue Ridge physiographic province coincides with the anticlinorium only along the narrow northern nose of the fold in Maryland and Pennsylvania. The Blue Ridge physiographic province in Virginia is formed by the western limb of the fold and in places by the western part of the core; the eastern part of the fold in Virginia is in the Piedmont physiographic province.

The stratigraphy of the northern part of the Blue Ridge anticlinorium is outlined in Table 1. The core of the fold as far north as southern Maryland is composed of an older Precambrian

* Publication authorized by the Director, U. S. Geological Survey.

basement complex which consists mostly of various plutonic gneisses, some of which are of charnockitic nature. Younger Precambrian gneiss and schist of graywacke composition are unconformable upon the basement complex. They are thickest and most persistent on the eastern limb of the fold (Lynchburg Formation, including Rockfish Conglomerate Member); a coeval unit on the western limb (Swift Run Formation) is tuffaceous, thin, and less widely distributed. Similar metasedimentary rocks are exposed in a narrow syncline about 60 mi (100 km) long in the center of the anticlinorium in the Mechum River area, Virginia. The Catoctin Formation of late Precambrian age, consisting mainly of metabasalt or greenstone, is widely distributed on both flanks of the fold. Metarhyolite occurs with metabasalt at the nose of the anticlinorium in Maryland and Pennsylvania. Small lenticular bodies of amphibolite and ultramafic rock intrude the basement complex and the Lynchburg Formation; the amphibolite bodies may represent feeders for metabasalt of the Catoctin Formation. Quartzite and phyllite of Early Cambrian and Early Cambrian(?) age (Chilhowee Group) occur along most of the western limb of the fold and in the northern part of the eastern limb. Schist, phyllite, marble, and greenstone of Paleozoic(?) age (Evington Group) are present in the southern part of the eastern limb. Downfaulted Triassic sedimentary rocks and diabase border the eastern side of the fold in its northern part.

The principal fold—the South Mountain anticline—at the northern end of the Blue Ridge anticlinorium was first recognized by Keith (1894)

TABLE 1.   Stratigraphic Relations in the Northern Part of the Blue Ridge Anticlinorium

West limb	Center	East limb
Chilhowee Group—Lower Cambrian and Lower Cambrian(?)		Chilhowee Group in northern part—Lower Cambrian and Lower Cambrian(?) Evington Group in southern part—Paleozoic(?)
- - - - - - - - - - - - - - -	Unconformity	- - - - - - - - - - - - - - -
		Ultramafic intrusions cut the upper part of the Lynchburg Formation—upper Precambrian(?)
Catoctin Formation, mostly metabasalt; mafic feeder dikes cut older rocks—upper Precambrian	Catoctin Formation and metarhyolite at north end of anticlinorium—upper Precambrian	Catoctin Formation, mostly metabasalt; mafic feeder dikes cut older rocks—upper Precambrian
Swift Run Formation—upper Precambrian	Mechum River metasedimentary rocks of Gooch (1958), equivalent in part to Lynchburg and Swift Run—upper Precambrian	Lynchburg Formation, including Rockfish Conglomerate Member at base—upper Precambrian
- - - - - - - - - - - - - - -	Unconformity	- - - - - - - - - - - - - - -

Basement complex—gneissic hypersthene granodiorite, quartz monzonite, granite, syenite, anorthosite, and para gneiss—older Precambrian

in his study of the northern Blue Ridge. He concluded that the granitic rocks in the core of the fold intruded the nonepidotic schists of the Catoctin but were older than epidotic schists of the Catoctin and the metarhyolite. This view that the Catoctin was intruded by the granitic gneisses prevailed for many years, and Jonas (1927, p. 845) expressed her opinion that the underlying Lynchburg Formation was also intruded by the granitic gneisses. Later, Nelson (1932) found a basal conglomerate on the Rockfish River beneath the Lynchburg Formation that contained pebbles and boulders of the underlying granitic gneisses. Jonas (1935), p. 47–49, 56–57) showed that the Catoctin Formation was unconformable upon gneissic granodiorite and that mafic dikes related to the metabasalt cut the gneiss, but she still thought that the Lynchburg Formation was intruded by granitic rocks. This last view was later revised (Jonas and Stose, 1939), and the current concept of the stratigraphy that is given in Table 1 was established. The most recent discussion of the Precambrian rocks in the Blue Ridge is by King (in press) in a review of the Precambrian of the Southeastern States. Recognition of an older Precambrian basement complex that is overlain unconformably by a thick series of upper Precambrian metasedimentary and metavolcanic rocks has been the critical key to the correct interpretation of Blue Ridge geology, particularly in the southern Blue Ridge where thrust faulting makes the structure much more complex than in the northern Blue Ridge.

PRECAMBRIAN BASEMENT COMPLEX

The gneissic rocks of the anticlinorium core have been given various names—injection complex (Jonas and Stose, 1939, p. 580–582), basement complex (Bloomer and Werner, 1955, p. 581), and Virginia Blue Ridge Complex (Brown, 1958, p. 7). Distinctive rock types—gneissic quartz diorite, hypersthene granodiorite, quartz monzonite, granite, syenite, and paragneiss—occur widely in the anticlinorium.

Some investigators have found gneisses that they believe to be sedimentary rocks that have been intruded by the granitic rocks. Jonas and Stose (1939, p. 581) stated that such paragneisses are widespread but did not mention specific localities. Paragneiss may be of limited occurrence in the basement complex of northern Virginia and Maryland because it is described only in Frederick County, Maryland (Stose and Stose, 1946, p. 16–18). In central Virginia, bodies of paragneiss are small and occur as numerous inclusions, as much as 10 ft (3 m) thick and several hundred feet long, in the granitic rocks (Bloomer and Werner, 1955, p. 581). Paragneiss may be more extensive farther south. In Bedford County, southwest of Lynchburg, Pegau (1932, p. 22–25) and Diggs (1955, p. 9–11) described gneiss composed of alternating layers of hornblende gneiss and biotite gneiss, which they called Moneta Gneiss and interpreted as an old paragneiss. On the James River near Lynchburg, Brown (1958, p. 12–17) described interlayered hornblende gneiss and leucogneiss, which he called the Reusens migmatite facies of Moneta Gneiss; the leucogneiss appears to intrude the hornblende gneiss. Recognition of old paragneiss in the basement complex may be a difficult matter in some areas; in places, interlayered younger rocks, such as Lynchburg Formation and hornblende gneiss equivalent to the Catoctin Formation, may be so highly deformed that they appear to be an older paragneiss.

Several types of intermediate and granitic gneisses are widespread in the Blue Ridge anticlinorium. Gneissic hypersthene syenite was described many years ago by Watson and Cline (1916) as being one of the most abundant rocks in the central Blue Ridge, occurring in every county between Floyd on the south and Warren on the north, a distance of more than 150 mi (240 km). These rocks occur mostly in the western part of the anticlinorium and form many of the high peaks. Jonas (1935, p. 50–55) later described the rock as hypersthene granodiorite. Bloomer and Werner (1955, p. 582) found these rocks to be an assemblage with wide range in composition—granite, granodiorite, syenite, quartz diorite, anorthosite, unakite—and proposed the name Pedlar Formation for these rocks. Later workers (Brown, 1958, p. 9–11; Allen, 1963, p. 18–23; Virginia Div. Mineral Resources, 1963; Werner, 1966, p. 7) have followed this nomenclature.

Typical hypersthene-bearing rocks are medium to coarse grained, green to dark green-gray, and are composed mainly of gray to blue quartz, plagioclase (An_{30-50}), potassium feldspar and perthite, hypersthene, augite, hornblende, and biotite. Watson and Cline (1916, p. 202–204) give five chemical analyses of these rocks, and Brown (1958, p. 10) and Allen (1963, p. 12) each give three modal analyses. Some hypersthene-bearing rocks are noritic in composition, have high contents of TiO_2 and P_2O_5, and may be related to nelsonite (see below) (Watson and Cline, 1916, p. 225–231). The Blue Ridge hypersthene-bearing gneisses are similar to pyroxene syenites of the Adirondacks and intermediate varieties of charnockite in Madras, India, according to Watson and Cline (1916, p. 212–220).

Unakite, a distinctive rock composed largely of blue quartz, plagioclase, pink potassium feldspar, and considerable epidote, is associated with the hypersthene rocks in places (Watson and Cline, 1916, p. 220–223). Jonas (1935, p. 50–51) stated that unakite occurs along the western side of the hypersthene granodiorite and is hydrothermally altered granodiorite.

Much of the eastern side of the anticlinorium is underlain by two varieties of granitic gneiss that have been called the Lovingston Granite Gneiss and the Marshall Gneiss of Jonas (Virginia Geol. Survey, 1928; Virginia Div. Mineral Resources, 1963). Both types are quartz monzonite in composition, but the Lovingston contains large porphyroblasts or augen of potassium feldspar, whereas the Marshall Gneiss of Jonas (Virginia Geol. Survey, 1928) has more uniform grain size. Brown (1958, p. 10) gives five modal analyses, and Allen (1963, p. 12) lists six. Plagioclase composition is An_{5-20}; biotite is abundant, and usually accompanied by a little muscovite, chlorite, and garnet. Both varieties of gneissic quartz monzonite are said to grade into the hypersthene-bearing rocks.

Two other types of granitic gneisses have been described in Greene and Madison Counties, Virginia. The Old Rag Granite of Furcron (1934, p. 406–407) is exposed on Old Rag Mountain and other high ridges in the vicinity (Reed, 1955, p. 876–877; Allen, 1963, p. 15–17). It is a coarse-grained rock composed mainly of blue quartz and potassium feldspar, accompanied by plagioclase, amphibole, and biotite. The other granitic gneiss was called the Robertson River Formation by Allen (1963, p. 24–25) and underlies an area more than 22 mi (35 km) long and several miles wide along the southeast side of the anticlinorium core. It is a hornblende–oligoclase granite, which contains two to four times as much potassium feldspar as plagioclase.

The Roseland Anorthosite forms an elongate body about 20 sq mi (52 sq km) in area (Fig. 1, index map) in charnockitic rocks and the Marshall Gneiss of Jonas (Virginia Geol. Survey, 1928) in Nelson and Amherst Counties (Watson and Taber, 1913, p. 68–90; Ross, 1941, p. 3–4; Bloomer and Werner, 1955, p. 582; Herz, in press). The rock is composed mainly of very coarse-grained andesine antiperthite and finer grained andesine–oligoclase and microcline in strongly granulated zones (Herz, in press). Blue quartz and clinozoisite are generally present, and, in places, hornblende, hypersthene, rutile, ilmenite, and apatite. Watson and Taber (1913, p. 76) give eight chemical analyses of the rock which they call syenite. Nelsonite, a rock rich in titanium (as either ilmenite or rutile) and phosphorus (apatite), forms dike-like bodies in the anorthosite or the adjacent gneiss (Watson and Taber, 1913, p. 100–106; Moore, 1940, p. 638–645; Ross, 1941, p. 20–23; Herz, in press).

Some recent investigators have found gradational contacts between different types of granitic gneisses and have concluded that these rocks have been formed by granitization of sedimentary rocks (Bloomer and Werner, 1955, p. 581–582; Brown, 1958, p. 17; Allen, 1963, p. 70–71). Little specific evidence is given to support this view of origin, however, with the exception of zircons showing possible detrital forms that Mertie (1956) found in hypersthene granodiorite, Lovingston Granite Gneiss, and the Marshall Gneiss of Jonas (Virginia Geol. Survey, 1928). Mertie believed these rocks to be paragneisses, at least in part. Extensive studies of similar rocks in other granulite terranes (the Adirondacks, for example, suggest that the plutonic gneisses in the Blue Ridge anticlinorium need much more study if their origin is to be understood.

One absolute age determination has been made of the basement complex. Lead isotope ages of zircon from the hypersthene granodiorite at the north end of Mary's Rock Tunnel on the Skyline Drive south of Thornton Gap, Virginia, range from 1,070 to 1,150 m.y., and biotite ages are 880 m.y. (Rb–Sr) and 800 m.y. (K–Ar) (Tilton and others, 1960, p. 4175–4176). Obviously, the significance of the zircon age depends upon whether the rock is an orthogneiss or paragneiss.

UPPER PRECAMBRIAN METAMORPHIC ROCKS

Rocks unconformably overlying the basement complex are mainly metasedimentary and include the Lynchburg Formation (and Rockfish Conglomerate Member) along the eastern limb of the anticlinorium, the Mechum River metasedimentary rocks of Gooch (1958) in a narrow syncline along the middle of the anticlinorium, and the Swift Run Formation along the western limb. The Catoctin Formation, mostly metabasalt or greenstone, is above these rocks on both limbs of the fold and is associated with metarhyolite at the northern end of the fold. All these rocks are here grouped together as an upper Precambrian sequence, for reasons given following the discussion of these formations.

The Lynchburg Formation, including the Rockfish Conglomerate Member, lies above the basement complex in a belt that extends southwest from northern Virginia for more than 150 mi (240 km). These beds are exposed across widths of as much as 6 mi (10 km) in the southern part of the region (Fig. 1, sections G-C' and D-D') but are much thinner in the vicinity of the Potomac River and to the north (Fig. 1, sections A-A' and B-B'). They consist mainly of metamorphosed graywacke (fine-grained biotite–

quartz–feldspar gneiss), accompanied by conglomerate, mica schist, and graphitic schist.

Thick beds of coarse conglomerate rest upon the basement complex in the central part of the region. Along the Rockfish River, Nelson County, the conglomerate is about 1,200 ft (365 m) thick, and has been considered by Nelson (1932; 1962, p. 17–19) to be a separate formation, the Rockfish Conglomerate. Bloomer and Werner (1955, p. 583–585), however, call these beds the Rockfish Member of the Lynchburg Formation. Similar conglomerate occurs in the Lynchburg Formation to the north (Allen, 1963, p. 28–31) and south (Brown, 1958, p. 21), but has not been mapped as separate beds.

In the Warrenton area of northern Virginia (Fig. 1, index map), east of the basement complex, Furcron (1939, p. 37–45, Pl. 1) mapped a belt of coarse arkose (which he correlated with the Loudoun Formation) and schist, slate, and marble (which he grouped together and named the Fauquier Formation); these beds are now generally believed to be equivalent to the Lynchburg Formation.

Along the western side of the anticlinorium, clastic and pyroclastic rocks (arkose, graywacke, conglomerate, slate, and phyllite), basalt, and thin marble lenses occur between the basement complex and Catoctin Formation at many places and are called the Swift Run Formation (Stose and Stose, 1946, p. 18–20; King, 1950, p. 9–12; Bloomer and Werner, 1955, p. 587–589; Reed, 1955, p. 880–881; Nickelsen, 1956, p. 243–244; Nelson, 1962, p. 22–24; Allen, 1963, p. 38–40; and Werner, 1966, p. 11–14; Brown, Chapter 23 this volume). Thickness of these beds usually ranges from less than 100 ft (30 m) to about 400 ft (120 m); Nelson (1962, p. 22–23) reported an unusual thickness of 1,380 ft (420 m) at one locality in Albemarle County.

A syncline (the Batesville syncline) of metasedimentary rocks about 60 mi (100 km) long and 0.5–2 mi (0.8–3.2 km) wide lies along the middle of the anticlinorium in north-central Virginia (Fig. 1, section C-C') (Gooch, 1958, p. 569–571; Nelson, 1962, p. 24; Allen, 1963, p. 32). These rocks are mainly phyllite and slightly metamorphosed graywacke, accompanied by metaconglomerate, and were called the Mechum River metasedimentary rocks by Gooch (1958, p. 571); he suggested that they are equivalent, at least in part, to the Swift Run and Lynchburg.

The Catoctin Formation is the most widespread formation in the region, extending along both flanks of the anticlinorium for about 200 mi (320 km) from near the James River in Virginia to

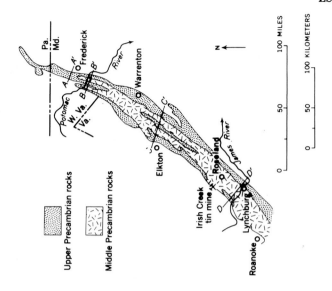

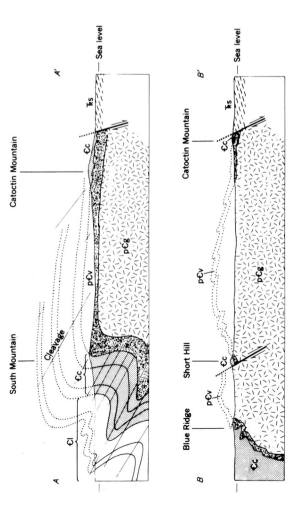

FIGURE 1. Structure section and index map of Blue Ridge anticlinorium. ℞s, Triassic sandstone and shale; O℄l, Cambrian and Ordovician limestone and dolomite; ℄l, Cambrian limestone, dolomite, and shale; ℄c, Cambrian and Cambrian(?) quartzite and phyllite (Chilhowee Group); Þze, Paleozoic(?) phyllite and schist (Evington Group); p℄v, upper Precambrian metabasalt, phyllite, and quartzite (Catoctin Formation and Swift Run Formation); p℄l, upper Precambrian mica gneiss and schist (Lynchburg Formation, including Rockfish Conglomerate Member, and metasedimentary rocks of Mechum River); p℄g, older Precambrian basement complex. Section A-A', near Frederick, Md., mainly from Cloos (1947, Pl. 13); east flank of fold

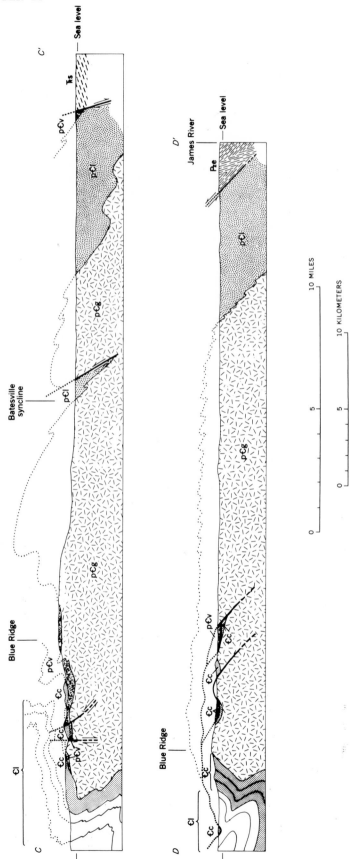

from Whitaker (1955, Pl. 1); Precambrian basement complex core interpreted from Jonas and Stose (1938). Section B-B', about 4 mi (6.5 km) south of Potomac River in Virginia, modified from Nickelsen (1956, Fig. 6). Section C-C', extending southeast from a point about 7½ mi (12 km) northeast of Elkton, Va.; part northwest of Blue Ridge from King (1950, Fig. 13); part southeast of Blue Ridge interpreted from Virginia Division of Mineral Resources (1963) and Allen (1963, Pl. 1). Section D-D', near Lynchburg, Va., modified from Bloomer and Werner (1955, Pl. 1); addition on east end from Espenshade (1954, Pl. 1).

southern Pennsylvania (see tectonic map at end of this volume). The Catoctin forms the northwest slope and locally the crest of the Blue Ridge and South Mountain, Maryland, on the west side of the fold, and Catoctin Mountain and other ridges farther south on the east side. Only selected references to the Catoctin are given here, because of the very voluminous literature. The formation was first described in the northern part of the fold by Keith (1894); its volcanic nature had been earlier recognized by Williams (1892, p. 490–492). The most thorough recent study of the Catoctin was that by Reed (1955) of an area in the Blue Ridge of northern Virginia. Here the Catoctin Formation is made up of a series of basalt flows, metamorphosed in the greenschist facies, but still showing such primary structures as amygdules, feldspar phenocrysts, flow breccias, columnar jointing, and sedimentary dikes. Individual flows are 150–270 ft (45–80 m) thick and are recognizable because of schistose zones formed in amygdular and breccia flow tops. Thin sedimentary beds, commonly phyllitic or sandy, and indistinguishable from the Swift Run Formation, are interlayered with the lava flows. Maximum total thickness of the Catoctin here is about 1,800 ft (550 m); it is evident that the flows were poured out on a surface that had a relief of at least 1,000 ft (300 m). The Catoctin farther south in the Blue Ridge was studied by Bloomer and Bloomer (1947).

The metabasalt or greenstone on the west side of the fold is composed mainly of albite, chlorite, epidote, and actinolite; relict pyroxene occurs locally (Reed, 1955, p. 887–890). Mineralogy of the greenstone on the east side of the fold is similar but reflects higher metamorphic grade, because biotite is present in places and pyroxene seems to be very rare (Brown, 1958, p. 26–27; Allen, 1963, p. 45).

The metabasalt was regarded by Bloomer and Werner (1955, p. 592) and Brown (1958, p. 24) as spilite that had formed under eugeosynclinal conditions. On the other hand, Reed (1955, p. 893–894) concluded that the Catoctin flows were definitely subaerial because of columnar jointing, absence of pillow structure, areal extent of individual flows, and thin breccia zones between flows, and suggested that the flows were tholeiitic basalts instead of spilites. Since then, Reed (1964) has presented four new chemical analyses that show that these rocks are of spilitic composition, but he suggests that their high soda content may be related to low-grade regional metamorphism of subaerial flows.

Metarhyolite associated with metabasalt of the Catoctin covers an area of more than 100 sq mi (260 sq km) at the north end of the anticlinorium in southern Pennsylvania and Maryland (Stose, G. W., 1932, p. 31–38; Jonas and Stose, 1938; Stose and Stose, 1946, p. 22–24; Gray and others, 1960). The metarhyolite has a variety of primary structures and textures — amygdules, spherulites, quartz and feldspar phenocrysts, breccia and flow structure—and some of it is evidently devitrified glass (Bascom, 1896, p. 36–67).

Dikes of amphibolite (metadiabase) occur widely in the basement complex, especially near the flanks of the fold; some of them were probably feeders for the basalt flows. Ultramafic intrusions (serpentinite, talc–chlorite schist, and metaperidotite) are also present near the top of the Lynchburg Formation for about 100 mi (160 km) northeast of Lynchburg (Bloomer and Werner, 1955, p. 593; Brown, 1958, p. 41–47; Nelson, 1962, p. 76–78; Allen, 1963, p. 35–37).

The age of the Swift Run and Catoctin has been assigned variously to the Precambrian, Precambrian and Cambrian, and Cambrian, because at some places these formations seem to grade upward into Lower Cambrian clastic sedimentary rocks having thin volcanic layers, and at other places Lower Cambrian beds are unconformable on the Catoctin, Swift Run and basement complex. In the Elkton area, Virginia (Fig. 1, index map), King (1949, p. 525–528; 1950, p. 14) assigned the Catoctin and Swift Run to the Precambrian because there is an unconformity above them and because Lower Cambrian rocks are different lithologically and have conglomerates containing Catoctin pebbles. Stose and Stose (1946, p. 28–29) thought there was an unconformity above the Catoctin in Maryland, and also concluded that the Catoctin is upper Precambrian. The Catoctin is here regarded as uppermost Precambrian on the basis of the above facts and the arguments of King (1949, p. 635–638).

ROCKS BORDERING THE ANTICLINORIUM

Clastic sedimentary rocks of the Chilhowee Group (Lower Cambrian and Lower Cambrian?) lie above the Catoctin Formation all along the west flank of the Blue Ridge anticlinorium to its nose in Pennsylvania, and make up the foothills along the west side of the Blue Ridge. The Chilhowee Group consists mainly of quartzite, sandstone, siltstone, and shale and varies somewhat from place to place in the nature and thickness of its lithologic units. In the Elkton area,

Virginia, it is about 2,800 to 3,400 ft (855 to 1,070 m) thick (King, 1950, p. 16–23). The Chilhowee has been mapped in places from south to north by Bloomer and Werner (1955, p. 594–597), King (1950, p. 14–24), Reed (1955, p. 878–879), Nickelsen (1956, p. 247–253), Jonas and Stose (1938), Stose and Stose (1946, p. 31–42), and Cloos (1941; 1951a, p. 28–39).

Proceeding southward on the east flank of the fold, the Catoctin is bordered for about 25 mi (40 km) by a downfaulted basin of Triassic conglomerate, sandstone, shale, and diabase, then by the Chilhowee Group for 50 mi (80 km) (Whitaker, 1955, p. 441–447; Toewe, 1966, p. 4–5), and again by Triassic rocks for about 60 mi (100 km).

Farther south, the east side of the fold is bordered by the Evington Group (phyllite, schist, marble, quartzite, and greenstone), which is strongly deformed by folding and faulting in the narrow James River synclinorium (Espenshade, 1954, p. 14–21; Brown, 1958, p. 28–38, also this volume; Smith and others, 1964, p. 7–12). The Evington is probably early Paleozoic in age and

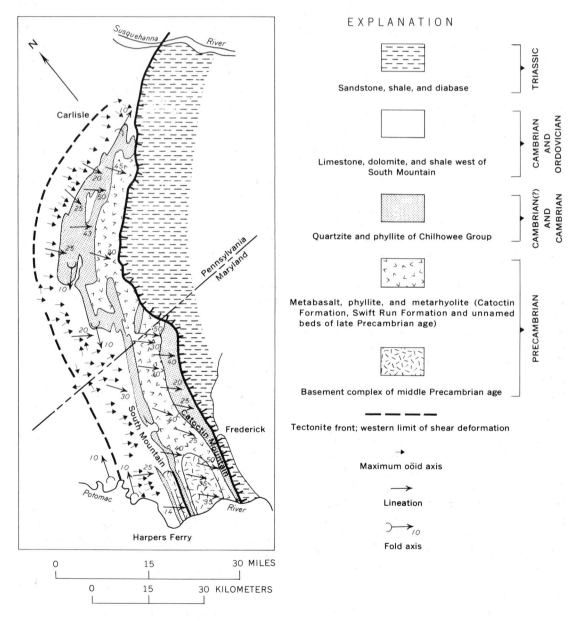

FIGURE 2. Structure map of the northern end of the Blue Ridge anticlinorium. Mainly from Cloos (1947, fig. 1) and Cloos and Hietanen (1941); older Precambrian gneiss core adapted from Jonas and Stose (1938).

may be partly equivalent to the Chilhowee, from which it is separated by the long basin of Triassic rocks. The Evington Group was formerly regarded as Precambrian and equivalent to the Glenarm Series by Furcron (1935, p. 22–37).

STRUCTURAL GEOLOGY

The structural features of the Blue Ridge anticlinorium are best known at its northern end, where they have been studied in detail by Cloos and his students (Fig. 2). The anticlinorium here is asymmetric, its western limb having overturned folds and the eastern limb dipping very gently eastward (Fig. 1, section A-A′). Structural details farther south in the anticlinorium are not so well known, and some interpretations shown in section B-B′, C-C′, and D-D′ of Figure 1 are rather diagrammatic.

In the northern part of the anticlinorium (Fig. 1), cleavage is present in all the rocks and generally dips 10–50 degrees eastward. Cleavage consists of foliation or schistosity in gneisses and other coarse-grained rocks and takes on a slaty character in fine-grained rocks. It gradually dies out westward in overlying calcareous beds of Cambrian and Ordovician age. Lineation in a down the cleavage dip is widespread, and lineations due to intersection of bedding and the principal cleavage (in b) and intersection of the principal cleavage and later slip cleavage are also present (Cloos, 1947, p. 895–897; 1950, p. 6–7; 1958, p. 13–17; Cloos and Hietanen, 1941, p. 80–84; Whitaker, 1955, p. 451–454; Nickelsen, 1956, p. 256–259). Lineation in a is represented by elongated mineral grains, pebbles, chloritic or epidotic amygdules, streaks, and by the long axes of oöids in calcareous oölites above the Chilhowee. Fold axes (b) are usually nearly horizontal.

The principal cleavage is essentially axial-plane cleavage, but is parallel to the axial plane only near the crests and troughs of the folds; elsewhere the dip of cleavage varies in a fanlike manner and diverges upward in anticlines and downward in synclines. Cleavage characteristically dips steeply where bedding dips gently, and gently where bedding is steep.

These various structural elements, illustrated diagrammatically in Fig. 3, have very uniform relations throughout an area of at least 300 sq mi (780 sq km) (Cloos, 1947, p. 913).

Deformation is most intense in the Precambrian rocks and the Chilhowee Group and decreases gradually for 5–8 mi (8–13 km) west of the crest of the anticlinorium, as shown by cleavage and elongated oöids in limestone and

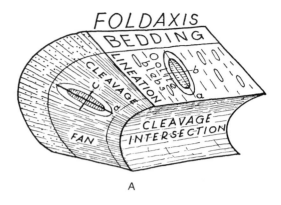

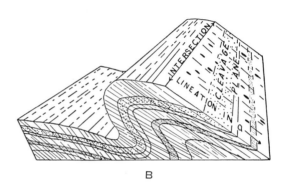

FIGURE 3. Diagrams showing relations of structural features in the northern end of the Blue Ridge anticlinorium. A. Western limb. (From Cloos 1951b); B, Eastern limb. (From Whitaker 1955.)

dolomite. Cloos (1947, p. 869–892) made an exhaustive quantitative study of deformation of the oöids (originally spherical and usually less than 1 mm in diameter) by measuring their three axes in thin sections of oriented specimens from many localities and calculating the average amount of deformation at each place. He concluded that the anticlinorium is a large shear fold in which deformation took place by laminar flow along cleavage planes, which caused much crustal shortening without important thrusting. The original thickness of beds must have been distorted greatly by the process of laminar flow, resulting in much thickening along the axes of folds and thinning on the flanks (Cloos, 1947, p. 901–911).

Similar deformation effects were found by Fellows (1943, p. 1415–1430, Pl. 2) in his petrofabric study of Appalachian quartzites; samples from the northern end of the anticlinorium and to the east all show recrystallization, elongation of quartz grains, and lattice orientation, whereas samples from farther west and north do not have these features.

Cloos (1957, 1966) recently investigated structural features farther south in the anticlinorium and found the same structural plan throughout the region. He also discovered that about 90 percent of all slickensides are in the *ac* plane, suggesting that late movement of blocks took place along this plane.

In the Elkton area, Virginia, King (1950, p. 41–52) found that the western limb of the anticlinorium is overturned westward in an asymmetric fold whose axial plane dips gently eastward (Fig. 1, section C-C'). Cleavage is widespread in the Catoctin and argillaceous beds of the Chilhowee. Deformation seems to have been by a shear mechanism in rocks of low competency, but by a flexure mechanism in massive quartzite of the Chilhowee. A few miles farther north in the Luray area, the anticlinorium is likewise overturned westward, and cleavage in the Catoctin has strong downdip lineation (Reed, 1955, p. 879–880).

About 10 mi (16 km) east of the Blue Ridge is the long narrow Batesville syncline of metasedimentary rocks (Fig. 1, section C-C'); the writer interprets this syncline as more tightly compressed and deeper than is shown in other interpretations (Gooch, 1958; Nelson, 1962, Pl. 1; Allen, 1963, Pl. 1).

The general structure of the anticlinorium in the southern part of the region, according to Bloomer and Werner (1955, p. 600–605), is shown in Figure 1, section D-D'. Foliation of Precambrian age is present in the basement complex; younger cleavage and downdip lineation occur in rocks of the Lynchburg Formation and Chilhowee Group and their equivalents. Cleavage and steep lineation are also conspicuous in rocks of the Evington Group on the east flank of the anticlinorium (Espenshade, 1954, p. 29–32; Brown, 1958, p. 57–60).

In the basement complex, cataclastic deformation and zones of strong shearing that produced phyllonite and mylonite are reported at various localities (Furcron, 1939, p. 29–30; Bloomer and Werner, 1955, p. 602; Nickelsen, 1956, p. 242–243; and Allen, 1963, p. 25). Deformation within the basement gneisses probably took place along these zones.

Faults are known at many places in the anticlinorium, and it was thought for many years that the anticlinorium was thrust westward along a profound fault that extended for hundreds of miles along the west front of the Blue Ridge (King, 1950, p. 47–50). Stose and Stose (1946, p. 104–113) postulated such a fault—the Harpers Ferry overthrust—in Maryland and northern Virginia, and regarded it as rising out of the basement core of the fold. Other workers (King, 1950, p. 47–50; Cloos, 1951b, p. 141–143; and Nickelsen, 1956, p. 263–265) have not found evidence for large-scale thrusting here. In the southern part of the region, a zone of small thrusts is present along the west side of the fold (Fig. 1, section D-D') (Bloomer and Werner, 1955, p. 602), and southwest of the James River the anticlinorium is bordered everywhere on the west by overthrusts (Virginia Div. Mineral Resources, 1963). Thus, according to recent views, the Blue Ridge anticlinorium appears to be autochthonous for about 200 mi (320 km) southwest of its nose, and becomes allochthonous farther southwest. However, Gwinn (Chapter 8 this volume) speculates that South Mountain may actually be underlain by a great detachment zone because the Paleozoic rocks that underlie the Great Valley, Valley and Ridge, and Alleghany Plateau to the west seem to be shortened about 50 mi. Thrusting in the northern Blue Ridge probably took place at about the same time (Late Paleozoic) as it did in the southern Blue Ridge (Bryant and Reed, Chapter 15 this volume; Rankin, Chapter 16 this volume).

Triassic sedimentary and igneous rocks are downfaulted along the eastern border of the anticlinorium for nearly 150 mi (240 km) south of the nose. These normal faults are the youngest known in the region and are considered to be Triassic. Faults of this age may be present within the anticlinorium but would be difficult to recognize. One possible Triassic fault of this nature (Nickelsen, 1956, p. 263) is the normal fault on the west side of Short Hill, Virginia (Fig. 1, section B-B').

METAMORPHISM

At least two periods of metamorphism may be recorded in the rocks here: a possible period of older Precambrian age and a definite middle Paleozoic period. The possible Precambrian metamorphism in the basement complex is suggested by the presence of paragneiss, ancient foliation, and by the occurrence of garnet at places where garnet is absent from younger rocks (Nickelsen, 1956, p. 243). Charnockitic rocks are widespread in the basement complex and presumably were formed under conditions of high temperature and pressure. If these rocks are of igneous origin, they were probably intruded about 1,100 million years ago, according to the lead-isotope age determination of zircon from hypersthene granodiorite cited earlier (Tilton and others, 1960, p. 4175–4176). This corresponds with ages of 1,000–1,100 m.y. for the earliest recorded plutonism and metamorphism in the southern Blue Ridge.

The effects of middle Paleozoic metamorphism are evident in all pre-Triassic rocks of the anticlinorium. These rocks in the northern part of the anticlinorium fold are in the greenschist facies; the basement granitic gneiss has chlorite, sericite, epidote, albite, and late biotite; the metabasalt of the Catoctin consists of albite, epidote, chlorite, and actinolite; and phyllitic beds in the Chilhowee have sericite and chlorite (Reed, 1955, p. 894–895; Whitaker, 1955, p. 451; Nickelsen, 1956, p. 255–256).

Jonas (1935, p. 50–51) concluded that the epidotization of hypersthene granodiorite to form unakite took place in Precambrian time because of the presence of unakite pebbles in Lower Cambrian conglomerates. Rankin (written communication, July 1967), however, finds evidence in the Mt. Rogers, Virginia, area that epidotization of pebbles took place within the conglomerate and thinks that this alteration is related to Paleozoic metamorphism.

In the southern part of the region near Lynchburg, Brown (1958, p. 9–11) described features in the basement gneiss—the formation of albite, sericite, and epidote, and replacement of hypersthene by biotite and hornblende—that may have been caused by Paleozoic metamorphism. Similar features in basement gneiss of the Roseland area are also attributed by Herz (in press) to Paleozoic metamorphism. The grade of metamorphism is somewhat higher in the southern part of the anticlinorium than in the northern part. Biotite and garnet are characteristic of the Lynchburg Formation, and kyanite and andalusite are locally present; the Catoctin has chlorite, actinolite, epidote, and some biotite (Brown, 1958, p. 20–28). Grade of metamorphism here increases eastward in the Evington Group flanking the anticlinorium (Espenshade, 1954, p. 38–43). In the Altavista area south of Lynchburg, Redden (1963, p. 84–86) described sillimanite, andalusite, staurolite, and cordierite in rocks of the Lynchburg Formation and Evington Group; these minerals are in part retrograded to muscovite, paragonite, and chlorite.

Biotite in Lynchburg Formation west of Lynchburg has given a date of 370 m.y. by the K–Ar method (Long and others, 1959, p. 595). Presumably this biotite was formed in the middle Paleozoic period of metamorphism, about the same time as the main metamorphism in the southern Blue Ridge (Bryant and Reed, Chapter 15 this volume; Rankin, Chapter 16 this volume).

PROBLEMS NEEDING INVESTIGATION

The older Precambrian gneiss core of the Blue Ridge anticlinorium is a major structural element of the Appalachian Mountain system that reappears at intervals for nearly 1,500 mi (2,400 km) to the northeast: in the Reading Prong–Highlands province, the Green Mountains, and the Long Range of western Newfoundland. This welt at many places separates Paleozoic miogeosynclinal rocks on the west from Paleozoic eugeosynclinal rocks on the east. Deep-seated rocks have been uplifted along a narrow belt hundreds of miles long. Is this a crustal shear of the first order? Rodgers (1964, p. 79) suggested that the west side of the anticlinorium marks an isotherm where the basement became warm enough to deform plastically; lower temperatures prevailed to the west. Surely it is significant that practically all Appalachian volcanic activity in late Precambrian and Paleozoic time occurred either along this welt or east of it; only very minor Paleozoic volcanism took place to the west. Did the welt exist during early Paleozoic time and separate a western miogeosyncline from an eastern eugeosyncline? King (1964, p. 17–18) also considered this problem and found little evidence for separate troughs in Virginia, but he pointed out that Middle Ordovician conglomerates in Tennessee suggest uplift in that region.

Gwinn (Chapter 8 this volume) presents evidence to suggest that the northern end of the Blue Ridge is overthrust a great distance westward, contrary to current opinion that this part of the Blue Ridge is rooted. Thus, the Precambrian core of the entire Blue Ridge may be allochthonous, as Drake (Chapter 19 this volume) believes to be the case for the Precambrian of the Reading Prong. Geophysical studies in the South Mountain region are needed to solve this problem, because careful geologic work has not found evidence for large-scale thrusting here.

The petrology, geochemistry, metamorphism, and structure of the basement complex core of the northern Blue Ridge anticlinorium, covering an area of about 2,000 sq mi (5,200 sq km), has received little attention from the point of view of modern concepts and methods. Charnockitic rocks underlie hundreds of square miles of this terrane and apparently closely resemble the charnockitic suite of rocks in the type locality of Madras, India (Subramaniam, 1959, p. 328–336), and similar rocks in many other areas (Howie, 1964). The origin of the basement gneisses needs careful study to determine if they are largely products of granitization, as recent workers have proposed, or if they are metamorphosed magmatic rocks, as appears to be the case for similar rocks in the Adirondacks (Buddington, 1952, p. 58–61).

Paleozoic granite and pegmatite intrude Blue Ridge rocks in North Carolina, but have not been recognized in the northern part of the anticlinorium. The possibility of their occurrence here must be kept in mind.

The age of the tin deposits (containing cassiterite, wolframite, beryl, and fluorite) that occur with greisen in pyroxene granodiorite at Irish Creek, Virginia (Koschmann and others, 1942, p. 279–285; Glass and others, 1958; Werner, 1966, p. 46), is a puzzle. Deposits of this nature are most unusual in rocks of the granulite facies, and the question arises whether these deposits may not be Paleozoic in age rather than Precambrian.

Radiometric age determinations have been published for only two localities in the part of the anticlinorium north of Roanoke, Virginia. Radiometric dating of carefully selected samples will certainly be of great aid to detailed studies of the igneous and metamorphic rocks of the anticlinorium.

The nature and distribution of the upper Precambrian sedimentary and volcanic rocks are known in a general way now, but detailed studies directed toward determining stratigraphic relations and depositional environment would undoubtedly be very fruitful.

REFERENCES

Allen, R. M., Jr., 1963, Geology and mineral resources of Greene and Madison Counties: Virginia Div. Min. Resources Bull. 78, 102 p.

Bascom, Florence, 1896, The ancient volcanic rocks of South Mountain, Pennsylvania: U. S. Geol. Survey Bull. 136, 124 p.

Bloomer, R. O., and Bloomer, R. R., 1947, The Catoctin Formation in central Virginia: Jour. Geology, v. 55, p. 94–106.

————, and Werner, H. J., 1955, Geology of the Blue Ridge region in central Virginia: Geol. Soc. America Bull., v. 66, p. 579–606.

Brown, W. R., 1958, Geology and mineral resources of the Lynchburg quadrangle, Virginia: Virginia Div. Min. Resources Bull. 74, 99 p.

Buddington, A. F., 1952, Chemical petrology of some metamorphosed Adirondack gabbroic, syenitic and quartz syenitic rocks: Am. Jour. Sci., Bowen Volume, pt. 1, p. 37–84.

Cloos, Ernst, 1941, Geologic map of Washington County: Baltimore, Maryland Geol. Survey, scale 1:62,500.

————, 1947, Oölite deformation in the South Mountain fold, Maryland: Geol. Soc. America Bull., v. 58, p. 843–918.

————, 1950, The geology of the South Mountain anticlinorium, Maryland: Johns Hopkins Univ. Studies in Geology, no. 16, pt. 1, 28 p.

————, 1951a, Stratigraphy of sedimentary rocks, *in* The physical features of Washington County. Maryland Dept. Geology, Mines, and Water Resources, Washington County [Rep. 14] p. 17–94.

————, 1951b, Structural geology of Washington County, *in* The physical features of Washington County: Maryland Dept. Geology, Mines, and Water Resources, Washington County [Rep.], p. 124–163.

————, 1957, Blue Ridge tectonics between Harrisburg, Pennsylvania, and Asheville, North Calolina: Natl. Acad. Sci. Proc., v. 43, no. 9, p. 834–839.

————, 1958, Structural geology of South Mountain and Appalachians in Maryland, Guidebooks 4–5: Johns Hopkins Univ. Studies in Geology, no. 17, 85 p.

————, 1966, Dating Blue Ridge deformation plan with slickensides and lineations [abs.]: Geol. Soc. America, Northeastern Section Ann. Mtg. Program, p. 17.

————, and Hietanen, Anna, 1941, Geology of the "Martic overthrust" and the Glenarm Series in Pennsylvania and Maryland: Geol. Soc. America Spec. Paper no. 35, 207 p.

Diggs, W. E., 1955, Geology of the Otter River area, Bedford County, Virginia: Virginia Polytechnic Inst. Bull., Eng. Exp. Sta. Series no. 101, 23 p.

Espenshade, G. H., 1954, Geology and mineral deposits of the James River–Roanoke River manganese district, Virginia: U. S. Geol. Survey Bull. 1008, 155 p.

Fellows, R. E., 1943, Recrystallization and flowage in Appalachian quartzite: Geol. Soc. America Bull., v. 54, p. 1399–1432.

Furcron, A. S., 1934, Igneous rocks of the Shenandoah National Park area: Jour. Geology v. 42, p. 400–410.

————, 1935, James River iron and marble belt: Virginia Geol. Survey Bull. 39, 124 p.

————, 1939, Geology and mineral resources of the Warrenton quadrangle, Virginia: Virginia Geol. Survey Bull. 54, 94 p.

Glass, J. J., Koschmann, A. H., and Vhay, J. S. 1958, Minerals of the cassiterite-bearing veins at Irish Creek, Virginia, and their paragenetic relations: Econ. Geology, v. 53, p. 65–84.

Gooch, E. O., 1958, Infolded metasedimentary rocks near the axial zone of the Catoctin Mountain–Blue Ridge anticlinorium in Virginia: Geol. Soc. America Bull., v. 69, p. 569–574.

Gray, Carlyle, and others, 1960, Geologic map of Pennsylvania: Harrisburg, Pennsylvania Geol. Survey, 4th ser., scale 1:250,000.

Herz, Norman, in press, The Roseland alkalic anorthosite massif, Virginia, *in* Isachsen, Y. W., ed., Origin of anorthosite: New York State Museum Mem.

Howie, R. A., 1964, Charnockites: Sci. Progress, v. 52, p. 628–644.

Jonas, A. I., 1927, Geologic reconnaissance in the Piedmont of Virginia: Geol. Soc. America Bull., v. 38, p. 837–846.

————, 1935, Hypersthene granodiorite in Virginia: Geol. Soc. America Bull., v. 46, p. 47–60.

————, and Stose, G. W., 1938, Geologic map of Frederick County and adjacent parts of Washington and Carroll Counties: Baltimore, Maryland Geol. Survey, scale 1:62,500.

————, 1939, Age relation of the pre-Cambrian rocks in the Catoctin Mountain–Blue Ridge and Mount Rogers anticlinoria in Virginia: Am. Jour Sci., v. 237, p. 575–593.

Keith, Arthur, 1894, Geology of the Catoctin belt. U. S. Geol. Survey 14th Ann. Rept., pt. 2, p. 285–395.

King, P. B., 1949, The base of the Cambrian in the southern Appalachians: Am. Jour. Sci., v. 247, p. 513–530; 622–645.

————, 1950, Geology of the Elkton area, Virginia: U. S. Geol. Survey Prof. Paper 230, 82 p.

————, 1964, Further thoughts on tectonic framework of the southeastern United States, in Lowry, W. D., ed., Tectonics of the southern Appalachians: Virginia Polytech. Inst., Dept. Geol. Sci. Mem. 1, p. 5–31.

————, in press, The Precambrian of the United States of America: Southeastern United States, in Rankama, Kalervo, ed., The Precambrian, v. 4: New York, Intersci. Pub. (John Wiley and Sons).

Koschmann, A. H., Glass, J. J., and Vhay, J. S., 1942, Tin deposits of Irish Creek, Virginia: U. S. Geol. Survey Bul. 936-K, p. 271–296.

Long, L. E., Kulp, J. L., and Ecklemann, F. D., 1959, Chronology of major metamorphic events in the southeastern United States: Am. Jour. Sci., v. 257, p. 585–603.

Mertie, J. B., Jr., 1956, Paragneissic formations in northern Virginia [abs.]: Geol. Soc. America Bull., v. 67, p. 1754–1755.

Moore, C. H., Jr., 1940, Origin of the nelsonite dikes of Amherst County, Virginia: Econ. Geology, v. 35, p. 629–645.

Nelson, W. A., 1932, [Rockfish conglomerate, Virginia]: Wash. Acad. Sci. Jour., v. 22, p. 456–457.

————, 1962, Geology and mineral resources of Albemarle County: Virginia Div. Mineral Resources Bull. 77, 92 p.

Nickelsen, R. P., 1956, Geology of the Blue Ridge near Harper's Ferry, West Virginia: Geol. Soc. America Bull., v. 67, p. 239–270.

Pegau, A. A., 1932, Pegmatite deposits of Virginia: Virginia Geol. Survey Bull. 33, 123 p.

Redden, J. A., 1963, Stratigraphy and metamorphism of the Altavista area, in Weinberg, E. L., and others, Geological excursions in southwestern Virginia: Virginia Polytechnic Inst., Eng. ext. Ser., Geol. Guidebook no. 2, p. 77–86.

Reed, J. C., Jr., 1955, Catoctin Formation near Luray, Virginia: Geol. Soc. America Bull., v. 66, p. 871–896.

————, 1964, Chemistry of greenstone of the Catoctin Formation in the Blue Ridge of central Virginia: U. S. Geol. Survey Prof. Paper 501-C, p. C69–C73.

Rodgers, John, 1964, Basement and no-basement hypotheses in the Jura and the Appalachian Valley and Ridge, in W. D. Lowry, ed., Tectonics of the southern Appalachians: Virginia Polytech. Inst., Dept. Geol. Sci. Mem. 1, p. 71–80.

Ross, C. S., 1941, Occurrence and origin of the titanium deposits of Nelson and Amherst Counties, Virginia: U. S. Geol. Survey Prof. Paper 198, 59 p.

Smith, J. W., Milici, R. C., and Greenberg, S. S., 1964, Geology and mineral resources of Fluvanna County: Virginia Div. Mineral Resources Bull. 79, 62 p.

Stose, A. J., and Stose, G. W., 1946, Geology of Carroll and Frederick Counties, in The physical features of Carroll County and Frederick County: Maryland Dept. Geology, Mines, and Water Resources, Carroll and Frederick Counties [Rep.], p. 11–131.

Stose, G. W., 1932, Geology and mineral resources of Adams County, Pennsylvania: Pennsylvania Geol. Survey, 4th ser., Bull. Cl, 153 p.

Subramaniam, A. P., 1959, Charnockites of the type area near Madras—a reinterpretation: Am. Jour. Sci., v. 257, p. 321–353.

Tilton, G. R., Wetherill, G. W., Davis, G. L., and Bass, M. N., 1960, 1,000-million-year-old minerals from the eastern United States and Canada: Jour. Geophys. Research, v. 65, p. 4173–4179.

Toewe, E. C., 1966, Geology of the Leesburg quadrangle, Virginia: Virginia Div. Mineral Resources Rept. Inv. 11, 52 p.

Virginia Div. Mineral Resources, 1963, Geologic map of Virginia: Charlottesville, scale 1:500,000.

Virginia Geological Survey, 1928, Geologic map of Virginia: Charlottesville, scale 1:500,000.

Watson, T. L., and Cline, J. H., 1916, Hypersthene syenite and related rocks of the Blue Ridge region, Virginia: Geol. Soc. America Bull., v. 27, p. 193–234.

————, and Taber, Stephen, 1913, Geology of the titanium and apatite deposits of Virginia: Virginia Geol. Survey Bull. 3-A, 308 p.

Werner, H. J., 1966, Geology of the Vesuvius quadrangle, Virginia: Virginia Div. Mineral Resources Rept. Inv. 7, 53 p.

Whitaker, J. C., 1955, Geology of Catoctin Mountain, Maryland and Virginia: Geol. Soc. America Bull., v. 66, p. 435–462.

Williams, G. H., 1892, The volcanic rocks of South Mountain in Pennsylvania and Maryland: Am. Jour. Sci., 3d ser., v. 44, p. 482–496.

Chapter 15

Structural and Metamorphic History
of the Southern Blue Ridge*

BRUCE BRYANT AND JOHN C. REED, JR.

INTRODUCTION

THE SOUTHERN BLUE RIDGE is defined for the purposes of this paper as that part of the Blue Ridge
geologic province southwest of Roanoke, Va. It
is bounded on the southeast by the Brevard fault
zone and James River synclinorium and on the
northwest by low-angle thrust faults that have
carried Precambrian plutonic and metamorphic
rocks of the Blue Ridge many miles northwestward across unmetamorphosed Paleozoic sedimentary rocks of the Unaka and Valley and
Ridge belts (King, 1955). To the northeast it is
continuous with the northern Blue Ridge (described by Espenshade in Chapter 14); to the
southwest it disappears beneath Cretaceous deposits of the Coastal Plain in central Alabama.

STRATIGRAPHY

The southern Blue Ridge consists of a complex
terrane of plutonic gneisses and intricately deformed and variously metamorphosed sedimentary and volcanic rocks of Precambrian age
invaded by a diverse array of intrusive rocks of
Precambrian and Paleozoic age.

The plutonic rocks comprise a wide variety of
granites, layered and nonlayered granitic gneisses
and migmatitic rocks that crop out chiefly along
the northwestern edge of the Blue Ridge in
western North Carolina and southwestern Virginia (Keith, 1904, 1905, 1907b; Stose and Stose,
1957; Bryant and Reed, 1962; Hadley and Goldsmith, 1963).

* Publication authorized by the Director, U. S. Geological Survey.

The more massive granitic-textured rocks in
the plutonic complex were mapped as Cranberry
and Max Patch Granites by Keith (1903, 1904).
They range in composition from quartz diorite to
granite, but most are granodiorites or quartz
monzonites. Their original textures ranged from
fine and even grained to coarsely porphyritic or
porphyroblastic, but subsequent shearing and recrystallization has reduced many of the rocks to
strongly foliated flaser and augen gneisses. These
granitic rocks pass transitionally into migmatites
and layered plutonic gneisses in which thin
granitic layers are thinly intercalated in all proportions with biotite and hornblende schist, amphibolite, and other nongranitic rocks. Although
some of the granitic rocks may be intrusive, the
field relations suggest that many are products of
granitization of preexisting sedimentary and volcanic rocks (Hamilton, 1960; Bryant, 1962;
Hadley and Goldsmith, 1963).

Lead–uranium isotopic ratios in euhedral zircons from the granitic gneisses indicate true ages
of 1,000–1,100 m.y. (Davis and others, 1962).
These ages are comparable to zircon ages from
the Baltimore Gneiss, from basement rocks in the
Blue Ridge in northern Virginia, and from rocks
of the Grenville province in Canada (Tilton and
others, 1960). These plutonic rocks are the oldest
dated rocks in the Blue Ridge and are considered
to be of older Precambrian age in terms of the
local informal usage of the U. S. Geological
Survey.

Much of the southeastern part of the Blue
Ridge in southwestern Virginia, North Carolina,
and Georgia is underlain by mica schist, mica

gneiss, and amphibolite, obviously derived from metamorphism of sedimentary and volcanic rocks. For many years all these rocks have been considered early or middle Precambrian in age because in several places they stratigraphically underlie upper Precambrian sedimentary rocks and locally they grade through migmatites into older Precambrian plutonic gneisses (Keith, 1904, 1905, 1907b; Brobst, 1962; Bryant and Reed, 1962; Hadley and Goldsmith, 1963). However, recent work by Hadley (Chapter 17 this volume) and Rankin (Chapter 16 this volume) has shown that a large part of this terrane consists of metamorphosed upper Precambrian rocks, although older rocks may be present in some areas.

In the northwestern part of the Blue Ridge, older Precambrian plutonic basement rocks are overlain unconformably by upper Precambrian clastic and volcanic strata, many of which are little metamorphosed. These younger Precambrian rocks comprise the Ocoee Series in North Carolina, Tennessee, and northern Georgia, the Mount Rogers Formation (Rankin, Chapter 16 this volume) and Lynchburg Formation in Virginia, and possibly part of the Talladega Slate in western Georgia and Alabama. Upper Precambrian rocks of low metamorphic grade are also exposed in the Grandfather Mountain window (lat. 36°00′; long. 81°45′) in northwestern North Carolina, where they have been assigned to the Grandfather Mountain Formation. The stratigraphy of these upper Precambrian rocks and their relations to the more metamorphosed rocks in the southeastern part of the Blue Ridge are discussed elsewhere in this volume by J. B. Hadley (Chapter 17) and by D. W. Rankin (Chapter 16).

Rocks southeast of the Dahlonega fault in Georgia and Alabama (see tectonic map at end of this volume) have not yet been studied in detail. For almost 200 mi these rocks are bounded on both the northwest and southeast by faults, so that the establishment of their stratigraphic relations to the flanking rocks is difficult. They have been considered Paleozoic in age, but future work may show that some or all of them are of late Precambrian age.

The Chilhowee Group of Early Cambrian(?) and Early Cambrian age is the basal clastic deposit of the Appalachian miogeosynclinal sequence northwest of the Blue Ridge (Colton, Chapter 2 this volume). These clastic rocks, and younger Cambrian and Ordovician strata, occur in thrust sheets in the Unaka belt along the northwest flank of the Blue Ridge and in windows beneath the Blue Ridge thrust sheet. The transition between Paleozoic miogeosynclinal and other facies occurred somewhere southeast of the present Blue Ridge, for allochthonous Lower Cambrian(?) and Lower Cambrian miogeosynclinal rocks in the Grandfather Mountain window must originally have been derived southeast of the present position of Brevard fault zone.

Other rocks of possible Paleozoic age occur in the Talladega belt along the northwestern edge of the Blue Ridge in Alabama and Georgia and in the Murphy syncline in northwestern Georgia and southwestern North Carolina (Keith, 1907a). These rocks are discussed by Hadley (Chapter 17 this volume).

INTRUSIVE ROCKS

Intrusive rocks within these various complexes are of diverse types and ages. Besides the gneissic granitic rocks previously discussed, the older Precambrian basement also contains large sharply bounded plutons of more felsic intrusive rocks—the Striped Rock Granite (Stose and Stose, 1957) in southwestern Virginia; the Beech Granite (Keith, 1903; Bryant, 1962) north and west of the Grandfather Mountain window; the Brown Mountain Granite (Reed, 1964b) in the Grandfather Mountain window; and the Whiteside Granite (Keith, 1907b) in western South Carolina and adjacent parts of North Carolina and Georgia. These rocks are largely medium to coarse grained, massive to rudely foliated light-colored granites or quartz monzonites; generally they occur in large sills or domes with contacts nearly concordant with foliation in the enclosing rocks. Because of the mineralogy and massive character of the rocks, effects of later shearing and metamorphism are commonly inconspicuous, but most display at least faint foliation and lineation produced during Paleozoic deformation.

The Striped Rock, Beech and Brown Mountain Granites invade only older Precambrian plutonic rocks. Rankin (1968) has noted petrographic similarities that indicate that the granites are related to volcanic rocks in the Grandfather Mountain and Mount Rogers Formations and has suggested that they are of late Precambrian age. This is in accord with radiometric zircon ages (Davis and others, 1962) which indicate a possible age of 800–900 m.y. for the Beech Granite. The Whiteside Granite locally invades rocks tentatively identified by A. E. Nelson (written communication, 1967) as belonging to the Ocoee Series. Because this granite does not display obvious effects of Paleozoic shearing and metamorphism, it has been considered of late Paleozoic age (Keith, 1907b;

Griffitts and Overstreet, 1952). It has yielded zircon with lead–alpha ages as old as 710 m.y. and monazite with ages ranging from 360 to 440 m.y. (Jaffe and others, 1959)). Overstreet and Bell (1965, p. 105) favor the interpretation that the Whiteside is of Paleozoic age and that the zircon is a relict detrital mineral, but they recognize the possibility that the granite is Precambrian and that the monazite ages are due to lead loss during Paleozoic metamorphism. The Whiteside may be composed in part of mobilized basement rock, as has been suggested by Hadley and Goldsmith (1963) for two smaller plutons east of the Great Smoky Mountains.

Swarms of dikes and small irregular plugs of variously metamorphosed gabbro, dolerite, and basalt are widespread in the Blue Ridge from the vicinity of Asheville, N.C., northeastward into Virginia. They are particularly abundant west of the Grandfather Mountain window where they were referred to as Bakersville Gabbro by Keith (1903). Although some of the mafic rocks are essentially unmetamorphosed, many have been converted to greenschist or amphibolite during Paleozoic regional metamorphism (Wilcox and Poldervaart, 1958). Mafic dikes of this group cut older and upper Precambrian rocks and are apparently the feeders for some of the basalt flows intercalated in the Mount Rogers Formation; they may also be the source of mafic sills and flows now represented by amphibolite layers in the Ashe Formation (Rankin, Chapter 16 this volume) and in the upper Precambrian rocks west and southwest of the Grandfather Mountain window (Hadley, Chapter 17 this volume). Similar mafic intrusions (Linville Metadiabase) in the Grandfather Mountain window may have fed the basalt flows in the Grandfather Mountain Formation (Bryant, 1962; Bryant and Reed, 1962).

A few dikes and sills of altered diorite and diabase cut rocks of the Ocoee Series in the central eastern parts of the Great Smoky Mountains (Hadley and Goldsmith, 1963). These bodies are considered to be of Paleozoic age by Hadley and Goldsmith (1963, p. 71) because they were apparently intruded after initial folding of the enclosing rocks but prior to the peak of Paleozoic regional metamorphism. Similar sills invade rocks below the Murphy Marble in the Murphy syncline (Van Horn, 1948; Fairley, 1965); they are probably no older than latest Precambrian and may be as young as middle Paleozoic. It is not clear whether these rocks are correlative with the basaltic rocks farther northwest or whether

they date from one or more entirely unrelated periods of intrusion.

The only intrusive rocks of well-established Paleozoic age in the southern Blue Ridge are the commercial mica-bearing pegmatites of Franklin–Sylva and Bryson City districts in southwestern North Carolina (lat. 35°05′ to 35°25′; long. 83°15′ to 83°35′) and similar pegmatites and associated stocks of leucogranodiorite ("alaskite" of many authors) in the Spruce Pine district in west-central North Carolina (lat. 36°00′; long. 82°05′). These pegmatites cut upper Precambrian rocks including the Bakersville Gabbro and have fairly consistent apparent radiometric ages of about 350 m.y. (Long and others, 1959). Although it is commonly assumed that these ages date the emplacement of the pegmatites some older ages are found for the pegmatite minerals in the Spruce Pine area (Davis and others, 1962), and wall rocks in the area contain micas with 350-m.y. ages. This suggests that the pegmatites there may have been emplaced somewhat earlier, prior to the climax of the metamorphic event reflected by the 350-m.y. ages. Micas from the wall rocks in the Franklin–Sylva district have older apparent ages than those in the pegmatites (Kulp and Eckelmann, 1961).

Small dikes of alkali granodiorite (trondjhemite) are found in a belt about 10 mi (16 km) wide and 50 mi (80 km) long, extending northeastward from near Waynesville, N.C. (lat. 35°30′; long. 83°00′) (Hadley and Goldsmith, 1963; Hadley, written communication, 1967). They occur chiefly in basement rocks, but locally cut upper Precambrian rocks of the Great Smoky Group and the Bakersville Gabbro. Bryant (1963) has mapped a swarm of similar granitic dikes and pegmatites near Deep Gap (lat. 36°15′; long. 81°30′) northeast of the Grandfather Mountain window. No radiometric ages are available for either of these groups of dikes. They may be of the same general age as the pegmatites near Spruce Pine, but because there is no significant overlap in their distribution the relation cannot be determined directly.

Rankin (1968) has suggested that the bodies of peraluminous granitic rocks near Stone Mountain (lat. 36°25′; long. 81°05′) and Mount Airy, N.C. (lat. 36°30′; long. 80°35′), resemble the granodiorites near Spruce Pine and may be part of the same magma series. As has been pointed out, some authors consider the Whiteside Granite to be of Paleozoic age. The quartz diorite of the Pinckneyville Granite (Prouty, 1923; Gault, 1945), the southwesternmost intrusive complex in the Blue Ridge (lat. 32°35′ to

33°10′; long. 85°50′ to 86°20′), is apparently syntectonic and probably of Paleozoic age.

Small bodies of metamorphosed ultramafic rocks are widespread in the southern Blue Ridge. Some intrude upper Precambrian strata, and at least one of those in the Spruce Pine district must be older than middle Paleozoic, for it is cut by a pegmatite at least 375 m.y. old (Kulp and Brobst, 1954; Davis and others, 1962). Thus, if all the ultramafic rocks are contemporaneous, they are probably of early Paleozoic age. However, it is quite possible that ultramafic bodies of several ages may be present.

METAMORPHISM

The older Precambrian basement rocks of the southern Blue Ridge were thoroughly metamorphosed at least once prior to deposition of the upper Precambrian Ocoee Series and its equivalents. The upper Precambrian rocks and the underlying basement were subsequently subjected to at least one and probably to several periods of deformation and regional metamorphism during the Paleozoic. Superposition of the effects of these various metamorphic episodes in many of the rocks and telescoping by large-scale low-angle thrusts (to be discussed later) has led to great difficulty in interpreting the complex structural and metamorphic history of the region.

The earliest recorded metamorphic event is the episode of plutonic metamorphism during which the 1,100-m.y. old basement rocks were formed. Textures and structures dating from this event have been largely obliterated, but the heterogeneous layered character of some of the gneisses and the bulk composition of many of the individual layers suggests that they were derived from sedimentary and volcanic rocks during this metamorphism (Bryant, 1962; Hadley and Goldsmith, 1963). The local occurrence of relic hypersthene in these rocks in northwestern North Carolina (Bryant, 1962) and Southwestern Virginia (Dietrich, 1959) and its more widespread occurrence in central and northern Virginia (Espenshade, Chapter 14 this volume), where the effects of Paleozoic metamorphism are less severe, suggests that granulite facies conditions may have at least locally prevailed during this metamorphism. In both the eastern Great Smoky Mountains and in the area northwest of the Grandfather Mountain window, these layered gneisses become more migmatitic toward the northwest and pass gradually into uniform granitic rocks apparently derived from them through feldspathization, partial melting, and local mobili-

zation. The same northwestward change from layered gneisses to more uniform granitic rocks is evident in the Grandfather Mountain window, but there the transition zone occurs at least 20 mi (32 km) southeast of the corresponding change in the surrounding rocks, all of which have subsequently been thrust northwestward across the older Precambrian rocks exposed in the window.

Nearly all the present metamorphic textures and mineral assemblages in the Blue Ridge rocks are products of regional dynamothermal metamorphism during the Paleozoic. In general, the grade of Paleozoic metamorphism increases southeastward (Fig. 1), in the opposite direction from the apparent gradient of the older Precambrian metamorphism (Hadley and others, 1955; Kulp and Eckelmann, 1961; Hadley and Goldsmith, 1963). Upper Precambrian rocks, some of which are essentially unmetamorphosed along the northwest edge of the Blue Ridge, pass progressively through the classic Barrovian zones, reaching the sillimanite zone in a narrow belt in the southeastern part of the Blue Ridge in northern Georgia and southwestern North Carolina (Hadley and Nelson, written communication, 1967). The underlying older Precambrian basement rocks were, of course, subjected to the same metamorphic gradient. They show obvious retrograde effects where the Paleozoic metamorphism was of low grade, but farther southeast the grade of Paleozoic metamorphism approaches that of the older Precambrian metamorphism, and the effects of the two become increasingly difficult to distinguish.

Evidence is widespread for at least two generations of superimposed folds and cleavages in the upper Precambrian rocks of the Blue Ridge (Hurst, 1955; Hamilton, 1961; Hadley and Goldsmith, 1963; King, 1964a; Neathery, 1965). The time relations between folding and Paleozoic metamorphism are not everywhere clear and consistent, but in most cases metamorphism apparently began during the earliest stages of folding and reached a climax during or somewhat after the formation of the second folds. In many areas, the climax of regional metamorphism was followed by shearing and retrogressive metamorphism which produced widespread cataclastic textures and structures and altered the older medium- and high-grade mineral assemblages. Much of this cataclasis and retrogression is evidently related to large-scale low-angle overthrusting during the Paleozoic, discussed below.

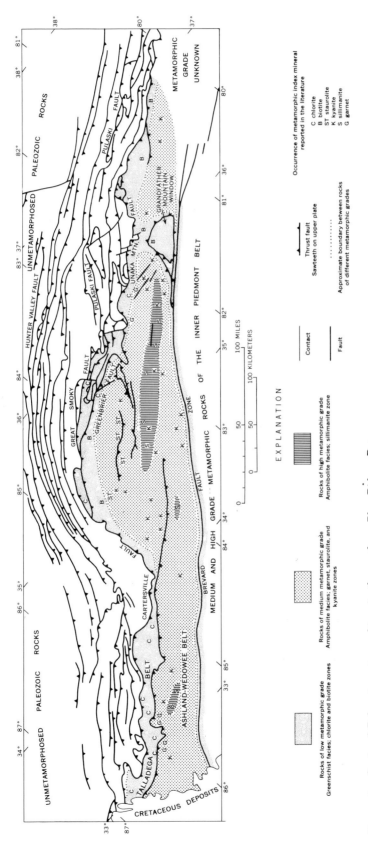

FIGURE 1. Grade of Paleozoic metamorphism in the southern Blue Ridge. Data from numerous published sources.

THRUST FAULTS

The large-scale low-angle overthrusts in the Paleozoic rocks of the Valley and Ridge belt have long been recognized (Hayes, 1891). Keith (1902, 1907c) described similar faults involving the crystalline rocks along the northwestern flank of the Blue Ridge, but it is only within the last few decades that it has been realized that the well-known thrust faults in the Valley and Ridge belt are matched by thrust faults of comparable magnitude and extent in the Blue Ridge. Everywhere along the northwest edge of the Blue Ridge, variously metamorphosed Precambrian and Paleozoic rocks are thrust northwestward over the imbricately faulted but unmetamorphosed Paleozoic miogeosynclinal rocks of the Valley and Ridge belt. The entire Blue Ridge belt is known to be allochthonous in the vicinity of the Grandfather Mountain window and may be allochthonous throughout the southern Appalachians (King, 1964b).

The Grandfather Mountain window extends nearly to the Brevard fault zone on the southeast side of the Blue Ridge (see map, fig. 2). Precambrian plutonic and medium-grade metamorphic rocks in the thrust sheet surrounding the window tectonically overlie upper Precambrian and Cambrian sedimentary and volcanic rocks in the window that have never been metamorphosed beyond low grade. At least 35 mi (55 km) of northwestward transport was required to bring the plutonic rocks of the northwest edge of the Blue Ridge thrust sheet from southeast of the Grandfather Mountain window (Bryant and Reed, 1962; Bryant, 1962, 1963; Reed, 1964a,b).

King (1964a) estimates postmetamorphic transport along the Great Smoky fault in the Great Smoky Mountains as more than 12 mi (19 km) and probably less than 24 mi (38 km), but we believe that the actual amount may have been greater. The great number of faults in the Valley and Ridge belt northwest of the northwestward bulge of the Blue Ridge may indicate that the maximum northwestward transport of the Blue Ridge rocks may have occurred in this segment.

The complex internal structure of the Great Smokies also suggests particularly great displacement there. The Great Smoky Group of the Ocoee Series moved northwestward (Hadley and Goldsmith, 1963; King, 1964a) on premetamorphic faults over the Snowbird Group, a different facies of the Ocoee. Hadley and Goldsmith (1963) suggest a minimum transport of 15 mi (25 km) on the Greenbrier fault, the principal premetamorphic fault.

At the southern end of the Blue Ridge, the stratigraphic and structural relations shown by Butts (1940) suggest that low-grade metamorphic rocks of the Talladega belt have been thrust at least 15 mi (25 km) northwestward over unmetamorphosed rocks of the Valley and Ridge belt along the Cartersville fault. Similar relations are suggested also by less detailed maps in northeast Alabama (Warman and Causey, 1962; Causey, 1965).

In southern Virginia, just south of Roanoke, the pattern of mapped faults (Woodward, 1932) indicates a minimum northwestward displacement of the Blue Ridge rocks of at least 8 mi (13 km).

The regional gravity low that extends along the Appalachian miogeosynclinal belt to central Virginia, cuts across the northwestward bulge of the Blue Ridge in southern Virginia, and in North Carolina it reaches the western edge of the Piedmont (Am. Geophys. Union and U. S. Geol. Survey, 1964). In Georgia the low becomes less pronounced and may be traced along the Brevard fault zone to the Coastal Plain.

If the regional gravity low of the Appalachians represents a continuous feature in the upper part of the earth's crust, as suggested by Griscom (1963), its change of position from the Valley and Ridge belt to the eastern Blue Ridge and western Piedmont may indicate as much as 80 mi (130 km) gross northwestward transport of the Blue Ridge rocks over the low. (Compare trend of gravity low with depth to Precambrian rocks (Colton, 1961, Fig. 4)) The possibility arises that the southern Blue Ridge may be underlain by a considerable thickness of rocks of early Paleozoic age, as suggested by King (1964b, p. 19). Negative anomalies are greatest where the Blue Ridge thrust sheet is thinnest, as around the Grandfather Mountain window, and are considerably less to the southwest where the thrust sheet is probably thicker and is composed of higher grade rocks with considerable amphibolite. This is probably due to density differences in the near-surface rocks.

Northwestward transport of the thrust sheets of the southern Blue Ridge is established by several lines of evidence. Around the Grandfather Mountain window, pervasive northwest-trending lineation (mineral streaking, grooving) is demonstrably in the direction of tectonic transport. Late minor folds in the thrust sheet are generally asymmetric toward the northwest. King (1964a) and Smith (1960) also report northwest mineral lineation along faults at the western margin of the Blue Ridge. Cloos (1957) found northwest lineations along the northwestern margin of the Blue Ridge from North Carolina to Pennsylvania, although thrust faults are absent in northern Virginia, Maryland, and Pennsylvania. The occurrence of windows containing unmetamorphosed lower Paleozoic miogeosynclinal rocks beneath metamorphosed upper Precambrian rocks in the Great Smoky Mountains (King, 1964a; Neuman and Nelson, 1965) and in the Hot Springs window (Oriel, 1950) shows that the Blue Ridge must have moved nothwestward, normal to the regional trend of the metamorphic zones (Fig. 1).

Clearly the thrust faults in the Blue Ridge did not cut steeply into the old continental crust, for the crystalline rocks in the thrust sheets are ordinary micaceous and hornblendic metamorphic and plutonic rocks, too silicic to have been derived from the lower crust or upper mantle. The thrust sheets have nevertheless traveled tens of miles northwestward over younger rocks. Thus the faults at depth must have been gently inclined or subhorizontal. This poses questions regarding the mechanism of thrusting. Perhaps the shearing originated in an essentially horizontal zone of transition between amphibolite and granulite facies rocks where conditions were favorable for dehydration and conversion of mica to feldspar and amphibole to pyroxene. Such a process might produce an increase in fluid pressure and a decrease in rock strength and furnish a mechanism to facilitate sliding of the sheet, as suggested by Heard and Rubey (1966).

In several places, fault zones near the exposed soles of thrust sheets in the Blue Ridge are marked by phyllonites, indicating the availability of water during shearing. These rocks are now far from any deep-seated environment, and some may not have been in such an environment at any time during the Paleozoic. The water required may have either migrated along fault planes from a deeper environment where dehydration was taking place or been derived by progressive metamorphism of upper Precambrian or lower Paleozoic sedimentary rocks beneath or in the upper part of the thrust sheet. In any case the formation of phyllonite from plutonic rock indicates a high fluid pressure, which could have been a factor contributing to the long travel of the various thrust sheets indicated by geologic evidence. The possibility of a genetic connection between these thrust faults and the Brevard fault zone is discussed in a separate paper (Reed, Bryant, and Myers, Chapter 18 this volume).

RELATIONS BETWEEN THE BLUE RIDGE THRUST SHEET AND THRUST SHEETS OF THE UNAKA BELT IN NORTH-EASTERN TENNESSEE

In northeastern Tennessee and southwestern Virginia the Blue Ridge is flanked on the northwest by a belt of mountains underlain by thrust sheets of upper Precambrian and Lower Cambrian rocks which King (1955) has termed the Unaka belt. The thrust sheets of the Unaka belt lie tectonically beneath the Blue Ridge thrust sheet, the principal thrust sheet surrounding the Grandfather Mountain window, and above the imbricate sheets of the Valley and Ridge belt. The structurally intermediate position of the Unaka sheets and stratigraphy of the rocks in them indicate that they probably originated somewhere between the initial sites of Valley and Ridge and Blue Ridge terranes. The structural analysis that follows indicates further that the Unaka sheets probably came from a site between the present Grandfather Mountain window and the Valley and Ridge terrane, although it is also possible that the sheets came from southeast of the window.

At least three major structural layers occur in the Unaka belt of northeastern Tennessee: the Mountain City window (including the Limestone Cove and Doe River inner windows), the Shady Valley thrust sheet, and the Buffalo Mountain thrust sheet (Fig. 2). All these layers are overridden by the Blue Ridge thrust sheet (Figs. 2 and 3. The thrust sheets of the Unaka belt are composed principally of unmetamorphosed Lower Cambrian(?) and Lower Cambrian rocks of the Chilhowee Group, Shady Dolomite, and Rome Formation, but they contain some upper Precambrian strata and older Precambrian basement rocks in their lower parts. In its upper part, the Shady Valley sheet contains rocks as young as the Sevier Shale of Middle Ordovician age.

King and Ferguson (1960, p. 79, 83) showed that the Shady Valley thrust sheet originated southeast of the Mountain City window. Rocks in the Shady Valley and Buffalo Mountain thrust sheets are unmetamorphosed and were not deformed prior to thrusting; clearly they were not subjected to the intense folding and low-grade metamorphism that affected rocks now exposed in the Grandfather Mountain window. Rocks of the Chilhowee Group in the Shady Valley thrust sheet are in stratigraphic contact with plutonic basement rocks. The thrust sheet, therefore, could not have originated in the area of the Grand-

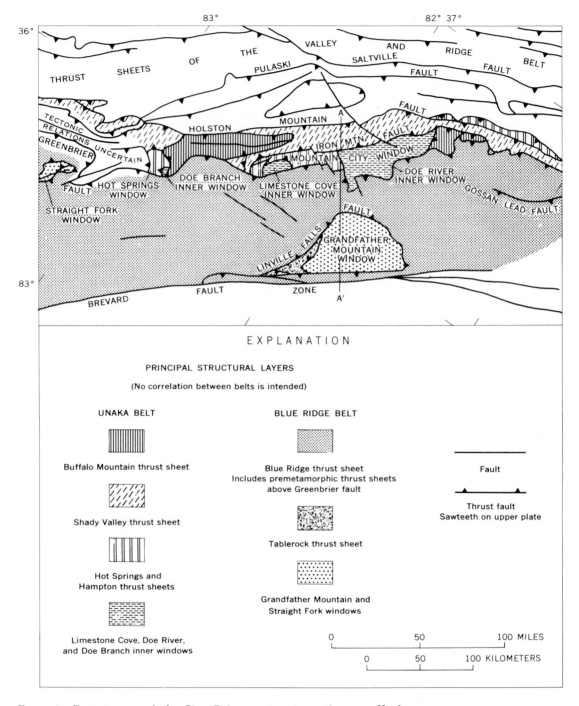

FIGURE 2. Tectonic map of the Blue Ridge province in northwestern North Carolina, northeastern Tennessee, and southwestern Virginia. Modified from P. B. King (*in* U. S. Geol. Survey and Am. Assoc. Petroleum Geologist, 1961). Northern boundaries of Mountain City window from D. W. Rankin (oral communication).

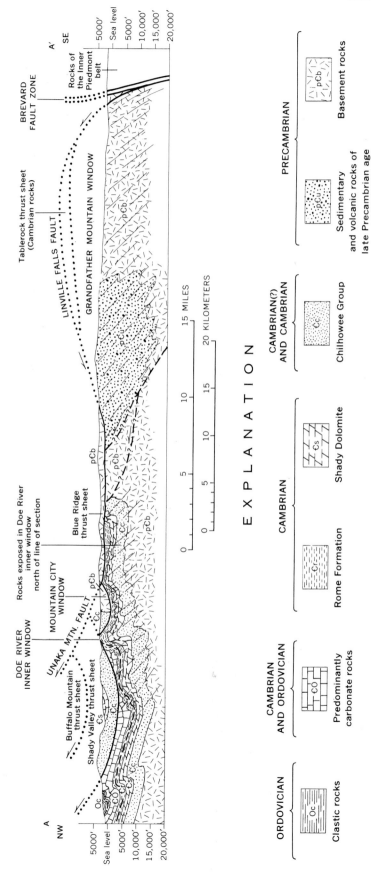

FIGURE 3. Geologic section of the Blue Ridge province in northwestern North Carolina and northeastern Tennessee, showing relations between structural layers of the Unaka belt, the Blue Ridge thrust sheet, and the Grandfather Mountain window. Line of section shown on Figure 2. Tectonic units shown in same pattern as Figure 2. Vertical scale same as horizontal.

EXPLANATION

ORDOVICIAN
Oc
Clastic rocks

CAMBRIAN AND ORDOVICIAN
CO
Predominantly carbonate rocks

CAMBRIAN
Cr
Rome Formation

Cs
Shady Dolomite

CAMBRIAN(?) AND CAMBRIAN
Cc
Chilhowee Group

PRECAMBRIAN
pCu
Sedimentary and volcanic rocks of late Precambrian age

pCb
Basement rocks

222   CHAPTER 15

father Mountain window, where a thick sequence of upper Precambrian rocks lies on the basement. The source of the sheet must, therefore, have been either between the Grandfather Mountain and Mountain City windows or southeast of the Grandfather Mountain window, beyond the southeastern edge of the late Precambrian basin of deposition. The Buffalo Mountain thrust sheet must have traveled at least 10 mi (16 km) farther than the Shady Valley thrust sheet in order to override it.

Southeast-dipping cleavage in the southeast part of the Buffalo Mountain thrust sheet is younger than thrusting (Ordway, 1959). Basement rocks in the thrust sheets display none of the pervasive cataclasis to conspicuous in the Blue Ridge thrust sheet and Grandfather Mountain window. The Shady Valley thrust sheet rests on unmetamorphosed rocks as young as Middle Ordovician that were folded prior to or during emplacement of the thrust sheet. The complicated folds and faults in the Doe River inner window contrast with simple broad folds of the Shady Valley thrust sheet (Fig. 3).

Derivation of the Shady Valley thrust sheet from southeast of the Grandfather Mountain window would require that rocks of the thrust sheet somehow escaped Paleozoic deformation and metamorphism, which strongly affected all known terranes in and southeast of the window. It, therefore, seems more probable that the source of the Shady Valley thrust sheet lay between the unmetamorphosed rocks of the Mountain City window and the metamorphosed rocks of the Grandfather Mountain window. If the Shady Valley thrust sheet originated in this interval, however, the original distance between the two windows must have been telescoped by a major thrust fault that is now hidden beneath the Blue Ridge thrust sheet (Fig. 3), for the present distance between the two windows is less than the exposed width of the Shady Valley thrust sheet. Thus, the rocks exposed in the Grandfather Mountain window are probably part of a thrust sheet beneath the Blue Ridge thrust sheet.

GEOCHRONOLOGY AND THE STRATIGRAPHIC RECORD OF TECTONISM

The Paleozoic metamorphic and tectonic events that have affected the southern Blue Ridge are nowhere recorded by contemporaneous sedimentary deposits within the region itself. The only means of dating these events involve isotopic dating of minerals from the metamorphic rocks and examination of the stratigraphic record in nearby

parts of the Appalachian miogeosyncline northwest of the Blue Ridge.

Radiometric ages of minerals furnish some of the principal clues to the metamorphic and structural history of the Blue Ridge, but radiometric age determinations on minerals from polymetamorphic rocks are fraught with uncertainties (compare Thomas, 1963, with Fullagar and Bottino, 1968). Hart (1964) found that K/Ar and Rb/Sr ratios are changed at temperatures as low as 200°C during contact metamorphism of crystalline rocks. The same temperature would be reached at a depth of only 4 mi (6.5 km) if the geothermal gradient was 30°C per km. As Hadley (1964) has suggested, mineral ages based on these ratios may merely record the last time the rocks cooled below some critical temperature rather than the date of crystallization or metamorphism.

Recognizing this uncertainty, Hadley (1964) compiled radiometric ages available in 1963 from the crystalline rocks of the southern Appalachians. He found a close correlation between Paleozoic K/Ar and Rb/Sr ages in the crystalline rocks of the Blue Ridge and Piedmont and the thicknesses of clastic deposits in the Appalachian basin.

The earliest frequency peak in the mineral ages is at about 430 m.y., or in the Early Silurian (Holmes, 1959; Kulp, 1961). Rapid deposition of clastics northwest of the Blue Ridge just prior to the recorded thermal event is indicated by a thick wedge of Middle Ordovician clastic rocks which rests disconformably on the Cambrian and Ordovician carbonate sequence in the southeastern part of the Valley and Ridge belt in Tennessee and adjacent parts of Georgia and Virginia. Pebbles in the Middle Ordovician conglomerates are composed of Cambrian and Ordovician carbonate rocks, of quartzite from the Chilhowee Group and Ocoee Series, and of vein quartz, feldspar, and volcanic rock (Kellberg and Grant, 1956). This detritus, which must have come from the southeast, indicates uplift of the Blue Ridge or of a terrane since overridden by the Blue Ridge thrust sheet. Perhaps uplift and erosion of the Blue Ridge crystalline rocks during the Middle Ordovician resulted in the cooling which is recorded by the radiometric dates in the Early Silurian. Sheets of rocks of the Ocoee Series, Chilhowee Group, and Cambrian–Ordovician carbonate section may have slid from the Ordovician tectonic high (Cooper, 1968). Possibly the premetamorphic faults, folds, and cleavage in the basement rocks of the Great Smoky Mountains (Hadley and Goldsmith, 1963), and early folds and cleavage in the Grandfather Mountain window date from this period of tectonic activity.

The next cluster of radiometric ages is in the range between 380 and 320 m.y. (Late Devonian to Late Mississippian according to Kulp, 1961), with a frequency peak at about 350 m.y. (Early Mississippian according to McDougall and others, 1966). Uplift and erosion of the Blue Ridge and Piedmont during this interval are indicated by a wedge of clastic rocks in the miogeosyncline. These Upper Devonian and Lower Mississippian deposits rest on a regional unconformity that bevels beds ranging in age from Early Devonian in northeastern Tennessee to as old as Early Ordovician in Georgia and Alabama (Rodgers, 1953).

Very few mineral ages from the southern Blue Ridge are younger than 320 m.y., although a number of ages from the eastern Piedmont cluster around 250 m.y., perhaps corresponding to the Permian clastic deposits in West Virginia, Ohio, and Pennsylvania. There seems to be no reflection in radiometric ages from the Blue Ridge of the wedge of Upper Mississippian and Pennsylvanian rocks in the Appalachian basin. These deposits thicken and coarsen southeastward and must have been derived from that direction. Other evidence (see below) suggests that the main episode of northwestward thrusting in the southern Blue Ridge may have occurred during this interval. Perhaps minerals whose radiometric ages reflect this episode are to be sought in the zones of sheared and retrogressively metamorphosed rocks along the postmetamorphic thrusts in the Blue Ridge.

The paucity of radiometric mineral ages in the southern Blue Ridge younger than 350 m.y. indicates that the climax of Paleozoic regional metamorphism occurred prior to the Early Mississippian. It is not clear, however, whether these were two main episodes of metamorphism, one prior to cooling in Early Silurian time and a second prior to cooling in Early Mississippian time, or whether metamorphism went on sporadically over a long interval, punctuated by two episodes of erosion, uplift, and cooling.

AGE OF THRUSTING

The oldest recognized thrust faults in the Blue Ridge are the Greenbrier and related faults in the Great Smoky Mountains. Hadley and Goldsmith (1963) have shown that metamorphic isograds cross the Greenbrier fault without offset. As the latest regional metamorphism in the Blue Ridge ended at least 350 million years ago, movement on the Greenbrier fault is assumed to have taken place before Early Mississippian time.

Most of the thrust faults along the northwestern edge of the Blue Ridge are postmetamorphic and clearly younger than the Greenbrier. The Great Smoky fault northwest of the Great Smoky Mountains involves rocks as young as Mississippian (Neuman and Nelson, 1965). Northeast of the Great Smoky Mountains, the Shady Valley thrust sheet apparently overrides the Pulaski fault, the southeasternmost of the imbricate thrusts in the Valley and Ridge belt (Fig. 2). The Pulaski involves rocks as young as Early Mississippian in Virginia and may be of the same general age as the thrusts farther northwest that involve Lower Pennsylvanian rocks. Thus, the Shady Valley thrust sheet and the tectonically higher Buffalo Mountain thrust sheet probably arrived in their present position in post-Early Mississippian and possibly in post-Early Pennsylvanian time. Latest movement of the Blue Ridge thrust sheet, which overrides both the Shady Valley and Buffalo Mountain sheets must have been even later.

An upper age limit for the major deformation is provided by a dike of unmetamorphosed diabase of probable Late Triassic age that extends without deflection from the Inner Piedmont belt across the Brevard fault zone and the Blue Ridge thrust sheet and into the Grandfather Mountain window (Reed, 1964a).

White (1950) has postulated Tertiary movements along a normal fault along the southeast side of the Blue Ridge physiographic province. Small young faults have been described (White, 1952; Conley and Drummond, 1965), but no major post-Triassic faulting has been proven in the southern Appalachians, although the region as a whole has no doubt been uplifted substantially since the Triassic (see Owens, Chapter 28 this volume).

REFERENCES

American Geophysical Union and U. S. Geol. Survey, 1964, Bouguer gravity anomaly map of the United States: scale 1:2,500,000.

Brobst, D. A., 1962, Geology of the Spruce Pine district, Avery, Mitchell, and Yancey Counties, North Carolina: U. S. Geol. Survey Bull. 1122-A, 26 p.

Bryant, B., 1962, Geology of the Linville quadrangle, North Carolina–Tennessee—A preliminary report: U. S. Geol. Survey Bull. 1121-D, 30 p.

————, 1963, Geology of the Blowing Rock quadrangle, North Carolina: U. S. Geol. Survey Geol. Quad. Map GQ-243.

————, and Reed, J. C., Jr., 1962, Structural and metamorphic history of the Grandfather Mountain area, North Carolina—A preliminary report: Am. Jour. Sci., v. 260, p. 161–180.

Butts, Charles, 1940, Description of the Montevallo and Columbiana quadrangles [Ala.]: U. S. Geol. Survey Geol. Atlas, Folio 226, 20 p.

Causey, L. V., 1965, Geologic rock-type map of Talladega County, Alabama: Alabama Geol. Survey, Map 38.

Cloos, Ernst, 1957, Blue Ridge tectonics between Harrisburg, Pennsylvania, and Asheville, North Carolina: Natl. Acad. Sci. Proc., v. 43, p. 834–839.

Colton, G. W., 1961, Geologic summary of the Appalachian Basin with reference to the subsurface disposal of radioactive waste solution: U. S. Geol. Survey TEI-791, 121 p., issued by U. S. Atomic Energy Comm. Tech. Inf. Service, Oak Ridge, Tenn.

Conley, J. F., and Drummond, K. M., 1965, Faulted alluvial and colluvial deposits along the Blue Ridge from near Saluda, North Carolina: Southeastern Geology, v. 7, p. 35–39.

Cooper, B. N., 1968, When was the Iron Mountain-Holston Mountain "thrust" Virginia and Tennessee formed? [abs.]: Geol. Soc. America Spec. Paper 101, p. 353–354.

Davis, G. L., Tilton, G. R., and Wetherill, G. W., 1962, Mineral ages from the Appalachian province in North Carolina and Tennessee: Jour. Geophys. Research, v. 67, p. 1987–1996.

Dietrich, R. V., 1959, Geology and mineral resources of Floyd County of the Blue Ridge Upland, southwestern Virginia; Va. Polytech. Inst. Bull. Eng. Expt. Sta. Ser., no. 134, 160 p.

Fairley, W. M., 1965, The Murphy syncline in the Tate quadrangle, Georgia: Georgia Dept. Mines, Mining and Geology Bull. 75, 71 p.

Fullagar, P. D., and Bottino, M. L., 1968, Comparison of Rb/Sr whole rock and mineral ages with K–Ar mineral ages of gneiss at Ore Knob, North Carolina [abs.]: Geol. Soc. America Spec. Paper 101, p. 74.

Gault, H. R., 1945, Petrography, structures and petrofabrics of the Pinkneyville quartz diorite, Alabama: Geol. Soc. America Bull. v. 56, p. 181–246.

Griffitts, W. R., and Overstreet, W. C., 1952, Granitic rocks of the western Carolina Piedmont: Am. Jour. Sci., v. 250, p. 777–789.

Griscom, Andrew, 1963, Tectonic significance of the Bouguer gravity field of the Appalachian system [abs.]: Geol. Soc. America Spec. Paper 73, p. 163–164.

Hadley, J. B., 1964, Correlation of isotopic ages, crustal heating, and sedimentation in the Appalachian region, in Lowry, W. D., ed., Tectonics of the southern Appalachians: Virginia Polytech. Inst., Dept. Geol. Sci. Mem. 1, p. 33–44.

————, and Goldsmith, Richard, 1963, Geology of eastern Great Smoky Mountains, North Carolina and Tennessee: U. S. Geol. Survey Prof. Paper 349-B, 118 p.

————, King, P. B., Neuman, R. B., and Goldsmith, Richard, 1955, Outline of the geology of the Great Smoky Mountains area, Tennessee and North Carolina, in Russell, R. J., ed., Guides to southeastern geology: New York, Geol. Soc. America, p. 390–411.

Hamilton, Warren, 1960, Description of the basement rocks, p. 13–17, in King, P. B., and Ferguson, H. W., Geology of northeasternmost Tennessee: U. S. Geol. Survey Prof. Paper 311, 136 p.

Hamilton, Warren, 1961 Geology of the Richardson cove and Jones cove quadrangles, Tennessee: U. S. Geol. Survey Prof. Paper 349-A, 55 p.

Hart, S. R., 1964, The petrology and isotopic-mineral age relations of a contact zone in the Front Range, Colorado: Jour. Geology, v. 72, p. 493–525.

Hayes, C. W., 1891, The overthrust faults of the southern Appalachians: Geol. Soc. America Bull., v. 2, p. 141–154.

Heard, H. C., and Rubey, W. W., 1966, Tectonic implications of gypsum dehydration: Geol. Soc. America Bull., v. 77, p. 741–760.

Holmes, Arthur, 1959, A revised geologic time scale: Edinburgh Geol. Soc. Trans., v. 17, p. 183–216.

Hurst, V. J., 1955, Stratigraphy, structure and mineral resources of the Mineral Bluff quadrangle, Georgia: Georgia Dept. Mines, Mining and Geology Bull. 63, 137 p.

Jaffe, H. W., Gottfried, David, Waring, G. L., and Worthing, H. W., 1959, Lead-alpha age determinations of accessory minerals of igneous rocks (1953–1957): U. S. Geol. Survey Bull. 1097-B, p. 65–148.

Keith, Arthur, 1902, Folded faults in the southern Appalachians [abs.]: Science (new ser.), v. 15, p. 822–823.

————, 1903, Description of the Cranberry quadrangle [N.C.–Tenn.]: U. S. Geol. Survey Geol. Atlas, Folio 90, 9 p.

————, 1904, Description of the Asheville quadrangle [N.C.–Tenn.]: U. S. Geol. Survey Geol. Atlas, Folio 116, 10 p.

————, 1905, Description of the Mount Mitchell quadrangle [N.C.–Tenn.]: U. S. Geol. Survey Geol. Atlas, Folio 124, 10 p.

————, 1907a, Description of the Nantahala quadrangle [N.C.–Tenn.]: U. S. Geol. Survey Geol. Atlas, Folio 143, 12 p.

————, 1907b, Description of the Pisgah quadrangle [N.C.–S.C.]: U. S. Geol. Survey Geol. Atlas, Folio 147, 8 p.

————, 1907c, Description of the Roan Mountain quadrangle [Tenn.–N.C.]: U. S. Geol. Survey Geol. Atlas, Folio 151, 12 p.

Kellberg, J. M., and Grant, L. F., 1956, Coarse conglomerates of the Middle Ordovician in the southern Appalachian Valley: Geol. Soc. America Bull., v. 67, p. 697–716.

King, P. B., 1955, A geologic section across the southern Appalachians—an outline of the geology in the segment in Tennessee, North Carolina, and South Carolina, in Russell, R. J., ed., Guides to southeastern geology: New York, Geol. Soc. America, p. 332–373.

————, 1964a, Geology of the central Great Smoky Mountains, Tennessee: U. S. Geol. Survey Prof. Paper 349-C, 148 p.

————, 1964b, Further thoughts on tectonic framework of the southeastern United States, in Lowry, W. D., ed., Tectonics of the southern Appalachians: Virginia Polytech. Inst., Dept. Geol. Sci. Mem. 1, p. 5–39.

————, and Ferguson, H. W., 1960, Geology of northeasternmost Tennessee. U. S. Geol. Survey Prof. Paper 311, 136 p.

Kulp, J. L., 1961, Geologic time scale: Science, v. 133, p. 1105–1114.

————, and Brobst, D. A., 1954, Notes on the dunite and the geochemistry of vermiculite at the Day Book dunite deposit, Yancey County, North Carolina: Econ. Geology, v. 49, p. 211–220.

————, and Eckelmann, F. D., 1961, Potassium-argon istopic ages on micas from the southern Appalachians: New York Acad. Sciences Annals, v. 91, p. 408–419.

Long, L. E., Kulp, J. L., and Eckelmann, F. D., 1959, Chronology of major metamorphic events in the southeastern United States: Am. Jour. Sci., v. 257, p. 585–603.

McDougall, I., Compston, W., and Bofinger, V. M., 1966, Isotopic age determinations on Upper Devonian rocks from Victoria, Australia: a revised estimate for the age of the Devonian–Carboniferous boundary: Geol. Soc. America Bull., v. 77, p. 1075–1088.

Neathery, T. L., 1965, Paragonite pseudomorphs after kyanite from Turkey Heaven Mountain, Cleborne County, Alabama: Am. Mineralogist, v. 50, p. 718–723.

Neuman, R. B., and Nelson, W. H., 1965, Geology of the western Great Smoky Mountains: U. S. Geol. Survey Prof. Paper 349-D, 81 p.

Ordway, R. J., 1959, Geology of the Buffalo Mountain–Cherokee Mountain area, northeastern Tennessee: Geol. Soc. America Bull., v. 70, p. 619–636.

Oriel, S. S., 1950, Geology and mineral resources of the Hot Springs window, Madison County, North Carolina: North Carolina Div. Mineral Resources Bull. 60, 70 p.

Overstreet, W. C., and Bell, Henry, III, 1965, The crystalline rocks of South Carolina: U. S. Geol. Survey Bull. 1183, 126 p.

Prouty, W. F., 1923, Geology and mineral resources of Clay County, with special reference to the graphite industry: Alabama Geol. Survey County Rept., no. 1, 190 p.

Rankin, D. W., 1968, Magmatic activity and orogeny in the Blue Ridge province in northwestern North Carolina and southwestern Virginia [abs.]: Geol. Soc. America Spec. Paper 115, p. 181.

Reed, J. C., Jr., 1964a, Geology of the Lenoir quadrangle, North Carolina: U. S. Geol. Survey Geol. Quad. Map GO-242.

————, 1964b, Geology of the Linville Falls quadrangle, North Carolina: U. S. Geol. Survey Bull., 1161-B, 53 p.

Rodgers, John, 1953, Geologic map of east Tennessee with explanatory text: Tennessee Div. Geology Bull. 58, 168 p.

Smith, J. W., 1960, Geology of the area along the Cartersville fault near Fairmont, Georgia [abs.]: Georgia Mineral Newsletter, v. 13, p. 107.

Stose, A. J., and Stose, G. W., 1957, Geology and mineral resources of the Gossan Lead district and adjacent areas: Virginia Div. Mineral Resources, Bul. 72, 233 p.

Thomas, H. H., 1963, Isotopic ages on coexisting hornblende, mica and feldspar [abs.]: Am. Geophys. Union Trans., v. 44, p. 110.

Tilton, G. R., Wetherill, G. W., Davis, G. W., and Bass, M. N., 1960, 1,000-million-year old minerals from the eastern United States and Canada: Jour. Geophys. Research, v. 65, p. 4173–4179.

U. S. Geological Survey and American Association of Petroleum Geologists, 1961, Tectonic map of the United States, exclusive of Alaska and Hawaii: 2 sheets, scale 1:2,500,000 [1962].

Van Horn, E. C., 1948, Talc deposits of the Murphy Marble belt: North Carolina Div. Mineral Resources Bull. 56, 54 p.

Warman, J. C., and Causey, L. V., 1962, Geology and ground-water resources of Calhoun County, Alabama: Alabama Geol. Survey County Rept. 7, 77 p.

White, W. A., 1950, Blue Ridge front—a fault scarp: Geol. Soc. America Bull., v. 61, p. 1309–1346.

————, 1952, Post-Cretaceous faults in Virginia and North Carolina: Geol. Soc. America Bull., v. 63, p. 745–748.

Wilcox, R. E., and Poldervaart, Arie, 1958, Metadolerite dike swarm in the Bakersville–Roan Mountain area, North Carolina: Geol. Soc. America Bull., v. 69, p. 1323–1367.

Woodward, H. P., 1932, Geology and mineral resources of the Roanoke area, Virginia: Virginia Geol. Survey Bull. 34, 172 p.

Chapter 16

Stratigraphy and Structure of Precambrian Rocks in Northwestern North Carolina*

DOUGLAS W. RANKIN

I have also discovered that many graduate students, both at my own university and elsewhere, have a warped attitude toward the study of metamorphic rocks. They think that petrography, physical chemistry, structural geology, and structural petrology will give answers that can be solved only by stratigraphy. I do not mean to discredit these other fields. . . . I merely regret that stratigraphy has been neglected.—Billings (1950).

INTRODUCTION

BY THE EARLY 1960's it was generally recognized that the Blue Ridge in northwestern Carolina consisted of a large allochthonous mass of crystalline rocks lying at least in part in the Blue Ridge thrust sheet. The Grandfather Mountain window, eroded through this sheet, exposes lower grade rocks ranging in age from older Precambrian through Early Cambrian. In the Blue Ridge thrust sheet, granitic rocks predominate to the northwest, as first recognized by Keith (1903, 1905, 1907) 60 years ago; they are structurally overlain by layered amphibolite and mica gneiss and schist to the southeast. Active controversy continues over whether any or all of these crystalline rocks are of late Precambrian age and over the origin and mutual relations of these rocks. Table 1 summarizes the major interpretations of the Blue Ridge crystalline rocks over the past 60 years.

Recent detailed and reconnaissance mapping in northwestern North Carolina and adjacent areas has shown that the only older Precambrian rocks are the granitic rocks along the northwest front of the Blue Ridge. This work has also demon-

* Publication authorized by the Director, U. S. Geological Survey.

strated that parts of the stratified sequence of upper Precambrian rocks near Mt. Rogers, Virginia, are in the Blue Ridge thrust sheet (Fig. 1). The older Precambrian granitic rocks form the core of a northeast-trending anticlinorium analogous to the Blue Ridge anticlinorium of northern Virginia (Espenshade, Chapter 14 this volume) and the Green Mountain anticline of Vermont. The granitic rocks in the core are overlain nonconformably on each side by upper Precambrian metasedimentary and metavolcanic rocks.

Metamorphic grade increases southeast across the anticlinorium, and a marked contrast in lithology of the upper Precambrian rocks on opposite limbs of the anticlinorium implies a facies change that, together with the distribution of metamorphic zones, places restrictions on the possible sites of origin for the rocks of various thrust sheets in the Blue Ridge and Unaka belts. The specific area from which the evidence is drawn is a sizable one, covering roughly 3,000 sq mi (7,800 sq km), and the conclusions reached have implications affecting a much larger area.

PREVIOUS WORK AND ACKNOWLEDGMENTS

Early reconnaissance mapping in northwestern North Carolina by Keith (1903, 1905, 1907) and by A. J. Stose (Jonas, 1932) has been summarized with considerable insight by King (1955). R. M. Hernon made a reconnaissance survey of some of the area, but his untimely death in 1965 forestalled the publication of his work. Some of his conclusions are much of the same as mine. In 1965, the U. S. Geological Sur-

TABLE 1. Major Contributions to or Interpretations of the Stratigraphy of the Crystalline Rocks of Northwestern North Carolina

	Keith (1903, 1905, 1907)	A. J. Stose & G. W. Stose, Jonas (1932) Jonas & Stose (1939) Stose & Stose (1957)	Spruce Pine Project. Kulp & Poldervaart (1956) Eckelmann & Kulp (1956) Brobst (1962)	Grandfather Mountain Project. Bryant & Reed (1962) Bryant (1962, 1963) Reed (1964a,b)	Hernon (unpublished manuscript dated 1961) Reconnaissance only	Rankin (Chapter 16 this volume)
Layered mica gneiss and schist	Carolina Gneiss. Metamorphic rock of unknown origin. Age: Archean (oldest unit).	Wissahickon Formation in early reports. Lynchburg Formation in later. Metasedimentary. Age: late Precambrian.	Carolina Gneiss (name abandoned by Brobst). Metasedimentary (clastic facies). Age: See below.	Biotite-muscovite schist and gneiss (unnamed). Metasedimentary. Age: older Precambrian.	Lynchburg Formation. Metasedimentary. Age: late Precambrian.	Ashe Formation (clastic facies). Age: late Precambrian.
Layered hornblende gneiss and amphibolite	Roan Gneiss. Igneous; intrusive into the Carolina Gneiss. Age: Archean.	Unnamed. Igneous. Intrusive into the Lynchburg Formation near Galax, Virginia. Wide area of similar rocks in North Carolina thought to be basalt flows.	Roan Gneiss (name abandoned by Brobst). Metasedimentary (carbonate facies). Age: See below.	Amphibolite and hornblende gneiss (unnamed). Origin not specified. Age: older Precambrian.	Hornblendic rocks (unnamed). Amygdaloidal metabasalt flows. Age: late Precambrian.	Ashe Formation. Metavolcanic facies. Metamorphosed mafic tuffs and basalt flows. Some mafic intrusive rocks. Age: late Precambrian.
Layered and nonlayered granitic gneiss	Cranberry Granite. Igneous; intrusive into Roan and Carolina. Age: Archean.	Cranberry Granite, Grayson granodiorite gneiss plus several other names. An injection complex. Age: older Precambrian.	Cranberry, Henderson Granite, Gneiss and Formation. Granitized sedimentary rocks. Age: all the above units thought to be part of a single sedimentary series of late Precambrian age that underwent an even later Precambrian metamorphism.	Cranberry Gneiss. Granitized amphibolite, hornblende gneiss, and mica schist and gneiss. Age: older Precambrian but younger than rocks assigned to the Ashe Formation by Rankin (Chapter 16 this volume).	Blue Ridge Complex. Plutonic igneous and metamorphic rocks. Age: older Precambrian.	Cranberry Gneiss. Mostly sheared quartz monzonite. Age: older Precambrian.

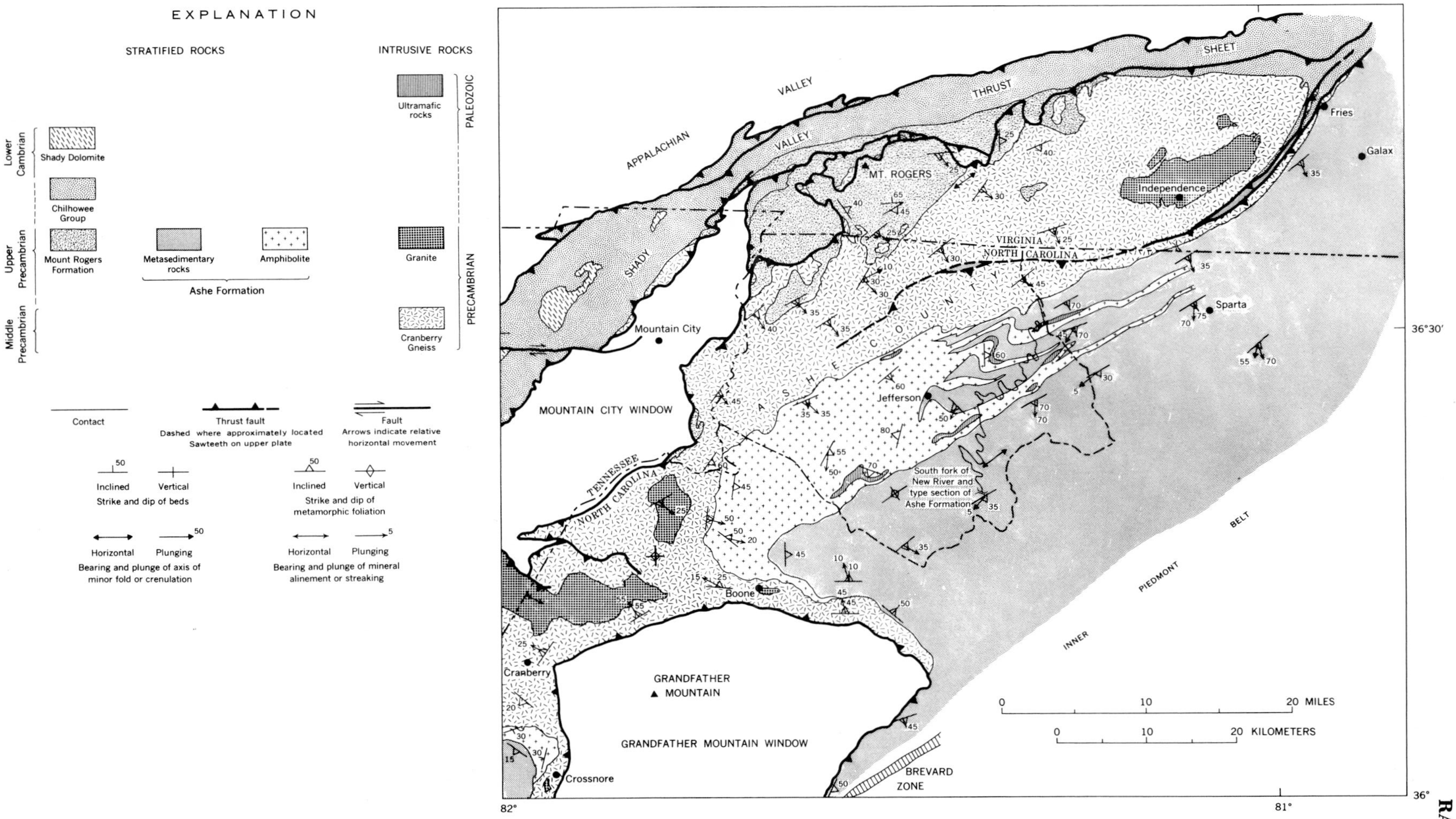

FIGURE 1. Geologic map of the Blue Ridge and Shady Valley thrust sheets in the west half of the Winston–Salem 2 degree quadrangle. Compiled from Bryant (1962, 1963), Virginia Division of Mineral Resources (1963), Espenshade (written communication, 1966), King and Ferguson (1960), Rankin (unpublished data), and Stose and Stose (1957).

vey began an experiment in the southern Appalachians of reconnaissance mapping of 1°×2° quadrangles at a scale of 1:250,000. The Winston-Salem quadrangle, covering the common corner of North Carolina, Tennessee, and Virginia (Fig. 1), is one of these. I have been working mostly in the Blue Ridge part of this quadrangle, trying to tie together earlier detailed work of King and Ferguson (1960) in northeasternmost Tennessee, of Bryant and Reed (1962) in the Grandfather Mountain area, of Stose and Stose (1957) in the Independence–Galax area, and my own earlier work in the Mt. Rogers area (Rankin, 1967a,b). Some of the results of this attempted synthesis are presented here.

In addition to the dependence upon previous work, I have benefited greatly from discussions with many geologists. I wish to acknowledge specifically, G. H. Espenshade of the U. S. Geological Survey, my colleague in the Winston-Salem project who has been mainly working in the Piedmont, and J. C. Reed, Jr., Bruce Bryant, J. B. Hadley, and E-an Zen, all of the U. S. Geological Survey.

MAJOR LITHOLOGIC UNITS IN THE BLUE RIDGE THRUST SHEET

Core of the anticlinorium

Cranberry Gneiss. The older granitic rocks of the Blue Ridge thrust sheet have been studied for a distance of more than 65 mi, from south of Cranberry, North Carolina, to Fries, Virginia. One is first impressed with the great variety of rocks, but much of the variation is accounted for by either metamorphism or younger intrusive rocks. Most of the plutonic rocks in the area between Cranberry and Fries may be mapped as a single lithologic unit. The name Cranberry Gneiss (Bryant, 1962) is in current usage for these rocks near Cranberry, North Carolina, and has precedence over various names used by Stose and Stose (1957) for granitic gneisses farther northeast near Fries. The name Cranberry Gneiss is used here.

The Cranberry Gneiss is least affected by Paleozoic metamorphism in the Shady Valley thrust sheet and along the northwest edge of the Blue Ridge thrust sheet (see Fig. 4A, B). In general, both shearing and recrystallization increase toward the southeast. Where least modified, the Cranberry Gneiss is an igneous plutonic rock that occurs in bodies of batholithic size.

Compositions range from diorite to granite; most rocks are quartz monzonite. Biotite is probably the primary dark mineral in the quartz

monzonite. Hornblende, with or without biotite, is the primary dark mineral in diorite and quartz diorite. Sphene is a common accessory mineral in all rocks. Textures range from fine to medium grained and equigranular to porphyritic. Phenocrysts of microcline are commonly as much as 1 in. (2.5 cm) long and rarely as much as 2½ in. (6.5 cm). The various textural varieties as well as compositional varieties may be present in a single outcrop. Commonly it is impossible to determine the mutual relations between various textural and compositional types. Biotite-bearing two-feldspar pegmatites, aplites, and quartz veins that are probably comagmatic are sparingly present and cut both quartz monzonite and granite. Mafic xenoliths are observed in a few outcrops.

Most of the Cranberry Gneiss is sheared and somewhat recrystallized, resulting in flaser gneisses and augen gneisses. Mineralogically the gneisses consist of quartz, microcline, and plagioclase in a matrix of sericite, shreddy biotite, stilpnomelane, epidote, and calcite. Augen gneiss and flaser gneiss are mutually gradational; the former is probably derived from porphyritic plutonic rocks by shearing and recrystallization. Farther southeast and closer to the overlying upper Precambrian rocks, recrystallization of the sheared rocks is more pronounced and results in distinctly gneissic rocks containing prominent megascopic muscovite, biotite, and, in some rocks, garnet.

The Wilson Creek Gneiss (Bryant, 1962), exposed in the Grandfather Mountain window, is probably equivalent to the Cranberry Gneiss. Both occupy the same stratigraphic position, that is, nonconformably beneath upper Precambrian stratified rocks, and both are about 1,050 million years old (Davis and others, 1962). Where least metamorphosed, the Wilson Creek Gneiss is a medium-grained biotite quartz monzonite that is virtually indistinguishable from the least metamorphosed Cranberry Gneiss. Bryant (1962, p. D-5 and D-13) noted that granitic gneiss of the Cranberry is commonly interlayered with mafic rocks, whereas that of the Wilson Creek is not. This difference was, in fact, a major reason why he mapped them separately. My interpretation is that most of the mafic layers in the Cranberry are younger intrusives (see below) and are not a valid criterion for differentiating two units composed dominantly of quartz monzonite. There are, however, scattered outcrops in the Cranberry terrane of layered gneisses cut by mafic dikes. These layered gneisses may represent scattered patches of older country rock into which the Cranberry was intruded. No areas of pre-Cranberry rocks are large enough to show on Figure 1, and no contact relations with the Cranberry have been observed.

The Blowing Rock Gneiss, an augen gneiss within the Grandfather Mountain window, may have developed from a porphyritic phase of the Wilson Creek in the same manner that augen gneiss in the Blue Ridge thrust sheet developed from a porphyritic phase of the Cranberry. The porphyritic and equigranular phases of the Cranberry Gneiss as well as the Wilson Creek and Blowing Rock Gneisses are thought to be cogenetic and together form the Elk Park plutonic group, formerly the Cranberry magma series (Rankin, 1967b), named for a village about a mile northwest of Cranberry, North Carolina.

Younger intrusive mafic rocks. Irregular bodies of gabbro intrude the Cranberry Gneiss, and diabase dikes cut rocks as young as late Precambrian near Mt. Rogers, Virginia. All are more or less metamorphosed. Keith (1903) called those rocks west of the Grandfather Mountain window, Bakersville Gabbro. Similar mafic rocks, called the Linville Metadiabase, cut the Wilson Creek Gneiss and overlying upper Precambrian stratified rocks within the Grandfather Mountain window. These mafic rocks are probably intrusive equivalents of the late Precambrian volcanic rocks. Some may be as young as basalt that occurs in the Lower Cambrian(?) Unicoi Formation to the northwest.

In places, the mafic intrusive rocks make up as much as 30 percent of the Cranberry terrane in areas as large as 3 sq mi. Where shearing is intense and mafic rocks abundant, crosscutting contacts are uncommon, and a layering of granitic gneiss and greenstone or amphibolite results. These layered gneisses have been interpreted by some as metasedimentary rocks (Eckelmann and Kulp, 1956; Bryant, 1962).

Younger intrusive felsic rocks. Felsic dikes and sills, though much less common than mafic dikes and sills, are found also in the Cranberry Gneiss and in upper Precambrian stratified rocks. Where present, quartz phenocrysts are commonly embayed, and alkali feldspar phenocrysts are highly perthitic and have a patch perthite texture typical of alkali feldspar phenocrysts in nearby upper Precambrian rhyolites. Where shearing is strong, these felsites are sericite phyllites and, like their mafic counterparts, tend to be semiconformable. Some recent workers interpret many of these as phyllonite and mylonite (Hamilton, *in* King and Ferguson, 1960, p. 13).

Several workers have recognized sizable bodies of granitic rocks that intrude the Cran-

berry Gneiss but not younger rocks. These include the Beech Granite west of Boone, North Carolina (Keith, 1903; Bryant, 1962), aegirine–augite granite near Crossnore and Boone, North Carolina (Eckelmann and Kulp, 1956; Bryant, 1962 and 1963), and the Striped Rock Granite near Independence, Virginia (Reiken, 1966). During the current study, a new large body (8 sq mi, 20 sq km) of aegirine–augite granite was found northwest of Boone (Fig. 1), as well as numerous smaller scattered bodies.

All these rocks are thought to be genetically related, and together with the Bakersville Gabbro, Linville Metadiabase, and the upper Precambrian volcanic rocks, they form the Crossnore plutonic–volcanic group, formerly the Mount Rogers magma series (Rankin, 1967b). Fluorite and zoned allanite are accessory minerals in all felsic members, and sodic amphibole occurs in the groundmass of some rhyolites. Isotopic ages confirm that the aegirine–augite granite at Crossnore and the Beech Granite are younger than the Cranberry Gneiss (Davis and others, 1962), but whether these granites are the same age as the upper Precambrian volcanic rocks is not known at present. Within the Grandfather Mountain window, the Brown Mountain Granite is younger than the Wilson Creek Gneiss and contains accessory allanite and fluorite (Reed, 1946a,b). It may also belong to the Crossnore plutonic–volcanic group.

Northwest flank of the anticlinorium

Mount Rogers Formation. Stose and Stose (1944, p. 410–411) named the interbedded volcanic and sedimentary rocks in southwestern Virginia the Mount Rogers Volcanic Series after exposures on Mt. Rogers, the highest mountain in the state. These rocks are readily subdivided lithologically, but the lithologies intertongue locally and are repeated throughout the unit. If any rocks, known or unknown, from another area are correlated with the Mount Rogers, the correlation will almost certainly be with the unit as a whole and not with a subdivision of it. The group status is herewith dropped and the name Mount Rogers Formation is substituted. Subdivisions of this unit proposed by Stose and Stose (1957) are poorly defined and are not useful as map units. A major thrust fault, unrecognized by the Stoses, further complicates their stratigraphic interpretation. The subdivisions Cinnamon Ridge Member, Flat Ridge Formation, and Cornett Basalt Member are therefore abandoned.

The Mount Rogers Formation is a sequence about 10,000 ft thick of interbedded and interfingering volcanic and sedimentary rocks. The volcanic rocks range in composition from basalt to rhyolite; flow rocks with intermediate compositions are uncommon. Thick masses of rhyolite are the most distinctive feature of the formation and make up about 50 percent of it.

The formation may be roughly divided into three parts. The lower part consists of interbedded sedimentary rock, basalt, and rhyolite. Thick masses of rhyolite make up the middle part, and sedimentary rock containing minor basalt and rhyolite makes up the upper part. Calcareous sandstone and calcareous shale are sparingly present in parts of the formation, but limestones are absent. Most of the sedimentary rocks were formed from poorly sorted, immature sediments.

Sedimentary rocks of the lower part of the formation are characterized by somewhat metamorphosed gray or greenish gray muddy-matrix conglomerate, gritty graywacke, laminated siltstone, shale, and minor calcareous sandstone. Most of the rocks are semischists, and the conglomerate contains stretched pebbles and boulders. Bedding is poorly defined in the conglomerate and graywacke.

Arkose, rhythmite, laminated pebbly mudstone and tillite, all maroon colored, characterize the upper part of the formation. Numerous examples of soft-sediment deformation are preserved. The arkose is locally crossbedded, and graded bedding is exceptionally well developed in the rhythmites. Some rhythmites contain outsized clasts of both plutonic and volcanic rocks. Carrington (1961) suggested that the rhythmites were tuffaceous and that the larger fragments in them were related to volcanic activity. Most of the rhythmites are, however, epiclastic. Because most of the clasts are Cranberry Gneiss, it is more likely that they were rafted into place. Because of the age of the beds and the size of the clasts (as much as about a meter across) ice is the most reasonable mechanism for rafting. Interbedded with and above the rhythmites are massive red matrix, poorly sorted, muddy, polymictic conglomerates— tillite. The close association of pebbly varvedlike rocks with the conglomerate suggests that they also may be of glacial origin. Harland and Rudwick (1964) have summarized evidence for a general late Precambrian glaciation.

Rhyolites are varied in texture and phenocryst content. In the massive rhyolite forming the middle part of the formation, phenocryst-poor sperulitic lava flows are overlain by porphyritic welded ash-flow sheets. These have an aggregate thickness of about 5,000 ft near Mount Rogers,

thought to be the approximate site of a volcanic center.

The age of the Mount Rogers is well established as late Precambrian (Jonas and Stose, 1939; Rankin, 1967a). The Mount Rogers rests nonconformably on the Cranberry Gneiss and is in stratigraphic contact with the overlying Unicoi Formation of Early Cambrian(?) age. Stose and Stose (1957, p. 57) thought that the latter contact was unconformable. I have followed the contact for 30 miles along strike in the Shady Valley thrust sheet and find no evidence for a structural unconformity, nor a metamorphic episode between the time of deposition of the Mount Rogers and the Unicoi. Evidence to the east in the Blue Ridge thrust sheet northwest of Fries, Virginia, suggests that there may have been a hiatus between the Mount Rogers and Unicoi perhaps involving broad epiorogenic movements and some erosion. The Cranberry Gneiss here contains abundant dikes of rhyolite and greenstone (Stose and Stose, 1957, Pl. 1). Rhyolite dikes are more abundant here than they are in any other region of the Cranberry Gneiss and are presumably of late Precambrian age. Both rhyolite and greenstone dikes extend to, but not across, the contact with overlying Unicoi Formation. No rhyolite flows or dikes have been observed in the Unicoi Formation. The rhyolite dikes extending to the base of the Unicoi suggest that volcanic activity occurred, but that the extrusive products were eroded prior to deposition of the Unicoi Formation.

Southeast flank of the anticlinorium

Ashe Formation. The names Roan Gneiss and Carolina Gneiss have been abandoned by the U. S. Geological Survey (Bryant, 1962; Brobst, 1962). The present study has demonstrated that the layered mica gneisses, schists, and amphibolites in central and southeastern Ashe County are a tremendously thick sequence of metavolcanic and metasedimentary rocks that are distinctive as a unit and rest nonconformably upon the Cranberry Gneiss. The name Ashe Formation is here proposed for these stratified rocks above the nonconformity; it is named for extensive exposures in Ashe County, North Carolina. The type section is designated as those exposures along the South Fork of the New River between the bridge along North Carolina Highway 163 and the nonconformity at the base of the formation 14.7 mi north–northeast along the Ashe-Alleghany County line (Fig. 1). The nonconformity may be located within 20 ft horizontally in the valley of a small stream flowing into the east side of the South

Fork (lat. 36°31.2′ N; long. 81°19.5′ W). The top of the formation has not been determined, but rocks thought to be in the Ashe Formation continue at least to the foot of the Blue Ridge escarpment, several miles southeast of the type section.

Keith (1903) and Bryant (1962, 1963), the only workers who have looked in detail at rocks both northeast and southwest of the Grandfather Mountain window, conclude that the layered amphibolites and mica gneisses and schists in the Spruce Pine district (Fig. 2B) are correlative with those north of Boone. On the basis of recent reconnaissance work, I concur with this correlation. The layered rocks in both areas are lithologically the same and occupy comparable structural positions. Therefore, the amphibolites and layered mica gneisses and schists of the Spruce Pine district are included in the Ashe Formation. The description that follows comes largely from my observations in the Blue Ridge between Boone, North Carolina, and Galax, Virginia.

Between Boone and the Virginia state line, rocks of the Ashe Formation lie within the kyanite–staurolite zone of regional metamorphism. The Ashe Formation consists dominantly of fine to medium grained biotite–muscovite gneiss and amphibolite. Biotite–muscovite schist, with or without garnet, kyanite, and staurolite, is interlayered with gneiss in many outcrops. Generally gneiss is dominant, and the schist occurs as partings between beds of gneiss. Some thick beds of schist could be mapped in a more detailed study. Gritty beds containing scattered 3–6 mm clasts of quartz and feldspar are common. Most beds are less than a foot thick, but gneiss beds 1–2 ft thick are common, and 10-ft beds are found in places. Both graded bedding and cross-bedding are rare. Much gneiss and schist is sulfidic, and some schists are rich in garnet and magnetite. Scattered lenses and beds of gneiss containing hornblende, epidote, garnet, and plagioclase resemble the pseudodiorites of the Great Smoky Group (see Hadley, Chapter 17 this volume). The schist and gneisses are metamorphosed sulfidic sandstones and grits containing shale partings and interbedded with thicker beds of shale. The calcium-rich beds and lenses (pseudodiorite) represent metamorphosed limy beds and lenses.

Amphibolite is present in the Ashe Formation in layers ranging in thickness from less than an inch to hundreds of feet. I interpret them as metamorphosed mafic volcanic and penecontemporaneous shallow intrusive rocks. Evidence for igneous origin is as follows. Some are relatively coarse-grained with relict gabbroic tex-

tures; these are probably analogous to the Linville Metadiabase in the Grandfather Mountain Formation. Other, finer grained amphibolites contain tabular pseudomorphed plagioclase phenocrysts (now largely mosaics of oligoclase–andesine with or without clinozoisite and garnet), closely resembling plagioclase phenocrysts in the metabasalts of the Mount Rogers Formation. These amphibolites are probably metabasalts. By analogy with other units with which the Ashe Formation is correlated, mafic igneous rocks are more likely to occur than extensive quantities of carbonate rock (including impure dolomites). Both instrusive and extrusive mafic rocks are present in the Mount Rogers Formation and in the Grandfather Mountain Formation, but carbonate rocks are atypical. Mafic dikes and small gabbro bodies cut the Cranberry Gneiss immediately beneath the Ashe Formation. Finally, near Fries, Virginia, where the metamorphic grade is lower, the Ashe Formation contains greenstone (quartz–albite–actinolite–epidote assemblage), but dolomitic beds have not been found. To be sure, it is impossible to demonstrate that all the amphibolites are igneous or pyroclastic, but I feel that most of them are.

Amphibolites are scattered throughout the known extent of the Ashe Formation, although they are mostly concentrated in a lens many thousands of feet thick near the base of the formation underlying the rugged mountain terrain west and southwest of Jefferson, North Carolina.

The thickness of the Ashe Formation is unknown because of the intense deformation it has undergone. The thickness must, however, be measured in miles.

CONTACT RELATIONS BETWEEN THE ASHE FORMATION AND CRANBERRY GNEISS

Stose and Stose (1957) interpreted the contact between the Ashe and Cranberry in southwestern Virginia as a wide fault zone between the Gossan Lead and Fries overthrusts. Bryant and Reed (1962, p. 164) thought that the Ashe Formation northeast of Spruce Pine (Fig. 2B) and north of Boone passed into the Cranberry through a zone of granitization. Bryant (written communication, 1967) notes that the Ashe Formation in North Carolina is of higher metamorphic grade than the Cranberry and suggests that the Ashe may be in a higher tectonic unit thrust over the Cranberry of the Blue Ridge thrust sheet along a hypothetical continuation of the Fries thrust. The same contact cannot be simultaneously a gradational one and a thrust fault. Eckelmann and Kulp (1956) thought that both the Cranberry and Ashe were metasedimentary units belonging to the same stratigraphic sequence and that the contact is a gradational stratigraphic one. Most of their data come from the Spruce Pine area, but their Figure 3 (Eckelmann and Kulp, 1956) clearly indicates that they also correlate the layered mica gneisses and amphibolites at Spruce Pine with those north of Boone. I interpret the contact as a nonconformity both in the Spruce Pine area and northeast of the Grandfather Mountain window.

The Ashe Formation does not grade into the Cranberry Gneiss. Where I have seen the contact in North Carolina, amphibolites are commonly interlayered with rocks on both sides of the contact and cannot be used as a mapping criterion. Sheared and recrystallized granitic rocks of the Cranberry Gneiss, although locally much like bedded gneisses of the Ashe Formation may, with luck, be distinguished from them. This may require examining a series of exposures rather than one or two.

Near Fries, Virginia, slate, metagraywacke, and greenstone of the Ashe rest directly on gneissic quartz monzonite. These stratified rocks are truly prograde rocks and not phyllonite and mylonite as Stose and Stose (1957) reported. Just east of the bridge carrying U. S. Highway 21 over the New River, layered gneiss of the Ashe Formation within 5 ft of the contact contains flattened cobbles of Cranberry. Further southwest, between Jefferson, North Carolina, and the Tennessee State line, where the amphibolites are more abundant, pelitic kyanite and staurolite schists commonly are the lowest exposed rock and contrast markedly with the underlying Cranberry. The pelitic schist may be a metamorphosed saprolite zone. There is no sharp change in metamorphic grade at the contact (see section on metamorphism).

There remains the problem of the Gossan Lead–Fries fault zone in southwestern Virginia. Stose and Stose (1957) mapped two major faults, the Fries overthrust on the northwest and the Gossan Lead overthrust on the southeast, separated by a wide zone of mylonitic rocks. The Gossan Lead overthrust is here interpreted not as a fault but as the northeastward continuation of the nonconformity between the Ashe Formation and the Cranberry Gneiss. Accumulating evidence, however, indicates that the Fries overthrust is a major fault or the northwest margin of a major fault zone.

Along the southern foot of the Iron Mountains northwest of Fries, Virginia, the Unicoi Formation rests nonconformably on the Cranberry Gneiss. The Unicoi Formation here contains some polymict conglomerate and maroon rhythmite similar to those in the Mount Rogers Formation, but most of the rocks are quartz pebble conglomerate and sandstone typical of the Unicoi Formation. The interbedded rocks similar to those in the Mount Rogers Formation may indicate an interfingering of the two units eastward from the main body of the Mount Rogers, or simply a repetition in time of lithology. In any case, there is no significant recognizable upper Precambrian unit preserved here between the Unicoi and Cranberry.

The nonconformity between the Unicoi and Cranberry has been observed at three localities. At each locality a quartz pebble conglomerate containing rounded quartz clasts as much as 9 in. long rests on quartz monzonite. The basal Unicoi beds wrap around the nose of an anticline, called the Elk Creek anticline by Stose and Stose (1957) and farther southwest are overridden by the Fries fault.[1] At Fries, Virginia, the nonconformity between the Cranberry Gneiss and Unicoi Formation is only about 3/4 mi (1 km) across strike from the unconformity between the Cranberry Gneiss and Ashe Formation (Fig. 1). Clearly these two nonconformities have been brought together from different geological environments by a major fault. Many thousands of feet of upper Precambrian rocks are missing between the Unicoi and Cranberry northwest of Fries that are present on top of the Cranberry just 3/4 mi (1 km) to the southeast.

The Fries fault has been extended about 10 mi into North Carolina in the Cranberry terrane on the basis of slivers of garnet gneiss and schist that may belong to the Ashe Formation (Fig. 1). This fault has not been recognized southwest of the Grandfather Mountain window.[2]

[1] Additional field work necessitates some revision of the map pattern near Fries, but only reinforces the conclusion that the Fries is a major thrust fault. (See Rankin, Espenshade, and Neuman, 1968, Geologic map of the western half of the Winston-Salem quadrangle, N.C.-Va.-Tenn.: U. S. Geol. Survey Open File Map.)

[2] The Fries fault is now thought to continue southward in the Cranberry, subparalleling the Ashe–Cranberry contact to the northeast corner of the Grandfather Mountain window where it probably merges with the Linville Falls fault bounding the window. A similar fault may be present in the Cranberry on the Spruce Pine side of the window. This fault pattern is similar to that proposed by Bryant (written communication 1967) but differs from it in that the fault is not located at the Ashe–Cranberry contact and in that this contact does not represent a metamorphic grade change.

AGE OF THE ASHE FORMATION AND REGIONAL CORRELATION

The Ashe Formation is clearly younger than the Cranberry Gneiss, which has been dated isotopically at about 1,050 m.y. (Davis and others, 1962). In the Spruce Pine district, the Ashe Formation is cut by pegmatites and alaskite dated at 350 m.y. (Long and others, 1959), placing an upper limit on the age.

Near Galax, Virginia, rocks here called the Ashe Formation were mapped by Stose and Stose (1957) as Lynchburg Formation, a unit of late Precambrian age in central Virginia (Brown, 1958 and Chapter 23 this volume). In Floyd County, Virginia, 25 mi (40 km) northeast of Galax, Dietrich (1959, p. 69) called the northeast strike extension of these rocks "Lynchburg Formation" because he was uncertain whether they were correlative with the type Lynchburg. On the basis of marked lithologic similarities between the type Lynchburg and the Ashe Formation and their comparable stratigraphic positions, I am convinced that the two formations are in part correlative. A new name is used because of distance separating Lynchburg, Virginia, and Ashe County, North Carolina, because much of the intervening country has not been mapped, and because the Ashe Formation may also include rocks correlative with the Swift Run and Catoctin Formations that overlie the Lynchburg.

Along the northwest limb of the anticlinorium, the Cranberry Gneiss is overlain nonconformably by the Mount Rogers Formation. Although there are significant lithologic differences between the Mount Rogers and Ashe Formations, these generally involve types of rocks that in many geologic terranes thicken and thin rapidly. These are mainly conglomerate and rhyolite, abundant in the Mount Rogers but generally absent in the Ashe. Metabasalts in both formations are most abundant in thick lenses near their respective bases. At least 5 mi of Cranberry Gneiss in the core of the anticlinorium currently separate the two formations and prior to movement on the Fries fault they must have been several miles farther apart. Between Boone, North Carolina, and Mountain City, Tennessee, small patches of Mount Rogers metabasalts and metasedimentary rocks are only 2 mi from the main outcrop belt of the Ashe. Not enough Mount Rogers is exposed there, however, to indicate how these rocks fit in the stratigraphic picture. The interpretation favored here is that the Mount Rogers and Ashe are correlative units of differing facies.

Hadley (Chapter 17 this volume) has briefly discussed the correlation of various upper Precambrian units, including the Mount Rogers and Lynchburg Formations, the Grandfather Mountain Formation, and the Ocoee Series, which underlies large areas southwest of Spruce Pine. He has also reported that rocks here included in the Ashe Formation near Spruce Pine pass southwest in a manner not yet known in detail into rocks lacking prominent amphibolites that are included by him in the Great Smoky Group of the Ocoee Series.

FACIES OF THE UPPER PRE-CAMBRIAN ROCKS

If we take 550 m.y. as the approximate date of the base of the Cambrian (Glaessner, 1963), a time interval roughly equivalent to the duration of Phanerozoic time is available for the deposition of upper Precambrian sediments in northwestern North Carolina. It may be unwarranted, albeit convenient, to consider them all coeval. Nevertheless, there are enough lithologic similarities to justify correlating, at least in part, the upper Precambrian rocks.

Without considering for the moment the subsequent complex deformation of this area, let us look at the three units of upper Precambrian stratified rocks in northwestern North Carolina: the Mount Rogers, Ashe, and Grandfather Mountain Formations (Figs. 2A, B and 3). The Mount Rogers and Ashe crop out in subparallel belts for at least 40 mi (64 km) along strike and offer interesting contrasts in lithologies. The Mount Rogers includes thick masses of rhyolite ash-flow sheets that are indicative of subaerial environment of deposition (Rankin, 1960). It contains massive conglomerates containing boulder-sized clasts of underlying basement rocks. It was deposited upon a Precambrian topography of sufficient relief that at one place or another nearly every unit within the formation from bottom to top is in contact with basement rocks. The formation thins to the northwest, as evidenced by its thinness in the Shady Valley thrust sheet as compared with that in the Blue Ridge thrust sheet. (See Fig. 4B for tectonic units.) The Mount Rogers does, in fact, thin to a featheredge laterally along the southeast margin of the Shady Valley thrust sheet. In northeasternmost Tennessee, rocks thought to be in the upper part of the Mount Rogers pinch out to the southwest, so that the Unicoi Formation rests directly upon the Cranberry Gneiss.* All factors indicate that the Mount Rogers Formation was deposited in a subaerial or shallow-water environment in a basin that deepened to the southeast.

A

Sedimentary and volcanic rocks,
interbedded

Felsic   Mafic
Volcanic rocks

Contact

Thrust fault
Dashed where approximately located
Sawteeth on upper plate

Fault

0   10   20 MILES

0   10   20 KILOMETERS

FIGURE 2. Geologic maps of northwestern North Carolina and adjacent parts of
Virginia and Tennessee. Sources of data, in addition to those given for figure 1,
are Brobst (1962), Reed (1964a,b), Rodgers (1953), and Tennessee Division of

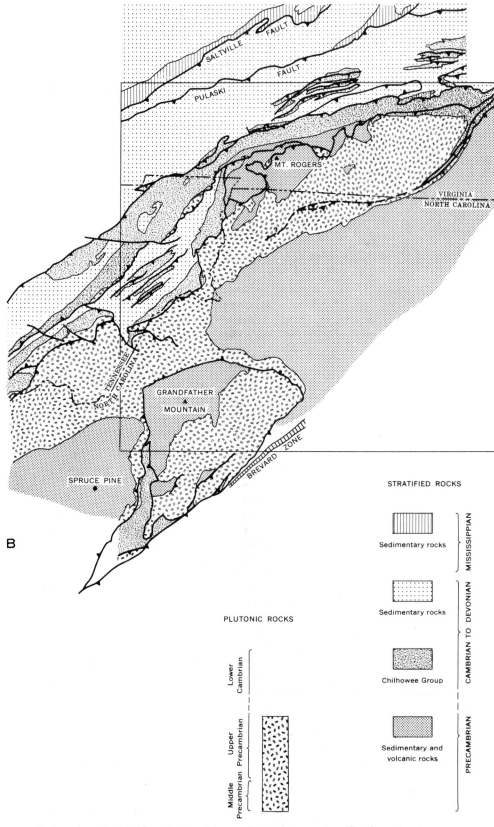

STRATIFIED ROCKS

Sedimentary rocks — MISSISSIPPIAN

Sedimentary rocks

Chilhowee Group — CAMBRIAN TO DEVONIAN

Sedimentary and volcanic rocks — PRECAMBRIAN

PLUTONIC ROCKS

Lower Cambrian

Upper Precambrian

Middle Precambrian

Geology (1966). (A) Distribution of upper Precambrian rocks. (B) Generalized geologic map.

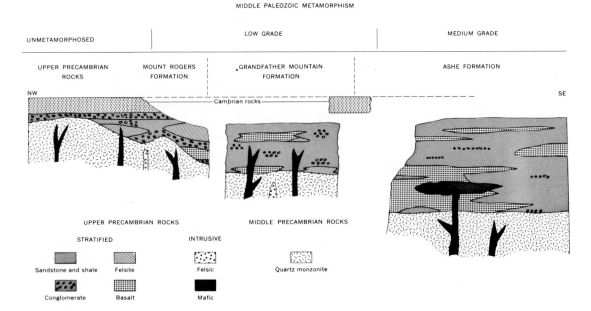

FIGURE 3. Sketch showing lithofacies in upper Precambrian rocks, prior to thrusting.

The Ashe Formation, on the southeast limb of the anticlinorium lacks rhyolites and coarse conglomerates. It contains a vastly thicker section of finer grained metasedimentary rocks. There is no evidence to date of significant topographic relief on the Cranberry basement beneath the Ashe Formation. These observations support the concept of a basin deepening to the southwest.

The Grandfather Mountain Formation (Bryant and Reed, 1962; Bryant, 1962; Reed, 1964b), is intermediate lithologically between these two units. It contains both metarhyolite and metabasalt, but more of the latter. Coarse conglomerates similar to those in the Mount Rogers are present, but less abundant. The metasedimentary rocks of the Grandfather Mountain have a gross stratigraphy reminiscent of the Mount Rogers. Metasandstones near the bottom commonly contain abundant metamorphic mica and are much like the graywacke of the Mount Rogers, but more recrystallized. Sandstones near the top of the Grandfather Mountain Formation tend to be feldspathic and have less clay matrix. This is again reminiscent of arkose most abundant in the upper part of the Mount Rogers. Further, pebbly mudstones and maroon tints are found in the upper part of the Grandfather Mountain Formation. On the other hand, at a higher metamorphic grade, the abundant micaceous metasandstones in

the lower part of the Grandfather Mountain Formation could become very satisfactory two-mica gneiss characteristic of the Ashe Formation.

PALEOZOIC METAMORPHISM

At least one episode of Paleozoic regional metamorphism, probably roughly the same age as the 350 m.y.-old pegmatites at Spruce Pine (Long and others, 1959), has left its imprint on northwestern North Carolina. There is a metamorphic gradient across the area from unmetamorphosed Paleozoic rocks of the Appalachian Valley to rocks of the kyanite and staurolite zones near Jefferson, North Carolina. Cataclastic deformation accompanied the mineralogic changes and also increased in intensity toward the southeast. The effects of shearing are most obvious, however, in the low-grade rocks in the core of the anticlinorium (see Figure 4A). Farther southeast, where the degree of recrystallization is greater, the rocks were deformed in a more homogeneous manner rather than along discrete macroscopic planes.

The record of progressive metamorphism is clear. Most workers have recognized that rocks of the Ashe Formation between Boone and Jefferson, North Carolina, are middle-grade metamorphic rocks. The metamorphic gradient has not been recognized, however, because pelitic rocks are not present in the Cranberry terrane to the northwest. Grade changes may be traced across the Cranberry in the core of the anti-

* The Mount Rogers Formation also pinches out to the east in the Blue Ridge thrust sheet, although here the pinchout may have involved some erosion (p. 232).

clinorium by mineral assemblages in younger mafic dikes.

Figure 4A shows the present distribution of middle Paleozoic metamorphic grades. Petrographic data are insufficient at present for a sophisticated treatment of the metamorphism. Most geologists would call rocks of the Appalachian Valley unmetamorphosed; much of the Mount Rogers Formation has been metamorphosed to low grade. Mafic rocks in the Mount Rogers are greenschists (albite–epidote–chlorite–tremolite assemblage), and many of the metagraywackes contain stilpnomelane.

The contact between low- and medium-grade rocks is drawn through the Cranberry terrane, where megascopic dark green amphibole becomes obvious in fine-grained greenstone dikes. This approximates the change from tremolite–actinolite to hornblende. This boundary is on the low-temperature side of the boundary between the greenschist facies and the almandine amphibolite facies as defined by Fyfe, Turner, and Verhoogen (1958, p. 217). To the northeast, this metamorphic boundary crosses the nonconformity between the Ashe and the Cranberry. The first occurrence of obvious megascopic dark-green amphibole in the mafic volcanic rocks of the Ashe northwest of Galax, Virginia, corresponds roughly to the first occurrence of garnet in the associated metasedimentary rocks. These garnets, however, are pale pink and may prove to be Mn-rich rather than typical almandine.

A comparable metamorphic gradient crosses the rocks exposed within the Grandfather Mountain window. A unit of layered gneiss and schist, that Bryant (1963) called "layered" Wilson Creek Gneiss, occurs within the window on its southeast side. These rocks are lithogically very similar to the Ashe Formation and are more probably of late Precambrian age. Near the window boundary fault they contain amphibolite and garnet gneiss and schist.

From the Tennessee–North Carolina state line to the Virginia–North Carolina line, kyanite-staurolite schist is commonly at or near the base of the Ashe Formation and is associated with garnet amphibolite. Because there are no pelitic rocks in the Cranberry, garnet, staurolite, and kyanite isograds cannot be mapped, but the metamorphic gradient must be quite steep here. More detailed work may show that the apparent steepness of the gradient is the result of telescoping of metamorphic zones by the Fries fault. Further northeast near Galax, it should be possible to sort out the metamorphic zones in more detail. Silli-

manite has not been observed in any rocks in the area mapped to date.

Bryant (1962) and Bryant and Reed (1962, and Chapter 15 this volume) attribute the broad area of low-grade Cranberry northwest of the Grandfather Mountain window to a pervasive retrogressive metamorphism after the Paleozoic metamorphism that produced kyanite and staurolite in the Ashe Formation. They further considered the Cranberry Gneiss bordering the Grandfather Mountain window on the north and southwest to be low grade, a result of the same pervasive retrogressive metamorphism.

New metamorphic minerals in the Cranberry Gneiss both northwest of the Grandfather Mountain window and in the core of the anticlinorium spacially between Mountain City, Tennessee, and Fries, Virginia, include biotite, albite, sericite, and epidote. New metamorphic minerals in mafic rocks in the same areas include albite, chlorite, actinolite, and epidote. My interpretation is that these minerals date from the progressive Paleozoic metamorphism. There is clear evidence that the mafic dikes in the northwest part of the Cranberry Gneiss have never been metamorphosed above their present low grade. Some contain pyroxene in relict ophitic textures. Relict pyroxene does not occur in higher grade amphibolitic rocks to the southeast. Thus, if these dikes are of late Precambrian or Early Cambrian age, the Cranberry Gneiss at the northwest margin of the Blue Ridge has never been metamorphosed beyond low grade in Phanerozoic time. The Paleozoic metamorphism of the Cranberry is retrogressive only in the sense that it affected Precambrian plutonic rock.

Mafic rocks in the Cranberry terrane bordering the Grandfather Mountain window on the north and southwest are amphibolites and garnet amphibolites. By my criterion, therefore, these are middle-grade metamorphic rocks. This interpretation is shown on Figure 4A and contrasts with the interpretation shown by Bryant and Reed (their Fig. 1, Chapter 15 this volume).

Bryant and Reed (Chapter 15 this volume) have summarized the evidence that progressive metamorphism in the southern Blue Ridge reached a peak during the middle of the Paleozoic Era. The thermal peak of the metamorphism occurred before the major thrusting, because isograds are offset by the thrusting. The pervasive cataclastic foliation of the lower grade rocks is an integral part of the regional metamorphism and therefore must also have developed before the major thrusting, as suggested by Hamilton (in King and Ferguson, 1960, p. 24–26), but for a

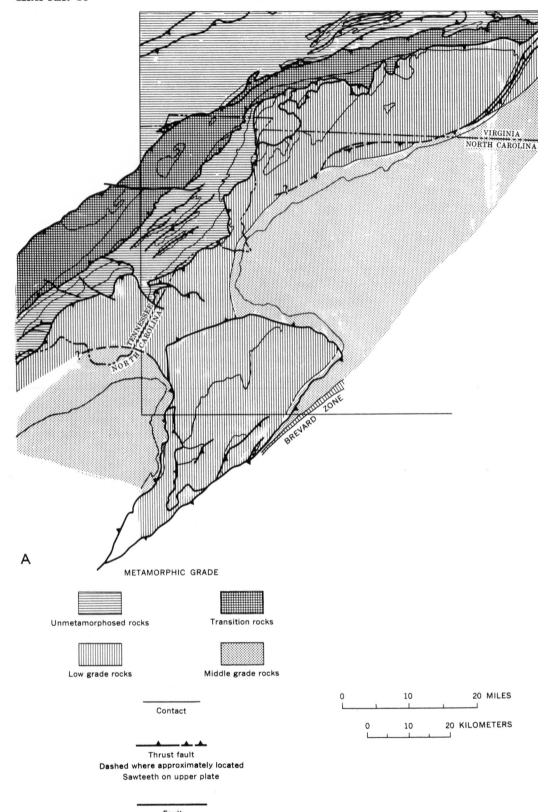

A

METAMORPHIC GRADE

Unmetamorphosed rocks

Transition rocks

Low grade rocks

Middle grade rocks

—————— Contact

0   10   20 MILES

0   10   20 KILOMETERS

━━▲━━▲━▲
Thrust fault
Dashed where approximately located
Sawteeth on upper plate

━━━━━━
Fault

FIGURE 4. Geologic maps of northwestern North Carolina and adjacent parts of
Virginia and Tennessee. (A) Map showing grade of middle Paleozoic metamorphism. (B) Tectonic map.

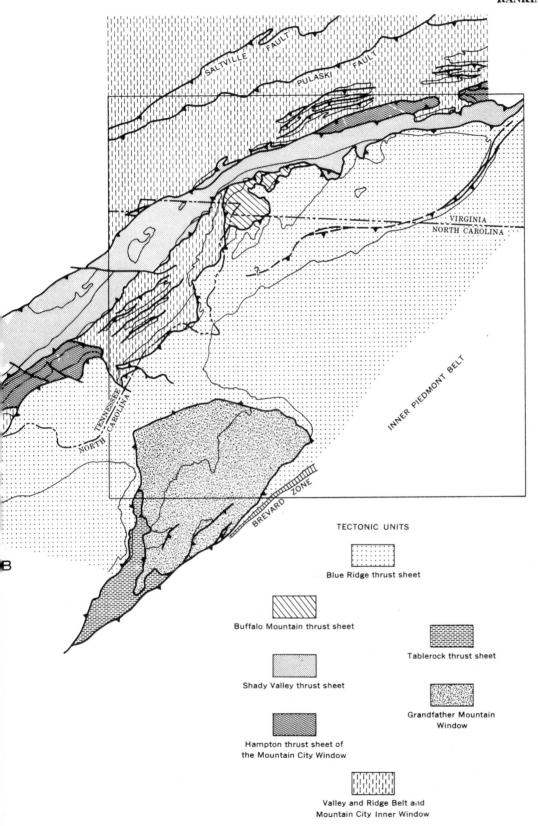

SALTVILLE FAULT

PULASKI FAULT

VIRGINIA
NORTH CAROLINA

INNER PIEDMONT BELT

NORTH CAROLINA
TENNESSEE

BREVARD ZONE

B

TECTONIC UNITS

Blue Ridge thrust sheet

Buffalo Mountain thrust sheet

Tablerock thrust sheet

Shady Valley thrust sheet

Grandfather Mountain
Window

Hampton thrust sheet of
the Mountain City Window

Valley and Ridge Belt and
Mountain City Inner Window

different reason. Southeast-plunging lineation (mineral streaking) and local shear zones probably date from the time of major thrusting.

TECTONIC UNITS

Certainly the dominant tectonic features of the southern Valley and Ridge province are the major thrust faults; this results in an imbricate structure. Many lines of evidence, such as mineral streaking, overturning of folds, and facies of Paleozoic sedimentary rocks indicate that the direction of the thrusting was toward the northwest. In northwestern North Carolina, the Grandfather Mountain window demonstrates that the crystalline rocks of the Blue Ridge are also allochthonous. A number of thrust sheets between the Blue Ridge thrust sheet and the Valley and Ridge province involve both crystalline basement rocks and Paleozoic stratified rocks. These sheets collectively form the Unaka belt of King (1955, p. 338).

Figure 4B portrays the succession and correlation of tectonic units from the Appalachian Valley to the Blue Ridge. The correlations of noncontiguous units are as much to indicate relative positions in the stacking of thrust sheets as they are to imply a continuity of sheets. It seems clear that the Blue Ridge thrust sheet is at the top of the stack and probably originated farthest southeast. It also seems clear that the Pulaski block of the Appalachian Valley (above the Pulaski fault) is at the bottom of the stack and is probably continuous with the inner Mountain City and Limestone Cove windows beneath the Shady Valley thrust sheets. In addition to these, however, there are at least three tectonic units in the Unaka belt intermediate in position between the Blue Ridge thrust sheet and the Pulaski block and two tectonic units in the Grandfather Mountain window. The mutual relations of these units is not immediately obvious, although we can place certain limits upon them.

In the thrust sheets of the Unaka belt, basal Chilhowee rocks rest directly on basement plutonic rocks with no intervening upper Precambrian rocks or with relatively thin intervening upper Precambrian rocks. In tectonic units now southeast of the Mountain City window, thick sections of upper Precambrian rocks rest on basement plutonic rocks. The Chilhowee in the Tablerock thrust sheet in the Grandfather Mountain window is not stratigraphically tied to basement rocks (Reed, 1964b) and does not influence the argument. Lithologic differences among upper Precambrian rocks suggest deposition of the Mount Rogers Formation farthest northwest, suc-

ceeded southeastward by the Grandfather Mountain Formation and the Ashe Formation. The Paleozoic metamorphic gradient used in conjunction with the facies of upper Precambrian rocks is also an aid in unshuffling the thrust sheets. In no place where the Chilhowee Group rests stratigraphically on plutonic basement is the metamorphic grade higher than it is where upper Precambrian rocks rest upon the basement. For example, the metamorphic grade of the Unaka belt is lower than that in the Grandfather Mountain window, including the Tablerock thrust sheet.

King and Ferguson (1960, p. 79, 83) concluded that the Shady Valley thrust sheet (and therefore the Buffalo Mountain thrust sheet) originated southeast of the Mountain City window. The most compelling evidence is that the closest source for the thick Chilhowee section in the Shady Valley thrust sheet is along the southeast edge of the Mountain City window. Bryant and Reed (Chapter 15 this volume) conclude that the source of the Shady Valley must have been either between the Mountain City and Grandfather Mountain windows or southeast of the Grandfather Mountain window. The combination of the sedimentary and metamorphic facies outlined above establishes with reasonable certainty the original relative position of the Shady Valley thrust sheet as between the Mountain City and Grandfather Mountain windows rather than southeast of the Grandfather Mountain window. The suggestion that the Shady Valley thrust sheet slid across the rocks exposed in the Grandfather Mountain window prior to the middle Paleozoic metamorphism requires a much more complex sedimentary history for the upper Precambrian rocks of the region.

Bryant and Reed (Chapter 15 this volume) point out that the present distance between the Mountain City and Grandfather Mountain windows is less than the exposed width of the Shady Valley thrust sheet. Furthermore, along the southeastern edge of the Shady Valley thrust sheet opposite the Grandfather Mountain window, Chilhowee rocks rest on basement rather than the Grandfather Mountain Formation. One is forced to conclude, as did Bryant and Reed, that the original distance between the two windows must have been telescoped by a major fault now hidden beneath the Blue Ridge thrust sheet. From this, it follows that all the rocks exposed in the Grandfather Mountain window are allochthonous.

We may conclude that the rocks of the structurally intermediate sheets of the Unaka belt have always been northwest of those in the structurally intermediate sheets exposed in the Grandfather Mountain window. Whether there are three intermediate sheets or more than five we do not know, nor do we know which sheet in the Grandfather Mountain window may correlate with which sheet (if any) in the Unaka belt.

AMOUNT OF THRUSTING

There can be no question that the major thrust sheets of the southern Appalachians have traveled many miles. Derivation of the Shady Valley thrust sheet from the southeast side of the Mountain City window requires a minimum displacement of 18 mi in the vicinity of long. 82° W. The Shady Valley thrust sheet need not have moved more than this relative to the Pulaski block or the Mountain City inner window.

Displacements of twice this, however, have been suggested for the Blue Ridge thrust sheet (Bryant and Reed, Chapter 15 this volume). My stratigraphic and metamorphic interpretations lead to more conservative figures, particularly for the Blue Ridge thrust sheet northeast of the Grandfather Mountain window. The suggested displacement is roughly the same relative to the Grandfather Mountain window as the suggested displacement of the Shady Valley thrust sheet relative to the Mountain City inner window. There are three lines of evidence for this:

First, correlation of the Cranberry and Wilson Creek Gneisses means that the minimum displacement of the Blue Ridge thrust sheet may be as little as 20 mi (30 km) rather than 35 mi (55 km). Second, the facies relations of the upper Precambrian rocks make it unlikely that the Mount Rogers Formation at the northwest edge of the Blue Ridge thrust sheet was deposited southeast of the Grandfather Mountain Formation. A northwest displacement of 20 mi for the Blue Ridge thrust sheet would place the original site of deposition of the Mount Rogers northeast of, but essentially on strike with the Grandfather Mountain Formation. Third, the rather crude metamorphic isograd between low and middle grades shown in Figure 4A is offset about 20 mi by the movement of the Blue Ridge thrust sheet.*

The minimum movement on the Fries fault is unknown but near Fries, Virginia, telescoping of sedimentary facies would suggest many miles of

movement. To the southwest, the Fries fault may continue unrecognized through the Cranberry terrane.

Finally, there is the problem of the amount of displacement of the Blue Ridge thrust sheet relative to the Shady Valley thrust sheet. There is no hard evidence for this in the northwest part of the Shady Valley sheet, the only part with which I am familiar. My feeling is that the amount of relative displacement is not more than the suggested 20-mi (30-km) movement of the Blue Ridge thrust sheet with reference to the Grandfather Mountain window. The Mount Rogers Formation is present in both the Shady Valley and Blue Ridge thrust sheets and in at least one significant intermediate tectonic unit. Although there are local differences, the same internal stratigraphy of the Mount Rogers occurs in adjacent parts of the three major tectonic units. Considering the predictable limited extent of some of these lithologic units, estimates of relative displacements should be conservative.

CONCLUSION

For the past 75 years or so geologists have gradually been whittling away at the area of Precambrian basement rocks in the Appalachians. Compare, for example, the 1893 geologic map of the United States (McGee, 1893), the 1961 tectonic map of the United States (U. S. Geol. Survey and Am. Assoc. Petroleum Geologists, 1961) and the tectonic map in the back of this volume. With the exceptions of gneiss domes in Maryland and New England, so far not recognized in the southern Appalachians, and a few areas in Rhode Island and eastern Massachusetts, the major axis of older Precambrian rocks in the northern and central Appalachians is at or very close to the west edge of the metamorphic terrane. In this paper I have attempted to show that the same pattern continues into the southern Appalachians, at least as far as northwestern North Carolina. As yet, in North Carolina there has been no successful correlation across the Brevard zone into the Piedmont. It will be interesting to see if the parallelism with New England is continued here by the presence of gneiss domes of older Precambrian rocks.

---

* The Blue Ridge thrust sheet northeast of the Grandfather Mountain window probably consists of an upper and lower plate separated by the Fries fault. (See footnote p. 234). The upper plate which includes the Ashe Formation has thus moved farther relative to the Grandfather Mountain window than the lower plate which includes most of the Cranberry Gneiss. This additional northwest transport of the Ashe makes the suggested

paleogeographic reconstruction of lithofacies even more reasonable. Further mapping is needed to better interpret the meaning of the offset metamorphic isograds. Southwest of the Grandfather Mountain window, isograds have probably been offset more than 20 mi. Some of this offset may have been along a fault analogous to or continuous with the Fries fault.

From Pennsylvania to the area herein discussed, the older Precambrian rocks consist of an assortment of granitic gneisses and granitoid rocks (see Drake, Chapter 19 this volume; Espenshade, Chapter 14 this volume). Farther southwest in the Great Smoky Mountains, Hadley and Goldsmith (1963, p. B6–B12) describe a heterogeneous assemblage of layered micaceous and hornblendic gneiss, mica schist, and amphibolite that they called Carolina Gneiss and considered to be part of the older Precambrian basement complex beneath the Ocoee Series. From their description, these rocks are very similar to the Ashe Formation and differ from the Ocoee Series largely in containing layered amphibolites. Similar rocks make up much of what has been called older Precambrian basement in the Blue Ridge of southwestern North Carolina. Hadley and Goldsmith (1963, p. B23) summarize their ideas concerning the relationship between the Carolina Gneiss and the granitic basement rocks in the eastern Great Smoky Mountains as follows:

"The older features of the complex indicate that a thick sequence of sandy, argillaceous, and in part calcareous or volcanic sedimentary rocks was metamorphosed and transformed progressively northwestward into more granitic rocks." These conclusions are essentially the same as those of Bryant and Reed (1962, p. 164) for the contact relations between the Cranberry Gneiss and Ashe Formation in northwestern North Carolina. Future work may well demonstrate that the Carolina Gneiss of Hadley and Goldsmith (1963) as well as much of the layered amphibolite, mica gneiss and schist in the Blue Ridge belt of southwestern North Carolina is of late Precambrian age. Such a reinterpretation would certainly be consistent with the occurrence to the northwest of a relatively narrow axis of older Precambrian granitic rocks.

Older Precambrian granitic rocks in northwestern North Carolina differ from those in the Blue Ridge in central Virginia, the Reading Prong in Pennsylvania, and the New Jersey Highlands in the rarity of hypersthene to the south (see Espenshade, Chapter 14 this volume; Drake, Chapter 19 this volume). Bryant (1962, p. D14) reports hypersthene in the Cranberry Gneiss on Little Hump Mountain west of the Grandfather Mountain window (lat. 36°8.6′ N.; long. 81°59′ W.). Dietrich (1959, p. 99) reports orthopyroxene in older Precambrian rocks of northeastern Floyd County, Virginia, 50 mi (80 km) northeast of Galax, Virginia. Hypersthene has not been reported elsewhere in the southern Blue Ridge. It is particularly significant that hypersthene does not occur in the Cranberry Gneiss along the northwest margin of the Blue Ridge thrust sheet and in the Unaka belt where the effects of Paleozoic metamorphism are least.

What is the significance of the general absence of hypersthene in the southern Blue Ridge as compared with its common occurrence farther northeast? Certainly granulite facies metamorphic conditions were not prevalent in the southern Blue Ridge at the time of formation of the Cranberry Gneiss. Several workers have suggested granulite facies conditions prevailed in central and northern Virginia (see Espenshade, Chapter 14 this volume) as well as in the Adirondacks and Reading Prong (Drake, Chapter 19 this volume) about a billion years ago. Is the basement to the northeast older, or are we seeing differing grades of Precambrian metamorphism beneath the overprint of Paleozoic metamorphism?

Finally, as our knowledge advances, we are finding that the volume of upper Precambrian stratified rocks in the central and southern Appalachians is tremendous. In addition to well-established extensive upper Precambrian units such as the Ocoee Series, Mount Rogers, Catoctin and Lynchburg Formations, consider for example, the rocks newly assigned to the upper Precambrian in the last 5 years (Hadley, Chapter 17 this volume; Bryant and Reed, 1962; Rankin, Chapter 16 this volume; Hopson, 1964). Darton and Keith (1901) seem to have been closer to the truth than they realized when they mapped the Wissahickon Formation near Washington, D.C. as "Carolina Gneiss."

REFERENCES

Billings, M. P., 1950, Stratigraphy and the study of metamorphic rocks: Geol. Soc. America Bull., v. 61, p. 435–448.

Brobst, D. A., 1962, Geology of the Spruce Pine district, Avery, Mitchell, and Yancey Counties, North Carolina: U. S. Geol. Survey Bull., 1122-A, 26 p.

Brown, W. R., 1958, Geology and mineral resources of the Lynchburg quadrangle, Virginia: Virginia Div. Mineral Resources Bull. 74, 99 p.

Bryant, B., 1962, Geology of the Linville quadrangle, North Carolina-Tennessee—A preliminary report: U. S. Geol. Survey Bull. 1121-D, 30 p.

————, 1963, Geology of the Blowing Rock quadrangle, North Carolina: U. S. Geol. Survey Geol. Quad. Map GQ-243.

————, and Reed, J. C., Jr., 1962, Structural and metamorphic history of the Grandfather Mountain area, North Carolina—A preliminary report: Am. Jour. Sci., v. 260, p. 161–180.

Carrington, T. J., 1961, Preliminary study of rhythmically layered, tuffaceous sediments near Konnarock, southwestern Virginia: Mineral Industries Jour. Virginia Polytech. Inst., v. 8, no. 2, p. 1–6.

Darton, N. H., and Keith, Arthur, 1901, Description of the Washington quadrangle: U. S. Geol. Survey Geol. Atlas, Folio 70.

Davis, G. L., Tilton, G. R., and Wetherill, G. W., 1962, Mineral ages from the Appalachian province in North Carolina and Tennessee: Jour. Geophys. Research, v. 67, p. 1987–1996.

Dietrich, R. V., 1959, Geology and mineral resources of Floyd County of the Blue Ridge upland, southwestern Virginia: Virginia Polytech. Inst. Bull., Eng. Expt. Sta. Ser., no. 134, 160 p.

Eckelmann, F. D., and Kulp, J. L., 1956, The sedimentary origin and stratigraphic equivalence of the so-called Cranberry and Henderson granites in western North Carolina: Am. Jour. Sci., v. 254, p. 288–315.

Fyfe, W. S., Turner, F. J., and Verhoogen, John, 1958, Metamorphic reactions and metamorphic facies: Geol. Soc. America Mem. 73, 259 p.

Glaessner, M. F., 1963, The dating of the base of the Cambrian: Geol. Soc. India Jour., v. 4, p. 1–11.

Hadley, J. B., and Goldsmith, Richard, 1963, Geology of eastern Great Smoky Mountains, North Carolina and Tennessee: U. S. Geol. Survey Prof. Paper 349-B, 118 p.

Harland, W. B., and Rudwick, M. J. S., 1964, The great Infra-Cambrian ice age: Sci. Am., v. 211, p. 28–36.

Hopson, C. A., 1964, The crystalline rocks of Howard and Montgomery Counties, in The geology of Howard and Montgomery Counties: Baltimore, Maryland Geol. Survey, p. 27–215, 270–336.

Jonas, A. I., 1932, Structure of the metamorphic belt of the southern Appalachians: Am. Jour. Sci., 5th ser., v. 24, p. 228–243.

————, and Stose, G. W., 1939, Age relations of the pre-Cambrian rocks in the Catoctin Mountain–Blue Ridge and Mount Rogers anticlinoria in Virginia: Am. Jour. Sci., v. 237, no. 8, p. 575–593.

Keith, Arthur, 1903, Description of the Cranberry quadrangle, [North Carolina-Tennessee]: U. S. Geol. Survey Geol. Atlas, Folio 90, 9 p.

————, 1905, Description of the Mt. Mitchell quadrangle [North Carolina–Tennessee]: U. S. Geol. Survey Geol. Atlas, Folio 124, 9 p.

————, 1907, Description of the Roan Mountain quadrangle [North Carolina–Tennessee]: U. S. Geol. Survey Geol. Atlas, Folio 151, 12 p.

King, P. B., 1955, A geologic section across the southern Appalachians—An outline of the geology in the segment in Tennessee, North Carolina, and South Carolina, in Russell, R. J., ed., Guides to southeastern geology: New York, Geol. Soc. America, p. 332–373.

————, and Ferguson, H. W., 1960, Geology of northeasternmost Tennessee: U. S. Geol. Survey Prof. Paper 311, 136 p.

Kulp, J. L., and Poldervaart, Arie, 1956, The metamorphic history of the Spruce Pine district: Am. Jour. Sci., v. 254, p. 394–403.

Long, L. E., Kulp, J. L., and Eckelmann, F. D., 1959, Chronology of major metamorphic events in the southeastern United States: Am. Jour. Sci., v. 257, p. 585–603.

McGee, W. J., compiler, 1893, Reconnaissance map of the United States: U. S. Geol. Survey 14th Annual Rept., Pt. 2.

Rankin, D. W., 1960, Paleogeographic implications of deposits of hot ash flows: Internat. Geol. Cong., 21st, Copenhagen 1960, Rept., Pt. 12. p. 19–34.

————, 1967a, The stratigraphic and structural position of the upper Precambrian Mount Rogers Volcanic Group, Virginia, North Carolina, and Tennessee [abs.]: Geol. Soc. America, Southeastern Section, Program for Annual Meeting, p. 49.

————, 1967b, Magmatic activity and orogeny in the Blue Ridge province of the southern Appalachian Mountain System in northwestern North Carolina and southwestern Virginia [abs.]: Geol. Soc. America, Program for 1967 Annual Meeting.

Reed, J. C., Jr., 1964a, Geology of the Lenoir quadrangle, North Carolina: U. S. Geol. Survey Geol. Quad. Map GQ-242.

————, 1964b, Geology of the Linville Falls quadrangle, North Carolina: U. S. Geol. Survey Bull. 1161-B, 53 p.

Reiken, C. C., 1966, Level of emplacement of the striped Rock Granite pluton, Grayson County, Virginia [abs.]: Geol. Soc. America, Southeastern Section, Program for Annual Meeting, p. 40.

Rodgers, John, 1953, Geologic map of east Tennessee with explanatory text: Tennessee Div. Geology Bull. 58, pt. 2, 168 p.

Stose, A. J., and Stose G. W., 1957, Geology and mineral resources of the Gossan Lead district and adjacent areas: Virginia Div. Mineral Resources Bull. 72, 233 p.

Stose, G. W., and Stose, A. J., 1944, The Chilhowee Group and the Ocoee Series of the southern Appalachians: Am. Jour. Sci., v. 242, p. 367–390, 401–416.

Tennessee Division of Geology, 1966, Geologic map of Tennessee: Nashville, scale 1:250,000.

U. S. Geological Survey and American Association of Petroleum Geologists, 1961, Tectonic map of the United States, exclusive of Alaska and Hawaii: 2 sheets, scale 1:2,500,000 (1962).

Virginia Division of Mineral Resources, 1963, Geologic map of Virginia: Charlottesville, scale 1:500,000.

Chapter 17

The Ocoee Series and Its Possible Correlatives*

JARVIS B. HADLEY

INTRODUCTION

THE CLASSIC GEOSYNCLINE of the central and southern Appalachians, a belt of folded and faulted Paleozoic sedimentary rocks ranging in age from Cambrian to Carboniferous, is bordered on the southeast by a wide belt of metamorphic rocks commonly known as the "crystalline Appalachians." Some of the rocks of this belt have long been known as older Precambrian rocks which are part of the basement on which the geosynclinal deposits were laid down. A few isolated patches of fossiliferous rocks have been recognized as metamorphosed rocks of early Paleozoic age surrounded by more intensely metamorphosed rocks whose age is unknown. Since the 1930's detailed work in several places in the western part of the Blue Ridge and Piedmont provinces of the crystalline Appalachians has shown that some of the metamorphic rocks represent very thick sequences of late Precambrian age and are the oldest deposits on the billion-year-old basement. It is now apparent that thick piles of metamorphosed sedimentary rocks of late Precambrian age are present throughout most of the length of the Appalachian chain southwest of Philadelphia. Their original lithologic character and stratigraphic arrangement, as well as their regional metamorphic alteration vary greatly from place to place, and they are involved with the underlying basement in complex folds and faults. This has made their study difficult and recognition of their significance slow. One of the principal bodies of late Precambrian rocks has long been known in the geologic literature as the Ocoee Series and is the principal subject of this essay.

* Publication authorized by the Director, U. S. Geological Survey.

THE OCOEE SERIES

The Ocoee Series is a large body of metasedimentary rocks which are exposed in an area nearly 50 mi (80 km) wide and 200 mi (300 km) long extending from northeastern Tennessee to northwestern Georgia. Throughout much of its extent the Ocoee Series is broken into as many as six major thrust sheets and many smaller fault slices which disrupt its stratigraphic continuity and make correlation within the series difficult (Fig. 1). Stratigraphic relations are best known in the vicinity of the Great Smoky Mountains in the western part of the Blue Ridge province, where three groups of formations total at least 40,000 ft (12,000 m) in thickness, comparable to the entire thickness of the Paleozoic rocks in most of the Appalachian region (Table 1). A short account of the stratigraphy of the Ocoee Series in the vicinity of the Great Smoky Mountains is given in King, Hadley, Neuman, and Hamilton (1958); detailed treatment of its stratigraphy, structure, and metamorphism will be found in U. S. Geological Survey Professional Paper 349, Geology of the Great Smoky Mountains, Tennessee and North Carolina (Hamilton, 1961; Hadley and Goldsmith, 1963; King, 1964; and Neuman and Nelson, 1965). The rocks of the series are everywhere regionally metamorphosed and range from lower greenschist facies in the northwest to upper amphibolite facies in the southeast. The purpose of this report, however, is to summarize the existing information on the lithologic character and stratigraphic relations of the Ocoee Series and other rocks that were probably deposited at about the same time in nearby parts of the Appalachian region. Therefore, the rocks will be described in sedimentary terms as much as possible.

TABLE 1. Stratigraphic units of the Ocoee Series, Tennessee and North Carolina. After King, Hadley, Neuman and Hamilton (1958, p. 955)

AGE	NORTH OF AND BELOW GREENBRIER FAULT				SOUTH OF AND ABOVE GREENBRIER FAULT		
CAMBRIAN AND CAMBRIAN(?)	CHILHOWEE GROUP	Cochran or Unicoi Formation and higher units *— DISCONFORMITY(?) —*		Correlation between these sequences not established	ROCKS OF MURPHY BELT	Nantahala Slate and higher units (PRECAMBRIAN(?) and early PALEOZOIC(?)) *LITHOLOGIC BREAK, BUT PROBABLY CONFORMABLE*	
LATER	OCOEE SERIES	WALDEN CREEK GROUP	Sandsuck Formation Wilhite Formation Shields Formation Licklog Formation *FAULT CONTACT, SEQUENCE UNCERTAIN* Unclassified formations		OCOEE SERIES	GREAT SMOKY GROUP	Unnamed higher strata Anakeesta Formation Thunderhead Sandstone Elkmont Sandstone
PRECAMBRIAN		SNOWBIRD GROUP	Metcalf Phyllite Pigeon Siltstone Roaring Fork Sandstone Longarm Quartzite Wading Branch Formation *— UNCONFORMITY —*			SNOWBIRD GROUP	Roaring Fork Sandstone Longarm Quartzite Wading Branch Formation *— UNCONFORMITY —*
EARLIER PRECAMBRIAN	Granitic and gneissic rocks				Granitic and gneissic rocks		

Snowbird Group

The oldest part of the Ocoee Series is the Snowbird Group, exposed only in the northern part of the Ocoee outcrop area from the foothills of the Great Smoky Mountains near Sevierville, Tennessee northeastward to the Nolichucky River. It lies unconformably on older Precambrian granitic rocks in the eastern Great Smoky Mountains and is overlain in part by the Walden Creek Group. Southwestward the Snowbird Group is cut off beneath large low-angle faults and its original extent in this direction is not known. The full thickness of the group can be seen only in the northeastern part of the Great Smoky Mountains where it is about 13,000 ft (4,000 m). Somewhat farther west near Sevierville, Tennessee, the group is at least 15,000 ft (4,500 m) thick even though the base is not exposed. Toward the southeast where the Snowbird is much thinned and overlapped by the Great Smoky Group, three of five formations in the Snowbird Group can be recognized, but their combined thickness is less than one-tenth of the thickness of the group farther northwest.

The Snowbird Group consists entirely of metamorphosed clastic sedimentary rocks derived largely from a basement complex of heterogeneous granitic and layered gneisses. The most abundant rocks are arkose, feldspathic sandstone, and argillaceous siltstone, with lesser amounts of graywacke and shale. A basal unit, found wherever the group lies directly on the basement, consists of metamorphosed quartzose shale and graywacke with sporadic thin lenses of quartz–pebble conglomerate.

All but the finest grained rocks in the group are highly feldspathic, the feldspar content commonly reaching or exceeding that of quartz. The feldspar is largely albite in the less metamorphosed siltstone and finer sandstone and microcline in the coarser sandstone and arkose. This textural segregation may be due in part to the size distribution of these minerals in the basement complex. Detrital heavy-mineral constituents, also recognizably similar to those in the basement rocks, are dominated by sphene, epidote, and magnetite–ilmenite. Strong concentrations of zircon occur locally in the lower part of the group. Detrital chlorite or biotite is abundant in many beds.

Most of the rocks are well bedded. Large to small scale current bedding is common, and cyclic graded beds are rare except in the basal graywacke. Sandstone ranges from thick to thin bedded; shale and siltstone are poorly bedded to thin bedded or laminated. Changes in grain size and composition are characteristically gradational from one bed to another in alternating repetition and bedding surfaces are commonly not well defined.

A strong westward facies-gradient prevails throughout the group, from coarser and cleaner in the eastern and stratigraphically lower parts to finer and muddier in the western and upper parts. Thus the Longarm Quartzite, which consists typically of light-colored arkose and feldspathic quartzite with abundant crossbeds and dominates all the eastern exposures of the group, passes westward into finer grained and darker rocks characteristic of the Roaring Fork Sandstone, Pigeon Siltstone, and Metcalf Phyllite. This gradient coincides with the dominant current directions indicated by crossbedding in the group and shows clearly that the sediments were transported by bottom traction from sources to the east and southeast into a westward deepening basin.

The mineral composition of the Snowbird Group indicates that it was derived from a crystalline terrane much like the older Precambrian basement complex exposed east and northeast of the Ocoee area. Whereas some of this terrane was granitic, the relatively high iron, magnesia, and lime content of the Snowbird as compared with the chemical composition of average granites shows that mafic rocks such as amphibolite, basalt, or gabbro were also abundant in the source areas. Some of the mafic material may have been contributed from volcanic sources, but there is no direct stratigraphic or petrographic evidence of important volcanic contributions. The abundance of relatively fresh feldspar, detrital epidote, and sphene shows that the source rocks were eroded by mechanical disintegration with little chemical decomposition. They were transported rapidly westward and southwestward by fluviatile and marine currents, and deposited in a presumably marine basin that eventually accommodated sediments at least 3 mi (4.8 km) thick.

Great Smoky Group

Rocks of the Great Smoky Group occupy more than three-fourths the known outcrop area of the Ocoee Series, including all the eastern, central, and southeastern parts. They make up the Great Smoky Mountains along the Tennessee–North Carolina state line and extend southwestward well into northern Georgia (Hayes, 1895; Keith, 1907; LeForge and Phalen, 1913; Bayley, 1928; Hurst, 1955; Fairley, 1965; Hernon, 1964). Recent reconnaissance mapping indicates that a large area east of Asheville, North Carolina, including all the Black Mountains is also underlain by rocks of the Great Smoky Group. Increasing metamorphism and invasion by granitic intrusive rocks makes recognition of the group difficult throughout the eastern and southeastern parts of the area, and its boundary there is only partly known (Fig. 1). The greatest thickness of the group in continuous, carefully studied sections is about 15,000 ft (4,500 m). Neither the top nor the base is exposed in these sections, however,

and the maximum overall thickness of the group may be as much as 25,000 ft (7,500 m).

The group consists dominantly of fine to coarse grained argillaceous feldspathic sandstone or graywacke, interbedded with siltstone, shale, and arkosic conglomerate. Typical sandstone is medium to thick bedded. Many beds 1–10 ft (0.3–3 m) thick are graded from coarse sandstone or fine conglomerate at the base to fine argillaceous sandstone, siltstone, or slate at the top and are interpreted as turbidites. Silty and argillaceous rocks form minor interbeds in the northwestern part of the outcrop area but thicken southeastward where they constitute units several hundred feet thick and may amount to one-third or one-half of the rocks exposed. Calcareous sandstone layers and calcareous lenses and concretions in normal sandstone are present in much of the group and are especially conspicuous in the more highly metamorphosed southeastern parts, where they appear as porphyroblastic hornblende and garnet-bearing granofels, referred to as "pseudodiorite" by Keith (1913) and by Emmons and Laney (1926, p. 19–21).

Some of the thicker and coarser sandstone beds in the northern and northwestern parts of the group contain intraformational chips and slabs derived from the interbedded pelitic rocks. Unusually coarse conglomerate composed of well-rounded cobbles of granite, gneiss, and minor quartzite forms beds as much as 30 ft (9 m) thick throughout a stratigraphic interval of about 2,000 ft (600 m) in the upper part of the group. Associated with the coarser sandstone and conglomeratic beds are restricted tongues and lenses of black, carbonaceous, and highly sulfidic shale forming units as much as 2,000 ft (600 m) thick and collectively described as the Anakeesta Formation. Locally they include a few thin beds of dark impure limestone or dolomite.

Basal beds of the Great Smoky Group are found only in the highly metamorphosed terrane in the northeastern part of its outcrop area and lie either on the Snowbird Group or unconformably on the basement complex (Fig. 1). Although now metamorphosed to mica schists, the basal beds were originally fine-grained argillaceous rocks and contain no basal conglomerate.

There is much intertonguing of rock-stratigraphic units within the Great Smoky Group, although their relations are less systematic than in the Snowbird Group and regional facies changes are less obvious. Coarse-grained and thick-bedded rocks are most abundant in the northern and northwestern parts of the group and decrease toward the south and southeast, where thinner bedded sandstone and pelitic rocks are more abundant. Intraformational breccia is principally associated with the thick, coarse sandstone beds in the northwestern outcrops, and coarse conglomerate beds are found only in a small area in the northernmost part of the Great Smoky Mountains. Fine arkosic conglomerate extends sparingly eastward into the Black Mountains and southwestward into northern Georgia, where both the conglomerate and associated sandstone are commonly less feldspathic than in areas farther north (Hurst, 1955, p. 32, 38, 57; Hernon, 1964). The carbonaceous and sulfidic shale facies is principally developed in lenses as much as 2,000 ft (600 m) thick that intertongue with the middle and upper parts of the Great Smoky Group in the northern part of the area; southward these lenses become fewer and thinner and none appear to persist south of the Georgia state line.

Mineralogical and chemical comparisons between the Great Smoky and Snowbird Groups indicate that the rocks of the Great Smoky Group were derived from dominantly granitic source areas in which mafic rocks of the type that supplied material to the Snowbird Group were much less abundant. This is recorded in higher proportions of alumina, and lower proportions of iron, magnesia, and lime in the finer grained rocks of the Great Smoky Group, whose bulk composition approximates that of average quartz monzonite. Notable also is the absence of detrital epidote and sphene in most of the Great Smoky Group, and the almost ubiquitous presence of detrital tourmaline, which is abundant also in the Walden Creek and Chilhowee Groups and in the rocks of the Murphy belt (Table 1) but is uncommon in the Snowbird Group and the adjoining basement rocks. Studies of detrital zircons in the tourmaline-bearing sandstones of the Great Smoky Group (Carroll, Neuman, and Jaffe, 1957; Stern and Rose, 1961) indicate that zircon populations in the Great Smoky Group range from 850 to 1,150 m.y. in age and suggest that younger tourmaline-bearing granites unexposed during the deposition of the Snowbird may have been exposed during deposition of the Great Smoky Group and younger rocks. The zircon studies do not indicate, however, a correlation between tourmaline content and the younger zircon.

Depositional features of the Great Smoky Group that distinguish it from the Snowbird Group are abundance of cyclic graded beds, poorer sorting, and paucity of crossbedding. The abundant parallel graded beds of the Great Smoky Group indicate that much of it may be resedimented material transported by turbidity currents from intermediate depositional sites of higher potential energy to sites of lower energy where sulfide and carbon-rich mud was also accumulating. Intraformational breccia, coarse grain size, and other indicators of high depositional energy in the northern and northwestern parts of the group suggest that at least the secondary source, and probably also the primary source, of the sediments lay north rather than east or south of the main depositional trough. The well-rounded cobbles and boulders in the northernmost outcrops may have originated as beach deposits on a shoreline of considerable local relief.

Walden Creek Group

The third group of formations in the Ocoee Series, the Walden Creek Group, is confined to a narrow, much faulted belt all along the northwest border of the Ocoee outcrop area. It is in stratigraphic sequence beneath the Chilhowee Group of Lower Cambrian and Lower Cambrian(?) age (Table 1) in many places throughout the belt but is detached from lower rocks everywhere except northeast of the Pigeon River, where it succeeds the Snowbird Group. The thickness of the group is poorly known because of structural and stratigraphic complexity but it is probably about 10,000 ft (3,000 m) in the thickest parts.

The Walden Creek Group is lithologically the most heterogeneous of the three groups. Much of it consists of dark silty to sandy argillite interbedded with siltstone and fine to coarse feldspathic sandstone. Sandstone and siltstone are commonly calcareous or ankeritic and much of the sandstone is coarser and less feldspathic than the average sandstone of the Snowbird Group. The rocks range from thin to thick bedded in many diverse types of parallel, cross-laminated, and graded strata. Flame structure, load cast, and slump structures are conspicuous locally.

Roundstone quartz–pebble conglomerate (Citico Conglomerate of Keith, 1895) is a characteristic rock of the Walden Creek Group. It consists of abundant well-rounded pebbles an inch or less in diameter together with intraformational fragments of calcareous sandstone, argillite, or impure limestone embedded in coarse carbonatic sandstone. The pebbles are reported (Hamilton, 1961, p. A19; King, 1964, p. C49, C59) as consisting largely of vein quartz, the remainder including feldspar anhedra, quartzite, dark siltstone, chert, or hornfels, and rare tourmalinized quartzite and granite. The conglomerate occurs in beds a foot (30 cm) or less thick to units more than 200 ft (60 m) thick in the upper and lower parts of the Walden Creek Group near Sevierville, Tennessee, and throughout the group farther southwest. It is rare in exposures of the group northeast of the Pigeon River.

The Walden Creek Group is the only part of the Ocoee Series that contains important amounts of carbonate rocks. These consist of dark, silty, sandy or argillaceous limestone, and dolomitic limestone, thin to thick bedded, occurring in units 1–150 ft (45 m) thick interbedded with feldspathic sandstone and various pelitic rocks. Rare oölitic beds occur but no organic or reeflike structures have been found. Some limestone beds are sedimentary breccias that consist of closely jumbled fragments of various types of limestone in a limestone matrix that commonly contains abundant angular quartz sand (Hamilton, 1961, p. A23–A24). Other limestone beds contain unusually well-rounded and uniformly sized quartz sand, either disseminated or concentrated in poorly defined beds. Northeast of the Pigeon River these sandy limestones are associated with thick-bedded calcareous quartz sandstone or orthoquartzite characterized by similar well-rounded and well-sorted grains suggestive of aeolian origin. Near Sevierville and throughout the Walden Creek belt to the northeast, the carbonate rocks are concentrated in a unit about 500 ft (150 m) thick near the middle of the group (Yellow Breeches Member of the Wilhite Formation). To the southwest they are more widely distributed through several thousand feet of argillite, sandstone, and conglomerate (Neuman and Nelson, 1965, p. D19–D20).

Rocks of the Walden Creek Group southwest of the Great Smoky Mountains have not been mapped or studied in detail, but the characteristic dark argillite, slate, limestone, and quartz–pebble conglomerate are described in older reports (Hayes, 1895, p. 2) as continuing at least as far as the Georgia state line. Northeast of the Pigeon River rocks of the Walden Creek Group lie in stratigraphic sequence beneath the Chilhowee group and above the Snowbird Group in various fault slices as far as the northern limit of the Ocoee Series in northeastern Tennessee (Fig. 1, localities 6, 7, 10, 11). Little detailed mapping has been done, however, and stratigraphic details, boundary relations and thickness of the three groups in these areas are not well known.

Sedimentary features of the Walden Creek Group in the western Great Smoky Mountains indicate source areas north of the depositional sites and record subaqueous deposition under varied conditions fluctuating between shallow, well-ventilated environments and deeper, less well ventilated ones (Neuman and Nelson, 1965, p. D67). According to Hamilton (1961, p. A27–A28), "Deposition of the Walden Creek Group can be pictured as a slow continuing sedimentation of pelitic debris, or infrequently of carbonate, with irregular periods of supply and reworking of coarse sandy and pebbly material. Perhaps deposition was near sea level for much of the material."

The rocks of the Ocoee Series have been telescoped by low-angle faults (Fig. 1) into a belt that is now much narrower than the basins in which these rocks were originally deposited. In order to understand their mutual stratigraphic relations it is necessary to restore these fault slices to approximately their original relative positions, as shown in Figure 2.

If the thrust sheets that contain the Walden Creek and Snowbird Groups in the northeastern part of the Ocoee area are pulled apart to their relative prefault positions, both groups appear to be thickest in the sheets that originate farthest southeast (group D of Fig. 2); they are thin in the intermediate sheets (groups B and C, Fig. 2) and absent in the northeastern parts of the lowermost sheets (group A, Fig. 2). They apparently thicken southwestward within some sheets, although the critical ones have not been mapped sufficiently to supply details. Currently available mapping shows that the Walden Creek and Snowbird Groups are overlapped northeastward by the Chilhowee Group, and that the dark siltstones of the upper part of the Snowbird are overlapped or replaced in the same direction by the Walden Creek. Much further work is needed, however, to determine the stratigraphic and tectonic significance of the northeastward thinning and eventual disappearance of the two groups.

The Great Smoky Group appears only in the southeasternmost and tectonically highest thrust sheets, which are underlain by the Greenbrier and Unaka Mountain–Stone Mountain faults and, presumably, by the fault bounding the Grandfather Mountain window (Bryant and Reed, 1962, p. 167). Within these sheets the Great Smoky Group overlaps the Snowbird Group eastward and southeastward, the zero isopach of the Snowbird lying a few miles west of Asheville and northwest of Waynesville. The thinning and disappearance of the Snowbird in this direction is largely depositional rather than erosional as shown by persistence almost to the zero isopach of three of the four principal lithostratigraphic subdivisions of the group. Moreover, the rocks of

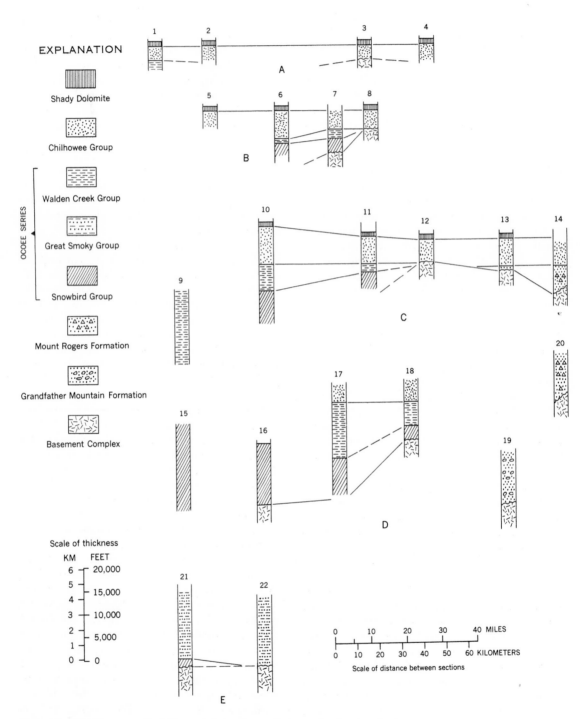

EXPLANATION

Shady Dolomite

Chilhowee Group

OCOEE SERIES

Walden Creek Group

Great Smoky Group

Snowbird Group

Mount Rogers Formation

Grandfather Mountain Formation

Basement Complex

Scale of thickness

KM	FEET
6	20,000
5	15,000
4	
3	10,000
2	5,000
1	
0	0

Scale of distance between sections

FIGURE 2. Distribution and thickness of the Ocoee Series and related rocks in thrust sheets of eastern Tennessee, western North Carolina, and southwestern Virginia. Sections in group A are in the structurally lowest sheets and the presumably autochthonous rocks of the Mountain City window. Groups B, C, D, and E represent higher sheets which originated successively farther southeast. (See Figure 1 for locations of numbered sections.)

the Greenbrier–Blue Ridge thrust sheets have moved northwestward relative to the overridden rocks by 15 mi (24 km) or more in the eastern Great Smoky Mountains (Hadley and Goldsmith, 1963, p. B80) and possibly more than 35 mi (55 km—the exposed width of the sheet—in the Grandfather Mountain area. Such movements suggest that the Great Smoky Group was deposited considerably southeast of the depositional basins of the other two groups, despite its apparent northerly source.

Since no Walden Creek appears in the highest thrust sheets, it was presumably overlapped southeastward by the Great Smoky Group and its southeastern limit is concealed beneath the Blue Ridge or Greenbrier thrust sheet. An alternative hypothesis is that the Walden Creek is equivalent to at least part of the Great Smoky Group, as suggested by Neuman and Nelson (1965, p. D63–D67). This idea derives support from the fact that coarse feldspathic sandstone and conglomerate of Great Smoky aspect (unclassified formations of Ocoee Series, Table 1) apparently succeed the Snowbird in various localities in the northern foothills of the Great Smoky Mountains (Hamilton, 1961, p. A15; Hadley and Goldsmith, 1963, p. B47). These coarser rocks, however, are unlike those of the Walden Creek Group, and few rocks of the Great Smoky Group, especially in its lower part, seem sufficiently similar to the Walden Creek Group to make direct correlation possible. A third alternative, discussed briefly in the following section, is that the Walden Creek may be younger than most of the Great Smoky Group and equivalent in age to the uppermost Great Smoky and overlying rocks of the Murphy belt.

CORRELATIVE UNITS

Grandfather Mountain Formation and Mount Rogers Formation

Three other thick sequences of rocks lie on the older Precambrian basement in nearby parts of the Blue Ridge province and appear to be contemporaneous with at least parts of the Ocoee Series. These are the Mount Rogers Formation (Rankin, Chapter 16 this volume) exposed in southwestern Virginia and adjacent areas in Tennessee and North Carolina, the closely related Grandfather Mountain Formation in northwestern North Carolina, and the rocks of the Spruce Pine area (Fig. 1).

The Grandfather Mountain Formation, described by Bryant (1962) and Reed (1964), consists of arkose and siltstone with interbedded metabasalt and other volcanic rocks, graywacke, and conglomerate. Much of the formation consists of light-colored arkose and feldspathic quartzite much like the Longarm Quartzite in composition and bedding features. The arkose is associated with plagioclase-rich siltstone and graywacke similar to parts of the Roaring Fork Sandstone and Pigeon Siltstone. Conglomeratic phases of the formation contain fragments of arkose, argillite, phyllite, basement granite, and felsic volcanic rocks. The conglomerate, however, and the metabasalt, rhyolite, and latite are quite unlike any rocks known in the Snowbird Group.

Recent detailed mapping of the Mount Rogers Formation by D. W. Rankin (Chapter 16 this volume) shows that it includes a lower heterogeneous unit consisting of conglomerate, graywacke, and basalt; a middle unit consisting of rhyolite lava and ash flows; and an upper unit of maroon tillite, arkose, rhythmite, shale, and minor basaltic pillow lava. The upper unit, with tillite at the top, underlies the Chilhowee Group of Early Cambrian and Early Cambrian(?) age with apparent conformity. Because several of these units from the oldest to the youngest lie directly upon older Precambrian granitic rocks, the Mount Rogers Formation must have been deposited on an area of basement rocks with considerable topographic relief.

Many stratigraphic and lithologic features common to the Mount Rogers, Grandfather Mountain, and Snowbird (summarized in Table 2) indicate that they may represent deposits produced under similar conditions of sediment supply and transport in a single basin or separate but contemporaneous ones. Comparison of differing characteristics of the three units suggests that the Mount Rogers Formation represents a partly subaerial or landward facies marked by maximum volcanism and nearby topographic relief, whereas the Grandfather Mountain Formation represents a region also near land but farther removed from volcanic centers. The rocks of the Snowbird Group 60 mi to the southwest may represent more seaward parts of this basin, in which dominantly detrital sediments derived from the basement rocks with little or no volcanic material came to rest.

Rocks of the Spruce Pine area

In the vicinity of Spruce Pine, North Carolina, at the northern end of the southeastern belt of Ocoee Series (Fig. 1) is a body of rocks described by D. A. Brobst as consisting of interlayered mica gneiss and schist, amphibole gneiss and schist, and a few beds of dolomite marble,

TABLE 2. Comparison of Stratigraphic and Lithologic Features of the Snowbird Group, Grandfather Mountain Formation, and Mount Rogers Formation

	Snowbird Group	Grandfather Mountain Formation	Mount Rogers Formation
Stratigraphic position	Lies on older Precambrian. Overlain by Walden Creek Group.	Same. Overlying rocks not known.	Same. Overlain by Chilhowee Group.
Rock types	Arkose and feldspathic sandstone low in sequence.	Arkose and feldspathic sandstone throughout.	Arkose prominent in upper part.
	Siltstone, dark, plagioclase-rich, in upper part.	Similar but less abundant.	Rhythmite in upper part.
	Polymictic conglomerate absent.	Polymictic conglomerate in middle and upper parts.	Tillite in upper part. Polymictic conglomerate in lower part.
	Volcanic rocks absent.	Volcanic rocks thin.	Volcanic rocks thick.
Bedding	Crossbeds common; graded beds rare.	Crossbeds local; massive beds common; graded beds rare.	Graded beds in rhythmites; crossbeds common in arkose.
Mineralogy	K-spar more abundant in coarser rocks.	Same.	
Detrital heavy minerals	Zircon, ilmenite, epidote, sphene (tourmaline lacking).	Zircon, ilmenite, sphene, rutile (tourmaline lacking).	Zircon (tourmaline lacking).
Thickness	15,000 ft (4,500 m) or more.	10,000 ft (3,000 m), possibly much more.	10,000 ft (3,000 m) approximately.
Paleogeography	Subaqueous to nearshore; some possibly fluvial. Low local relief.	Subaerial or nearshore; in part fluvial.	Large part subaerial. High local relief.

all of Precambrian age (Brobst, 1962, p. A5). These rocks are invaded by many large and small bodies of muscovite-rich granite pegmatite and variably coarse-grained oligoclase granodiorite deficient in mafic constituents.

On the basis of recent reconnaissance in this area, I have concluded that most of the layered gneisses are of mixed volcanic and epiclastic origin. The most characteristic rocks are layered hornblende–feldspar and biotite–quartz–feldspar gneisses of varied composition, commonly in layers an inch to several feet (2.5 cm to 1 m) thick. They are highly feldspathic and appear to represent interbedded pyroclastic rocks ranging from basaltic to more felsic compositions. Nonlayered amphibolite of fairly constant composition forms bodies 10–50 ft (3–15 m) thick which probably represent shallow intrusives or flows. Quartz-rich micaceous gneiss and schist, either interbedded with the metavolcanic rocks or in thicker units, apparently represent metamorphosed feldspathic sandstone and shale. Formation of the layered amphibole-bearing gneisses by metamorphism of dolomitic sedimentary rocks is considered unlikely because very few carbonate rocks are found in the assemblage and these are not associated with mafic gneisses. The total thickness of the metavolcanic and associated stratified rocks is not known but must be thousands of feet.

The rocks of the Spruce Pine area form a complexly folded, grossly synclinal mass which overlies older granitic (basement) gneisses and, in part at least, underlies thick-bedded metasandstone and micaceous schist belonging to the Great Smoky Group of the Ocoee Series. The map boundary between the volcanic sequence and the Great Smoky is complex and probably results from folding and intertonguing in a manner not yet known in detail. The contact with the underlying granitic rocks is obscured by mafic intrusives (metamorphosed Bakersville Gabbro) in the granitic gneisses and probably in the lower part of the overlying sequence and by strong shearing of all the rocks involved. Because volcanic rocks are not known in the Ocoee Series to the southwest, correlation of the Spruce Pine rocks with the type Ocoee is uncertain. They are probably equivalent to thick amphibolite interbedded with metasedimentary gneisses northeast of the Grandfather Mountain window described by Rankin (Chapter 16 this volume).

Rocks of the Murphy belt

An unnamed group of rocks extending from southwestern North Carolina to northern Georgia is not classed as part of the Ocoee Series but is considered here because it includes the only rocks known to overlie the Great Smoky Group and therefore is an important key to the stratigraphic and regional relations of the group. These rocks lie in a synclinal belt 100 mi (160 km) long called the Murphy belt, and are flanked on both sides by the Great Smoky Group. The rocks on the southeast flank of the syncline are of considerably higher metamorphic grade than those on the northwest and were assigned in the earlier reports to the older Precambrian basement complex. Structure within the Murphy belt is varied and complex. In the northern part, in North Carolina, the stratigraphic units are repeated by large recumbent folds and a low-angle thrust fault, all of which have been refolded. Farther southwest the units have been thinned presumably by tectonic squeezing and faulting. Minimum stratigraphic thickness of the rocks overlying the Great Smoky Group in the northern part of the belt is estimated at about 6,000 ft (1,800 m).

The rocks of the Murphy belt before regional metamorphism were largely thin-bedded argillaceous siltstone, shale, and fine-grained sandstone. The basal unit (Nantahala Slate, Table 3) consists of dark, variably silty or sandy argillite or slate, commonly interlaminated with fine-grained, light-colored, moderately feldspathic quartzite. The quartzite beds locally thicken and increase in proportion to argillite, forming quartzite units several tens of feet thick, varying in number and in stratigraphic position within the formation. The uppermost quartzite is commonly the most conspicuous and was called the Tusquitee Quartzite by Keith (1907, p. 4) who regarded it as succeeding the dark argillite and thought that its repetition resulted from folding.

The two rock types, however, are interbedded throughout much of the Nantahala and are more feasibly considered as a single formation. Although numerous and persistent throughout the northern part of the belt, the quartzite beds diminish southwestward and no important quartzite units are reported in the Nantahala in the Tate quadrangle near the southern end of the belt (Fairley, 1965, p. 19).

The rocks overlying the Nantahala–Tusquitee include at least two distinctly different types. One is dark, argillaceous but generally nonfissile metasiltstone, thinly but indistinctly bedded, rich in iron and alumina and characterized by metamorphic garnet, staurolite, and black mica. The other type is also thin-bedded but lighter colored, sandy, and locally distinguished by lenticular–laminate bedding or small-scale crossbedding. The darker, iron-rich beds were assigned to the Brasstown Formation by Keith (1907, p. A5) and the sandier metashale to the overlying Valleytown Formation. Keith's mapping, however, indicates that these rock types intertongue to a considerable degree and subsequent workers have generally found Keith's stratigraphic division unworkable (Hurst, 1955, p. 49; Fairley, 1965, p. 19).

The Murphy Marble which overlies the Brasstown–Valleytown is the least exposed but the most thoroughly investigated unit in the belt because of its economic importance. It consists of quite pure to highly impure micaceous calcite and dolomitic marble, whose thickness is hard to estimate because of structural complexity and poor exposures. It ranges between 150 and 500 ft (50 and 150 m) in the northern part of the Murphy syncline according to Keith (1907, p. 5) and from 75 to 250 ft (25–75 m) near the Georgia state line according to Hurst (1955, Pl. 1). The marble is in turn overlain by quartz–

TABLE 3. Formations in the Murphy Syncline, Southwestern North Carolina and Northwestern Georgia

Nantahala quadrangle, North Carolina and Tennessee (Keith, 1907)	Mineral Bluff quadrangle, Georgia (Hurst, 1955)	Tate quadrangle, Georgia (Fairley, 1965)
	Mineral Bluff Formation	
Nottely Quartzite	Nottely Quartzite	
		Andrews Schist
Andrews Schist	Andrews Formation	
		Marble Hill Schist
Murphy Marble	Murphy Marble	Murphy Marble
Valleytown Formation		
	Brasstown Schist	Brasstown Schist
Brasstown Schist		
Tusquitee Quartzite	Tusquitee Quartzite	
Nantahala Slate	Nantahala Slate	Nantahala Schist
Great Smoky Conglomerate	Dean Formation	Great Smoky Formation

sericite schist and slate or phyllite of the Andrews Schist, whose basal beds are calcareous and ferruginous and grade to marble below. The Nottely Quartzite, 75–150 ft (25–50 m) thick at the top of Keith's sequence, in part resembles the quartzite units in the Nantahala Slate, although in places it is coarser and faintly cross-bedded.

The upper part of the Great Smoky Group in the Murphy syncline, described by Hurst as the Dean Formation (Hurst, 1955, p. 40–45) also contains prominent beds of iron-rich aluminous metapelite (porphyroblastic biotite and staurolite–mica schist) as well as one or more beds of quartz–feldspar conglomerate. These rocks can be recognized throughout the northwest limb of the Murphy syncline and in the northern part of the southeastern limb. They were included by Keith (1907), LaForge and Phalen (1913), and Fairley (1965, p. 17) in the lower part of the Nantahala slate but, because of the typical Great Smoky aspect of the conglomerate and the easily traceable lithostratigraphic boundary at the base of the Nantahala, Hurst's classification seems the better one. In many places the boundary between the Nantahala and Dean Formation is gradational over a few tens of feet, although minor lithologic variations in the uppermost beds of the Dean Formation are mentioned by Hurst (1955, p. 43–45). Regionally the two units appear to be essentially conformable.

The sandstones of the Murphy belt are generally less feldspathic (Fairley, 1965, p. 14–15) than comparable rocks in either the Snowbird or Great Smoky Groups farther north, a feature which, together with the abundance of iron and alumina in pelitic beds, presumably reflects greater chemical weathering and less rapid erosion in the source areas. In these characteristics they resemble those of the Walden Creek Group previously described. Hurst (1955, p. 56–57) points out, however, that essentially similar conditions also existed during sedimentation of parts of the Great Smoky Group in the Murphy syncline and regards the transition between the two groups as recording no important changes in source areas or conditions of deposition.

The rocks overlying the Great Smoky Group in the Murphy syncline have long been tentatively correlated with the Chilhowee Group and younger rocks that overlie the Walden Creek Group (Keith, 1907, p. 11; Hurst, 1955, p. 72; Neuman and Nelson, 1965, p. D65), a deduction based largely on correlation of the Murphy Marble with the Shady Dolomite of Early Cambrian age, the oldest major carbonate unit in the Paleozoic sequence northwest of the Blue Ridge. For this correlation to be correct, however, some explanation is required for the large differences in rock types and stratigraphic arrangement that distinguish the Chilhowee Group from the rocks underlying the Murphy Marble. Possibly the differences can be explained as changes in sedimentary facies southeastward from the interbedded orthoquartzite and micaceous shale characteristic of the upper part of the Chilhowee Group, although recognition of these rock types beneath Shady Dolomite in thrust sheets in the Grandfather Mountain window (Reed, 1964, p. B14–B18) indicates that such facies changes do not exist across the Chilhowee Group farther northeast.

Recent mapping by Hernon (1964) and reconnaissance by Hurst (1962) indicates that rocks of Walden Creek aspect stratigraphically overlie the Great Smoky Group in southeastern Tennessee (Fig. 1) and may be equivalent to at least part of the rocks of the Murphy belt. W. M. Fairley (1965) has made a similar suggestion on the basis of mapping and reconnaissance northeast of Cartersville, Georgia. Hayes (1895) and Salisbury (1961, p. 40–41), however, concluded that the finer grained rocks of Walden Creek aspect in the same belt and its extension just south of the Georgia state line are stratigraphically beneath the Great Smoky Group. The structural and stratigraphic relations of the rocks in southeastern Tennessee are not well documented by modern mapping; moreover the structure shown in Figure 1 indicates that they may lie south of the Greenbrier fault and, therefore, in a very different structural block from all other rocks of the Walden Creek Group.

Talladega Slate

A belt of rocks in western Georgia and eastern Alabama, extending southwest along the strike of the Murphy belt and known as the Talladega Slate, has long been considered to contain equivalents of the Ocoee Series. The lower part of the Talladega in Alabama is about 13,000 ft (4,000 m) thick according to Butts (1926, p. 49–61) and consists largely of low-grade bluish or greenish gray, locally purplish slate and phyllite with subordinate sandstone, arkose, conglomerate, quartzite, and limestone. The rocks are unfossiliferous and their age is unknown. They are overlain by fossiliferous Devonian and Carboniferous rocks at the southwestern end of the Talladega belt but stratigraphically underlying rocks are unknown. The Talladega Slate was correlated by Crickmay (1936) with the rocks of the Murphy belt and rocks now recognized as the more metamorphosed parts of the Ocoee Series in North Carolina. No specific units of the Ocoee Series or of the Murphy belt have been recognized in the Talladega, however, and its outcrop width where it joins the Murphy and Ocoee belts near Cartersville, Georgia, is less than a mile wide so that even the continuity is questionable. Little additional information has become available since the work of Butts, but on the basis of a field review of the type Talladega in Alabama, Rodgers and Shaw (1963) concluded that all the rocks in it are Paleozoic.

Northern Correlatives of the Ocoee Series

Thick sequences of clastic metasedimentary rocks analogous to the Ocoee Series occur along the eastern flank of the Blue Ridge anticlinorium (King, 1951, p. 128, Fig. 33) as far north as Pennsylvania (see tectonic map at the end of this volume). Chief among these are the Lynchburg Gneiss of central Virginia and the Wissahickon Formation of southeastern Pennsylvania and central Maryland.

The Lynchburg Gneiss, described by W. R. Brown (1958 and this volume), consists of 10,000 or more ft (3,000 m) of metamorphosed arkosic metasandstone or graywacke that is interbedded with graphitic and alumina-rich schist and much resembles the metamorphosed sandstones of the Great Smoky Group. It lies directly on older Precambrian granitic gneisses and migmatite along the east side of the Blue Ridge and contains thick lenses of granite–cobble conglomerate at and near its base. The upper part of the formation contains intercalations of mafic metavolcanic rocks or intrusive sheets and is overlain by metabasalt flows correlated with the Catoctin Formation of late Precambrian(?) age. This is in turn overlain by the Evington Group of Paleozoic(?) age, made up of metashale, quartzite, and marble having many lithologic features in common with the rocks of the Murphy belt.

The Lynchburg Gneiss is the oldest sedimentary deposit on the older Precambrian basement complex in central Virginia and represents an immense eastward thickening of clastic deposits stratigraphically beneath the Lower Cambrian(?) units of the Chilhowee Group which lie directly on the basement a few miles to the west. As most of the Lynchburg has been metamorphosed to amphibolite grade there is little internal evidence of the direction or manner of transport, but the restriction of the coarsest deposits to the west side of the outcrop belt suggests eastward transport from highlands to the west.

Farther northeast, along the strike of the Lynchburg Gneiss, the Wissahickon Formation and the closely related Sykesville Formation of former usage and Laurel Formation of Cloos and Broedel (1940) in central Maryland have been intensively studied by Hopson (1964, p. 70–117). These formations constitute a thick body of schists and gneisses interpreted by Hopson as representing shale, graywacke and chaotic submarine slide deposits, at least 20,000 ft (6,000 m) thick, of pre-Ordovician and probably of late Precambrian age. The formations consist of upward-coarsening metapelite and sandstone, including turbidites, that intertongue eastward with thickening wedges of feldspathic sandstone containing chaotically distributed fragments of various intraformational and exotic rock types. These wedges, individually as much as 2,000 ft (600 m) thick in places, appear to represent huge submarine slide masses shed westward from a sharply rising ridge of basement rocks at the eastern edge of the depositional basin. Westward the slide wedges grade into a thick sequence of rhythmically bedded fine-grained feldspathic graywacke and metashale.

The Wissahickon Formation lies on older Precambrian granitic gneiss now exposed only in small domical uplifts in the eastern part of the outcrop area. Intervening between the Wissahickon and the basement, however, is a well-defined limestone unit (Cockeysville Marble) underlain by quartzite (Setters Formation). These formations are not found beneath the late Precambrian clastics farther south, but resemble, except in thickness, the basal Cambrian quartzite and overlying limestone that rest on the older Precambrian basement a few miles to the northwest. Despite this resemblance, Hopson presents cogent arguments that the Wissahickon and related rocks are probably late Precambrian in age and correlates them with the Lynchburg Gneiss and Ocoee Series.

CONCLUSIONS

In trying to assess the regional significance of the Ocoee Series and related deposits of late Precambrian age in the central and southern Appalachians, the following points seem pertinent:

All the known deposits represent an abrupt eastward thickening of Precambrian clastic sedimentary rocks older than the basal Cambrian deposits of the classic Appalachian geosyncline. They must mark the hinge of major downwarping of the continental margin in late Precambrian time.

In Maryland and western North Carolina, older Precambrian basement rocks east of the downwarped basins served as sources of the sediments that filled them, indicating that the basins lay within the Precambrian continental margin rather than on one-sided continental shelves. In central Virginia no basement rocks have been identified or inferred east of the Precambrian sedimentary rocks, which may have been continental shelf deposits. It should be noted, however, that the wide but poorly known belt of metamorphic rocks farther east in the Piedmont of Virginia and the Carolinas may prove to contain other belts of late Precambrian rocks and possibly welts of basement.

The known late Precambrian sedimentary rocks were laid down in deep orogenic basins or troughs partly bordered by highlands of considerable relief. Probably several basins were separated by tectonically active ridges marginal to the Precambrian continent. Depositional conditions in these troughs varied markedly from place to place. Some were filled with bottom-traction deposits, deposited perhaps at shallow depths. Many, however, were filled with the products of turbidity flows or other gravity-powered mechanisms. Several reveal concurrent volcanism within the troughs or marginal to them.

The western limit of these troughs lies in the vicinity of the Blue Ridge anticlinorium (King, 1951, p. 128, Fig. 33 and p. 103, Fig. 21) in which older Precambrian basement rocks are exposed from New England to northern Georgia. In New England the anticlinorium marks approximately the boundary between western miogeosynclinal and eastern eugeosynclinal belts of Paleozoic rocks with different stratigraphic sequences. In the central and southern Appalachians it marks the hinge line or western limit of early crustal downwarping that initiated the complex geosynclinal and orogenic developments of the Paleozoic Era.

REFERENCES

Bayley, W. S., 1928, Geology of the Tate quadrangle, Georgia: Georgia Geol. Survey Bull. 43, 170 p.

Brobst, D. A., 1962, Geology of the Spruce Pine district, Avery, Mitchell, and Yancey Counties, North Carolina: U. S. Geol. Survey Bull. 1122-A, p. A1–A26.

Brown, W. R., 1958, Geology and mineral resources of the Lynchburg quadrangle, Virginia: Virginia Div. Mineral Resources Bull. 74, p. 1–99.

Bryant, B., 1962, Geology of the Linville quadrangle, North Carolina–Tennessee: a preliminary report: U. S. Geol. Survey Bull. 1121-D, p. 1–30.

————, and Reed, J. C., Jr., 1962, Structural and metamorphic history of the Grandfather Mountain area, North Carolina: a preliminary report: Am. Jour. Sci., v. 260, p. 161–180.

Butts, C., 1926, The Paleozoic Rocks, in Geology of Alabama: Alabama Geol. Survey, Spec. Rept. No. 14, p. 41–230.

Carroll, Dorothy, Neuman, R. B., and Jaffe, H. W., 1957, Heavy minerals in arenaceous beds in parts of the Ocoee Series, Great Smoky Mountains, Tennessee: Am. Jour. Sci., v. 255, p. 175–193.

Cloos, E., and Broedel, C. H., 1940, Geologic map of Howard and adjacent parts of Montgomery and Baltimore Counties: Maryland Geol. Survey.

Crickmay, G. W., 1936, Status of the Talladega Series in southern Appalachian stratigraphy: Geol. Soc. America Bull., v. 47, p. 1371–1392.

Emmons, W. H., and Laney, F. B., 1926, Geology and ore deposits of the Ducktown mining district, Tennessee: U. S. Geol. Survey Prof. Paper 139, 114 p.

Fairley, W. M., 1965, The Murphy syncline in the Tate quadrangle, Georgia: Georgia Geol. Survey Bull. 75, 71 p.

Hadley, J. B., and Goldsmith, Richard, 1963, Geology of the eastern Great Smoky Mountains, North Carolina and Tennessee: U. S. Geol. Survey Prof. Paper 349-B, 118 p.

Hamilton, Warren, 1961, Geology of the Richardson Cove and Jones Cove quadrangles, Tennessee: U. S. Geol. Survey Prof. Paper 349-A, 55 p.

Hayes, C. W., 1895, Description of the Cleveland sheet (Tennessee): U. S. Geol. Survey Geol. Atlas, Folio 20, 4 p.

Hernon, R. M., 1964, Preliminary geologic maps and sections of the Ducktown, Isabella, and Persimmon Creek quadrangles, Tennessee and North Carolina: U. S. Geol. Survey open-file report, 4 sheets, scale 1:24,000.

Hopson, C. A., 1964, The crystalline rocks of Howard and Montgomery Counties, in The geology of Howard and Montgomery Counties: Baltimore, Maryland Geol. Survey, p. 27–337.

Hurst, V. J., 1955, Stratigraphy, structure and mineral resources of the Mineral Bluff quadrangle, Georgia: Georgia Geol. Survey Bull. 63, 137 p.

————, 1962, Ocoee metasediments, north central Georgia and southeast Tennessee: Georgia Dept. Mining and Geol., Guidebook No. 3, 28 p.

Keith, Arthur, 1895, Description of the Knoxville sheet (Tennessee–North Carolina): U. S. Geol Survey Geol. Atlas, Folio 16, 6 p.

————, 1907, Description of the Nantahala quadrangle (North Carolina and Tennessee): U. S. Geol. Survey Geol. Atlas, Folio 143, 11 p.

————, 1913, Production of apparent diorite by metamorphism: Geol. Soc. America Bull., v. 24, p. 684–685.

King, P. B., 1951, The tectonics of middle North America: Princeton, New Jersey, Princeton Univ. Press, 203 p.

————, 1964, Geology of the central Great Smoky Mountains; Tennessee: U. S. Geol. Survey Prof. Paper 349-C, 148 p.

————, Hadley, J. B., Neuman, R. B., and Hamilton, Warren, 1958, Stratigraphy of the Ocoee Series,

Great Smoky Mountains, Tennessee and North Carolina: Geol. Soc. America Bull., v. 69, p. 947–966.

LaForge, Laurence, and Phalen, W. C., 1913, Description of the Ellijay quadrangle (Georgia–North Carolina, Tennessee): U. S. Geol. Survey Geol. Atlas, Folio 187, 18 p.

Neuman, R. B., and Nelson, W. H., 1965, Geology of the western Great Smoky Mountains, Tennessee: U. S. Geol. Survey Prof. Paper 349-D, 81 p.

Reed, J. C., Jr., 1964, Geology of the Linville Falls quadrangle, North Carolina: U. S. Geol. Survey Bull. 1161-B, p. 1–53.

Rodgers, J. and Shaw, C. E., 1963, Age of the Talladega Slate of Alabama [abs.]: Geol. Soc. America Spec. Paper 73, p. 226–227.

Salisbury, J. W., 1961, Geology and mineral resources of the northwest quarter of the Cohutta Mountain quadrangle: Georgia Geol. Survey Bull. 71, 61 p.

Stern, T. W., and Rose, H. J., Jr., 1961, New results from lead-alpha age measurements; Am. Mineralogist, v. 46, p. 606–612.

Chapter 18

The Brevard Zone: A Reinterpretation*

JOHN C. REED, JR., BRUCE BRYANT, AND W. B. MYERS

INTRODUCTION

THE BREVARD ZONE is a narrow belt of low-grade metamorphic rocks that marks the southeastern edge of the Blue Ridge belt from northern North Carolina to the edge of the Coastal Plain in northwestern Alabama (see tectonic map at end of this volume). The known length of the zone is more than 325 mi (525 km); its width is generally less than 5 mi (8 km).

Most rocks in the zone are intensely sheared. Many are phyllonites and blastomylonites derived by retrogressive metamorphism of the flanking rocks. Some, however, are progressively metamorphosed sedimentary rocks that locally retain primary textures and structures. Cleavage in the low-grade rocks is commonly steeply dipping or vertical, but in some places dips as gentle as 30° SE are prevalent. Perhaps the most notable features of the Brevard are its great length, its continuity, and its trace, which is remarkably straight and independent of the structural salients in the Blue Ridge and Valley and Ridge belts. Throughout much of its length the zone is marked by a distinct topographic lineament, by an abrupt structure break, and by a distinct change in aeromagnetic patterns (for example, see Philbin, Petrafeso, and Long, 1964). The zone is essentially coincident with the foot of the broad gravity gradient on the northwest side of the Piedmont gravity high. The crest of this regional gravity high lies along the southeastern edge of the Piedmont from Alabama through northern

Virginia (see Am. Geophys. Union, Spec. Comm. Geophys. and Geol. Study Continents, 1964). Both gravity and seismic data suggest that the zone marks the junction between two crustal blocks of quite different character.

Much of the Blue Ridge belt northwest of the Brevard zone is composed of interlayered mica schist, mica gneiss, amphibolite, and hornblende gneiss of medium or high metamorphic grade. Many of the micaceous rocks are recognizable as metamorphosed upper Precambrian sedimentary rocks of the Ocoee Series and its stratigraphic equivalents (Hadley, Chapter 17 this volume), but some are clearly of pre-Ocoee age. Some of the amphibolitic rocks are metamorphosed sedimentary or volcanic rocks, and others are metamorphosed basaltic dikes and sills. Many of them probably are of late Precambrian age, but some are older.

Similar schists, gneisses, and amphibolite are widespread in the Inner Piedmont belt southeast of the Brevard zone. The age and stratigraphic relations of the layered rocks in the Inner Piedmont remain to be established; some or all of the rocks may be of late Precambrian age, but some may well be of Paleozoic age.

The lithologic similarity between layered metamorphic rocks of the Blue Ridge and those of the Inner Piedmont, coupled with uncertainties as to their ages, make it difficult to establish any fundamental differences between these belts. The few detailed geologic maps spanning the Brevard zone show gross contrasts on opposite sides, such as differences in metamorphic grade, in structural pattern, in relative proportions of schist, gneiss,

* Publication authorized by the Director, U. S. Geological Survey.

and amphibolite, and in character and abundance of granitic rocks. The Henderson Gneiss, a distinctive coarse-grained augen gneiss, occurs interleaved with Inner Piedmont rocks along the southeast side of the Brevard zone for many miles in Georgia and the Carolinas, but nowhere does it occur northwest of the zone. Similarly, the Toluca Quartz Monzonite and pegmatites related to it invade schists and gneisses of the Inner Piedmont belt throughout much of North and South Carolina and northern Georgia, but it is not found on the opposite side of the Bervard zone. Moreover, granitic basement gneisses that yield isotopic zircon ages of 1,000–1,100 m.y. are widely distributed in the Blue Ridge, but nowhere have rocks of comparable age been recognized southeast of the Brevard zone.

Whatever its origin, the Brevard zone is clearly a major structural element—probably one of the fundamental tectonic features of the crystalline belt of the southern Appalachians.

PREVIOUS INTERPRETATIONS OF THE BREVARD ZONE

Arthur Keith first recognized the distinctive rocks in the Brevard zone during his survey of the Pisgah and Mt. Mitchell quadrangles, North Carolina (Keith, 1905, 1907). He named the low-grade rocks in the zone the "Brevard schist" and interpreted them as a narrow syncline of Cambrian rocks infolded into an older, higher grade metamorphic terrane. His maps clearly show the remarkable continuity of the zone and the striking contrast between the rocks on opposite sides, but he failed to explain how these features could be reconciled with his synclinal interpretation of the Brevard. Sloan (1908) traced the Brevard zone across South Carolina, where he termed it the Chauga zone. He advanced no new interpretation of the zone, but his work has led to the suggestion that the feature might better be named "Sloan's lineament."

Jonas (1932) recognized the retrogressive character of many of the rocks in the Brevard zone and the continuity and essential unity of the belt. She suggested that the zone is the sole of a great overthrust, the Brevard overthrust, which she believed extended northeastward to connect with the Martic overthrust in Maryland and Pennsylvania.

Reed and Bryant (1964) reviewed previous descriptions and interpretations of the Brevard and described the rocks and structures along the zone in the Grandfather Mountain area, North Carolina, in some detail. They suggested that the zone marks a large-scale strike-slip fault because

of its straight trace, the generally steep dips within it, and the strong subhorizontal cataclastic lineation in the sheared rocks along it. The contrast in rock types across the zone, especially the absence of the Henderson Gneiss northwest of the zone anywhere in North Carolina, South Carolina, and northern Georgia, indicates that total lateral displacement has been at least 135 mi (220 km). On the basis of their interpretation of the relations between lineations in the Blue Ridge thrust sheet and those along the Brevard zone in the Grandfather Mountain area (Reed and Bryant, 1964, p. 1188), they concluded that the displacement was right lateral and that it had occurred after northwestward movement of the Blue Ridge thrust sheet.

King (1964, p. 12–14) has pointed out that large-scale strike-slip faults parallel to the structural grain of mountain belts have now been recognized in many parts of the world. These faults are generally interpreted as later than and unrelated to the folds and thrust faults commonly attributed directly or indirectly to lateral compression. He suggests that the association of "compressional" structures with strike-slip faults is too widespread to be entirely fortuitous and that there may well be a closer genetic relation between these types of structures than has previously been recognized. Burchfiel and Livingston (1967) have recently called attention to the close analogy between the Brevard zone and the root zones of the Alpine nappes such as the Urseren zone, the Pusteria–Insubric line, and others. They infer that the thrust sheets of the Blue Ridge are somehow rooted in the Brevard zone rather than being truncated by it as we previously suggested.

These suggestions have led us to reexamine our interpretation of the structures along the Brevard zone in the Grandfather Mountain area. This area is particularly critical, because there the entire Blue Ridge belt is clearly allochthonous, and the fault at the sole of the Blue Ridge thrust sheet crops out along the southeast side of the Grandfather Mountain window less than 2 mi from the Brevard zone. It is also one of the few areas in which the geology of an extensive segment of the Brevard zone and the flanking rocks has been studied in some detail.

GRANDFATHER MOUNTAIN AREA

General Geology

The Grandfather Mountain window, 45 mi (70 km) long and 20 mi (32 km) wide, exposes rocks beneath the Blue Ridge thrust sheet (Fig. 1).

In the window, granitic gneisses and granites about 1,100 m.y. old (Davis and others, 1962) are unconformably overlain by the Grandfather Mountain Formation, a thick sequence of sedimentary and volcanic rocks of late Precambrian age (Bryant 1962; Reed, 1964). In the southwestern part of the window the granitic basement and the upper Precambrian rocks are separated from the tectonically overlying rocks of the Blue Ridge thrust sheet by the Tablerock thrust sheet (Fig. 1, outline map). The Tablerock is a thin but extensive tectonic slice composed chiefly of quartzite and phyllite of the Chilhowee Group of Early Cambrian and Early Cambrian(?) age and, locally, the Shady Dolomite of Early Cambrian age.

The lower part of the Blue Ridge thrust sheet surrounding the Grandfather Mountain window consists chiefly of layered and nonlayered granitic gneiss and granite that in part resemble plutonic basement rocks in the window and yield zircon with similar isotopic ages. The tectonically higher part of the Blue Ridge thrust sheet is composed largely of mica schist, mica gneiss, and amphibolite. In some places these rocks are in sharp contact with the underlying granitic rocks, and the contact may be a fault or an unconformity, as interpreted by Rankin (Chapter 16 this volume) north of the Grandfather Mountain area. Elsewhere they grade into the layered granitic gneisses through a zone of mixed rocks as much as 3 mi wide.

The Inner Piedmont belt in the Grandfather Mountain area is composed predominantly of thinly layered fine-grained biotite–plagioclase gneiss interleaved with mica schist, amphibolite, coarse augen gneiss (Henderson Gneiss), and sillimanite schist. All these rocks are invaded by irregular semiconcordant bodies of granitic rock (quartz diorite to quartz monzonite) of early or middle Paleozoic age. The layered rocks of the Inner Piedmont differ from those of the Blue Ridge thrust sheet in bulk composition, mineralogy, texture, and in the relative proportions of the various rock types, although some similar rock types occur in both terranes. No rocks comparable to the Henderson Gneiss or the granitic rocks of the Inner Piedmont are found in either the Grandfather Mountain window or the Blue Ridge thrust sheet.

The segment of the Brevard zone in the Grandfather Mountain area is less than a mile (1.6 km) wide and is marked by a discontinuous zone of silicified blastomylonite, locally as much as 2,000 ft (600 m) wide, and by a narrow but more continuous belt of thinly interlayered phyllonitic paragonite-bearing schist and gneiss. The latter rocks are unlike any of the flanking schists and gneisses, and lack pegmatites and granitic rocks of any kind, although pegmatites are widespread in adjacent rocks of both the Inner Piedmont and the Blue Ridge thrust sheet.

G. H. Espenshade (personal communication, 1966) has found that the belt of paragonite-bearing phyllonitic schist and gneiss can be traced northeastward into southern Virginia where it apparently connects with phyllite, schist, and metagraywacke of the Evington Group of early Paleozoic(?) age in the southwestern end of the James River synclinorium. In the Grandfather Mountain area, the belt of paragonitic rocks is apparently an exotic tectonic slice along the Brevard fault zone.

A single dike of unmetamorphosed diabase crosses the Inner Piedmont belt, the Brevard zone, the Linville Falls fault, and extends without deflection into the Grandfather Mountain window. The dike rock is petrographically identical with similarly oriented dikes that are widespread in other parts of the Inner Piedmont. Some of these dikes cut Triassic rocks, and they are generally believed to be of Late Triassic age, although some may be as young as Early Cretaceous (King, 1961).

Structure and Metamorphism

All the granitic basement rocks in the Grandfather Mountain window have been sheared and retrogressively metamorphosed to greenschist grade. At the same time, the overlying upper Precambrian sedimentary and volcanic rocks were progressively metamorphosed under similar conditions. Both the basement rocks and the overlying rocks display an intense and pervasive cleavage or cataclastic foliation which generally strikes northeast and dips steeply or moderately southeast (Fig. 1). This cleavage is parallel to the axial planes of shear folds in the sedimentary and volcanic rocks. In the northern part of the window, cleavage strikes north or northwest and dips east or northeast, parallel to axial planes of a second generation of folds superimposed on the earlier northeast-trending set.

Rocks of the Tablerock thrust sheet are also of low metamorphic grade, but they do not display the same cleavage pattern as the rest of the rocks in the window. Cleavage is parallel to axial planes of recumbent isoclines and transects bedding only in the fold noses. It dips gently westward in the western part of the thrust sheet and gently to moderately southeastward in the southeastern part. It is essentially parallel to the

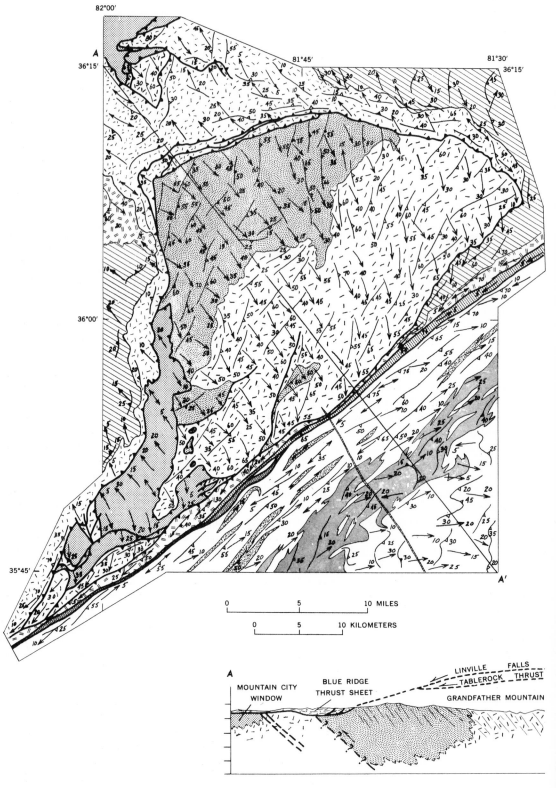

FIGURE 1. Generalized geologic map of the Grandfather Mountain window and vicinity.

EXPLANATION

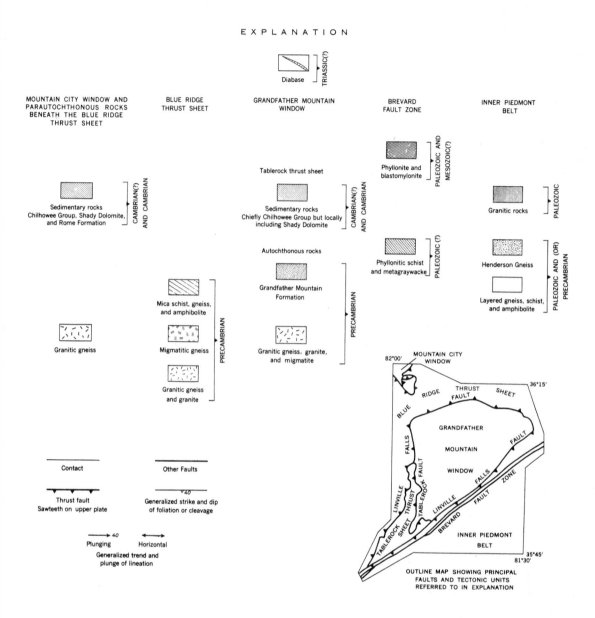

Diabase

TRIASSIC(?)

MOUNTAIN CITY WINDOW AND
PARAUTOCHTHONOUS ROCKS
BENEATH THE BLUE RIDGE
THRUST SHEET

BLUE RIDGE
THRUST SHEET

GRANDFATHER MOUNTAIN
WINDOW

BREVARD
FAULT ZONE

INNER PIEDMONT
BELT

Sedimentary rocks
Chilhowee Group, Shady Dolomite,
and Rome Formation

CAMBRIAN(?) AND CAMBRIAN

Tablerock thrust sheet

Sedimentary rocks
Chiefly Chilhowee Group but locally
including Shady Dolomite

CAMBRIAN(?) AND CAMBRIAN

Phyllonite and
blastomylonite

PALEOZOIC AND MESOZOIC(?)

Granitic rocks

PALEOZOIC

Mica schist, gneiss,
and amphibolite

Migmatitic gneiss

Granitic gneiss
and granite

PRECAMBRIAN

Autochthonous rocks

Grandfather Mountain
Formation

Granitic gneiss, granite,
and migmatite

PRECAMBRIAN

Phyllonitic schist
and metagraywacke

PALEOZOIC(?)

Henderson Gneiss

Layered gneiss, schist,
and amphibolite

PALEOZOIC AND (OR) PRECAMBRIAN

Granitic gneiss

Contact

Other Faults

Thrust fault
Sawteeth on upper plate

Generalized strike and dip
of foliation or cleavage

Plunging   Horizontal

Generalized trend and
plunge of lineation

OUTLINE MAP SHOWING PRINCIPAL
FAULTS AND TECTONIC UNITS
REFERRED TO IN EXPLANATION

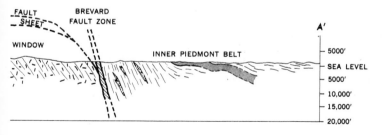

Tablerock fault which forms the sole of the thrust sheet and is generally at a high angle to cleavage in the rocks beneath.

The granitic rocks in the lower part of the Blue Ridge thrust sheet have been sheared and retrograded under low-grade conditions similar to those that affected the basement rocks in the Grandfather Mountain window. They display intense and pervasive cataclastic foliation that dips away from the window at gentle to moderate angles along the west and north sides and at moderate to steep angles along the southeast side.

The mica schist, mica gneiss, and amphibolite that compose most of the remainder of the Blue Ridge thrust sheet are mostly of medium metamorphic grade and show no evidence of extensive retrogression. Foliation in them dips gently to moderately in various directions, producing a complex map pattern that is locally discordant with foliation trends in the retrograded basement rocks beneath. Along the western and northern edges of the Grandfather Mountain window, the Linville Falls fault is sharply discordant with cleavage and foliation in the window rocks. Along the southeastern edge, the fault is parallel to foliation in the window and in the Blue Ridge thrust sheet.

The rocks of the Inner Piedmont belt more than 5 mi (8 km) southeast of the Brevard zone are all of the high metamorphic grade. Foliation parallel to layering (relict bedding?) is openly flexed; folds that plunge gently eastward were apparently synchronous with sillimanite-grade metamorphism and emplacement of the Paleozoic granitic rocks. Closer to the Brevard zone, polymetamorphic textures appear and become increasingly apparent as the Brevard is approached. Sillimanite is absent; relict staurolite and kyanite indicates that the rocks were originally of medium grade. Recrystallization in the southeastern part of the belt was also at medium grade, but closer to the Brevard it was low grade (albite–chlorite), and the original assemblages were largely destroyed. Foliation dips moderately in the southeastern part of the belt but steepens to nearly vertical close to the Brevard zone. The abrupt transition from gentle dips and variable strikes in the high-grade rocks to consistent northeast strikes and southeast dips takes place where polymetamorphic textures first become apparent.

The rocks in the Brevard zone are all of low metamorphic grade. The blastomylonites and phyllonites have been thoroughly recrystallized and almost all vestiges of their original textures and structures destroyed. The phyllonitic schist contains abundant porphyroclasts of muscovite and garnet and locally contains relicts of staurolite, indicating that it is derived by retrogression of an original medium-grade rock. Foliation in the rocks of the Brevard zone is generally parallel to the trend of the belt and dips steeply southeast. Locally it is vertical or even dips steeply west.

Lineations

Almost all the low- and medium-grade rocks in the Grandfather Mountain window display conspicuous lineations marked by aligned mineral grains and aggregates, streaking and grooving on foliation or cleavage planes, and by elongated porphyroclasts, pebbles, or amygdules. In the basement and late Precambrian rocks of the Grandfather Mountain window, the lineation lies on cataclastic foliation and cleavage planes and plunges southeastward, maintaining a surprisingly uniform trend (Fig. 1). Where the cleavage strikes northeast, the lineation plunges down the dip and is approximately normal to fold axes. Where the cleavage strikes north or northwest, the trend of the lineation remains southeast, but plunges are correspondingly gentler. Virtually without exception, lineation is perpendicular to the trace of the Brevard zone.

Lineation in the rocks of the Blue Ridge thrust sheet lies on foliation planes that are parallel to compositional layering in the gneissic rocks. It is best developed in the strongly sheared and retrograded plutonic rocks near the sole of the thrust sheet but is also found locally in the overlying medium-grade metamorphic rocks. Trends are parallel to trends of lineation in the window, but north and west of the window, plunges are generally northwest. Orientation of lineation in the Tablerock thrust sheet is the same as in nearby parts of the Blue Ridge thrust sheet.

Near the southeastern edge of the Grandfather Mountain window, the lineation swings abruptly clockwise, changing from southeasterly plunges and trends perpendicular to the trace of the Brevard zone to subhorizontal with trends parallel to the zone. This transition takes place in a belt ranging in width from less than half a mile (0.8 km) to as much as 3 mi (5 km). The swing in lineation trend is independent of lithology and involves all tectonic units from the basement rocks of the window to the Brevard zone itself.

Lineation in the Brevard zone and in the polymetamorphic rocks to the southeast is subhorizontal, parallel to the trace of the zone and perpendicular to lineation within the Blue Ridge. The orthogonal relationship of lineation across the northwestern margin of the Brevard zone is believed to be a key to the overall movement pattern.

Lineation is of a different character in the high-grade rocks of the Inner Piedmont, where polymetamorphic textures are absent. It is formed by alignment of sillimanite needles and fibers, by hornblende needles, and by long dimensions of mica flakes and quartz–feldspar aggregates. It is parallel to crenulations and minor folds in individual outcrop and to axes of larger open folds inferred from map patterns and statistical studies of attitudes of layering and foliation. This lineation apparently formed at the same time as folding and sillimanite-grade regional metamorphism. It trends east–west and generally plunges gently eastward.

Interpretation of the Lineation Pattern

The strong development of cataclastic lineation in the retrogressively metamorphosed rocks near the sole of the Blue Ridge thrust sheet indicates that the lineation was formed during movement of the thrust sheet. The similarity in character and striking parallelism in trend between the lineation in the Blue Ridge thrust sheet and that in the Grandfather Mountain window indicates that the lineation in the window also formed during thrusting. The latter lineation is normal to the axes of shear folds in the upper Precambrian rocks, and the offset of bedding planes in the sedimentary rocks and of pegmatite dikes in the basement rocks shows that movement on the cleavage and foliation planes has been parallel to the lineation, demonstrating that the lineation is in the *a* direction. The similar northwest-trending lineations in the Blue Ridge and Tablerock thrust sheets are also thought to be *a* lineations that mark the direction of thrusting.

The identification of the cataclastic lineation in the Blue Ridge thrust sheet as an *a* lineation is confirmed by petrofabric analyses (Reed and Bryant, 1964, p. 1188) and by regional geologic relations (King, 1964, p. 18–19; Bryant and Reed, Chapter 15 this volume) which require at least 35 mi (55 km) of northwestward transport of the crystalline rocks of the Blue Ridge across rocks of the Grandfather Mountain window.

The cataclastic lineation in the Brevard zone and in the polymetamorphic rocks of the Inner Piedmont is similar in appearance and petrographic character to the northwest-trending cataclastic lineation in the Blue Ridge thrust sheet and the Grandfather Mountain window. Petrofabric diagrams of quartz axes in blastomylonites with northwest-trending lineation along the Linville Falls fault are similar in symmetry and pat-

tern to equivalent diagrams from blastomylonites with subhorizontal northeast-trending lineations along the Brevard zone (Reed and Bryant, 1964, p. 1188). The lineations in the Brevard zone and in the sheared rocks to the southeast are, therefore, also believed to be in *a*.

Originally Reed and Bryant interpreted the subhorizontal *a* lineation along the southeast side of the Brevard zone as younger than the northwest–southeast-trending *a* lineation produced during northwestward thrusting of the Blue Ridge rocks. They attributed the clockwise swing in lineation trend along the northwest side of the Brevard to drag during right-lateral strike-slip movement and therefore concluded that the principal movement along the Brevard was later than and unrelated to thrusting. They suggested that the parallelism of the Brevard zone and Linville Falls fault along the southeast side of the Grandfather Mountain window was due to reactivation and rotation of the Linville Falls fault during later movement along the Brevard.

It now appears that this interpretation is unlikely for several reasons. First, the close physical similarity between the lineation in the Blue Ridge thrust sheet and Grandfather Mountain window and the lineation along the Brevard zone, and the smooth swing in trend from one to the other, indicate that the structures are of the same age. Nowhere are cataclastic lineations with different trends superimposed, as might be expected if they were of different ages. Second, rotation of an older northwest–southeast-trending lineation by drag would require a corresponding rotation of the planes on which the lineation lies. No indication of such rotation is evident in the map pattern or in statistical diagrams of poles of cleavage and foliation. Such rotation of planar structures would be obvious unless it occurred around axes everywhere normal to the diverse planes involved. In the zone where the lineation swing occurs, foliation planes in the Blue Ridge thrust sheet and in the basement rocks of the Grandfather Mountain window dip 50°–60° SE. Bedding and cleavage in the Tablerock thrust sheet dip 15°–35° SE. Both of these sets of planes could not have been rotated around the same axis to produce the observed lineation swing. We therefore must conclude that the subhorizontal northeast-trending *a* lineation along the southeast side of the Brevard zone is of the same age as the northwest-trending *a* lineation in the rocks of the Blue Ridge thrust sheet and Grandfather Mountain window, and that northwestward movement of the Blue Ridge and Tablerock thrust sheets was contemporaneous with and intimately related to strike-slip movement along the Brevard zone. Evidently, the Blue Ridge thrust sheet is rooted along the northwest side of the zone, as suggested by Burchfiel and Livingston (1967), rather than being cut off by the Brevard as we have previously suggested. The lineation pattern indicates that during strike-slip movement along the Brevard, material moved northward and upward along the northwest side and then northwestward across the Grandfather Mountain window.

The cataclastic lineation in the polymetamorphic rocks along the southeast side of the Brevard zone is quite different from the lineation in the high-grade rocks of the Inner Piedmont farther southeast. The former is clearly related to movement along the Brevard, whereas the later is an older structure dating from an earlier episode of folding and regional metamorphism. Locally, there is some evidence of counterclockwise rotation of the older lineation near the southeast edge of the belt of polymetamorphic rocks, but generally the lineation in the high-grade rocks maintains its east–west trend until it is obliterated by shearing and replaced by the northeast-trending lineation in the polymetamorphic rocks.

SPECULATION ON THE MECHANISM OF CONCURRENT STRIKE-SLIP FAULTING AND THRUSTING

If strike-slip faulting along the Brevard and northwestward thrusting in the Blue Ridge are related, what was the mechanism involved? How was the vast volume of material in the Blue Ridge thrust sheet, which must have originated southeast of the Grandfather Mountain window, derived from the narrow belt between the southeastern edge of the window and the Brevard zone?

The most likely explanation of concurrent strike-slip and thrust faulting of this magnitude must involve the resolution of forces resulting from horizontal movement of crustal blocks of at least subcontinental dimensions. Figure 2 illustrates a possible mechanism.

The Brevard is envisioned as a transcurrent fault developed in response to relative northward drift of the crustal block southeast of the fault, in the general direction of the heavy arrows on the top surface of the block (Fig. 2). The original position of the fault (A-B) is assumed to have been southeast of its present location (A'-B'). Relative northward drift of the block by an amount (M, M') would result in left-lateral strike-slip movement (s, s') along the Brevard zone and would require northwestward migration of the entire zone by an amount (t, t'). North-

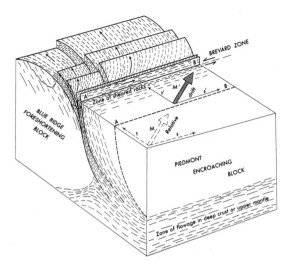

FIGURE 2. Diagramatic sketch illustrating possible relation between strike-slip faulting and thrusting. Top of block is approximately at present level of erosion.

westward migration of the zone would be accomplished by downbuckling and thickening of the block assumed to be stationary and by northwestward thrusting of material from along the southeast edge of the rising thickened prism (the Tablerock, Blue Ridge, and possibly higher thrust sheets). Once these thrust sheets reached the crest of the tectonic high northwest of the Brevard, further northwestward movement would be facilitated by gravity sliding. The orthogonal requirement of thrust and strike-slip motion in the model is rigorously met by the regular perpendicularity of lineation between the rocks of the Blue Ridge and the polymetamorphic rocks of the Inner Piedmont, near the Brevard zone. The swing in the trend of the cataclastic lineations in the rocks immediately northwest of the Brevard would then be due to the change from predominately strike-slip motion near the zone to predominately thrust motion farther to the northwest. The trend of the lineation at any point (light lines on the fault surfaces, Fig. 2) is the vector sum of the local strike-slip (s, s') and thrust (t, t') components.

In order to calculate the amount of relative northwestward migration of the Brevard zone necessary to explain the thrusting in the Grandfather Mountain area, it would be necessary to know the total cross-sectional area of the thrust sheets, the amount of thickening of the foreshortening block, and the depth to the base of the encroaching block. The thrust sheets in the Blue Ridge northwest of the Grandfather Mountain window are about 60 km wide and probably at least 5 km thick (see figs 2 and 3 in Bryant and

Reed, Chapter 15 this volume). If the depth to the base of the encroaching block is assumed to be 20 km and thickening of the foreshortening block is neglected, only 15 km of relative northwestward migration of the Brevard is required to explain the thrusting; if the encroaching block extends to the base of the crust (about 35 km), less than 9 km of migration would be necessary.

The model yields no information on the amount of strike-slip displacement because the azimuths of the orthogonal movement components are determined by the trace of the join between the crustal blocks and not by the direction of relative drift. Drift may well have been more nearly parallel to the Brevard zone than the arbitrary direction illustrated (Fig. 2).

This model of simultaneous strike-slip movement and thrusting requires *left-lateral displacement* along the Brevard, rather than right-lateral as we previously suggested. There is as yet no conclusive evidence as to the sense of movement along the zone, but the above model seems more nearly compatible with the observed structural relations in the Grandfather Mountain area than was our previous interpretation.

An interesting corollary to this hypothesis is that the diabase dike that crosses the Brevard in the Grandfather Mountain area lies in the expected orientation of the rotated shear set conjugate to the left-lateral strike-slip fault (Cloos, 1955, p. 244–245). Swarms of similarly oriented diabase dikes occur southeast of the Brevard in Alabama, Georgia, and South Carolina, (King, 1961); some cut the Brevard zone, and several extend entirely across it, but the dike swarms are absent northwest of the zone. Perhaps these dikes filled fractures that originated during the closing stages of movement along the Brevard. Note also that the old lineation and fold axes in the Inner Piedmont are approximately normal to the assumed direction of movement of the Piedmont block.

If the above interpretation of the structural relation between the Brevard fault and the northwestward thrusting of the Blue Ridge thrust sheet is correct, most of the movement of the Brevard fault would have been concomitant with the thrusting and could have started in the middle Paleozoic after the climax of regional metamorphism about 350 m.y. ago (Bryant and Reed, Chapter 15 this volume). However, the Brevard fault may have been active even during early Paleozoic, and some of the sedimentary rocks now

present in tectonic slices along the Brevard may have been deposited in elongate basins developed in a manner analogous to development of the Gulf of California along the San Andreas fault (Hamilton, 1961). Movement along the zone must have ended prior to or was concurrent with the emplacement of the dike, presumably in the Late Triassic.

REFERENCES

American Geophysical Union, Special Committee for the Geophysical and Geological Study of the Continents, 1964, Bouguer gravity anomaly map of the United States U. S. Geol. Survey Spec. Map, 2 sheets, scale 1:2,500,000.

Bryant, Bruce, 1962, Geology of the Linville quadrangle, North Carolina–Tennessee—a preliminary report: U. S. Geol. Survey Bull. 1121-D, 30 p.

Burchfiel, B. C., and Livingston, J. L., 1967, Brevard zone compared to alpine root zones: Am. Jour. Sci., v. 265, p. 241–256.

Cloos, Ernst, 1955, Experimental analysis of fracture patterns: Geol. Soc. America Bull., v. 66, p. 241–256.

Davis, G. L., Tilton, G. R., and Wetherill, G. W., 1962, Mineral ages from the Appalachian province in North Carolina and Tennessee: Jour. Geophys. Research, v. 67, p. 1987–1996.

Hamilton, W. B., 1961, Origin of the Gulf of California: Geol. Soc. America Bull., v. 72, p. 1307–1318.

Jonas, A. I., 1932, Structure of the metamorphic belt of the southern Appalachians: Am. Jour. Sci., 5th ser., v. 24, p. 228–243.

Keith, Arthur, 1905, Description of the Mount Mitchell quadrangle [N.C.–Tenn.]: U. S. Geol. Survey Geol. Atlas Folio 124, 10 p.

————, 1907, Description of the Pisgah quadrangle [N.C.–S.C.]: U. S. Geol. Survey Geol. Atlas Folio 147, 8 p.

King, P. B., 1961, Systematic pattern of Triassic dikes in the Appalachian region: U. S. Geol. Survey Prof. Paper 424-B, p. 93–95.

————, 1964, Further thoughts on tectonic framework of the southeastern United States, in Lowry, W. D., ed., Tectonics of the southern Appalachians: Virginia Polytech. Inst., Dept. Geol. Sci. Mem. 1, p. 5–31.

Philbin, P. W., Petrafeso, F. A., and Long, C. L., 1964, Aeromagnetic map of the Georgia Nuclear Laboratory area, Georgia: U. S. Geol. Survey Geophys. Inv. Map GP-488.

Reed, J. C., Jr., 1964, Geology of the Linville Falls quadrangle, N. C.: U. S. Geol. Survey Bull. 1161-B, 53 p.

————, and Bryant, Bruce, 1964, Evidence for strike-slip faulting along the Brevard zone in North Carolina: Geol. Soc. America Bull., v. 75, p. 1177–1196.

Sloan, Earle, 1908, Catalogue of the mineral localities of South Carolina: South Carolina Geol. Survey Bull. 2, 4th ser., 505 p.

Chapter 19

Structural Geology of the Reading Prong*

AVERY ALA DRAKE, JR.

INTRODUCTION

THE READING PRONG is a prominent physiographic and geologic feature that extends from east of the Hudson River in New York to Reading, Pennsylvania (Fig. 1). Geologically, it also includes Little South Mountain, an outlying mountain mass about 10 mi (17 km) west of Reading. In New York, the Prong is known as the Hudson Highlands, in New Jersey, as the New Jersey Highlands, and in Pennsylvania, as the Reading and Durham Hills.

The Prong consists of ridges underlain by Precambrian igneous and metamorphic rocks and intermontane valleys underlain by carbonate and pelitic rocks of Cambrian and Ordovician age. In northeastern New Jersey and New York, clastic rocks of Silurian and Devonian age also crop out within the Prong. The Prong is bounded on the north by the Great Valley, which is underlain by the same Cambrian and Ordovician carbonate and pelitic rocks that crop out in the intermontane valleys. In New Jersey and Pennsylvania, the Triassic border fault forms most of the south border of the Prong. Small areas of Paleozoic rocks locally occur between the Precambrian and Triassic rocks. In New York, the Highlands are largely in fault contact on the south with the New York City Group of uncertain but probable early Paleozoic age, although some rocks of the Great Valley sequence of unquestioned early Paleozoic age are present along the Hudson River.

The regional geology of the Prong is known mostly from reconnaissance mapping in New York (Broughton and others, 1962), reconnais-

sance studies in New Jersey (Bayley, 1941; Bayley and others, 1914; Darton and others, 1908; Spencer and others, 1908), and county studies in Pennsylvania (Miller and others, 1939, 1941; Willard and others, 1959).

In recent years detailed mapping has been concentrated in three separate areas: (1) northeastern New Jersey and adjoining New York (Baker, 1956; Buddington and Baker, 1961; Dodd, 1962; Hague and others, 1956; Hotz, 1952; Sims, 1958); (2) the Delaware Valley of eastern Pennsylvania and western New Jersey (Davis and others, 1967; Drake, 1965, 1967a,b; Drake, Davis, and Alvord, 1960; Drake and Epstein, 1967; Drake, Epstein, and Aaron, 1969; Drake, McLaughlin, and Davis, 1961, 1967); and (3) the Boyertown–Reading area, Pennsylvania (Buckwalter, 1959, 1962; Geyer and others, 1963). The Delaware Valley is the only area where detailed geologic mapping has been carried across the entire Prong and the adjoining Paleozoic rocks.

This work has shown that the fundamental problem of the area is the geometric and tectonic relation of the Precambrian rocks to the sedimentary rocks of the Great Valley and the smaller intra-Prong valleys. Many geologists, myself included, believe that the Precambrian rocks are a far-traveled allochthonous mass.

I would like to thank J. M. Aaron, R. E. Davis, and J. B. Epstein of the U. S. Geological Survey, my co-workers in the Delaware and Lehigh Valleys, for data and many years of thoughtful discussions, the late V. E. Gwinn, and G. H. Wood, Jr., of the U. S. Geological Survey for help in crystallizing many of the ideas presented herein, and J. M. Aaron and D. L. Southwick of the U. S. Geological Survey for most helpful technical reviews of earlier drafts of the paper.

* Publication authorized by the Director, U. S. Geological Survey.

271

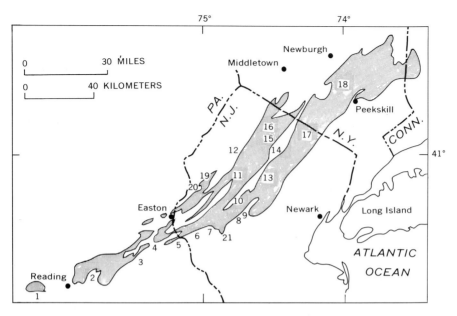

FIGURE 1. Sketch map of the Reading Prong showing some places mentioned in text. (1) Little South Mountain; (2) Oley Valley; (3) Limeport; (4) Saucon Valley and Bethlehem; (5) Rattlesnake Hill window and Monroe; (6) West Portal window; (7) Clinton, Jutland, High Bridge, and Annandale area; (8) Pottersville Falls; (9) Peapack Valley; (10) German Valley; (11) Stanhope; (12) Andover; (13) Dover; (14) Green Pond syncline; (15) Edison area; (16) Franklin–Sterling area; (17) Sterling Lake–Ringwood area; (18) Popolopen Lake area; (19) Jenny Jump Mountain; (20) Buttzville; (21) McPherson.

STRATIGRAPHY AND PETROLOGY

Precambrian Rocks

The Precambrian rocks of the Prong are chiefly high-grade iron- and soda-rich metasedimentary and metavolcanic rocks that are interlayered with smaller amounts of amphibolites and marble and are plutonized and migmatized by sodic granitic rocks, hornblende granite, and alaskite. All these rocks are in the hornblende granulite subfacies of the granulite facies of regional metamorphism (Fyfe and others, 1958). Except for minor amounts of charnockite, rocks of this suite are not known in the core of the northern Blue Ridge anticlinorium to the south (Espenshade, Chapter 14 this volume) or along the Green Mountain axis to the north. These rocks are, however, very similar to those of the Adirondacks and the Grenville province of Canada and New York and have been called Grennoble by Engel (1956). A sequence of lower grade interbedded metasedimentary and metavolcanic rocks overlies the older metamorphic and plutonic rocks.

In New Jersey and Pennsylvania, these Precambrian rocks have been divided into four principal units: Franklin Limestone, Pochuck Gneiss (all the mafic rocks), Losee Gneiss (feldspar dominantly plagioclase), and Byram Gneiss (feldspar dominantly K-feldspar); the last unit is roughly equivalent to the Storm King Granite (Lowe, 1950) of New York. In addition, Bayley (1941) recognized a fifth group of rocks near the Delaware River, the Pickering Gneiss, and Fraser (in Miller and others, 1939, 1941) mapped a unit in eastern Pennsylvania which he called the Moravian Heights Formation. The Pochuck, Losee, and Byram were generally thought to have had an intrusive origin while the Pickering Gneiss, Moravian Heights Formation, and Franklin Limestone were thought to be metasedimentary and metavolcanic rocks. Although these units were useful in reconnaissance mapping, they include diverse lithologies and rocks of different origins, and most modern workers have abandoned them in favor of more precise lithologic units. The units from seven areas within the Prong are compared and correlated in Table 1.

It can be seen that similar rocks crop out throughout the Highland area. Sodic rocks are absent and marble is very sparse in the western part of the Prong. Minor amounts of other rock types have been mapped locally, but they are not considered herein.

Paleozoic Rocks

Sedimentary rocks of Cambrian and Ordovician age flank the Prong on the northwest, crop out within intermontane valleys, and are sparingly present in Pennsylvania and New Jersey between the Precambrian and Triassic rocks to the southeast. These rocks were deposited in the Paleozoic geosyncline and include, from southeast to northwest and oldest to youngest, orthoquartzite–carbonate shelf deposits (Lower Cambrian through Lower Ordovician), deeper water neritic carbonate deposits (Middle Ordovician), and graywacke–shale flysch deposits (Middle and Upper Ordovician). The stratigraphy of these rocks is better known in some areas than others, but more or less similar rocks are present throughout the Prong. Correlations and reported thicknesses are given in Table 2. In my opinion, the reported thicknesses of the Martinsburg Shale in New Jersey and the Snake Hill Shale in New York are grossly underestimated. These rocks are unconformably overlain to the northwest by molasse deposits of Silurian age, but these and younger rocks will not be considered herein.

At many places in the Appalachians, a Precambrian core or welt separates miogeosynclinal rocks from eugeosynclinal rocks. This is not the case, however, in the Prong, where autochthonous rocks of eugeosynclinal origin have not been recognized and the same rock units crop out on both sides of the Precambrian highlands.

In New York, the relations are more complex, as the Precambrian rocks are in fault contact on the south with rocks of the New York City Group of Prucha (1956). These rocks have been considered at one time or another to be Precambrian (Berkey, 1907; Berkey and Rice, 1919; Prucha, 1956, 1959) or lower Paleozoic and equivalent to rocks of the Great Valley (Balk, 1936; Norton, 1959; Isachsen, 1964). This thorny problem is yet to be resolved, although currently the consensus is that the bulk of the New York City Group is of Cambrian and Ordovician age. It is difficult, however, to rationalize the relation on the Hudson south of the Prong near Peekskill, New York (Broughton and others, 1962), where relatively unmetamorphosed Cambrian and Ordovician rocks are almost in juxtaposition with their supposed metamorphic equivalents. It may well be that a major structure here has so far escaped detection. If not, the New York City Group may

be equivalent to the Glenarm Series of Pennsylvania and Maryland.

In 1946, Stose recognized varicolored pelites and other exotic rock types within the Martinsburg outcrop belt between the Susquehanna and Lehigh Rivers and attributed their presence to a far-traveled Hamburg klippe. Subsequent work has turned up faunal evidence that at least some of these rocks are older than Trenton and therefore are older than the Martinsburg (Field Conference Pennsylvania Geologists, 1966). It is now thought that these rocks occur in several klippen that were emplaced during Martinsburg deposition. Basaltic volcanic rocks crop out near Jonestown, Pennsylvania, (southwest of the area shown in Figure 1) within the allochthonous pelite and are also probably allochthonous. This assemblage is a typical eugeosynclinal suite. The source of these klippen is not known, but it may be the Cocalico Shale which crops out south of the Triassic basin.

Pelitic rocks in thrust contact with Lower and Middle Cambrian carbonate rocks have been mapped as Martinsburg in the Lebanon Valley (Gray and others, 1958; Geyer and others, 1963). In addition, thermally metamorphosed pelite that crops out along the Cambrian carbonate–Triassic boundary has been mapped as "Mine Hill Slate" (Geyer and others, 1958). The geographic and tectonic position of these rocks suggests that they also may be klippen of the Hamburg type. One other small body of possible allochthonous pelite lies on Allentown Dolomite near Limeport, Pennsylvania (Miller and others, 1941). The relation there, however, is conventionally interpreted as an unconformity.

Rocks similar to those in the Hamburg klippen crop out between the Precambrian rocks of the Prong and the Triassic basin near Clinton and Jutland, New Jersey. These rocks contain Normanskill graptolites and were recognized by Bayley and others (1914) as being older than Martinsburg. More recent work by Harry Dodge and R. B. Neuman (in U. S. Geological Survey, 1964, p. A83) has established that graptolites of both Early (Deepkill) and Middle (Normanskill) Ordovician age are present in these rocks, which overlie Jacksonburg Limestone of Middle Ordovician (Trenton) age. Here again, older rocks overlie younger rocks and can be presumed allochthonous.

One other area of pelite in the Peapack Valley of New Jersey has been mapped as Martinsburg, although these rocks do not resemble the Martinsburg of the Great Valley. They are, however, like the rocks of the Clinton–Jutland area. The

TABLE 1. Comparison, Probable Correlation, and Interpreted Origin of Major Precambrian Rock Units in Seven Reading Prong Areas

Reading area, Penna. (Buckwalter, 1959, 1962)	Delaware Valley, N. J.–Penna. (Drake, 1967a,b; Drake and others, 1961, 1967, and 1969)	Dover district, N.J. (Sims, 1958)	Franklin area, N.J. (Hague and others, 1956)	Edison Area, N.J. (Buddington and Baker, 1961)	Sterling Lake, N.Y.– Ringwood, N.J. area (Hotz, 1952)	Popolopen Lake quadrangle, N.Y. (Dodd, 1962)
Granitic gneiss. Mostly intrusive.	Hornblende granite, microperthite alaskite, and biotite granite. Mostly intrusive.	Hornblende granite, alaskite, and related rocks. Intrusive.	Byram gneiss. Intrusive.	Hornblende granite and alaskite. Intrusive.	Hornblende granite and related facies. Intrusive.	Hornblende granite, leucogranite, and biotite granite. Intrusive.
—	Microantiperthite alaskite. Intrusive.	Microantiperthite granite. Intrusive.	—	Pyroxene alaskite, pyroxene granite, and pyroxene syenite. Intrusive.	—	—
—	Albite-oligoclase granite and related albite pegmatite. Rheomorphic, probably anatectic, with attendant granitization.	Albite-oligoclase granite and albite-quartz pegmatite. Anatectic.	Granodiorite gneiss. Intrusive.	Albite alaskite. Uncertain origin.	Quartz–oligoclase gneiss (part). Intrusive.	Plagioclase–quartz leucogneiss (part). Probably metavolcanic.
Quartz diorite gneiss. Migmatitic variant of granitic gneiss. (Correlation doubtful).	Quartz diorite. Plutonic metamorphism of sodic tuff and mafic volcanic rock.	Quartz diorite. Intrusive.	—	Hypersthene–quartz-oligoclase gneiss. Metasedimentary.	Quartz–oligoclase gneiss (part). Intrusive.	Hypersthene–quartz-oligoclase gneiss. Metavolcanic.
—	Oligoclase-quartz gneiss. Metavolcanic.	Oligoclase–quartz–biotite gneiss. Metasedimentary.	Losee gneiss and oligoclase gneiss. Intrusive.	Quartz–oligoclase gneiss. Uncertain origin.	Quartz–oligoclase gneiss (part). Intrusive.	Plagioclase–quartz leucogneiss (part). Probably metavolcanic.

Graphitic gneiss and quartz–biotite–feldspar gneiss. Metasedimentary.	Quartz–feldspathic gneiss, including biotite–quartz–plagioclase gneiss, sillimanite-bearing gneiss, and potassic feldspar gneiss. Metasedimentary.	Biotite–quartz–feldspar gneiss. Metasedimentary.	Microcline gneiss, garnet gneiss, biotite gneiss, and graphitic gneiss. Metasedimentary.	Biotite–quartz–plagioclase gneiss, and quartz–microcline granulite. Uncertain origin. Quartz–potassium feldspar gneiss, quartz–plagioclase gneiss, epidote-scapolite–quartz gneiss, and garnetiferous gneiss. Metasedimentary.	Garnetiferous quartz–biotite gneiss and quartzite. Metasedimentary.	Rusty and nonrusty biotite–quartz–feldspar gneiss. Metasedimentary.
Marble and lime silicate gneiss. Metasedimentary.	Marble and its alteration products, including skarn. Metasedimentary.	Skarn and related rocks. Metasedimentary.	Marble. Metasedimentary.	Marble. Metasedimentary.	Marble and skarn. Metasedimentary.	Marble and pyroxenite. Metasedimentary.
Hornblende gneiss. Metasedimentary and metaigneous(?).	Amphibolite. Metaigneous, metavolcanic, and metasedimentary.	Amphibolite. Metasedimentary and metaigneous.	Hornblende gneiss. Metasedimentary, metaigneous, and metavolcanic.	Amphibolite. Uncertain origin.	Pyroxene amphibolite and amphibolite. Metasedimentary.	Amphibolite, pyroxene–hornblende–plagioclase gneiss, and biotite–hornblende–quartz–feldspar gneiss. Metavolcanic.
—	Pyroxene gneiss. Metasedimentary.	—	Pyroxene gneiss and related rocks. Metasedimentary.	Pyroxene gneiss. Metasedimentary.	—	—
Migmatite and closely interbedded gneiss. Injection gneiss.	Migmatite: microperthite alaskite neosome in amphibolite paleosome, albite–oligoclase granite neosome in amphibolite paleosome.	Hornblende granite and alaskite, and albite–oligoclase granite containing more than 20 percent metasedimentary rocks. Injection gneiss.	—	—	—	—

TABLE 2.   Correlation of Lower Paleozoic Rocks that Border the Reading Prong

Age	Reading area, Penna. (Geyer and others 1963)	Delaware Valley, N.J.–Penna. (Drake and others, 1961; Drake, 1965; Drake and Epstein, 1967)	New Jersey (Lewis and Kummel, 1915)	Eastern New York (Broughton and others, 1962)
Middle and Upper Ordovician	Martinsburg Formation (?)	Martinsburg Formation 12,000 ft (3,658 m)	Martinsburg Shale 3,000 ft (914 m)	Snake Hill Shale 3,000 ft (914 m)
Middle Ordovician	Hershey Formation 400 ft (122 m)	Jacksonburg Limestone 500–800 ft (152–244 m)	Jacksonburg Limestone 125–300 ft (38–91 m)	Balmville Limestone <100 ft (30 m)
	Myerstown Formation 150 ft (46 m)			
	Annville Formation 120 ft (37 m)			
Lower Ordovician	Beekmantown Group 2,950 ft (900 m) +	Beekmantown Group 1,450 ft (442 m)	Kittatinny Limestone 2,500–3,000 ft (762–914 m)	Stockbridge Group 4,000 ft (1,219 m)
Upper Cambrian	Conococheague Group 3,700 ft (1,158 m) +	Allentown Dolomite 1,700 ft (518 m)		
Middle Cambrian	Buffalo Springs Formation 500 ft (152 m) +	Leithsville Formation 1,000 ft (305 m)		
	Tomstown Formation 500 ft (152 m) +			
Lower Cambrian	Hardyston Formation 250–600 ft (76–183 m)	Hardyston Quartzite 100 ft (30 m)	Hardyston Quartzite 5–200 ft (2–60 m)	Poughquag Ortho-quartzite 250 ft (76 m)

Jacksonburg Limestone is absent in this area, and the pelite overlies probable Beekmantown carbonate (Minard, 1959). This may be another klippe. The similarity of the relations mentioned above to those in the Taconic region (see Zen, 1967) is obvious.

Clastic rocks of Silurian through Late Devonian age are infolded and infaulted with the Prong rocks in New Jersey and New York for a distance of about 53 mi (88 km) (Lewis and Kümmel, 1910-12; Broughton and others, 1962). These rocks are in the Green Pond syncline, which ranges from less than a mile (1.6 km) to slightly more than 4 mi (7 km) in width, and they lie unconformably on older rocks. Generally they are in contact with Precambrian rocks, but at places they lie on rocks as young as the Martinsburg Formation (Kümmel and Weller, 1902). Apparently there is an unconformity within this sequence, and rocks of Onondaga age lie on rocks of Late Silurian age. The rocks have a thickness of about 6,000 ft (1,829 m) (Lewis and Kümmel, 1940) and presumably are a near-shore and on-shore facies of rocks of equivalent age that crop out in the folded Appalachians to the northwest.

Minard (1959) has mapped a lens of apparent Silurian orthoquartzite and quartzite conglomerate in the Peapack Valley of New Jersey. As the lithology resembles that of the Shawangunk Conglomerate of the main Silurian outcrop belt, Minard correlates it with that unit rather than with the equivalent Green Pond Conglomerate of the Green Pond sequence. This rock is in sedimentary contact with allochthonous(?) Ordovician pelite on the south and is in fault contact with Precambrian rocks on the north. This is a very important exposure as no rocks of this age have heretofore been found south of the Precambrian highlands. Does this remnant mean that there was once a syncline containing Silurian and perhaps Devonian rocks south of the Prong in the area now covered by Triassic rocks?

The metamorphic grade of the Cambrian and Ordovician rocks progressively increases to the northeast from essentially unmetamorphosed rocks near Harrisburg to probable lower greenschist facies at the Delaware River, to biotite zone at the Hudson River, to sillimanite zone where the Prong plunges out in New York. This progression is largely based on the mineralogy of the Martinsburg Formation and equivalents, because carbonate rocks are not sensitive indicators of low-grade regional metamorphism. The metamorphic grade at the northeast end of the Prong is well established (Balk, 1936), as is the relatively unmetamorphosed character of the rock at the southwest end. The grade in the area between is a subject of controversy.

In the Pennsylvania slate belt, almost all authors, especially Behre (1927, 1933) have considered the slate to be the result of metamorphic recrystallization, the principal metamorphic minerals being sericite (or muscovite) and chlorite. Bates (1947) found that much of the mica in the commercial slate (Pen Argyl Member of the Martinsburg Formation) is illite rather than sericite, although both sericite and chlorite are present in lesser quantities. He thought that the illite was a metamorphic mineral. More recently, McBride (1962) reported that shales and slates from the Martinsburg contain both 2M muscovite and chlorite that seems to be metamorphic. He also reported that these minerals have recrystallized in metamorphosed graywacke. Similar relations have been noted in the Delaware Valley. Maxwell (1962), on the other hand, believes that the mineralogy of the Martinsburg in the Delaware Valley results from diagenesis rather than metamorphism and provides abundant references that both degraded and well-crystallized illite, as well as chlorite, can form at low temperatures and pressures. It is difficult, if not impossible, to differentiate between metamorphic recrystallization and authigenic recrystallization. It should be pointed out, however, that just because a mineral *can form* at low temperatures and pressures does not necessarily mean that it *has* so formed. Because of these conflicting data, evidence from other rocks must be considered.

Maxwell (1962) also reports that the carbonate rocks that lie between the Martinsburg and the Precambrian show no signs of plastic deformation or recrystallization. This is in error. All writers from the time of the Second Pennsylvania Geological Survey (Lesley and others, 1883) have described and illustrated abundant examples of flowage and recrystallization in these rocks (see Miller and others, 1939, 1941; Gray, 1951, 1952, 1959; Gray and others, 1958; Geyer and others, 1958, 1963; Prouty, 1959; Drake and others, 1960; Sherwood, 1964). Recrystallization is common parallel to cleavage, and in less competent

beds, flowage parallel to cleavage has produced a tectonic layering. In such rocks, the only bedding seen is tiny fold hinges marked by thin silt layers. Abundant examples have been illustrated where dolomite has been boudined out into lenses that now float in limestone that has flowed around them (see Miller and others, 1941). In other places, folds are defined by dolomite beds, whereas the limestones have deformed plastically and have only a flow fabric. In areas of extreme deformation, the Jacksonburg Limestone contains veins of columnar quartz and calcite which parallel cleavage and thrust faults. The Martinsburg Formation contains similar veins of quartz. These veins are obviously secondary and probably are of intraformational origin. Their restriction to areas of deformation suggests that the movement of material and recrystallization resulted from metamorphism rather than simple cold-water percolation, as plentiful open space in other areas is not filled.

Pelitic beds within the carbonate sequence, as well as in the Hardyston Quartzite, all contain recrystallized mica and are now phyllites. It is not known, however, what the mica mineral is.

All these data support the contention that the rocks at the Delaware River have undergone low-grade regional metamorphism rather than only diagenesis.

The Silurian and Devonian rocks in the Green Pond syncline are cleaved, but no data are available on their metamorphic grade.

Triassic Rocks

The Prong, except in New York, is bordered on the southeast by clastic sedimentary rocks of the Brunswick Formation (Gettysburg Shale west of Reading) of the Newark Group of Triassic age. These rocks are primarily red beds and are coarser and more arkosic near the Precambrian contact, and therefore obviously had a northern provenance. Fanglomerates that occur all along the Precambrian border and in the gap between the west end of the Prong and Little South Mountain are of special interest.

Two types of fanglomerate have been mapped along the border: (1) those that have a relatively large areal extent and contain clasts predominantly of Silurian and Devonian quartzite, and (2) those of limited areal extent that contain clasts of Cambrian and (or) Ordovician carbonate, Ordovician pelite, Precambrian gneiss, or some other local rock type. Along the Delaware River, fossiliferous clasts from the quartzite fanglomerate have been identified as probably stemming from the Green Pond Conglomerate and

Decker Limestone (Drake and others, 1961). Both these formations crop out in the Green Pond syncline to the northeast, east of the Delaware Valley; it seems that the larger fans fed into the Newark basin from the northeast rather than normal to the present Precambrian–Triassic border, as conventionally envisioned. The shape of the larger fans (see Drake and others, 1961) supports this observation. The smaller fans that contain clasts of local origin, on the other hand, seem to have been fed down a slope about normal to the present border.

The large body of coarse clastic rock between Reading and Little South Mountain consists of conglomerate containing clasts of Hardyston Quartzite interbedded with arkosic sandstone. These rocks are of especial interest in that no rock now exposed could have been their source. McLaughlin (1938) believes that they stemmed from a klippe that was subsequently completely eroded away, and that idea is accepted here.

The Brunswick Formation is about 3,400 ft (1,036 m) thick in the Delaware Valley (Drake and others, 1961) and is thought to be 6,000–8,000 ft (1,829–2,438 m) in New Jersey. Near Little South Mountain in Pennsylvania, the Gettysburg Shale is about 9,400 ft (2,865 m) thick (Geyer and others, 1963). These rocks are generally considered to have been deposited in a basin. A border fault has been mapped throughout New Jersey (Lewis and Kümmel, 1910-12), the Delaware Valley (Drake 1967b; Drake and others, 1961, 1967), and eastern Pennsylvania (Gray and others, 1960). In east-central Pennsylvania, however, no such fault has been mapped at all places (Gray and others, 1960); the faults there may have been buried by later sedimentation.

Sheets of diabase and basalt flows occur within the sedimentary rocks; in one place diabase is in contact with Precambrian rocks. Dikes of diabase, presumably of Triassic age, cut rocks of the Prong at various places.

STRUCTURAL GEOLOGY

The Reading Prong traditionally has been considered a first-order anticlinorium cored with Precambrian rock. It is part of a major Appalachian structural element that includes the Blue Ridge anticlinorium (Espenshade, Chapter 14 this volume; Bryant and Reed, Chapter 15 this volume) to the south and the Green Mountain anticlinorium of Vermont and the Long Range anticlinorium of western Newfoundland to the north. The rocks in the core of the anticlinorium were deformed plastically during the Precam-

brian, have been through the Taconic, Acadian, and Appalachian orogenies, and were again deformed during the Triassic. Their structure and current position is the cumulative result of this long and complicated history of deformation.

Internal Precambrian Structure

Lithologic boundaries and crystallization foliation in the Precambrian rocks of the Reading Prong trend northeast from the southwesternmost exposures near Reading (Buckwalter, 1962) to the northeasternmost extremity of the Prong (Broughton and others, 1962), although local variations are common. Contacts and foliation, as well as compositional layering, commonly dip southeast, although locally the rocks are vertical or dip northwest. Most authors think that the compositional layering is relict bedding and have found that layering and foliation are essentially parallel. They also find that foliation is generally poorly developed in fold noses. Baker (1956) reports that in fold noses in the Edison, New Jersey, area, foliation parallels the axial surfaces of folds rather than layering which follows lithologic contacts. Elsewhere in the Prong, foliation is apparently absent from fold noses, lineation being the only minor structure. Layering is probably largely parallel to relict bedding, but possibly it is parallel to cleavage as well; if so, the lithologic units are the limbs of lensed-out or sheared-out isoclinal folds. This seems much more likely when the evidence from highly deformed but nonmetamorphic or low-grade metamorphic terranes is considered. Crystallization foliation in the plutonic rocks parallels the layering in the metamorphic rocks and probably is largely flow structure.

Lineation is poorly to well developed throughout the Prong and is commonly marked by mineral alinements or streaks, crinkles, rods, or fluted surfaces. Less common lineations include grooves, ridges, boudinage, and minor fold axes. So far as is known, virtually all lineation related to the Precambrian plastic deformation is parallel to the tectonic b axis. Sims (1958) reports an exception in the Dover, New Jersey, district related to intense deformation and overturning, which he believes to be lineation in a. A few minor fold axes in the Delaware Valley also may be a lineations, but they are sparse. Lineation has been well studied in the Prong because it has been long recognized that the magnetite ore bodies plunge parallel to the linear elements in the enclosing rocks. Regionally, plunges are gentle to moderate to the northeast, statistical averages ranging from N. 49° E. to N. 70° E. in the Delaware Valley to

N. 39° E. in southwestern New York (Dodd, 1962). Lineation has not been studied southwest of the Delaware River.

The Precambrian rocks are well jointed, and in most areas, statistical studies show that cross, longitudinal, and diagonal joints can be tectonically related to the lineation. Many other joints related to later deformation are abundant throughout the Prong.

The distribution of lithologic units and the attitudes of foliation indicate that the Precambrian rocks are complexly folded throughout the Prong. The appalling lack of exposure south of the Wisconsin terminal moraine, however, has made detailed structural studies impossible in the southwestern area. For example, Buckwalter (1959, 1962) found 50 outcrops in the Boyertown 15 min quadrangle and 69 outcrops in the Reading 15-min quadrangle and therefore could map no folds. Exposure is slightly better in the Delaware Valley, although Drake and others (1961) found only 13 outcrops in the Frenchtown 7½-min quadrangle.

Folds range in magnitude from a few inches in wavelength and amplitude to as much as 7 mi. long parallel to the axis and a mile in width. All are in foliation which may or may not be parallel to bedding. Because the stratigraphic relations are not known, the terms antiforms and synforms are preferred to anticlines and synclines for most of these folds. The folds range from upright and open to isoclinal overturned or more rarely isoclinal recumbent; isoclinal folds overturned to the northwest are most common. Most folds plunge northeast essentially parallel to the regional lineation, although southwest–plunging folds have been mapped in the Delaware Valley. All folds seem to have limited vertical extents. The maximum amplitude recorded in the Delaware Valley is 2,600 ft. No other authors give amplitude estimates. Most of the folding was disharmonic and was accomplished largely by flexural slip and attendant flowage of the less competent rock. The most competent units, primarily amphibolite (or migmatite), have broken and have been pulled apart in sites of strongest deformation.

Cross folds have been recognized by Hague and others (1956), Sims (1958), Drake (1967a,-b), and Drake and others (1961, 1967). These folds are all minor and almost certainly were formed by a vector quantity of the northwest-directed stress that caused the major folding.

No direct evidence has so far been found for more than one episode of folding during Precambrian time. All major and minor structures can be related to one period of plastic folding in the Precambrian or to post-Precambrian deformation. It should be pointed out, however, that the plutonic rocks (840 million years old) are generally thought to have been syntectonically emplaced into a suite of rocks that dates from 1,200 m.y. ago (Long and Kulp, 1962). If this is so, at least some of these sedimentary and volcanic rocks were metamorphosed and therefore probably folded at that time. Two lines of vague and inconclusive evidence bear on this problem. In the Delaware Valley, a sheet of metasomatic albite–oligoclase granite contains several skialiths of isoclinally folded amphibolite. Obviously, the amphibolite was folded prior to its replacement, but the time of deformation is uncertain. The lineation in the amphibolite is within the regional variation, so nothing conclusive can be determined.

In Scotts Mountain, near the Delaware River, several metasedimentary and metavolcanic layers are so related spatially that with imagination one could reconstruct them into a refolded recumbent fold (Drake, 1967a,b; Drake and others, 1969). Unusually steep dips and plunges and odd plunge directions in this area adds supporting evidence, but the exposure is such that this interpretation is open to debate. There is evidence, therefore, for only one period of Precambrian deformation, although I personally believe that this deformation obliterated most evidence of any earlier deformation.

Relation of the Precambrian to the Paleozoic Rocks

The principal structural problem in the Prong is the relation of the Precambrian rocks to the Cambrian and Ordovician rocks of the Great Valley and the smaller intermontane valleys. Early geologists in Pennsylvania (Rogers, 1858; Lesley and others, 1883; Miller, 1925) believed that the Precambrian rocks formed anticlinal ridges that were locally overturned and (or) thrust faulted, and that the intermontane valleys underlain by Paleozoic rocks were synclines and (or) grabenlike blocks downdropped by normal faults. The Paleozoic rocks of the Great Valley were considered to be largely in normal contact with the Precambrian rocks.

In New Jersey, Lewis and Kümmel (1940) recognized that at many places Precambrian rocks were thrust on Paleozoic rocks and believed that both were subsequently folded. In their interpretation, the northern boundaries of the intermontane valleys are high-angle normal faults that formed later than the thrusting and folding, probably during Triassic time.

In New York, all early workers considered the Precambrian rocks of the Hudson Highlands to be autochthonous, although Balk (1936) drew an almost continuous fault around the Highlands. In addition, he drew faults on all sides of several small Precambrian masses along which the Precambrian rocks were upthrust en masse.

In 1935, Stose and Jonas proposed that the Precambrian rocks and Hardyston Quartzite of the mountain area were part of a large, flat, far-traveled thrust sheet and that the inlying limestone valleys were tectonic windows. The Stose-Jonas hypothesis was vigorously denied by B. L. Miller and his co-workers (Miller and Fraser, 1936; Fraser, 1938), whose comments were answered by Stose and Jonas (1939). Shortly before his death, Miller (1944) presented specific evidence that clearly showed that the flat-thrust theory was completely untenable.

Miller's work was unquestioned until recently, when detailed geologic mapping began in the Delaware Valley. Concurrent with this mapping, an aeromagnetic survey was made of the Precambrian outcrop belt in Pennsylvania, and later, gravity studies were made in the southwestern part of the Prong in Pennsylvania and on Pohatcong Mountain in New Jersey. The geologic studies first led to the recognition that thrust faults and overturned folds were more of a factor in the distribution of Precambrian rocks than supposed by Miller and that some of the crystalline ridges were probably klippen (Field Conference Pennsylvania Geologists, 1961). Preliminary interpretation of the aeromagnetic data (Bromery and others, 1960) suggested that several Precambrian bodies were relatively thin and a few had no magnetic expression whatsoever. These magnetic data combined with the available geologic data suggested that the ridges of Precambrian rocks may indeed be detached parts of a single thrust sheet (A. A. Drake, Jr., and J. B. Epstein, *in* U. S. Geol. Survey, 1962). Later analyses of aeromagnetic and gravity data (R. W. Bromery, *in* U. S. Geol. Survey, 1964) indicated that the crystalline rocks at the west end of the Prong occurred at both the surface and at a deeper level. A still later regional gravity analysis (R. W. Bromery, *in* U. S. Geol. Survey, 1966) agreed with the prior interpretation of aeromagnetic and gravity data and confirmed that the Precambrian in the western part of the Prong was thin.

About the same time of the above work, geologists of the Pennsylvania Geological Survey (Gray and others, 1960; Socolow, 1961; Geyer and others, 1963) recognized that Little South Mountain and other small blocks of Precambrian rocks and Hardyston Quartzite were klippen. These interpretations meshed nicely with their finding that the Great Valley was recumbent and inverted.

Isachsen (1964), in a survey paper on the Precambrian in the northeastern United States, interpreted the Reading Prong as being a gigantic klippe composed of both Precambrian and lower Paleozoic rocks. In his interpretation, these rocks were emplaced as an immense allochthonous plum in a Martinsburg pudding, the thrust relation being obscured by later Martinsburg sedimentation.

Most geologists currently working in the Prong have accepted an allochthonous interpretation, and recently published maps (Geyer, 1963; Davis and others, 1967; Drake, 1967a,b; Drake and others, 1967, and 1969) show nonrooted Precambrian rocks. Buckwalter (1959, 1962) does not accept the allochthonous interpretation, although his maps show the Precambrian as being almost completely fault-bounded. Many other geologists probably do not accept the allochthonous interpretation, and as evidence is far from obvious and clear-cut it will be summarized below.

Geophysical evidence. An allochthonous origin is compatible with the regional gravity data (American Geophys. Union Spec. Comm. Geophys. and Geol. Study Continents, 1964). The thrust-faulted but parautochthonous Green Mountain anticlinorium in New England lies on the axis and steep western gradient of the Appalachian gravity high and is bordered on the west by a major gravity low which is coincident with a known basement depression, the Middlebury synclinorium. To the south, the Berkshires continue on the gravity gradient, but in New York the Prong cuts across the gravity gradient and at the Hudson is off the gradient entirely. From north-central New Jersey to Reading, Pennsylvania, the Prong lies dead center in the gravity low (Figure 2). To the south, the autochthonous Blue Ridge–South Mountain anticlinorium (Espenshade, Chapter 14 this volume) lies on the gravity gradient and follows it southwest to the vicinity of Roanoke, Virginia, where the Blue Ridge becomes allochthonous. From this point southwest, the allochthonous Blue Ridge (see Bryant and Reed, Chapter 15 this volume) lies in the trough of the gravity low. The comparison of the Prong to the southern Blue Ridge is obvious. As has been suggested by Woollard (1943), this gravity trough probably represents downbuckled crust beneath the foot of the known thrusts.

The conclusions of Bromery's regional gravity work have already been mentioned. In addition, he did a gravity survey of Pohatcong Mountain (Bloomsbury quadrangle), New Jersey. No gravity anomaly was associated with the mountain (R. W. Bromery, written communication, 1964). Wollard (1943) had earlier suggested that Mus-

conetcong Mountain in the same quadrangle was a thrust sheet having an average thickness of 500 ft (152 m).

Most of the Prong has been mapped aeromagnetically, and regional relations can best be seen on maps by Henderson and others (1966) and Bromery and Griscom (1967). The Precambrian

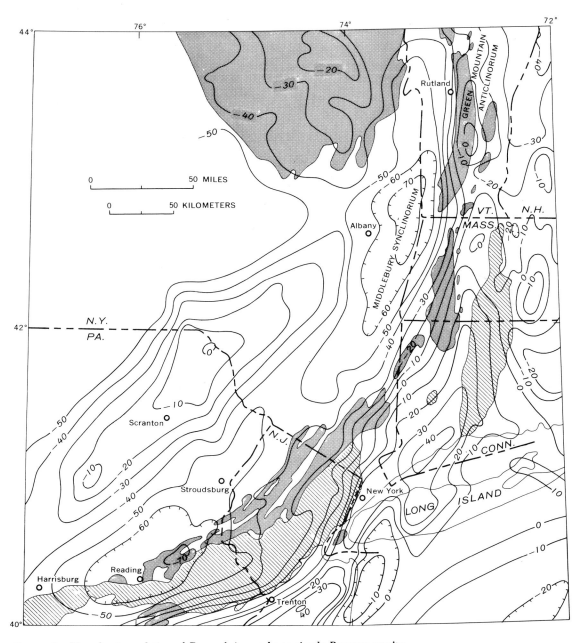

FIGURE 2. Map showing relation of Precambrian rocks to simple Bouguer gravity anomalies in the northeastern United States. Map modified from the Tectonic map of the United States (U.S. Geol. Survey and Am. Assoc. Petroleum Geologiests, 1961 and Bouguer gravity anomaly map of the United States (Am. Geophys. Union, Spec. Comm. Geophys. and Geol. Study Continents, 1964). Precambrian rocks, stippled; Triassic rocks, hatchered; Paleozoic rocks and Coastal-Plain deposits, unpatterned. Contour interval 10 milligals.

rocks have a characteristic "birds-eye maple" magnetic pattern. In the Delaware Valley and to the west, the Precambrian–Paleozoic interface also has a characteristic pattern. As is shown on Figure 3, the Precambrian ridges are bounded on the north by strong negative anomalies and on the south by gradients that drop off into the trench of the negative anomaly north of an adjacent Precambrian ridge. This relation suggests that Precambrian rock is absent beneath Paleozoic rocks on the north sides of the ridges but is buried beneath Paleozoic rocks on the south sides. The suggested model is a series of southeast-dipping, rootless Precambrian bodies that have nonmagnetic material in front, below, and behind.

Negative anomaly basins also occur at the ends of the Precambrian ridges. If the Precambrian were rooted, these ridge ends would be plunging noses of anticlines, and the magnetic rocks should be expressed for some distance along plunge beneath the cover. This relation is not seen, however, and it seems that the Precambrian rocks spoon out rather than plunge under. This relation is well shown where the Precambrian outcrop ends near Reading (Bromery and Griscom, 1967).

Geophysical work also has been done within the Paleozoic outcrop belt that flanks the Prong. Wood and Carter (in press) report that basement lies at a depth greater than 25,000 ft (7,620 m) at the Blue Mountain structural front as interpreted from aeromagnetic data. In addition, Wood and Carter (in press) report that oil-company seismic surveys indicate depths to basement between 30,000 and 45,000 ft (9,144–13,716 m) beneath the Great Valley near the Blue Mountain structural front. All these data are in harmony with the regional gravity data reported above.

Thrust-fault Evidence. Rather than describe all the thrust faults in the Prong or present evidence for others that have not been previously recognized, I will discuss two specific examples in order to demonstrate the tectonic style.

Along the north border of the Prong a thrust has been mapped from near Bethlehem, Pennsylvania, to near Buttzville, New Jersey, a distance of about 30 mi (50 m). Minimum horizontal transport seems to have been about 3 mi (5 km) although the maximum stratigraphic separation is less than 3,000 ft (914 m). Jenny Jump Mountain in New Jersey, a thrust block which is probably related to this thrust, lies to the north and is in contact with rocks as young as Martinsburg Formation. No other faults have been mapped along the northern Precambrian–Paleo-

zoic interface in New Jersey, although Baum (1967) has recently described thrusting from near Andover, New Jersey. In Kittatinny Valley there are several klippen of Kittatinny Limestone lying on Martinsburg, as well as two small gneiss klippen (Bayley and others, 1914). These klippen suggest a minimum horizontal transport of 3 mi (5 km) and occupy synforms in the folded thrust.

In New York, a series of Precambrian klippen lie on Snake Hill Shale between the state line and the Hudson and are believed to be the remnants of one thrust sheet (Broughton and others, 1962). Isachsen (1964) believes that Stissing Mountain in the Taconic terrane 25 mi (42 km) north of the Prong probably had the same origin as these klippen. The data presented above, should convince even the most skeptical reader that there has been a good deal of horizontal transport along the north border of the Prong.

Near the south border of the Prong in New Jersey, a thrust fault has been mapped along the front of the Musconetcong–Schooley Mountain mass between Stanhope and Bloomsbury, a distance of about 26.5 mi (44.5 km) (Lewis and Kümmel, 1910-12). Recently, it has been extended another 12 mi (20 km) into Pennsylvania where this Precambrian mass spoons out (Drake, 1967a,b; Drake and others, 1967). Tectonic windows have been mapped near West Portal, New Jersey (Drake, 1967b), and Monroe, Pennsylvania (Drake and others, 1967), and a small klippe sits on carbonate rocks near West Portal, indicating minimum transport of 2 mi (3.3 km). Pohatcong Mountain to the north, however, is completely surrounded by Allentown Dolomite so that transport had to be much greater. What is almost certainly the same thrust, however, comes to the surface to the south between the Precambrian mass and the Triassic rocks, and the whole Musconetcong–Schooley Mountain ridge is a klippe (Drake and others, 1967). This interpretation recently has been substantiated by diamond drilling on Rattlesnake Hill in Pennsylvania (Fig. 3) where inverted Leithsville was cut at a depth of 392 ft (119 m). It can be seen that this is an extremely well-established thrust, but nowhere in a length of nearly 40 mi (67 km) is the stratigraphic separation much more than 1,000 ft (305 m)! This relation will be pursued further later in this paper.

Two excellent examples of Precambrian rocks thrust onto Paleozoic rocks occur in the tail of Precambrian rocks that extends south from High Bridge, New Jersey (Lewis and Kümmel, 1910-12). One example is along the brook about 0.45 mi (0.75 km) east of Annandale, New Jersey,

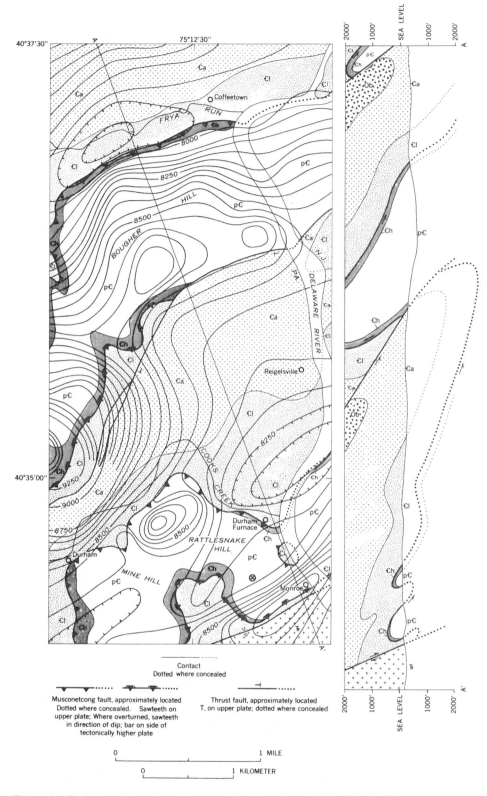

FIGURE 3. Geologic and aeromagnetic map and section of part of the Riegelsville quadrangle, Pennsylvania–New Jersey, modified from Drake and others, 1967, and Bromery and others, 1960. pЄ, Precambrian rocks; Єh, Hardyston Quartzite; Єl, Leithsville Formation; Єa Allentown Dolomite; Or, Rickenbach Dolomite; Oe, Epler Formation; Tʀ, Triassic rocks; ⊗, diamond drill hole. Magnetic contour interval 50 gammas.

where amphibolite lies on probable Allentown Dolomite. The other is a dolomite quarry about half a mile southwest of McPherson, New Jersey, where Precambrian gneiss clearly overlies dolomite, especially in a cave that has dolomite walls and a gneiss roof!

A study of the Geologic map of Pennsylvania (Gray and others, 1960) suggests that other thrusts come to the surface along the south border of the Prong. Evidence is especially compelling in the Reading area (Buckwalter, 1959, 1962) where faults have been mapped almost completely around the Precambrian. These faults faithfully follow the topography and suggest at least moderate dips; stream reentrants suggest that at some places the faults dip under the gneiss, not away. In addition, many of the small intermontane valleys that figured so prominently in the Stose-Miller controversy are shown by Buckwalter (1959, 1962) to be underlain by carbonate rocks and in many instances to be fault-bounded. These areas are, without exception, coextensive with negative magnetic anomalies. Buckwalter did not interpret these relations. A thrust-fault interpretation is currently (Oct. 1967) being tested by diamond drilling.

In New York, Isachsen (1964) reports that the fault bounding the Prong in the south also dips north beneath the Precambrian rocks. He tends, however, to negate this because of the probable influence of later deformation.

Evidence for steep faults. In New Jersey, steep faults bound the north sides of many of the valleys underlain by Paleozoic rocks. These faults traditionally were considered to be of Triassic age and to be related to the Triassic border fault. Therefore, the valleys were thought to be grabens bounded on the south, in some cases at least, by thrust faults of Paleozoic age and on the north by normal faults of Triassic age. Similar relations were noted in Pennsylvania (Miller and others, 1939, 1941), though there the valleys are not nearly as grabenlike in appearance. Detailed mapping in the Delaware Valley revealed that many ridges apparently are upthrust on the north side and bounded on the south by a normal fault. Stratigraphic displacement apparently was not large on either fault. On ridge noses the two faults come together and the supposed Triassic fault does not displace the older Paleozoic fault; this seems to indicate that the north and south borders are actually the same single fault. Balk (1936) apparently found the same relation in New York where he concluded that Precambrian blocks were popped into the Paleozoic pile by upthrusts on all sides. I once applied the same

reasoning to relations in the Delaware Valley to a considerable lack of critical acclaim.

Some steep faults, however, are indeed separate faults and have been so mapped (Drake, 1967a,b; Drake and others, 1967, and 1969). These faults bring rocks as young as Martinsburg Formation into contact with the Precambrian rocks. Such a fault bounds the north side of the Saucon Valley in Pennsylvania and brings rocks as young as Jacksonburg Limestone into contact with Precambrian rocks (Miller and others, 1941). Three holes drilled here by the New Jersey Zinc Company proved conclusively that the Precambrian did not lie in flat thrust contact with the Paleozoic rocks (Miller, 1944). The fault dips about 70° SE (R. B. Hoy, written communication, 1958). Similar faults are present all through the Prong to the southwest and have been especially noted by Buckwalter (1962). It is not necessary, however, that these faults be normal faults. Minor structures indicate that some are steep thrusts, and sedimentary structures indicate that in many places the rocks in contact with the Precambrian are overturned (Fig. 4). Buckwalter attempted to explain this by faulting out the normal limb of synclines. This is not, however, very convincing mechanically. Strong negative aeromagnetic anomalies are coextensive with such faults and suggest that nonmagnetic rocks are present at depth rather than magnetic basement, which would be the case if the Hardyston were in shallow synclines.

In New Jersey, two water wells were drilled at Bunnvale, about 1¼ mi (2.1 km) east of the fault-bounded German Valley. Both wells cut dolomite at less than 150 ft (38 m). Although no fault was mapped by Bayley and others (1914) the north-bounding fault in German Valley has probably brought carbonate rock nearly to the surface here as well.

Fault patterns farther northeast in New Jersey are difficult to interpret, as detailed stratigraphy has not been established. It is interesting, however, that in discussing the Zero fault, which bounds the north side of a carbonate valley in the Franklin area, Hague and others (1956) note that a study of "shears" indicated that the east side of the fault was up although *it was known that the east side was down.* One speculates whether this fact was known, or assumed because of the conventional normal fault interpretation. The Silurian and Devonian rocks have probably been faulted down into the Precambrian, although no data are at hand.

Several steep faults of this type have been mapped in New York. Such faults there were

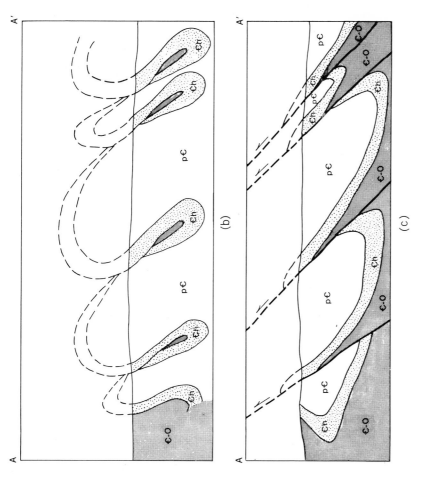

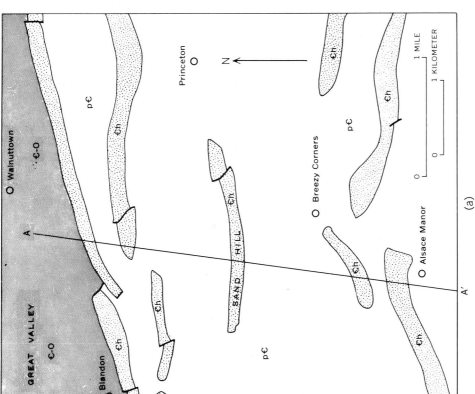

FIGURE 4. Geologic map and sections of part of the Reading quadrangle, Pennsylvania. (a) and (b) from Buckwalter (1962). (a), Geologic map; (b), Buckwalter's interpretation of sheared normal limbs of synclines; (c) inbricated nappe interpretation. pЄ, Precambrian rocks; Єh, Hardyston Quartzite; Є-O, Cambrian and Ordovician carbonate rocks. Vertical scale same as horizontal.

interpreted by Isachsen (1964) as resulting from imbrication of a thrust sheet by reverse faults.

Regional evidence from the Great Valley. Studies in the Great Valley in Pennsylvania between Harrisburg and Little South Mountain have shown that the Cambrian and Ordovician rocks therein are regionally overturned and are in the inverted limb of a nappe (Field Conference Pennsylvania Geologists, 1966; Geyer and others, 1958, 1963; Gray, 1959; Gray and others, 1958). The Precambrian rocks of Little South Mountain are involved in this structure (Geyer and others, 1963). To the northeast in the Delaware Valley, a nappe has also been recognized (Davis and others, 1967; Drake, 1967a; Drake and others, 1969). Here, however, most of the rocks in the valley lie in the brow and normal limb of the structure.

Although the intervening country has not been mapped, it is not unreasonable to infer that both areas are on the same structure and that the regional northeast plunge has brought the normal limb to the surface in the Delaware Valley. In any case, it is preposterous to suggest, as some have, that the regionally inverted rocks in the Lebanon Valley could pass into a barely crumpled upright monocline in the same belt of outcrop. The presence of the nappe is in harmony with the geophysical evidence previously cited which suggests that at least a double, if not a triple thickness of lower Paleozoic rock is present. The recently published Basement map of North America (American Assoc. Petroleum Geologists and U. S. Geol. Survey, 1967) also shows that the Prong and adjoining Great Valley lie in the deepest part of the Appalachian basin.

The lower Paleozoic rocks that lie south of the Precambrian are best exposed in the Oley Valley, which was interpreted by Stose and Jonas (1935) as a window. Recent reconnaissance in that area by V. E. Gwinn (written communication, 1966) has found that rocks of the Beekmantown Group are generally inverted and domed; younger rocks are exposed in antiforms. This indicates that the same type of structure prevails both north and south of the Precambrian highlands and that the Precambrian is at a higher tectonic level.

All these observations are in agreement with those previously cited and are more in line with an allochthonous rather than an autochthonous interpretation of the Prong.

Nappe interpretation. The geologic relations in any given quadrangle, or even two quadrangles, can be rationalized with a little thrusting here, a little folding there, and a little normal faulting somewhere else. One is reminded, however, of Bailey's (1935) comments on the discovery of the Hohe Tauern window: "Accordingly, to ease the situation, a motley group of local faults and thrusts was quietly introduced. Meanwhile it escaped notice that immense thrusting was required to explain the observable tectonic relations." In the Prong, most regional data point to grand tectonics involving allochthonous Precambrian rock. As Miller (1944) demolished the flat-thrust hypothesis, some other relation of Precambrian rock to cover must be sought. Geophysical and geologic relations allow two possible interpretations: (1) Several southeast-dipping imbricate thrust sheets, or (2) a nappe de recouvrement.

The imbricate-thrust hypothesis is attractive in that very large distances of tectonic transport can be avoided. Regional stratigraphic relations, however, make this interpretation implausible, as each Precambrian body has nearly normal stratigraphic relations with the Paleozoic rocks that surround it. Except for the greater transport along the north edge of the Prong, the maximum stratigraphic separation recognized to date is about 1,500 ft. It is difficult to visualize thrust sheets that can be traced for more than 40 mi along strike having such paraconformable relations with autochthonous rocks, and even more difficult when the thrust sheets have similar displacements along normal faults on their south sides.

The nappe de recouvrement interpretation is favored. The nappe is visualized as embracing all rocks up to and including the Martinsburg Formation, the Precambrian rocks forming the core. The geometric relations seen at the present level of erosion are the result of refolding and faulting of the core and inverted limb of the nappe, so that Precambrian rocks occupy the troughs of synforms and stratigraphically youngest rocks lie in the cores of antiforms. Shearing between the Precambrian and cover rocks produced the stratigraphic separations on the "thrust faults"; the apparent normal faults on the south sides of the ridges are actually thrust faults developed after emplacement of the nappe. The greater transport along the north border is the result of the core shearing through the cover and riding out on the normal limb. Figure 5 is an interpretation of the relations in the Delaware Valley where the structure has been called the Musconetcong nappe.

The brow of the nappe is in the Martinsburg Formation in the Delaware Valley (Davis and others, 1967; Drake and others, 1969), and

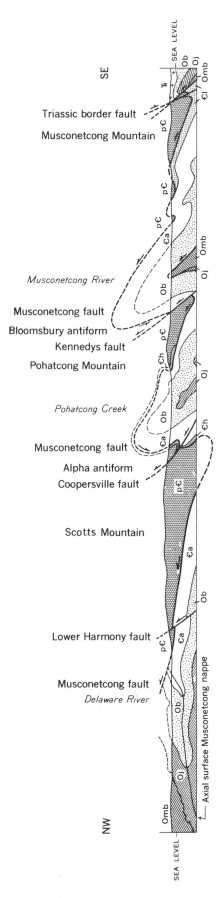

SE

Triassic border fault

Musconetcong Mountain

Musconetcong River

Musconetcong fault
Bloomsbury antiform
Kennedys fault
Pohatcong Mountain

Pohatcong Creek

Musconetcong fault
Alpha antiform
Coopersville fault

Scotts Mountain

Lower Harmony fault

Musconetcong fault
Delaware River

Axial surface Musconetcong nappe

NW

FIGURE 5. Geologic section across the Reading Prong in the Delaware Valley modified from Drake (1967b) and Drake and others 1969. Symbols: Ƚ℟, Triassic rocks; Omb, Martinsburg Formation; Oj, Jacksonburg Limestone; Ob, Beekmantown Group; Єa, Allentown Dolomite; Oj, Jacksonburg Limestone; Єl, Leithsville Formation; Єh, Hardyston Quartzite; pЄ, Precambrian rocks.

as it plunges northeast, the inverted limb is exposed in Pennsylvania and the normal limb in New Jersey. In the Delaware Valley, however, there seems to be a plunge culmination from which synforms spill both to the northeast and southwest. The passage from the upper limb to the lower limb around the core is exposed as Little South Mountain (Geyer and others, 1963). Geophysical evidence suggests a similar relationship north of Reading, Pennsylvania, and throughout most of New Jersey. Unfortunately, so little is known in detail of New Jersey geology that the relations there are difficult to interpret.

There may be a transition between allochthonous and autochthonous Precambrian in New Jersey similar to that between the southern and northern Blue Ridge in Virginia. If so, it is not readily apparent. It is just as likely that much of the lower Paleozoic rock in New Jersey and adjoining New York is upper limb, for the Silurian and Devonian rocks in the infolds are unconformable on all rocks from Precambrian through Ordovician. If this were true, the Precambrian would appear to be rooted although in actuality it is floating. This would go a long way toward reconciling the thoughts of those who have worked to the northeast with those who have worked in Pennsylvania, as upper limb infolds should be preserved in the least eroded part of the structure. The Precambrian rock is much more cataclastically deformed to the southwest and contains tectonic windows and isolated klippen. This may indicate that the sole of the core is near and that the body of Precambrian rock is relatively thin. The maximum thickness of the Precambrian core is not known. Unconfirmed reports that the New Jersey Zinc Company has drilled at least 8,000 ft (1,829 m) in the Franklin–Sterling Hill Mine in nothing but Precambrian rock do not unequivocally prove that the Precambrian is rooted. The Blue Ridge thrust sheet is estimated to be as much as 5 km thick (J. C. Reed, Jr., oral communication, 1967), so it is not unreasonable to assume that the nappe core is as much as 2 km thick. In New York, the Precambrian core probably thins before it spoons out east of the Hudson.

The magnitude of the nappe is tremendous. The Precambrian core crops out over a width about 9 mi (15 km) in the Delaware Valley, the total width of the structure being about 20 mi (33 km). The root zone is unknown, but almost certainly is in the Triassic basin, in parts of which, the aeromagnetic reflection of Precambrian rocks can be seen through the Triassic cover. Finding the root may not be this simple,

however, as Woodward (1964) has postulated that a right lateral wrench fault along which there has been as much as 80 mi (134 km) movement lies buried beneath the Triassic rocks. Bromery (*in* U. S. Geol. Survey, 1966, written communication, 1967) has found good geophysical evidence that such a fault does exist about on the line Doylestown–Birdsboro–Mount Hope, Pennsylvania, and has a displacement about as visualized by Woodward. These data suggest that the nappe may be laterally displaced from its root zone. It is possible, however, that this fault is the root zone of the nappe as well as a wrench fault as visualized by Reed and others (Chapter 18 this volume) for the Brevard and Burchfiel and Livingston (1967) for several Alpine nappes and the Brevard. If this fault is the root zone, in the Delaware Valley the Precambrian rock has been displaced about 30 mi (50 km) from its root. To the southwest, displacements apparently were less. No data are available to interpret the relations in New Jersey and New York.

Origin of the nappe. It is difficult to formulate an origin for a structural feature as poorly understood as the Musconetcong nappe. It probably developed high on the Appalachian geanticline by the upthrusting of basement into cover rocks. The movement of the Precambrian rocks into the core of the structure was largely mechanical, as there is no sign of post-Precambrian plastic deformation in the crystalline rocks. Shear is manifested along Precambrian–Paleozoic interfaces by envelopes of mylonite or cataclasite in the Precambrian rocks 1 ft (0.3 m) to several tens of feet thick. In a few places, the crystalline rock has been refoliated parallel to the contact, but only for a few inches. Carbonate rocks in the contact zone are commonly shattered, silicified, and granulated; bedding is absent. Thin-bedded shaly material has been converted to phyllite and in most places is now a mass of sheared-out isoclines.

It is postulated that as the nappe continued to grow under the influence of northwest-directed stress, it was unable to support its own weight and collapsed. Gravity probably played a large part in the northwestward transport of the structure from the geanticlinal high toward the basin. The skeptic might ask why, if this is so, do most of the structural elements now dip southeast? In part this is due to later folding, but a solution borrowed from Gordon Wood (Wood and Carter, in press) seems applicable. He points out that the Newark Triassic basin has subsided several thousand feet subsequent to the Appalachian deformations, probably owing to regional collapse of the geanticline. Such subsidence tilted this part of the Appalachian region to the southeast and thereby rotated the original northwest-sloping surfaces to the southeast.

Much of the northwest transport under the influence of gravity was probably accomplished by shearing within incompetent units, particularly the Martinsburg Formation, Jacksonburg Limestone, and Leithsville Formation in the lower limb of the nappe. To date, definite location of such faults in outcrop has been impossible, although the subsurface Stockertown fault (Davis and others, 1967) may be such a feature. This fault and associated imbricates crop out west of the area currently mapped in the Delaware Valley (J. B. Epstein, oral communication, 1967). Other possible glide zones are the Fish Hatchery fault (Drake, 1967b) and unnamed faults in the Jacksonburg Limestone in the Easton quadrangle (Drake, 1967a). Aside from shearing, cataclasis, and flowage in the purer limestones, the nappe emplacement seems to have been atectonic, as all penetrative features can now be related to later deformation. Such situations are common in Alpine nappes, and the sedimentary pile was probably only poorly consolidated when the nappe was emplaced.

Later deformation resulted in the antiforms and synforms now seen. Probably much of this folding is actually the result of imbrication of the Precambrian core of the nappe. The Precambrian rocks contain a cataclastic foliation that is marked with both a and b lineations. This foliation is best developed and most common near the borders of Precambrian bodies and is geometrically related to the cleavage in the Paleozoic rocks. Joint sets are related to the cataclastic deformation. Steep, northwest-striking cross joints are best developed; most are slickensided and grooved or have fault steps suggesting that there has been movement along them. The Precambrian rocks have not been folded themselves, but have been accommodated in the structures by the relative incompetence of the Paleozoic rocks and by movement on cataclastic foliation, cross joints (ac planes), and fractures.

Time of emplacement. The nappe was probably emplaced during the Taconic orogeny immediately after or during the waning stages of Martinsburg flysch deposition, as no younger rocks are involved. The allochthonous pelitic rock which occurs in the lower limb probably heralded the event as it was shed into the basin before the orogeny rolled into the site of the nappe's root zone.

No isotopic dates are available for the area that has been most intensely studied. However, Long and Kulp (1962) report that metamorphism in the Hudson River pelite dates from 450 m.y. ago, a Taconic date. Geologists of the Pennsylvania Geological Survey (Field Conference Pennsylvania Geologists, 1966) believe that the allochthon in the Harrisburg area was emplaced during the Taconic, and that later major thrusting occurred during the Acadian orogeny. The folding of the nappe core and lower limb cannot be directly dated. The Acadian event from the Harrisburg area could be extrapolated, but there is no evidence of deformation of this age in the younger rocks. In New York, however, isotopic dates indicate a metamorphism at 360 m.y., an Acadian event. J. B. Epstein (Field Conference Pennsylvania Geologists, 1967) has recently shown that the cleavage in the Martinsburg Formation in the upper limb of the nappe along the Blue Mountain structural front is a result of folding during the Appalachian orogeny, rather than of diagenesis as postulated by Maxwell (1962). Because this cleavage can be traced south into the core area and related to the refolding, it is concluded that the refolding is a result of the Appalachian orogeny.

Determining which features formed during which deformation is exceedingly difficult in this region because all deformation during the Paleozoic was essentially homoaxial. Structural trends apparently diverge no more than 20 degrees. This is a regional problem of the greatest magnitude.

REFERENCES

American Association of Petroleum Geologists and U. S. Geological Survey, 1967, Basement map of North America between latitudes 24 degrees and 60 degrees N: Washington, D.C., U. S. Geol. Survey, scale 1:5,000,000.

American Geophysical Union, Special Committee for the Geophysical and Geological Study of the Continents, 1964, Bouguer gravity anomaly map of the United States (exclusive of Alaska and Hawaii): U. S. Geol. Survey, Spec. Map, 2 sheets, scale 1:2,500,000.

Bailey, E. B., 1935, Tectonic essays, mainly Alpine: Oxford, Clarendon Press, 200 p.

Baker, D. R., 1956, Geology of the Edison area, Sussex County, N. J.: U. S. Geol Survey open-file rept., 275 p.

Balk, Robert, 1936, Structural and petrologic studies in Dutchess County, New York: Geol. Soc. America Bull., v. 47, p. 685–774.

Bates, T. F., 1947, Investigation of the micaceous minerals in slate: Am. Mineralogist, v. 32, p. 625–636.

Baum, J. L., 1967, Stratigraphy and structure of an anomalous area in the vicinity of Andover, Sussex County, N. J. [abs.]: Geol. Soc. America, Northeast Section Meeting, 1967, Program, p. 14.

Bayley, W. S., 1914, The Precambrian sedimentary rocks in the Highlands of New Jersey: Internat. Geol. Cong., 12th, Toronto, 1913, Comptes rendus, p. 325–334.

————, 1941, Pre-Cambrian geology and mineral resources of the Delaware Water Gap and Easton quadrangles, New Jersey and Pennsylvania: U. S. Geol. Survey Bull. 920, 98 p.

————, Salisbury, R. D., and Kümmel, H. B., 1914, Description of the Raritan quadrangle [New Jersey]: U. S. Geol. Survey Geol. Atlas, Folio 191.

Behre, C. H., Jr., 1927, Slate in Northampton County, Pennsylvania: Pennsylvania Geol. Survey, 4th ser., Bull. M9, 308 p.

————, 1933, Slate in Pennsylvania: Pennsylvania Geol. Survey 4th ser., Bull. M16, 400 p.

Berkey, C. P., 1907, Structural and stratigraphic features of the basal gneisses of the Highlands: New York State Mus. Bull. 107, p. 361–378.

————, and Rice, Marion, 1919, Geology of the West Point quadrangle, New York: New York State Mus. Bull. 225–226, 152 p.

Bromery, R. W., and Griscom, Andrew, 1967, Aeromagnetic and generalized geologic map of southeastern Pennsylvania: U. S. Geol. Survey Geophys. Inv. Map GP-577.

————, and others, 1960, Aeromagnetic map of part of the Riegelsville quadrangle, Bucks and Northampton Counties, Pennsylvania, and Hunterdon and Warren Counties, New Jersey: U. S. Geol. Survey Geophys. Inv. Map GP-236.

Broughton, J. G., and others, 1962, Geologic map of New York, 1961: New York State Mus. and Sci. Service, Geol. Survey, Map and Chart Ser. no. 5, 5 map sheets, scale 1:250,000, and text.

Buckwalter, T. V., 1959, Geology of the Precambrian rocks and Hardyston Formation of the Boyertown quadrangle: Pennsylvania Geol. Survey, 4th ser., Geol. Atlas 197, 15 p.

————, 1962, The Precambrian geology of the Reading 15′ quadrangle: Pennsylvania Geol. Survey, 4th ser., Prog. Rept. 161, 49 p.

Buddington, A. F., 1959, Granite emplacement with special reference to North America: Geol. Soc. America Bull., v. 70, p. 671–747.

————, and Baker, D. R., 1961, Geology of the Franklin and part of the Hamburg quadrangles, New Jersey: U. S. Geol. Survey Misc. Geol. Inv. Map I-346.

Burchfiel, B. C., and Livingston, J. L., 1967, Brevard zone compared to Alpine root zones: Am. Jour. Sci., v. 265, p. 241–256.

Darton, N. H., and others, 1908, Description of the Passaic quadrangle (N. J.-N. Y.): U. S. Geol. Survey Geol. Atlas, Folio 157.

Davis, R. E., Drake, A. A., Jr., and Epstein, J. B., 1967, Geologic map of the Bangor quadrangle, Pennsylvania–New Jersey: U. S. Geol. Survey Geol. Quad. Map GQ-665.

Dodd, R. T., 1962, Precambrian geology of the Popolopen Lake quadrangle, southeastern New York: Unpub. Ph.D. thesis, Princeton Univ., 178 p.

Drake, A. A., Jr., 1965, Carbonate rocks of Cambrian and Ordovician age, Northampton and Bucks Counties, eastern Pennsylvania and Warren and Hunterdon Counties, western New Jersey: U. S. Geol. Survey Bull. 1194-L, 7 p.

————, 1967a, Geologic map of the Easton quadrangle, New Jersey–Pennsylvania: U. S. Geol. Survey Geol. Quad. Map GQ-594.

————, 1967b, Geologic map of the Bloomsbury quadrangle, New Jersey: U. S. Geol. Survey Geol. Quad. Map GQ-595.

————, Davis, R. E., and Alvord, D. C., 1960, Taconic and post-Taconic folds in eastern Pennsylvania and western New Jersey: U. S. Geol. Survey Prof. Paper 400-B, p. B180–B181.

————, and Epstein, J. B., 1967, The Martinsburg Formation (Middle and Upper Ordovician) in the Delaware Valley, Pennsylvania–New Jersey: U. S. Geol. Survey Bull. 1244-H, 16 p.

————, Epstein, J. B., and Aaron, J. M., 1969, Geologic map and sections of parts of the Portland and Belvidere quadrangles, New Jersey–Pennsylvania: U. S. Geol. Survey Misc. Geol. Inv. Map I-552.

————, McLaughlin, D. B., and Davis, R. E., 1961, Geology of the Frenchtown quadrangle, New Jersey–Pennsylvania: U. S. Geol. Survey Geol. Quad. Map GQ-133.

————, 1967, Geologic map of the Riegelsville quadrangle, Pennsylvania–New Jersey: U. S. Geol. Survey Geol. Quad. Map GQ-593.

Engel, A. E. J., 1956, Apropos the Grenville, in Thomson, J. E., ed., The Grenville problem: Royal Soc. Canada Spec. Pub., no. 1, p. 74–96.

Field Conference of Pennsylvania Geologists, 26th, Bethlehem, 1961, Structure and stratigraphy of the Reading Hills and Lehigh Valley in Northampton and Lehigh Counties, Pennsylvania. Edited by J. Donald Ryan: Bethlehem, Pa., Lehigh Univ., 82 p.

Field Conference of Pennsylvania Geologists, 31st, Harrisburg, 1966, Comparative tectonics and stratigraphy of the Cumberland and Lebanon Valleys, by D. B. MacLachlan and S. I. Root: Harrisburg, Pa., Pennsylvania Geol. Survey, 90 p.

Field Conference of Pennsylvania Geologists, 32d, East Stroudsburg, 1967, Geology in the region of the Delaware to Lehigh Water Gaps, by J. B. Epstein and A. G. Epstein: Harrisburg, Pa., Pennsylvania Geol. Survey, 89 p.

Fraser, D. M., 1938, Contributions to the geology of the Reading Hills, Pa.: Geol. Soc. America Bull., v. 49, p. 1199–1212.

Fyfe, W. S., Turner, F. J., and Verhoogen, John, 1958, Metamorphic reactions and metamorphic facies: Geol. Soc. America Mem. 73, 259 p.

Geyer, A. R., and others, 1958, Geologic map of the Lebanon quadrangle, Pennsylvania: Pennsylvania Geol. Survey, 4th ser., Atlas 167C.

————, 1963, Geology and mineral resources of the Womelsdorf quadrangle: Pennsylvania Geol. Survey, 4th ser., Atlas 177C, 96 p.

Gray, C., 1951, Preliminary report on certain limestones and dolomites of Berks County, Pennsylvania: Pennsylvania Geol. Survey, 4th ser., Prog. Rept. 136, 85 p.

————, 1952, The high calcium limestones of the Annville belt in Lebanon and Berks Counties, Pennsylvania: Pennsylvania Geol. Survey, 4th ser., Prog. Rept. 140, 17 p.

————, 1959, Nappe structures in Pennsylvania [abs.]: Geol. Soc. America Bull., v. 70, no. 12, pt. 2, p. 1611.

————, Geyer, A. R., and McLaughlin, D. B., 1958, Geologic map of the Richland quadrangle, Pennsylvania: Pennsylvania Geol. Survey, 4th ser., Atlas 167D.

————, and others, 1960, Geologic map of Pennsylvania: Pennsylvania Geol. Survey, 4th ser., scale 1:250,000.

Hague, J. M., and others, 1956, Geology and structure of the Franklin–Sterling area, New Jersey: Geol. Soc. America Bull., v. 67, p. 435–474.

Henderson, J. R., Andreasen, G. E., and Petty, A. J., 1966, Aeromagnetic map of northern New Jersey and adjacent parts of New York and Pennsylvania: U. S. Geol. Survey Geophys. Inv. Map GP-562.

Hotz, P. E., 1952, Magnetite deposits of the Sterling Lake, N. Y.–Ringwood, N. J., area: U. S. Geol. Survey Bull. 982-F, p. 153–244.

Isachsen, Y. W., 1964, Extent and configuration of the Precambrian in northeastern United States: New York Acad. Sci. Trans., ser. 2, v. 26, no. 7, p. 812–829.

Kümmel, H. B., and Weller, Stuart, 1902, The rocks of the Green Pond Mountain region: New Jersey Geol. Survey Ann. Rept. State Geologist for 1901, p. 3–51.

Lesley, J. P., and others, 1883, The geology of Lehigh and Northampton Counties: Pennsylvania Geol. Survey, 2d., Rept. D3, v. 1, 283 p.

Lewis, J. V., and Kümmel, H. B., 1910-12, Geologic map of New Jersey: New Jersey Geol. Survey; revised 1931 by H. B. Kümmel; 1950 revision by M. E. Johnson, pub. as New Jersey Dept. Conserv. and Econ. Devel., Atlas Sheet 40.

————, 1940, The geology of New Jersey: New Jersey Dept. Conserv. and Devel., Geol. Ser. Bull. 50, 203 p.

Long, L. E., and Kulp, J. L., 1962, Isotopic age study of the metamorphic history of the Manhattan and Reading Prongs: Geol. Soc. America Bull., v. 73, p. 969–996.

Lowe, K. E., 1950, Storm King Granite at Bear Mountain, New York: Geol. Soc. America Bull., v. 61, p. 137–190.

Maxwell, J. C., 1962, Origin of slaty and fracture cleavage in the Delaware Water Gap area, New Jersey and Pennsylvania, *in* Petrologic studies—A volume in honor of A. F. Buddington: New York, Geol. Soc. America, p. 281–311.

McBride, E. F., 1962, Flysch and associated beds of the Martinsburg Formation (Ordovician), central Appalachians: Jour. Sed. Petrology, v. 32, p. 39–91.

McLaughlin, D. B., 1938, A great alluvial fan in the Triassic of Pennsylvania: Michigan Acad. Sci., Arts, and Letters, Papers, v. 24, p. 59–74.

Miller, B. L., 1925, Mineral resources of the Allentown quadrangle, Pennsylvania: Pennsylvania Geol. Survey, 4th ser., Atlas 206, 195 p.

————, 1944, Specific data on the so-called "Reading Overthrust": Geol. Soc. America Bull., v. 55, p. 211–254.

————, and Fraser, D. M., 1936, Comment *on* Highlands near Reading, Pennsylvania; an erosion remnant of a great overthrust sheet, by G. W. Stose and A. I. Jonas: Geol. Soc. America Bull., v. 46, p. 2031–2038; Reply by G. W. Stose and A. I. Jonas, p. 2038–2040.

————, Fraser, D. M., and Miller, R. L., 1939, Northampton County, Pennsylvania: Pennsylvania Geol. Survey, 4th ser., Bull. C48, 496 p.

————, and others, 1941, Lehigh County, Pennsylvania: Pennsylvania Geol. Survey, 4th ser., Bull. C39, 492 p.

Minard, J. P., 1959, The geology of Peapack–Ralston Valley in north central New Jersey: Unpub. M. S. thesis, Rutgers Univ. (New Brunswick), 103 p.

Norton, M. F., 1959, Stratigraphic position of the Lowerre quartzite: New York Acad. Sci. Annals, v. 80, art. 4, p. 1148–1158.

Prouty, C. E., 1959, The Annville, Myerstown and Hershey Formations of Pennsylvania: Pennsylvania Geol. Survey, 4th ser., Bull. G31, 47 p.

Prucha, J. J., 1956, Stratigraphic relationships of the metamorphic rocks in southeastern New York: Am. Jour. Sci., v. 254, p. 672–684.

————, 1959, Field relationships bearing on the age of the New York City Group of the Manhattan Prong: New York Acad. Sci. Annals, v. 80, art. 4, p. 1159–1169.

Rogers, H. D., 1858, The geology of Pennsylvania, a government survey: Philalelphia, 2 v.

Sherwood, W. C., 1964, Structure of the Jacksonburg Formation in Northampton and Lehigh Counties, Pennsylvania: Pennsylvania Geol. Survey, 4th ser., Bull. G45, 64 p.

Sims, P. K., 1958, Geology and magnetite deposits of Dover district, Morris County, New Jersey: U. S. Geol. Survey Prof. Paper 287, 162 p.

Socolow, A. A., 1961, Geologic interpretation of certain aeromagnetic maps of Lancaster, Berks, and Lebanon Counties, Pennsylvania: Pennsylvania Geol. Survey, 4th ser., Inf. Circ. 41, 19 p.

Spencer, A. C., and others, 1908, Description of the Franklin Furnace quadrangle [N. J.]: U. S. Geol. Survey Geol. Atlas, Folio 161.

Stose, G. W., 1946, The Taconic sequence in Pennsylvania: Am. Jour. Sci., v. 244, no. 10, p. 655–696.

————, and Jonas, A. I., 1935, Highlands near Reading, Pa., an erosion remnant of a great overthrust sheet: Geol. Soc. America Bull., v. 46, no. 5, p. 757–779.

————,1939, Discussion of the geology of the Reading Hills, Pa.: Am. Jour. Sci., v. 237, no. 4, p. 281–286.

U. S. Geological Survey, 1962, Geological Survey research 1962: U. S. Geol. Survey Prof. Paper 450-A, 257 p.

————, 1964, Geological Survey research 1964: U. S. Geol. Survey Prof. Paper 501-A, 367 p.

————, 1966, Geological Survey research 1966: U. S. Geol. Survey Prof. Paper 550-A, 385 p.

U. S. Geological Survey and American Association of Petroleum Geologists, 1961, Tectonic map of the United States, exclusive of Alaska and Hawaii: 2 sheets, scale 1:2,500,000 [1962].

Willard, Bradford, and others, 1959, Geology and mineral resources of Bucks County, Pennsylvania: Pennsylvania Geol. Survey, 4th ser., Bull. C9, 243 p.

Wood, G. H., Jr., and Carter, M. D., ————, Tectonics of the Anthracite region, Pennsylvania: Pennsylvania Geol. Survey, 4th ser., Bull. ————, in press.

Woodward, H. P., 1964, Central Appalachian tectonics and the deep basin: Am. Assoc. Petroleum Geologists Bull., v. 48, p. 338–356.

Woollard, G. P., 1943, Geologic correlation of aerial gravitational and magnetic studies in New Jersey and vicinity: Geol. Soc. America Bull., v. 54, p. 791–818.

Zen, E-an, 1967, Time and space relationships of the Taconic allochthon and autochton: Geol. Soc. America Spec. Paper 97, 107 p.

THE PIEDMONT

Chapter 20

Introduction

GEORGE W. FISHER

THE APPALACHIAN PIEDMONT is a gently rolling plain, bounded on the west by the hills of the Blue Ridge, and overlapped on the east by the sedimentary rocks of the Coastal Plain. It is underlain by deformed and metamorphosed eugeosynclinal rocks of late Precambrian to early Paleozoic age, mantling older Precambrian (1,100 m.y.) basement gneiss, and dotted here and there with basins of red Triassic sedimentary rocks. Intrusive rocks are common, and include preorogenic mafic sheets, syn- and postorogenic granitic plutons, and postorogenic diabase dikes.

Except along the major stream valleys, the rocks of the Piedmont are veiled by a thick mantle of saprolite. The poor exposures, lack of distinctive stratigraphic units, and scarcity of fossils, coupled with the effects of repeated deformation and variable metamorphism, have combined to make the Piedmont one of the least understood, and yet most fascinating parts of the Appalachians.

Seen on a map, the Piedmont has the shape of a wild duck swimming gracefully northward. From its beak in central New Jersey to its tail in Alabama, it stretches 840 mi (1,350 km). At its neck in northern Viriginia, it is only 10 mi (16 km) across, but at its belly in the Carolinas it widens to nearly 150 mi (240 km).

The older Precambrian basement rocks appear first in the northern tip of the beak, in a series of anticlinoria, and recumbent folds. Southward, they reappear in a series of elongate domes along the throat in Maryland, and northeastern Virginia, and also in a nearly continuous strip along the spine of the Blue Ridge, from the head in Maryland nearly to the tail in northeast Georgia. These basement rocks are predominantly quartzofeldspathic gneisses and migmatites (e.g., Hopson, 1964), but they include anorthosite and hypersthene granodiorite, reminding one of the Precambrian rocks of the Andirondacks and the Grenville Province of Canada. Radiometric dating has clearly established that these rocks are a southward extension of the Grenville Province (Tilton and others, 1960; Tilton and others, Chapter 29 this volume).

Mantling these basement rocks is a thick pile of variably metamorphosed sedimentary rocks, eugeosynclinal for the most part, but including some miogeosynclinal and transitional rocks near the Blue Ridge. Because fossils are exceedingly rare, most of these rocks are of uncertain age, but at least two distinct sequences can be recognized.

Along the eastern front of the Blue Ridge, a thick, upper Precambrian pelite–turbidite sequence (Lynchburg Formation in Virginia; Ocoee Series in the Carolinas; Great Smoky Group in Georgia) rests unconformably on the older Precambrian basement and underlies upper Precambrian volcanic rocks (Catoctin Formation) and lower Paleozoic clastic rocks (Chilhowee Group). Nearly 15,000 ft (4,600 m) thick in central Virginia, these rocks thin rapidly to the west and to the north, and disappear altogether in northern Virginia. Still farther north, in southern Pennsylvania, the Catoctin also wedges out, and the Chilhowee Group rests directly on the basement.

Near the eastern edge of the Piedmont, a thick pile of metamorphosed shales, turbidites, and andesitic to rhyolitic volcanic rocks crops out in the Carolina Slate Belt. Although the exact age of these rocks is debatable (Cambrian fossils have been found in them, but zircons from a felsic tuff suggest an Ordovician age), they appear to be definitely younger than the Lynchburg–Ocoee sequence, and are probably time equivalent to at least some of the lower Paleozoic miogeosynclinal rocks in the Valley and Ridge Province.

In the James River Synclinorium of central Virginia, paragonite phyllite, quartzite, and marble (Evington Group) lie conformably above the Lynchburg and Catoctin Formations. The Evington has long been considered Paleozoic, because it is in the same stratigraphic position as the lower Paleozoic Chilhowee Group to the west, and has yielded trilobite fragments, unfortunately unidentifiable (Jonas, 1927). The lithology of these rocks suggests that they are an off-shelf facies transitional from the miogeosyncline of the Valley and Ridge to the eugeosyncline of the Slate Belt.

Slates in the Arvonia, Quantico, and Peach Bottom synclines of the central Piedmont have yielded Ordovician fossils, and may represent a still younger sequence, or may correlate with some of the Slate Belt rocks.

Many of the other rocks in the Piedmont are also metamorphosed shales and turbidites, but it is not yet clear to which of the above sequences they belong, or whether they represent a still different cycle of sedimentation. The central problems of Piedmont geology revolve about the age and correlation of these rocks. Which ones represent the upper Precambrian turbidite sequence, and what was the extent of the basin in which they were deposited? Which ones are eugeosynclinal equivalents of the Paleozoic rocks in the Valley and Ridge Province, and what was the nature of the transition from eugeosyncline to miogeosyncline? From what source did these rocks come? Did they come from the craton, as recently suggested by Dietz (1963)? From a landmass of granitic rocks somewhere to the east, long ago dubbed Appalachia by Schuchert (1910)? Or did they come from a series of volcanic islands built on a thin oceanic crust within the eugeosyncline?

The papers in this section attack these problems from several points of view. Brown discusses the rocks of the western Piedmont, and Blue Ridge in Virginia. He shows that the miogeosynclinal clastic rocks of the Chilhowee Group were transported from the west and northwest across a shallow shelf, and suggests that the Evington Group of the Piedmont is a deep marine, off-shelf equivalent, derived partly from the west, and partly from the east. The upper Precambrian Lynchburg Formation appears to represent an earlier turbidite sequence, also derived partly from the west, and partly from the east.

Farther north, in Pennsylvania, the transition from the miogeosynclinal rocks to the phyllites and argillaceous marbles of the northwestern Piedmont occurs approximately at the controversial Martic Line, discussed in the paper by Wise. He shows that the transition appears to involve a facies change from a shelf environment northwest of the Martic Line to a deeper water environment southeast of the line, combined with local imbricate thrusting, clearly demonstrated by the classic mapping of Cloos (Cloos and Hietanen, 1941).

Fisher describes a well-exposed section through the metamorphosed pelite–turbidite sequence of the Maryland Piedmont (Glenarm Series). Abundant cobbles and boulders of basement gneiss in a huge submarine slump breccia suggest that these rocks were derived from a nearby, easterly source area of crystalline rocks. The Glenarm Series appears to underlie the phyllites, metamorphosed volcanic rocks, and marbles of the western Maryland Piedmont, which are probably equivalent to the upper Precambrian Catoctin volcanics and/or the lower Paleozoic Evington Group of Virginia. If so, the Glenarm Series may be a northeastern extension of the upper Precambrian Lynchburg–Ocoee sequence, as suggested by Hopson (1964).

Alternatively, the Glenarm could be lower Cambrian, a eugeosynclinal equivalent of the Chilhowee Group. Tilton and others report early Cambrian ages (550 m.y.) for zircons from the soda-rich metamorphosed tuffs formerly included in the Baltimore Gneiss (Baltimore paragneiss of Hopson, 1964). If these tuffs are a part of the Glenarm Series, a Cambrian age for the Series as a whole would appear likely.

Tilton and others also report new zircon data which confirm an older Precambrian age (1,100–1,300 m.y.) for the migmatites and granitic rocks of the Baltimore Gneiss. These results prove that the overlying Glenarm Series was deposited on crystalline rocks of the continental crust, not on oceanic crust, and that no large amount of continental accretion occurred in Maryland during growth of the Appalachians.

Problems of southern Piedmont geology are discussed in the papers by Sundelius, Overstreet, and Hurst. Sundelius summarizes the results of recent detailed mapping in the lower Paleozoic rocks of the Carolina Slate Belt, chiefly metagraywackes and metamorphosed mafic to felsic volcanic flows and pyroclastic deposits. He suggests that these rocks were derived from nearby volcanic islands and tectonic ridges, and deposited in the intervening basins.

Overstreet suggests that the rocks of the Carolina Slate Belt connect westward with the more intensely metamorphosed rocks of the Kings Mountain Belt in South Carolina, and overlie the variably metamorphosed pelites, graywackes, and arkoses of the Inner Piedmont Belt, and the southern part of the Charlotte Belt. This lower sequence is probably upper Precambrian or Cambrian, and may be in part equivalent to the Ocoee Series of the Blue Ridge.

The rocks of the Carolina Slate Belt can be traced southward into the Georgia Piedmont, discussed by Hurst. Between the Slate Belt Rocks (Little River Series) and the upper Precambrian rocks in the Blue Ridge lies a vast area of mica schist, gneiss of medium to high metamorphic grade, and a few quartzite horizons. Attempts to interpret the age and stratigraphic significance of these rocks have so far been frustrated by the presence of several strike faults of unknown displacement. Marble crops out the Brevard zone in a thin, nearly continuous strip from central Georgia northeast to the South Carolina state line and beyond. Marble is virtually absent in the rest of the Georgia Piedmont, and it seems possible that the beds in the Brevard zone are a downfolded remnant of overlying Paleozoic rocks (cf. Burchfiel and Livingston, 1967).

The sedimentary history of the Piedmont, however, is only part of its story. Other problems involve the origin of the plutonic rocks, and the tectonic and metamorphic history of the entire crystalline complex.

The plutonic rocks of the southern Piedmont are described in the papers by Overstreet, Hurst, and Sundelius. They include concordant early orogenic intrusions ranging from serpentine and gabbro through quartz monzonite, and younger cross-cutting plutons of gabbro, quartz monzonite, granite pegmatite, and syenite.

In the northern Piedmont, the plutonic rocks can be divided into two comagmatic series: an early gabbroic series, and a later granitic series (Hopson, 1964, p. 131–193). Southwick's paper in this section describes the huge gabbro–peridotite complex in northern Maryland and southern Pennsylvania, part of the gabbroic series. This complex appears to have been emplaced partly as a stratiform sheet, prior to regional deformation, and partly as a semisolid crystal mush, during regional deformation. Closely associated with the gabbro, and probably differentiated from it, are leucocratic soda-rich quartz diorite and albite granite. Relative to the gabbro, these rocks are strongly enriched in quartz, and impoverished in mafic minerals, but they show scarcely any increase in potash feldspar.

The rocks of the granitic series receive scant attention in this section. Hopson (1964) reports that they range from early synorogenic hornblende–biotite quartz diorite (Cambrian) through granodiorite to late cross-cutting quartz monzonite (Silurian). The younger rocks are progressively poorer in iron and magnesium, and richer in alkalis, notably potassium. In contrast to the late differentiates of the gabbroic series, they are greatly enriched in potash feldspar, and show a slight decrease in quartz content. Late pegmatites appear to have crystallized from anatectic melts squeezed out of the rising domes of basement gneiss, and may represent a third magmatic series.

Later igneous activity was confined to the emplacement of Triassic diabase dikes (see below), and Jurassic rocks ranging in composition from peridotite to nepheline syenite. The Jurassic rocks appear to be an eastern extension of a widespread province of alkalic rocks in the central United States (Zartman and others, 1967).

The tectonic and metamorphic evolution of the Piedmont appears to have been very complex, and only the broad outlines have been worked out. The papers in this section show that two periods of metamorphism, and several phases of deformation can be recognized at many places. In the northern Piedmont, local thrusting occurred prior to metamorphism, and at least two generations of folds developed near the peak of metamorphism. Much later, shearing and retrograde metamorphism were locally intense. In the southern Piedmont, the structural and metamorphic picture is further complicated by several strike faults of unknown displacement. Until more detailed mapping is done in the intervening areas, and until more radiometric dates are available, detailed correlation of the phases of deformation and recrystallization cannot be attempted. Nevertheless, evidence from several areas suggests that the main period of metamorphism and the principal stages of deformation occurred in Ordovician time, and that the upper Precambrian rocks were not regionally deformed or metamorphosed prior to deposition of the Paleozoic sedimentary rocks.

The upper Triassic sedimentary rocks of the Piedmont set a younger limit on the age of this deformation and metamorphism, but unfortunately they are not discussed in any of the papers of this section. Space permits only a brief mention here; for more detail, see the review by Reeside and others (1957, p. 1486–1498).

Most of the Triassic rocks are red, fluvial arkoses, siltstones, and shales, but lacustrine deposits and some coal beds occur locally. All are completely unmetamorphosed, and overlie the crystalline rocks of the Piedmont with profound unconformity; they are faulted, and broadly warped, but otherwise undeformed.

These rocks occupy a series of downfaulted basins extending from Nova Scotia to the Carolinas, and probably beyond; similar rocks have been found in deep wells in the Coastal Plain as far south as Florida. The chain of basins running southwest from the Newark Basin in New Jersey to the Danville Basin in North Carolina are all bounded on the west by normal faults with the southeast side downthrown, and beds in all dip predominantly northwest, toward the faults. Fanglomerates are present along the border faults in many places, and intertongue with the rocks throughout the section, showing that the faults were active during deposition.

The Deep River Basin in South Carolina has opposite symmetry; it is faulted on the east, and the beds all dip east, toward the fault. With the Danville Basin, it defines a large graben, or a symmetrical pair of half-grabens. A basin in Connecticut is also faulted on the east, and, with the Newark Basin, defines a second graben. There is no proof that the rocks ever formed a continuous blanket between these basins, but in the Newark–Connecticut graben, the presence of three lava flows at similar stratigraphic positions in both basins suggests a connection. The opposite dip of the beds in these two basin pairs shows post-Triassic arching along a northeast–southwest axis near the center of the Piedmont. The grabens may be keystone blocks dropped into place during the initial stages of this arching.

The faults bounding the Triassic basins closely follow structural trends in the older rocks, but the numerous dikes of Triassic or post-Triassic diabase do not; in any given area, the trend of the dikes is highly consistent, but it bears no systematic relation to the structures in the older rocks, or to the Triassic faults. The fractures in which the dikes solidified probably extend to the mantle, presumably the source of the diabase, and their pattern may reflect the orientation of stresses at a deeper level than those which produced the grabens (King, 1961).

The final chapter in the evolution of the Piedmont is partly recorded in the post-Triassic sedimentary rocks of the Coastal Plain, discussed in the paper by Owens. Up to 25,000 ft (7,600 m) of sediment accumulated in a series of radial tectonic troughs and longitudinal basins. Judging from the distribution of quartzose clastic rocks and carbonate rocks in these basins, post-Cretaceous uplift was greatest in the central Appalachians, and decreased southward.

Out of this discussion, and out of the papers in this section, two main points seem to emerge. First, we really know very little about the Piedmont rocks. The outlines of their history seem to be coming into focus, but detailed work has been done in only a few, widely scattered places, and the intervening areas of ignorance are vast. In most of the Piedmont, the problems remaining far outnumber the solutions. The second point is that these problems can be solved. Despite their metamorphism and intense deformation, the Piedmont rocks can be subdivided into mappable units. By combining detailed field work with modern petrologic, geophysical, and radiometric dating techniques, the history of the Piedmont can be pieced together. This realization has infused new life into Piedmont geology, and there is hope that in a decade or two a much clearer picture of the Piedmont will begin to emerge. The ideas developed in this section represent a beginning, not an end.

REFERENCES

Cloos, E., and Hietanen, A., 1941, Geology of the "Martic Overthrust" and the Glenarm Series in Pennsylvania and Maryland: Geol. Soc. America Spec. Paper 35, 207 p.

Burchfiel, B. C., and Livington, J. L., 1967, Brevard Zone compared to Alpine Root Zones: Am. Jour. Sci., v. 265, p. 241–256.

Dietz, R. S., 1963, Collapsing continental rises; an actualistic concept of geosynclines and mountain building: Jour. Geology, v. 71, p. 314–333.

Hopson, C. A., 1964, The crystalline rocks of Howard and Montgomery Counties, in The Geology of Howard and Montgomery Counties: Maryland Geol. Survey, p. 27–215.

Jonas, Anna I., 1927, Geologic reconnaissance in the Piedmont of Virginia: Geol. Soc. America Bull., v. 38, p. 837–846.

King, P. B., 1961, Systematic Pattern of Triassic Dikes in the Appalachian region: U. S. Geol. Survey Prof. Paper 424, p. B93–B95.

Reeside, J. B., and others, 1957, Correlation of the Triassic formations of North America exclusive of Canada: Geol. Soc. America Bull., v. 68, p. 1451–1514.

Schuchert, C., 1910, Paleogeography of North America: Geol. Soc. America Bull., v. 20, p. 427–606.

Tilton, G. R., Wetherill, G. W., Davis, G. L., and Bass, M. N., 1960, 1000-million-year-old minerals from the eastern United States and Canada: Jour. Geophys. Research, v. 65, p. 4173–4179.

Zartman, R. E., Brock, M. R., Heyl, A. V., and Thomas, H. H., 1967, K–Ar and Rb–Sr ages of some alkalic intrusive rocks from central and eastern United States: Am. Jour. Sci., v. 265, p. 848–870.

Chapter 21

The Metamorphosed Sedimentary Rocks
along the Potomac River
near Washington, D.C.

GEORGE W. FISHER

INTRODUCTION

MOST OF THE Maryland Piedmont is underlain by
the Glenarm Series, a thick sequence of metamorphosed geosynclinal sediments, mantling the
Precambrian Baltimore Gneiss (Fig. 1). Much
has been written on the Glenarm Series but its
significance in Appalachian geology remains controversial. Some geologists (e.g., Swartz, 1948)
have argued that it is equivalent to the lower
Paleozoic geosynclinal section in the Valley and
Ridge Province; others (e.g., Hopson, 1964)
have suggested that it is an upper Precambrian
sequence, underlying the Paleozoic rocks.

Mapping the Glenarm rocks is difficult because
of their complex sedimentary facies relationships,
their variable metamorphic grade, their intense
and repeated deformation, and the lack of distinctive lithologic units. But the most frustrating
obstacle to understanding these rocks is the
scarcity of good exposures. The bluffs along the
Potomac River provide an exceptionally good
cross section of Glenarm rocks from Washington,
D.C. to Seneca, Md., about 20 mi (32 km) to
the west (Fig. 2). Cloos (Cloos and Anderson,
1950) provided a tantalizing glimpse into the
complexities of these rocks by mapping a small
area just southeast of Great Falls in great detail.
In the hope that a careful study of the entire
section would provide a useful framework for a
more extensive study of the Piedmont rocks, I
described the Potomac section in a Ph.D. disser-

tation at Johns Hopkins University (Fisher,
1963). Reed and Jolly (1963) also studied the
rocks just southeast of Great Falls, and Hopson
(1964) described many features of these rocks.
Taken together, these studies have revealed many
areas of agreement, and have uncovered several
unsolved problems. This paper is an attempt
to summarize what we now know about these
rocks and their significance in Appalachian geology, and to single out the unsolved problems, in
hopes of stimulating others to study these very
challenging rocks.

While studying these rocks, I have benefited
greatly from discussions and field trips with many
geologists among whom Ernst Cloos, C. A. Hopson, E. H. Hanson, and D. L. Southwick deserve
special mention. D. W. Elliott, J. C. Reed, Jr.,
and C. A. Hopson read the manuscript, and
offered several helpful suggestions.

THE WISSAHICKON FORMATION

The Wissahickon Formation is a thick sequence
of metamorphosed graywackes, shales, and submarine slump deposits. It overlies thin deposits
of metamorphosed quartz sandstone, potassic
shale, and carbonate rocks, called the Setters
Formation and the Cockeysville Marble; together,
these three formations comprise the Glenarm
Series.

Southwick and Fisher (1967) have subdivided
the Wissahickon Formation into five informal

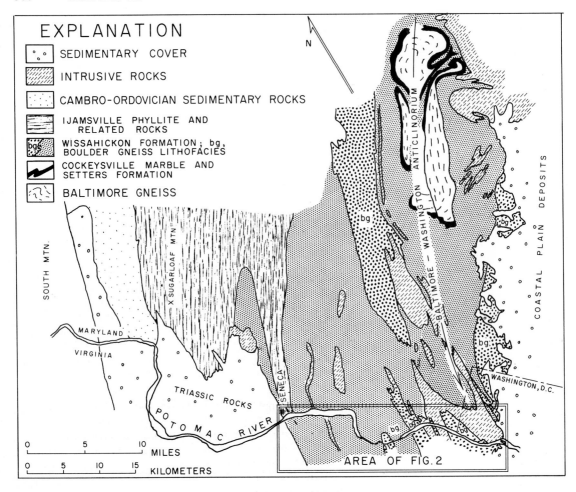

EXPLANATION

- SEDIMENTARY COVER
- INTRUSIVE ROCKS
- CAMBRO-ORDOVICIAN SEDIMENTARY ROCKS
- IJAMSVILLE PHYLLITE AND RELATED ROCKS
- WISSAHICKON FORMATION; bg, BOULDER GNEISS LITHOFACIES
- COCKEYSVILLE MARBLE AND SETTERS FORMATION
- BALTIMORE GNEISS

FIGURE 1. Generalized geologic map of the Appalachian Piedmont in southern Maryland, showing area described in this paper. Adapted from published maps of the Maryland Geological Survey, and from Bennison and Milton (1950) and Hopson (1964).

lithofacies, defined on the basis of original lithology. Three of these lithofacies—the metagraywacke, boulder gneiss and upper pelitic schist—crop out along the Potomac, and have been studied in detail. The lower pelitic schist was examined along Rock Creek in Washington, D.C., and is described with the upper pelitic schist. The metaconglomerate lithofacies is a restricted unit, which crops out only in northern Maryland, and is not discussed further.

The Metagraywacke Lithofacies

The metagraywacke lithofacies* is characterized by psammitic beds of metamorphosed graywacke and subgraywacke, rhythmically interbedded with pelitic rocks. The pelitic interbeds

* Includes the Peters Creek Formation of Jonas and Knopf (1921) and part of the Western Sequence of the Wissahickon Formation of Hopson (1964, p. 87–99).

vary from chlorite–sericite phyllites, near Seneca, to coarsely segregated sillimanite–mica schists, southeast of Great Falls. The psammitic beds vary from chlorite–sericite–quartz granofels to garnet–mica–quartz gneiss. Rare, thin layers of amphibolite probably represent basaltic tuffs, and local pods and layers of epidote-rich granofels probably formed from calcareous concretions and thin beds of calcareous sand.

The limits of the metagraywacke lithofacies on the geologic map (Fig. 2) were arbitrarily drawn where the proportion of psammitic rocks fell below 25 percent, but the psammitic beds are interlayered with pelitic rocks in all proportions, and isolated beds and zones of psammitic rock too thin to map are common in the pelitic schist zones.

Despite their intense metamorphism, the psammitic rocks retain many sedimentary structures.

Bedding is well preserved, even at the highest grades of metamorphism (Fig. 3A). Individual beds average 8 in. (20 cm) thick, and persist laterally with little change in thickness. Many of the psammitic beds are graded. In most, the grading is preserved only as a gradual upward diminution in quartz content (Fig. 3A). But a few of the thickest, least deformed beds contain relict detrital grains of quartz and feldspar which are up to 0.2 in. (5 mm) in diameter at the base, and decrease in size upward (Fig. 3B,C).

Many of the graded beds are massive at the base, and grade upward into a zone where thin micaceous laminations parallel to bedding alternate with quartzose layers. In some beds, the lower unlaminated zones pinch out along strike, and protrude into the underlying pelitic beds in lumpy protuberances resembling load casts (Fig. 3A). Other beds, which are not graded, are laminated throughout. These laminations are similar to structures in unmetamorphosed turbidites (Bouma, 1962; Lombard, 1963) and many may be relict sedimentary laminations, perhaps slightly accentuated by metamorphic differentiation. However, very similar laminations in other beds pass undeflected from a pelitic bed into the base of the next overlying psammitic bed (Fig. 3B). Crushing of relict detrital grains in the psammitic bed clearly shows that these laminations formed by segregation of micas along a tectonic cleavage. Probably both tectonic cleavage and sedimentary lamination are present, but it is exceedingly difficult to draw a clear distinction between them in all cases.

In some beds, the laminations are intensely contorted, but the beds themselves are not visibly deformed; the structure is very reminiscent of the type of convolute bedding ascribed to liquefaction of unconsolidated, water-saturated sand (Williams, 1960). The "balls" of unlaminated sand at the base of some graded beds also suggest liquefaction. Hopson (1964), p. 90–93; Pl. 15–18) has eloquently described many other features suggestive of soft-sediment deformation including clastic dikes, beds with fragments of pelitic material engulfed in a matrix of structureless, soupy sand (Fig. 3D), and larger zones of chaotic deformation.

The close resemblance of these features to structures produced by soft-sediment deformation in unmetamorphosed rocks elsewhere, the incompetent behavior of the psammitic beds, and the fact that these features are not restricted to migmatites, but occur in rocks of a wide range of metamorphic grade in the Piedmont (Hopson, 1964, p. 93) strongly suggest that they formed

prior to consolidation. However, some of the chaotic zones southeast of Great Falls are intimately associated with migmatitic schists containing quartz–albite exudation veinlets, and may have formed as a part of the migmatization; more detailed study of these structures is required to determine which stage of the deformation they represent.

Chemical analyses of the psammitic rocks show the excess of Na_2O over K_2O, and the high total iron plus magnesium content characteristic of graywackes and subgraywackes (Hopson, 1964).

Many features of the metagraywackes suggest deposition by turbidity currents: their rhythmic interlayering with shale; their graded bedding; their upward increase in number of laminations; their load casts; their slump features; and their chemical immaturity. The ungraded, laminated beds may have been deposited by less dense clouds of suspended material advancing in front of turbidity currents (Lombard, 1963). The thin, interbedded pelitic beds presumably represent periods of quiet deposition between turbidity currents. This assemblage of rock types, and the immensity of the Wissahickon Formation, strongly suggest deposition in a deep, marine flysch environment.

The Upper Pelitic Schist Lithofacies

The upper pelitic schist lithofacies* is a thick sequence of metamorphosed shales. Interbedded metagraywackes occur locally throughout the section, both as isolated beds, and as groups of beds too thin to map.

Near Seneca, Maryland, the pelitic rocks are chlorite–muscovite phyllites; eastward, they grade into coarse grained mica schist, locally containing garnet, staurolite, andalusite, and sillimanite. Chemically, the rocks are normal marine shales with an excess of K_2O over Na_2O and a high aluminum content (Hopson, 1964, p. 78).

Their association with the turbidites of the metagraywacke lithofacies suggests that the pelitic rocks also accumulated in a deep, marine basin.

The Boulder Gneiss Lithofacies

The boulder gneiss lithofacies† is a remarkably uniform sequence of metamorphosed sandy mudstones, containing scattered quartz granules, pebbles and rock fragments. It crops out in two

* Equivalent to part of the Western Sequence of the Wissahickon Formation of Hopson (1964).
† Equivalent to the Sykesville and Laurel Formations of Hopson (1964) and Cloos and Cooke (1953), and the chlorite–muscovite gneiss of Reed and Jolly (1963).

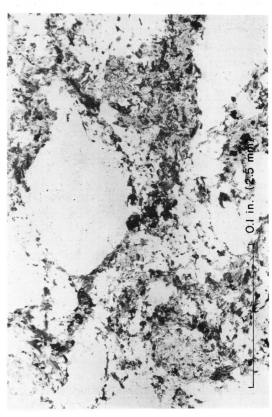

converging belts, which outline the limbs of the Baltimore–Washington Anticlinorium (Fig. 1).

The most common pebbles and rock fragments are:

1. rounded granules and pebbles of quartz, and of quartz and feldspar, ranging from sand size to 4 in. (10 cm) in diameter;

2. flattened mica schist fragments ranging from ½ in. (12 mm) blebs to slabs 15 ft (5 m) long;

3. angular cobbles and irregular, hook-shaped fragments of metagraywacke and concretionary calcareous siltstone;

4. cobbles of schistose amphibolite, biotite-quartz gneiss, and granite derived from the Baltimore Gneiss.

The smaller pebbles commonly lie parallel to schistosity in the matrix, but many larger blocks are randomly oriented. The larger rock fragments are locally concentrated in ill-defined zones, 50–100 ft (15–30 m) thick, which define a crude layering, but normal bedding is very rare.

The matrix is a granular garnet–oligoclase-mica–quartz gneiss with a chemical composition nearly identical to the average of the metagraywacke and the pelitic schist (Fisher, 1963, p. 42). Like the coarse-grained metagraywackes, it locally retains a relict clastic texture. Large, rounded grains of quartz and detrital feldspar are set in a fine-grained paste of quartz, recrystallized feldspar, micas, and chlorite.

The contacts of the boulder gneiss with the upper pelitic schist are gradational over 200–500 ft (60–150 m). Approaching the contact from the boulder gneiss side, the large rock fragments disappear first, then the smaller quartz pebbles and micaceous blebs decrease in number, and finally disappear altogether; the matrix becomes finer grained, gradually loses its granular, sandy texture, and passes gradationally into mica schist. In mapping, the contact was arbitrarily placed where the small mica schist fragments and coarse detrital quartz granules disappear.

The boulder gneiss was originally thought to be an intrusive granite, because of its massive appearance and its abundant cobbles and boulders, interpreted as xenoliths (Stose and Stose, 1946). However, several facts indicate that the rock is a metasediment: (1) its distribution shows that it is a tabular lens interstratified with the enclosing sediments, and folded across the Baltimore–Washington anticlinorium; (2) its contacts with the upper pelitic schist and with the metagraywacke lithofacies in Carroll County (Hopson, 1964, p. 118) suggest sedimentary gradation and lateral change of facies, not intrusion; (3) it contains abundant normative quartz and corundum, common in sedimentary rocks, but rare in igneous rocks (Hopson, 1964, p. 110); (4) its relict sandy textures show that the matrix was originally a sandy mudstone, not an igneous rock.

Hopson (1964, p. 108–112) considered various possible origins for the boulder gneiss, and concluded that it had probably formed by repeated slumping and mixing of Wissahickon graywacke and shale, boulders of basement rock, and quartz sand and pebbles. This explanation best accounts for the presence of randomly oriented boulders of metamorphic rock and contorted slabs of partly consolidated sediments in an unsorted, poorly bedded mixture of graywacke and shale, closely associated with deep marine turbidites. But, as Hopson noted, there remain several puzzling aspects of the boulder gneiss. First is its immense size. It appears to be as much as 15,000 ft thick in places, and can be traced along strike for nearly 100 mi. Though many submarine slide deposits have been described from other geosynclines (Ksiazkiewicz, 1958), none are as large. Second is the juxtaposition of coarse boulders and near-shore debris with deep-water flysch sediments. Hopson tentatively suggested that this could be explained by seaward migration of the shore line concurrent with rapid deepening of the basin, leading to cannibalization and slumping of unconsolidated sediments, together with simultaneous input of boulders and coarse sand derived

FIGURE 3. (A) Graded metagraywackes, interbedded with pelitic schist. The middle graywacke bed is laminated at the top, and massive at the base; the massive base is "balled up" into load casts. Hammer handle is 14 in (35 cm) long. Central Bear Island. (B) Relict detrital sand grains at the base of a graded metagraywacke bed, overlying fine-grained metagraywacke. The sand grains are stretched into augen, flattened parallel to a weak cleavage lamination, which cuts across bedding. Northern Bear Island. (C) Photomicrograph of relict detrital sand grains set in a matrix of fine-grained micas and quartz, at the base of a graded metagraywacke bed. Northern Bear Island, plane light. (D) Slabs of contorted mica schist in a matrix of "soupy" metagraywacke; deformed zone is confined to a single bed, traceable along strike for 40 ft (12 m). Beds above (not visible in photograph) and below are undisturbed. Central Bear Island.

from the uplifted basement. However, a final explanation of this peculiar unit awaits a more detailed study of its relations with the surrounding rocks, and a better understanding of flysch sedimentation.

Stratigraphic Relations between the Lithofacies

The direction of grading in metagraywacke beds, the relative orientation of bedding and axial plane schistosity, and the attitude of minor folds suggest that the mapped metagraywacke zones represent two different stratigraphic units, much repeated by folding, but gradually becoming younger to the west. The upper zone crops out in central Bear Island, in a south-plunging syncline with inward-facing graded beds at both contacts (Fig. 2). A different zone of distinctive, thick-bedded metagraywacke 1,000 ft (300 m) to the west contains graded beds facing predominantly east, and therefore lies stratigraphically below the synclinally folded zone.

Farther west, the thick-bedded metagraywackes pass stratigraphically downward into pelitic schist, which in turn passes into metagraywacke, also thick-bedded, but with graded beds facing west. Therefore, these two zones of thick-bedded metagraywacke appear to be parts of a single stratigraphic unit, on opposite limbs of an anticline. From this point, the same metagraywacke zone can be followed north along the Potomac to Bealls Island, where it swings south in a syncline. The orientation of bedding and axial plane cleavage, and the attitude of scattered graded beds suggest that the pelitic schist just west of Bealls Island occupies the core of an anticline, and that the same metagraywacke crops out west of the schist zone.

The orientation of graded beds and minor folds in the metagraywacke zone west of Sycamore Island suggest that this zone overlies the thick-bedded metagraywacke, and may correlate with the zone on central Bear Island. The metagraywacke zone just west of Offutt Island contains a few west-facing graded beds, and it may pass under the syncline on Bear Island, and connect with the thick-bedded unit to the west.

On Bear Island the lower metagraywacke zone is about 1,000 ft (300 m) thick. The thickness of the upper zone is indeterminate there, because its top is not exposed. The zone west of Sycamore Island, which may be equivalent, appears to be about 2,000 ft (600 m) thick, allowing for repetition by minor folds. The original thickness of the pelitic rocks is difficult to estimate, due to the intense deformation; but if the correlation of metagraywacke zones outlined above is correct, about 4,000 ft (1,200 m) of pelitic rocks now underlie the lowermost metagraywacke zone, 1,000 ft (300 m) lie between the two metagraywacke zones, and 6,000 ft (1,800 m) overlie the upper metagraywacke zone.

The two main belts of boulder gneiss converge and almost come together in Washington, D.C. (Figs. 1 and 2), and it seems certain that they are parts of a single stratigraphic unit, folded across the anticlinorium. The boulder gneiss zone at Cabin John nearly merges with the main belt along the south bank of the Potomac, and must be at almost the same stratigraphic level. The small boulder gneiss zone at Offutt Island is entirely separate from the main belt, and appears to be a small lense within the pelitic schist.

Structural and stratigraphic relations in northern Montgomery and Howard Counties (Hopson, 1964) leave little doubt that the main boulder gneiss belt overlies the pelitic schists to the east (Fig. 1). As already described, the metagraywackes west of the boulder gneiss appear to become younger to the west, though much repeated by folding. Therefore, the metagraywackes and pelitic schists west of the boulder gneiss probably overlie the gneiss in a simple homocline. However, graded beds are scarce in the critical area just west of the boulder gneiss, and it is conceivable that the gneiss occupies the trough of a local syncline, the southern continuation of the Peach Bottom Syncline in Pennsylvania. If so, however, the entire thickness of boulder gneiss must disappear by intertonguing with metagraywacke or pelitic schist to the west, since it does not reappear in the section farther west; although possible, this seems unlikely.

THE IJAMSVILLE PHYLLITE

A thin sliver of chlorite–muscovite phyllite crops out along the Potomac, just before the Wissahickon Formation passes under the Triassic rocks at Seneca (Fig. 2). Hopson (1964, p. 87–88, 121–124) correlated this phyllite with the Ijamsville Phyllite, one of several phyllites which are interbedded with volcanic rocks in the western Piedmont. The contact between the Ijamsville and the Wissahickon is poorly exposed, and may be complicated by Triassic faulting; further mapping along strike is needed to clarify the relations between these units. Provisionally, however, the orientation of graded beds and the pattern of folds in the Wissahickon Formation east of the contact suggest that the sequence is overturned to the west, and that the Wissahickon is older than the Ijamsville (Fig. 2, Section A-A').

SUMMARY OF SEDIMENTARY HISTORY

The Glenarm metasediments record a complex sedimentary history which began with deposition of quartz sand and mud (Setters Formation) and carbonate sediments (Cockeysville Marble), during a period of slow subsidence, and stable tectonic conditions (Hopson, 1964, p. 128–131). The overlying shales, graywackes, and slump deposits of the Wissahickon record the development of a deep, marine trough and the transition to flysch sedimentation. In the lower part, shales predominate, and sedimentation was probably relatively quiet. Higher in the section, turbidites are more abundant, suggesting a gradual buildup of tectonic activity, which reached its climax during deposition and slumping of the boulders and coarse, sandy sediments of the boulder gneiss.

As pointed out by Hopson, the boulders of basement rock in the boulder gneiss almost certainly came from the east, indicating that transport was predominantly westward. In detail, however, the direction of transport probably varied locally; the lateral intertonguing of boulder gneiss with metagraywacke and pelitic schist suggests that the boulder gneiss accumulated as huge submarine fans, intertonguing laterally with finer sediments. The relation between lithologies may have been much like that in the southern California basins today, where graded turbidites, deposited by density currents discharged from submarine canyons, interfinger laterally with slump deposits accumulating at the base of the continental slope between canyons (Gorsline and Emery, 1959).

The metagraywackes and pelitic schists above the boulder gneiss record a return to more normal sedimentation. The two metagraywacke zones probably reflect separate spasmodic pulses of tectonic activity and rapid sedimentation. The local slide deposits within the metagraywacke zones indicate recurrence of the conditions which produced the boulder gneiss, but on a much smaller scale. The zones of pelitic schist between the metagraywacke units probably record periods of relative stability.

The phyllites, quartzites and volcanic rocks of the western Piedmont appear to reflect the transition back to relatively shallow sedimentation as the geosynclinal basin was filled (Hopson, 1964, p. 128).

TECTONIC STRUCTURES

The structures along the Potomac reveal a long, complex history of deformation (Table 1). The slump deposits and convolute folds described earlier record the initial stages of deformation, by sliding parallel to bedding while the sediments were still soft. Later, during metamorphism, regional deformation produced three sets of folds and related planar structures, described below.

S_2 and F_2

The earliest schistosity (S_2) is best displayed in the competent metagraywackes and amphibolites southeast of Great Falls. In the metagraywackes, it is defined by thin micaceous laminae alternating with quartzose layers; in the amphibolites by layers of aligned hornblende crystals interleaved with thin plagioclase–epidote laminae. In most outcrops, the S_2 parallels bedding and wraps around younger fold hinges (Fig. 4B,D; see also Cloos and Anderson, 1950; and Reed and Jolly, 1963, p. H11); but at the hinges of a few, early folds (F_2) it parallels the axial plane, and cuts across bedding (Fig. 4A).

TABLE 1.   Structural Elements Along the Potomac River

Planar structures		Folds		Approximate time of development
S_1	Bedding in metasedimentary rocks.	F_1	Convolute folds without axial plane cleavage or schistosity.	During diagenesis.
S_2	Schistosity; most parallels S_1, but some is axial to minor folds.	F_2	Small isoclinical folds in S_1 with S_2 parallel to axial plane.	During rising metamorphism.
S_3	Axial plane cleavage or schistosity, slip cleavage, and axial plane of plications.	F_3	Large and small folds in S_1 and S_2, with S_3 parallel to axial plane.	Near peak of metamorphism.
S_4	Slip cleavage or schistosity, locally axial to minor folds.	F_4	Small folds in S_1, S_2, and S_3, with S_4 parallel to axial plane.	During retrograde metamorphism.
Steep faults				After metamorphism.

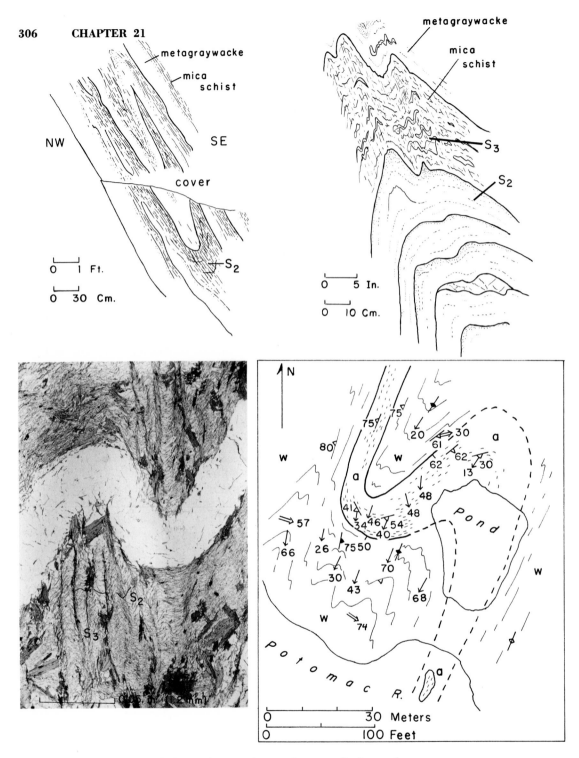

FIGURE 4. (A) Isoclinal F_2 fold, seen in a section nearly perpendicular to the fold axis (azimuth 75°, plunge 30° NE). S_2 schistosity parallels the axial plane of the fold, and cuts bedding (S_1) at the crest of the fold. Sketched from photograph. Central Bear Island. (B) Mica schist interbedded with metagraywacke in F_3 fold, sketched from photograph. S_2 parallels bedding; S_3 plications and slip cleavage parallel the axial plane, and are best developed in the pelitic schist at the fold crest. Loose block on northern Bear Island. (C) Photomicrograph of S_3 slip cleavage cutting across S_2 schistosity; note that the quartz veinlet parallel to S_2 is plicated, but is not ruptured. North bank of the Potomac River, near Sycamore Island. Plane light. (D) Amphibolite sill (a) and Wissahickon metagraywacke (w) in F_3 anticline, with minor F_2 folds (double arrows), F_3 folds and lineation in amphibolite (single arrows), S_2 (open symbols) and S_3 (solid symbols). Central Bear Island.

Most F_2 folds are essentially isoclinal, with greatly thickened beds at the hinge (Fig. 4A). Unfortunately, few measurements of the F_2 fold axes have been made, because these folds were not recognized until late in this study, and because their axes are difficult to measure in the polished outcrops along the river. The few that have been measured trend more or less east–west, and plunge down the dip of S_2, folded by later structures (Fig. 4D).

Some of the S_2 schistosity appears to have formed mimetically. Many of the slumped metagraywacke beds contain crumpled and rotated fragments of pelitic schist, with schistosity parallel to bedding. If these beds slumped prior to consolidation, and there is abundant evidence that they did, the coarse micas now defining the schistosity must have grown mimetically, parallel to an original bedding fissility.

However, three lines of evidence show that much of the S_2 formed by recrystallization during penetrative deformation: (1) the metagraywackes southeast of Great Falls preserve all stages in the development of S_2 by progressive granulation of originally coarse-grained rocks with no preferred orientation of micas (Fig. 3C), through slightly more deformed rocks with an incipient schistosity (Fig. 3B), to rocks with well-developed S_2, where all the detrital grains have been crushed into thin stringers, parallel to the schistosity; (2) the thin amphibolite sills and the margins of the thick sills contain well-developed S_2 (Fig. 4D); but the central parts of the thick sills are unfoliated, even in relict chill zones within composite sills. Therefore, it is most unlikely that the margins had any primary foliation, and the schistosity there must have formed by penetrative deformation, not by mimetic recrystallization; (3) the S_2 schistosity at F_2 fold hinges cuts across bedding, and must have formed as a result of deformation.

This deformation must have acted in a plane nearly parallel to bedding, presumably horizontal, because the opposite limbs of most F_2 folds are very different in length; one may be 10 or 20 times as long as the other. Flattening in the plane of S_2, in addition to slip parallel to the schistosity, is clearly indicated by numerous ptygmatically folded quartz veins. This deformation must have occurred during high grade metamorphism, because amphibolite assemblages (hornblende-oligioclase) are developed along S_2 on Bear Island; but it occurred prior to the peak of metamorphism, because andalusite and kyanite(?) have grown athwart S_2 in the pelitic beds.

No large scale F_2 folds have yet been recognized in the southern Maryland Piedmont, and it is not clear whether this deformation involved major horizontal transport, or simply flattening in the plane of bedding, combined with local slip along S_2 schistosity.

S_3 and F_3

Three structures commonly deform S_2: axial plane schistosity or cleavage, slip cleavage, and sets of plications with parallel axial planes. These three structures are collectively termed S_3 because they are parallel, and grade into each other over short distances.

The plications are similar folds with amplitudes of 0.2–2 in. (5–50 mm) and slightly thickened hinges containing micas parallel to S_2 (Fig. 4B). The slip cleavage is represented by sets of minute crinkles in S_2. Along the crinkle hinges, micas parallel S_2, but along the limbs, micas are bent toward the slip cleavage planes, and some new micas have grown parallel to the cleavage (Fig. 4C). The axial plane schistosity or cleavage is defined by micas which parallel the cleavage throughout the rock, not merely along discrete planes. In the phyllites and chlorite–sericite schists near Seneca, the S_3 cleavage is a smooth, lustrous surface made up of very fine micas in near-perfect parallel orientation. Farther east, it is a schistosity, defined by laminae of coarse, parallel micas, alternating with quartz veinlets.

The chlorite–biotite schists west of Great Falls preserve all stages in the gradual conversion of slip cleavage to axial plane cleavage. In intermediate stages, the slip-cleavage planes have evolved into thin, micaceous laminae. Nearly all of the micas parallel these laminae, but in a few of the intervening quartz-rich layers, bent micas oriented across the cleavage preserve remnants of the old crinkle hinges.

The F_3 folds are larger scale versions of the plications and crinkles; they are similar folds which deform bedding and S_2, and contain S_3 parallel to their axial planes. Most folds along the Potomac are of this type, including minor folds visible in a single outcrop (Fig. 4B), larger folds mappable in the field (Fig. 4D) and probably the Baltimore–Washington Anticlinorium (Fig. 1). F_3-fold axes trend uniformly north–northeast, but their plunge varies, possibly reflecting the earlier east–west F_2 folds (Fig. 2). The axial planes of F_3 folds (S_3) dip east near Seneca, steepen and become vertical near Great Falls, and dip west near Washington, D.C. (Fig. 2).

In the low-grade schists and phyllites of the western Piedmont, the F_3 folds appear to have formed mainly by flow or slip parallel to the cleavage; they contain well-developed S_3 throughout, and both pelitic rocks and metagraywackes are greatly thickened at fold hinges. In the folds near Great Falls, pelitic rocks in the fold hinges contain prominent S_3 schistosity, and are greatly thickened (Fig. 4B); apparently, they were deformed mainly by movement parallel to S_3. But on the fold limbs, and in the interbedded metagraywackes and amphibolites, folding apparently involved some slip parallel to bedding; beds are not greatly thickened, and S_3 is weak, or absent (Fig. 4B). It appears that the F_3 folds evolved by a combination of flexural folding and movement parallel to cleavage. The predominant mechanism was determined by regional variations in metamorphic grade, and by local variations in lithology, position within the fold, and perhaps tightness of folding.

The F_3 folds must have formed near the peak of metamorphism, because (1) the mineral assemblages developed along S_3 reflect the highest grade of metamorphism in each rock; (2) hornblende crystals in the amphibolites are aligned parallel to F_3-fold axes (Fig. 4D); (3) migmatitic quartz–albite veinlets parallel to S_2 in the rocks at Bear Island are folded, but new veinlets have formed parallel to S_3. Recrystallization must have outlasted deformation, however, because helicitic biotite and staurolite have grown across S_3 cleavage in many rocks, and because aluminum silicates have grown athwart the S_3 schistosity in a few places.

S_4 and F_4

In many rocks, S_3 is crinkled by a late slip cleavage (S_4). Near Stubblefield Falls, S_4 grades into an irregular, steeply dipping schistosity parallel to the axial planes of rare F_4 folds, which trend north–northeast (Fig. 2). Much of the S_4 has chloritized earlier biotite, and at Stubblefield Falls it has reduced veined schists and gneisses to phyllonites, showing that it formed well after the peak of metamorphism.

Faults

The youngest structures along the Potomac are two steep faults, just southeast of Great Falls, which offset an F_3 anticline, and two unmetamorphosed lamprophyre dikes (Fig. 2). Farther South, in Virginia, Bennison and Milton (1950) have demonstrated fault movement along the Triassic dike which passes near Seneca, but none could be proven at the Potomac.

METAMORPHISM

The Piedmont rocks gradually increase in metamorphic grade from phyllites in the west to kyanite- and sillimanite-bearing schists flanking the gneiss domes near Baltimore. Along the Potomac, this increase is reflected by the successive appearance of chlorite, biotite and garnet in the pelitic rocks; but this simple regional gradient is modified by two nodes of staurolite, kyanite(?) and sillimanite-bearing rocks, and by local zones of intense retrograde metamorphism (Fig. 2).

Mineral Assemblages and Metamorphic Reactions

The fine-grained schists and phyllites with S_3 cleavage near Seneca, and the retrograde phyllonites with S_4 cleavage near Stubblefield Falls are the lowest grade metamorphic rocks along the Potomac. Both contain the assemblage quartz–muscovite–chlorite–albite–epidote–magnetite–hematite (Table 2). The muscovite is a greenish 2M sericite; the chlorite is an aluminous 14 Å variety, with intermediate Mg/Mg + Fe'' (Fig. 5A).

About 1 mi (1.6 km) east of Blockhouse Point, scattered flakes of greenish biotite appear in a few rocks. Farther east, the biotite gradually becomes coarser grained and more abundant, and begins to form thin segregation laminae parallel to S_3. All of the chlorite zone minerals persist into the biotite zone, and appear to have been in equilibrium with biotite (Table 2; Fig. 5B).

Many rocks originally containing biotite have been retrogressively altered to the chlorite zone assemblage. This retrograde reaction shows that the development of biotite cannot have been caused by differences in original bulk composition alone; and it also shows that biotite was unstable in the chlorite zone, and that its failure to form there cannot be attributed solely to unfavorable reaction kinetics. The gradual appearance of biotite, its increase in abundance to the east, and the failure of any chlorite zone mineral to disappear in the biotite zone, suggest that the biotite formed because of a systematic change in the composition of one or more solid solution minerals. The chlorite compositions appear to be independent of metamorphic grade (Fig. 5), and the most likely explanation is that the biotite formed from celadonite, progressively exsolved from the sericite (cf. Velde, 1965). If so, the exact position of the biotite "isograd" depends partly on bulk composition, in addition to metamorphic conditions (cf. Thompson, 1957, p. 856).

About 3.5 mi (5.6 km) east of Blockhouse Point, almandine-rich garnet suddenly appears, and hematite disappears (Table 2, Fig. 5C,D). The garnet appears to have been in approximate equilibrium with coarse flakes of chlorite and biotite, because: (1) the three occur in sharp mutual contact, with no textural indication of replacement; (2) the assemblages satisfy the mineralogical phase rule; and (3) the tie lines plotted in Fig. 5C,D cross only slightly. Garnets from 15 Wissahickon pelites in northern Maryland are composed chiefly of almandine, pyrope, and grossularite, with only minor amounts of spessartite (D. L. Southwick, unpublished analyses). On this basis, spessartite was assumed negligible in garnets from the rocks along the Potomac (Table 2).

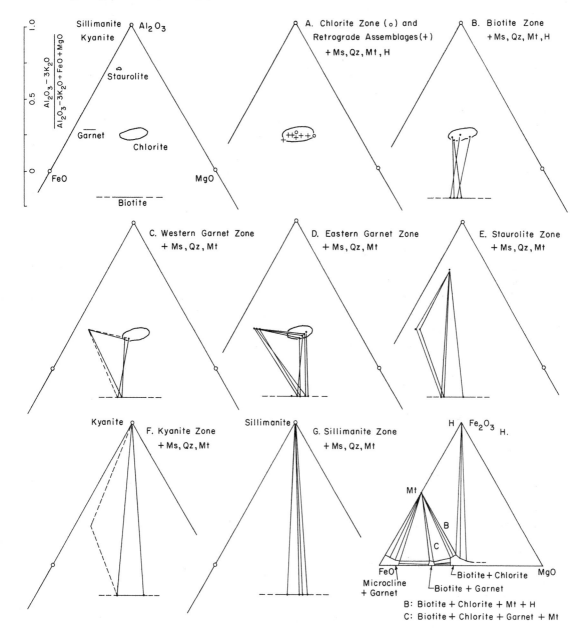

FIGURE 5. A through G show mineral assemblages in the system Al_2O_3–K_2O–FeO–MgO (AKFM) inferred to have been in equilibrium in each metamorphic zone at the peak of metamorphism; projected from muscovite (ms), quartz (qz) and H_2O onto the plane Al_2O_3–FeO–MgO, shown in key. H shows relations between the assemblages along the line FeO–MgO in B and C, when Fe_2O_3 is considered as a component.

TABLE 2.   Mineral Assemblages in Rocks Containing Muscovite and Quartz

	Chlorite—(Mg,Fe,Al)$_6$(Si,Al)$_4$O$_{10}$(OH)$_8$						Biotite		Garnet					Plagioclase		Oxides	Aluminum Silicates
	n ±0.001	B ±0.0005	Mg	Fe	Al	Si	n_z ±0.001	100 Mg/Fe″ + Mg	a_0 ±0.005Å	n ±0.005	Al	Py	Gr	n ±0.001	% An		
Ch Zone																	
108-A	1.615	+0.0044	2.9	1.8	2.6	2.7								1.531	5	H, Mt ($a°$ = 8.391 Å)	
109-B	1.628	+0.0033	2.3	2.3	2.8	2.6								nd		Mt	
Bi Zone																	
83-A	1.629	+0.0026	2.2	2.4	2.8	2.6	1.637	44						nd		Mt, H	
88-A	1.631	−0.0008	2.1	2.7	2.4	2.8	1.634	47						nd		Mt, H	
103-A	1.627	−0.0008	2.2	2.7	2.2	2.9	1.629	52						nd		Mt, H	
104-B	1.617	+0.0035	2.7	2.1	2.6	2.7	1.634	47						nd		Mt, H	
G Zone																	
20-A	1.627	+0.0010	2.3	2.5	2.4	2.8	1.638	42		nd				nd		Mt	
71-A*							1.630	52	11.655	1.785	54	5	41	nd*		Mt	
75-A							1.634	47								Mt	
102-A	1.617	+0.0038	2.8	1.7	2.6	2.7	1.626	55		nd				1.545	32		
102-B	1.623	+0.0030	2.5	2.2	2.6	2.7	1.630	52	11.562	1.797	49	16	15			Mt ($a°$ = 8.394 Å)	
102-C	1.625	+0.0048	2.4	2.2	2.8	2.6	1.632	51	11.566	1.800	72	13	16	1.546	34	Mt	
120-A	1.630	−0.0011	2.1	2.8	2.2	2.9	1.625	57	11.610	1.795	65	8	27	nd		Mt	
175-A	1.623	+0.0024	2.4	2.3	2.6	2.7	1.637	43	11.562	1.805	75	10	15	nd		Mt	
188-A							1.626	56	11.571	1.802	73	11	16	nd		Mt ($a°$ = 8.391 Å)	
St Zone																	
174-A							1.639	41	11.550	1.810	81	10	9	nd		Mt	St
174-C							1.639	41	11.547	1.809	80	12	8	nd		Mt	St
182-B							1.630	51	nd	nd							

Sample	Fisher (1963, Fig. 33)							Smith (1960, Pl. 12)			
										Mt	Ky/Si
Ky Zone											
79-A	1.625	56						nd		Mt	Ky
173-A	1.639	41						nd		Mt	Ky
Si Zone											
11-B*	1.637	47						1.536*	14	Mt	Si
15-A	1.626	56						1.533	18	Mt	Si
85-A	1.627	54			nd			nd		Mt, H	Si
95-A	nd		11.600	1.800	70	6	24	nd		Mt	Si
162-B	nd		11.590	1.800	70	9	21	nd		Mt	Si
178-A	1.628	53			nd			nd		Mt	Si
Retrograded Rocks											
3-E	1.625	+0.0029	2.4	2.3	2.6	2.7		1.534	11	Mt	
28-A	1.624	+0.0030	2.4	2.3	2.6	2.7		nd		Mt, H	
29-A	1.619	+0.0041	2.7	2.0	2.6	2.7		nd		Mt	
59-A	1.627	+0.0018	2.3	2.4	2.6	2.7		nd		Mt, H	
63-A	1.626	+0.0020	2.3	2.4	2.6	2.7		nd		Mt	
87-A	1.634	−0.0024	1.9	3.0	2.2	2.9		nd		Mt	

Column study references: Fisher (1963, Fig. 33); Sriramadas (1957, Fig. 8); Smith (1960, Pl. 12).

Determinative Method: Hey (1954, p. 284)

Notes

Space blank indicates mineral absent.
nd: mineral present, composition not determined.
* microcline $(2\theta_{130} - 2\theta_{1\bar{3}0} = 0.85°)$ also present.
H = hematite.
Ky = kyanite.
Mt = magnetite.
Si = sillimanite.
St = staurolite.

When plotted in a phase diagram, the minerals in the western part of the garnet zone do not overlap those in the biotite zone (Fig. 5B,C,H). The two assemblages could coexist at the same P, T, and a_{H_2O} in rocks of slightly different bulk composition; therefore the appearance of garnet may reflect an original difference in Fe′′′/Fe′′′ + Fe″ of two stratigraphically distinct units, or progressive reduction during metamorphism of rocks with high initial Fe′′′/Fe′′′ + Fe″. More detailed work on this boundary is needed to distinguish between these two possibilities. But whichever is correct, the appearance of garnet in these rocks clearly depends on bulk composition, in addition to metamorphic conditions.

Disregarding the grossularite in the garnet, these rocks contain four minerals in the four component AKFM system, so that all the minerals should have fixed compositions under fixed metamorphic conditions. Therefore, the differences in composition of biotite and chlorite coexisting with garnet in the eastern and western parts of the garnet zone (Fig. 5C,D) probably reflect variations in P, T, or a_{H_2O} within the garnet zone, and may be a useful indicator of metamorphic grade.

Two nodes of higher grade metamorphic rocks occur within the broad expanse of garnet zone rocks (Fig. 2). The node centered on Bear Island culminates in sillimanite-bearing migmatites, flanked by staurolite, and kyanite(?)-bearing schists. The node at Stubblefield Falls also contains migmatites, but the mapped zone boundaries are highly schematic, because the original minerals have been obliterated by pervasive retrograde metamorphism, and shearing along S_4. The following discussion is based entirely on the rocks near Bear Island.

The staurolite zone pelites lack the coarse-grained chlorite of the garnet zone rocks, and contain large, oval porphyroblasts of sieved staurolite, largely altered to a sericite–chlorite schimmer aggregate. The staurolite encloses biotite and garnet, and appears to have been in equilibrium with them (Table 2, Fig. 5E). The staurolite–biotite tie line in Figure 5E intersects the garnet–chlorite tie line in Figure 5C, suggesting that the staurolite formed by the reaction*: 3 chlorite + garnet + 3 muscovite $\rightleftharpoons$ staurolite + 3 biotite + 8 quartz + $10H_2O$. This reaction involves five phases in the four-component AKFM system, and should therefore depend only on the external conditions of metamorphism, not

on the bulk composition of the participating rocks (cf. Thompson, 1957, p. 856).

About 0.5 mi (0.8 km) west of Great Falls, staurolite disappears, and large bladed schimmer aggregates resembling altered kyanite coexist with garnet and biotite (Table 2). The configuration of tie lines in Fig. 5F suggests that the kyanite(?) formed by the reaction: 3 staurolite + muscovite + 6 quartz $\rightleftharpoons$ biotite + 2 garnet + 20 kyanite + $4H_2O$ In order for this reaction to proceed stably, the reaction staurolite + 6 chlorite + 6 muscovite $\rightleftharpoons$ 6 biotite + 20 kyanite + 6 quartz + $20H_2O$ should occur within the staurolite zone, but no rocks with the composition required to show this reaction were found.

On Bear Island, minute sillimanite needles fringe both the kyanite (?) pseudomorphs and andalusite, and occur in felted mats replacing muscovite. Segregation veinlets of quartz and albite are locally abundant parallel to S_3, and small bodies of aplite and pegmatite have intruded and locally feldspathized the schist. Hopson (1964) suggested that a rise in temperature accompanying the intrusion of these pegmatites might have produced the sillimanite in a separate period of recrystallization well after the main period of metamorphism. However, if the pegmatites alone had produced the sillimanite, sillimanite should be associated with all the pegmatites, and it is not. Numerous small pegmatites intrude the lower grade metamorphic rocks, but sillimanite occurs only near the pegmatites which intrude rocks containing andalusite and kyanite(?). This association strongly suggests that the rocks were already undergoing high temperature metamorphism when the pegmatites were intruded; heat accompanying the pegmatites may have been the final factor which pushed the rocks into the sillimanite field, but it cannot have been the sole cause of the sillimanite, and the sillimanite should be viewed as the culmination of the main period of metamorphism.

In addition to sillimanite, these rocks contain biotite, sparse grossularite-rich garnet (Table 2), and altered cordierite (Reed and Jolly, 1963, Fig. 6). Cordierite is not plotted in Figure 5G, because its paragenesis is uncertain. A few rocks contain microcline in contact with sillimanite, while others contain muscovite and albite apparently in equilibrium; these two assemblages are incompatible, and should not coexist at the same metamorphic grade (cf. Evans and Guidotti, 1966, Fig. 9). These different assemblages may reflect local differences in a_{H_2O}, caused by in-

---

* Coefficients in this and the following reactions were calculated using the mineral compositions in Figure 5, then rounded to one significant figure.

flux of magmatic fluids from the pegmatite intrusions.

In the eastern part of Bear Island, sillimanite fringes andalusite, and appears to have grown at its expense; but in the western part, it surrounds kyanite(?) pseudomorphs, and appears to have formed from kyanite. The juxtaposition of these two reactions suggests that the rocks entered the sillimanite field near the triple point of the Al_2SiO_5 system (Fig. 6). Most of the sillimanite occurs as felted masses of randomly oriented needles embedded in muscovite, apparently formed by the sliding equilibrium:

sodic muscovite + quartz $\rightleftharpoons$ potassic muscovite
+ sillimanite + albite + H_2O

(cf. Evans and Guidotti, 1966). Therefore, the rocks were in the part of the sillimanite field below the breakdown of pure muscovite plus quartz (Fig. 6). The intrusive pegmatites contain books of coarse, magmatic muscovite, and must have crystallized in the area between the muscovite plus quartz breakdown and the granite minimum (Fig. 6). Taken together, these facts indicate that the sillimanite zone rocks were metamorphosed at a total pressure near 4 kb, with water pressure at least locally equal to total pressure, and at a temperature between 500 and 650°C, but probably closer to the latter.

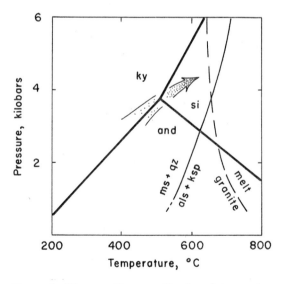

FIGURE 6. Pressure ($P_{total} = P_{H_2O}$) and temperature conditions of sillimanite zone rocks, inferred from equilibrium relations of (1) andalusite (and), kyanite (ky) and sillimanite (si) (heavy lines; after Newton, 1966); (2) breakdown of muscovite plus quartz (ms + qz) to aluminum silicate plus potash feldspar (als + ksp) (light line, after Evans, 1965); (3) minimum melt in "granite" system (dashed line, after Luth and others, 1964).

Interpretation of the Metamorphism

The progressive development of chlorite, biotite, and almandine in pelitic rocks along the Potomac records an increase in metamorphic grade eastward, toward the core of the Baltimore–Washington anticlinorium. This increase is in the direction of stratigraphically lower rocks, and increasing depth of burial probably contributed to the metamorphic gradient. But the isograds near Baltimore closely parallel the gneiss domes (Doe and others, 1965, Fig. 2), and unless the metamorphism there predated the doming, a major factor in the metamorphism must have been influx of heat from the Baltimore Gneiss, or possibly loss of water to the gneiss. The local nodes of sillimanite-zone metamorphic rocks superimposed on this gentle regional gradient were probably accentuated by intrusion of pegmatites; but the pattern of the nodes predates the pegmatites, and probably reflects local areas of high heat flow, perhaps from subjacent mafic intrusions (see Griscom and Peterson, 1961).

Local zones of intense retrograde metamorphism coincide with the areas where S_4 is strongly developed, and reflect late shearing and influx of water.

AGE OF METAMORPHISM AND DEFORMATION

The ages of the different stages of deformation and recrystallization which affected the Glenarm Series can be bracketed fairly closely by a combination of geologic and radiometric methods. The S_2 schistosity may have begun to form early in the history of these rocks. However, in southeastern Pennsylvania, the Wissahickon Formation contains a schistosity which is virtually identical in style and orientation with S_2 in the rocks along the Potomac, and which can be traced westward into Ordovician limestones (Freedman and others, 1964). Long-range structural correlations are extremely tenuous, but if S_2 in southern Maryland is synchronous with that in Pennsylvania, it must have continued to form until Ordovician time. S_3 and the associated F_3 folds are younger than S_2, but they cannot be much younger, because radiometric dates on zircon and feldspar from postorogenic pegmatites in the Maryland Piedmont show that the major deformation had ended by Late Ordovician or Silurian time (Wetherill and others, 1966). These pegmatites are similar to those which accompanied the sillimanite at Bear Island, and probably the metamorphism reached its peak at about this time.

S_4 is later than S_3, and the large difference in metamorphic grade between the high-grade assemblages developed along S_3 and the retrograde assemblages developed along S_4 suggest that there was a major time interval between them. However, S_4 must have formed prior to deposition of the unmetamorphosed Triassic rocks in the western Piedmont. Possibly the many Carboniferous dates given by Piedmont micas (Lapham and Basset, 1964; Wetherill and others, 1966) reflect this period of deformation and recrystallization.

The faults along the Potomac must be later than S_4, because they offset unmetamorphosed lamprophyre dikes; and they may be synchronous with the Triassic and post-Triassic faults in the Western Piedmont (cf. Cloos and Cooke, 1953).

In short, all of these stages of deformation and recrystallization appear to be parts of a single, complex, but essentially unified orogeny which occurred mainly in Paleozoic time.

AGE AND CORRELATION OF THE GLENARM SERIES

The age and the correlation of the Glenarm Series have long been controversial. At times, the weight of geologic opinion has favored a Precambrian age; at other times, an early Paleozoic age. Recent radiometric data show that the Glenarm must be younger than 1,100 m.y., the age of the basement gneiss; and it must be older than 550 m.y., the age of early orogenic intrusions cutting the Wissahickon Formation (Hopson, 1964, p. 203–204; Steiger, Hopson, and Fisher, manuscript in preparation). Within these limits, the Glenarm Series could be either (1) late Precambrian, a northeastern extension of the thick upper Precambrian turbidite sequence of the southern Appalachians (the Lynchburg Formation in Virginia, the Ocoee Series in the Carolinas, and the Great Smoky Group in Georgia); or (2) early Cambrian, a eugeosynclinal equivalent of the miogeosynclinal rocks in the Valley and Ridge Province (Chilhowee Group). A Cambrian age would require a rapid facies change just east of South Mountain. Both possibilities are consistent with the absence of Precambrian deformation in the Glenarm Series, because nowhere in the Appalachians is there evidence that the Lynchburg–Ocoee Series was deformed before deposition of the Chilhowee Group.

The stratigraphic relations between the Wissahickon Formation and the Ijamsville phyllite bear directly on this problem. The contact is poorly exposed, at least along the Potomac, and the succession is not yet firmly established; but both the orientation of graded beds near the contact, and the inferred stratigraphic relations between the lithofacies of the Wissahickon suggest that the Wissahickon is older than the Ijamsville. The detailed stratigraphic relations between the Ijamsville and the associated metavolcanics, marbles, and phyllites of the western Piedmont are debatable (see review in Hopson, 1964, p. 120–121); but nearly all students of these rocks agree that the thick quartzite interbedded with the phyllites at Sugarloaf Mountain can be correlated with the lower Cambrian Weverton Formation of the Chilhowee Group, and that the metavolcanics are either of similar age, or are equivalent to the upper Precambrian Catoctin Formation in South Mountain (e.g., Scotford, 1951; Stose and Stose, 1951). Therefore, if the Wissahickon underlies these rocks, it is probably older than the Chilhowee Group, and is most likely late Precambrian.

The metamorphosed andesitic tuffs of the northeastern Maryland Piedmont (James Run Gneiss, and Baltimore paragneiss of Hopson, 1964) also bear on this question. Tilton and others (Chapter 29 this volume) have shown that these rocks are of early or middle Cambrian age (550 m.y.). If they are a part of the Glenarm Series, these results would favor a Cambrian age for the series as a whole.

This conflicting evidence leaves the exact age of the Glenarm Series open to question. But it also suggests a very promising approach to the problem: determining more accurately the stratigraphic relations between the Wissahickon Formation, the phyllites and metavolcanic rocks of the western Piedmont, and the metamorphosed tuffs in northeastern Maryland. These rocks appear to hold the key to the age of the Glenarm, one of the major riddles of Piedmont geology.

REFERENCES

Bennison, A. P., and Milton, C., 1950, Preliminary geological map of the Fairfax, Virginia and part of the Seneca, Virginia–Maryland quadrangles, U. S. Geol. Survey Open File Map, scale 1:62,500.

Bouma, A. H., 1962, Sedimentology of some flysch deposits: Amsterdam, Elsevier, 168 p.

Cloos, E., and Anderson, J. L., 1950, The geology of Bear Island Potomac River, Maryland: The Johns Hopkins Univ. Studies in Geology, No. 16, Pt. 2, 13 p.

————, and Cooke, C. W., 1953, Geologic map of Montgomery County and the District of Columbia: Md. Geol. Survey, scale 1:62,500.

Darton, N. H., 1947, Sedimentary formations of Washington, D. C. and Vicinity, U. S. Geol. Survey, scale 1:31680.

Doe, B. R., Tilton, G. R., and Hopson, C. A., 1965, Lead isotopes in feldspars from selected granitic rocks associated with regional metamorphism: Jour. Geophysical Research, v. 70, p. 1947–1968.

Evans, B. W., 1965, Application of a reaction-rate method to the breakdown equilibria of muscovite and muscovite plus quartz: Am. Jour. Sci., v. 263, p. 647–667.

————, and Guidotti, C. V., 1966, The sillimanite-potash feldspar isograd in western Maine, U.S.A.: Contr. Mineral. and Petrol., v. 12, p. 25–62.

Fisher, G. W., 1963, The petrology and structure of the crystalline rocks along the Potomac River near Washington, D. C.: Unpub. Ph.D. dissertation, The Johns Hopkins Univ., Baltimore, 241 p.

Freedman, J., Wise, D. U., and R. D. Bentley, 1964, Pattern of folded folds in the Appalachian Piedmont along the Susquehanna River: Geol. Soc. America Bull., v. 75, p. 621–638.

Gorsline, D. S., and Emery, K. O., 1959, Turbidity-current deposits in San Pedro and Santa Monica Basins off southern California: Geol. Soc. America Bull., v. 70, p. 279–290.

Griscom, A., and Peterson, D. L., 1961, Aeromagnetic, aeroradioactivity, and gravity investigations of Piedmont rocks in the Rockville quadrangle, Maryland: U. S. Geol. Survey Prof. Paper 424-D, p. 267–271.

Hey, M. H., 1954, A new review of the chlorites: Mineralogical Mag., v. 30, p. 277–292.

Hopson, C. A., 1964, The crystalline rocks of Howard and Montgomery Counties, *in* The Geology of Howard and Montgomery Counties: Md. Geol. Survey, 1964, p. 27–215.

Jonas, A. I., and Knopf, E. B., 1921, Stratigraphy of the metamorphic rocks of southeastern Pennsylvania and Maryland: Wash. Acad. Sci. Jour., v. 11, p. 446–447.

Ksiazkiewicz, Marian, 1958, Submarine slumping in the Carpathian flysch: Pologne Soc. Geol. Annales, v. 28, p. 123–152.

Lapham, D. M., and Bassett, W. A., 1964, K–Ar dating of rocks and tectonic events in the Piedmont of southeastern Pennsylvania: Geol. Soc. America Bull., v. 75, p. 661–668.

Lombard, A., 1963, Laminites: A structure of flysch-type sediments: Jour. Sed. Petrol., v. 33, p. 14–22.

Luth, W. C., Jahns, R. H., and Tuttle, O. F., 1964, The granite system at pressures of 4 to 10 kilobars: Jour. Geophysical Research, v. 69, p. 759–773.

Newton, R. C., 1966, Kyanite–Andalusite equilibrium from 700° to 800°C: Science, v. 153, p. 170–172.

Reed, J. C. Jr., and Jolly, Janice, 1963, Crystalline rocks of the Potomac River Gorge near Washington, D. C.: U. S. Geol. Survey Prof. Paper 414-H, 16 p.

Scotford, D. M., 1951, Structure of the Sugarloaf Mountain area, Maryland as a key to Piedmont stratigraphy: Geol. Soc. America Bull., v. 62, p. 45–76.

Smith, J. R., 1960, Optical properties of low-temperature plagioclase, *in* Hess, H. H., Stillwater igneous complex, Montana; Geol. Soc. America Memoir 80, p. 191–220.

Southwick, D. L., and Fisher, G. W., 1967, Revision of stratigraphic nomenclature of the Glenarm Series, in Maryland: Md. Geol. Survey, Report of Investigations 6, 19 p.

Sriramadas, A., 1957, Diagrams for the correlation of unit cell edges and refractive indices with the chemical composition of garnets; Am. Mineralogist, v. 42, p. 294–298.

Stose, A. J., and Stose, G. W., 1946, Geology of Carroll and Frederick Counties, *in* The Physical Features of Carroll County and Frederick County: Md. Geol. Survey, 1946, p. 11–131.

————, 1951, Structure of the Sugarloaf Mountain area, Maryland, as a key to Piedmont stratigraphy: Geol. Soc. America Bull., v. 62, p. 697–699.

Swartz, F. M., 1948, Trenton and sub-Trenton of outcrop areas in New York, Pennsylvania, and Maryland: Am. Assoc. Petroleum Geologists Bull., v. 32, p. 1493–1595.

Thompson, J. B. Jr., 1957, The graphical analysis of mineral assemblages in pelitic schists: Am. Mineralogist, v. 42, p. 842–858.

Wetherill, G. W., Tilton, G. R., Davis, G. L., Hart, S. R., and Hopson, C. A., 1966, Age measurements in the Maryland Piedmont: Jour. Geophysical Research, v. 71, p. 2139–2155.

Velde, B., 1965, Phengite micas: Synthesis, stability and natural occurrence: Am. Jour. Sci., v. 263, p. 886–913.

Williams, E., 1960, Intrastratal flow and convolute folding: Geol. Mag., v. 97, p. 208–214.

Chapter 22

Multiple Deformation, Geosynclinal Transitions and the Martic Problem in Pennsylvania*

DONALD U. WISE

INTRODUCTION

THE MARTIC PROBLEM in Southeastern Pennsylvania is a combination of several problems centering around the nature of transition from the flysch-like, locally volcanic, highly metamorphic (eugeosynclinal?) core of the Appalachians to the bordering early Paleozoic (miogeosynclinal?) carbonates on the northwest. This line of transition for early Paleozoic time, now termed the Martic Line, is ordinarily drawn along the junction of known Cambro-Ordovician rocks (chiefly the Conestoga "Limestone") on the northwest (Fig. 1) with pelitic schists of the Glenarm Series, ranging from late Precambrian through early Paleozoic, on the southeast (Fig. 1). Like many regions of eugosynclinal–miogeosynclinal transition, the Martic zone involves very rapid facies changes, combined with structural telescoping, and followed by overprinting of several periods of folding. The Martic area, in addition to all these problems, lies squarely on the sharpest curve of the Appalachians, the so-called Susquehanna Orocline, resulting in differing strikes for the several deformations.

The Martic problem was originally defined by Knopf and Jonas (1929) in the type area of Martic Township in southern Lancaster County, Pennsylvania, and by Stose and Jonas (1935) as

* Part of an open file manuscript (February, 1968), Geological Survey of Pennsylvania. Published with permission of the Director.

a great northwestward thrust of "Precambrian" Glenarm Series over "Ordovician" Conestoga "Limestone." The supposed thrust was subsequently drawn for hundreds of miles along the Appalachians as the boundary of the inner and outer Piedmont. This view was challenged by Miller (1935) who interpreted the contact as an unconformity. In the classic study of the region, Cloos and Hietanen (1941) proved the existence of imbricate thrusting in the type area and provided structural data as a foundation for all subsequent work. The present paper is an attempt to synthesize details of this complex structural zone which have begun to emerge in the quarter century since Cloos' original publication.

SEGMENTS OF THE MARTIC LINE

The Martic Line of Pennsylvania is a complex polygenetic feature made up of distinct segments as named on Figure 2. The segments reflect differing degrees of influence of several major structural events: segments C,D,E and F were strongly influenced by early imbricate thrusting; all segments were subjected to the first period of major isoclinal flowage or deformation (D_1), the fold pattern of which controls the irregular shape of the Line in segments C and E; the last major deformation (D_2) involved uplift of the elongate basement block of Mine Ridge, sharply upturning the older structures along its edges to form seg-

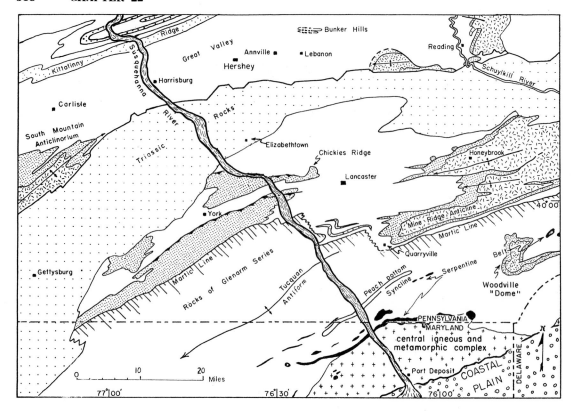

FIGURE 1. Index map to the Martic region, Pennsylvania and Maryland. From
Wise and Kaufmann (1960).

ments B and D. In the extreme east, the Line
merges with the Cream Valley–Huntingdon Valley fault zone on which McKinstry (1961) suggests the possibility of 15–20 mi (24–32 km) of
right lateral displacement. Segment G, the Magnetic Martic Line, is the location of a sharp break
in pattern on aeromagnetic maps of Bromery
et al. (1959) and is also the location of the Marburg Schist–Wissahickon Schist boundary west
of the Susquehanna River. In that area, the Magnetic Martic Line may be a better place to draw
the Martic contact than the traditional Martic
Line.

The point of this discussion of segmentation is
that *"The* Martic Line" does not exist as a single
structure but instead is a complicated array of
structures which must be understood separately
and sequentially in order to explain the overall
character of the Martic contact as a complex
polygenetic tectonic zone passing through the
Pennsylvania Piedmont, and as a fundamental
boundary within the early Paleozoic geosyncline.

EARLY SEDIMENTARY HINTS OF
THE MARTIC ZONE

Despite its diverse and segmented nature, the
Martic Zone location was clearly foreshadowed in

the Cambro-Ordovician sedimentary record. The
Conestoga "Limestone" (marble and calcareous
phyllite) of Ordovician and/or Cambrian age
rests on progressively older and older Cambrian
formations to the south as illustrated in Figure 3.
The contacts of Conestoga "Limestone" against
the underlying formations are marked by different symbols and connected into a map pattern
showing the formations on which the Conestoga
rests, band by band, progressively older to the
south. Parallelism with the future Martic Zone
is clear.

This relationship has been interpreted traditionally by Stose and Jonas (1923) as an unconformity representing uplift and erosion of Cambrian formations on the south followed by subsidence and deposition of the euxinic "Ordovician"
Conestoga "Limestone." Recently, Rodgers (in
press) has proposed that the relationship represents progressive northwestward advance of a
deep-water euxinic Conestoga carbonate bank
forming on the edge of a deep-water basin to
the south. Deeper or southward in the basin
would be the site of deposition of eugeosynclinal
sediments of the Glenarm Series continuing on
into the Paleozoic from late Precambrian times.

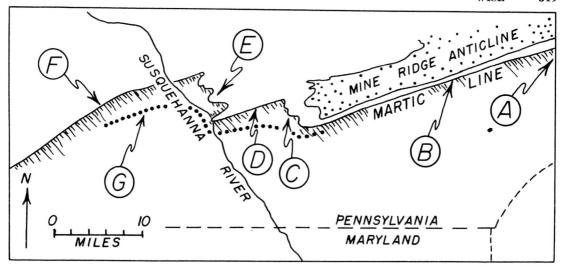

FIGURE 2. Segments of the Martic Line. (A) Cream Valley segment; (B) Coatesville segment; (C) New Providence segment; (D) Martic Forge segment; (E) Turkey Hill segment; (F) York County segment; (G) Magnetic Martic Line.

In either the Stose or Rodgers model, structural disruptions were present along the Martic Zone in Cambro-Ordovician time to help localize the next event in its history, that of imbricate thrusting to the northwest. In the Stose model the older lines of weakness associated with uplift of the unconformity helped localize thrusting along the zone. In the Rodgers model the thrust zone would reflect the thick sediments of the Glenarm Series riding northwestward up the basement slope and out across the lip of the carbonate covered shelf. In either model the sedimentary evidence indicates that the Martic Zone is not simply a chance concentration of late Paleozoic structures, but rather that the zone trend and general location had been defined by early Paleozoic time.

EARLY BEDDING PLANE THRUSTING

A complex series of imbricate thrusts is the first major disturbance which can be recognized clearly in the Martic Zone. This was established by Cloos' (Cloos and Hietanen, 1941) mapping of at least five repetitions of sheets containing the Cambro-Ordovician sequence from oldest to youngest: Antietam quartzitic schist, Vintage Dolomite, and Conestoga "Limestone" (Fig. 4). Remapping by students of Franklin and Marshall College in recent years has confirmed almost all details of Cloos' map picture of a series of thin sheets which were subsequently folded. Additional details of the thrusting can be deciphered viewing the map, Figure 4, down the plunge of the fold axes to approximate a cross-sectional view (Fig. 5). Several master thrusts appear in

the cross section, splaying into lesser faults at their distal ends. Details of small folds at the ends of these splays (Fig. 6) indicate that these folds are older than the main cleavage superimposed on the region and are most likely related to rolling of the thrust edges.

As traditionally mapped (Fig. 4) the Wissahickon Schist rests upon the uppermost of the thrust sheets with several patches of Conestoga "Limestone" in a band just south of the Martic Line. The cross section (Fig. 5) shows the difficulty of explaining this contact as the junction of the uppermost thrust sheet of Antietam Schist with the Wissahickon Schist. It seems more reasonable to consider the traditional Martic Line as one more thrust with a thin sheet of Conestoga "Limestone" on its back (Fig. 4), accounting for the "fensters" in the Martic "Thrust" interpretation of Knopf and Jonas (1929). The Magnetic Martic Line (Figs. 2 and 4) marked by a sharp break in magnetic pattern 1 mi (1.6 km) to the south, may be a better place to draw the Martic boundary between the nearly indistinguishable Antietam and Wissahickon Schists. West of the Susquehanna River, the Magnetic Martic Line merges with the traditionally mapped boundary of Wissahickon and Marburg Schists suggesting that the Marburg is an Antietam–Chickies equivalent. Weaver (1954) came to this same conclusion on lithologic and structural grounds.

Shifting the Martic contact southward does not change the basic Martic problem; it merely shifts the focus of structural interest to a new set of outcrops. It does strengthen the case for a facies change from the Antietam Formation of the mio-

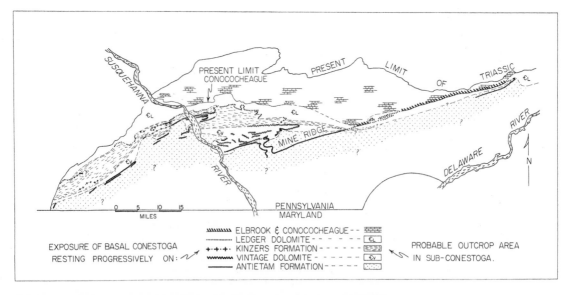

FIGURE 3. Sub-Conestoga Geologic Map. All contacts are taken from the Geologic Map of Pennsylvania (1960). The similarity in trends to the present Martic Line is foreshadowed by these lower Paleozoic overlap relationships.

geosyncline southward to the Glenarm lithologies of the eugeosyncline, also suggested by cross sections of the Maryland Piedmont drawn by Hopson (1964) and by the southeastward transition along the Susquehanna River from Chickies orthoquartzite to Harpers–Chickies "slate" (Stose and Stose, 1939). This facies change makes the old debate of thrusting or nonthrusting of the precise Martic contact in this one area become less important if only minor displacements were required to bring adjacent facies into juxtaposition.

The overall thrusting in the Martic Zone must have taken place before any severe folding of the area in order for large sheetlike thrust behavior to occur. The first major potential décollement plane above basement in the area east of the Susquehanna River is the Harpers Phyllite, commonly mapped with the Antietam Schist. The associated overlying Antietam quartzitic schist would thus be the logical unit to form the base of the observed thrust repetitions. West of the Susquehanna, Catoctin-type volcanics separate basement from the Cambrian quartzites, changing the potential décollement level and possibly accounting for the change in thrust pattern.

The direction of thrusting is unknown but is most likely to have been subperpendicular to the local edge of the basin as indicated by the trends of the sub-Conestoga contacts (Fig. 3). Using that model for thrust direction, the cumulative displacement on the early bedding plane thrusts need not have exceeded 10 mi (16 km).

MULTIPLE FOLD HISTORY

The overprint of two major systems of folds followed by minor or local folding, kinking, and jointing brought the imbricate thrusts of the Martic area into their present configuration. Each of the two major or regional fold systems had a distinctive orientation and style: the earlier (F_1) involved isoclinal folding, regional flowage and recrystallization; the second (F_2) disrupted these isoclinal folds by movement of relatively rigid basement blocks producing strain-slip cleavage and local folding in the overlying metasediments. The sequence of events is illustrated in Figure 7 and the terminology employed for the multiple fold systems is discussed in the caption to Figure 8, following the usage of Freedman, Wise, and Bentley (1964).

Isoclinal F_1 Fold System

The first clearly recognizable regional fold system superimposed on the Martic region is termed F_1 with its axial planes and schistosity as S_1. There was an earlier period of local folding associated with distal ends of imbricate thrusts ($F_{0 \times 0.5}$ axes on Fig. 10) as discussed previously for Figure 6. There may have been earlier structural/metamorphic events as suggested by thin section evidence of Cloos and Hietenen (1941) of early biotites being disorted and resorbed by micas of the main schistosity. To date no larger scale evidence of these early metamorphic events has been recognized.

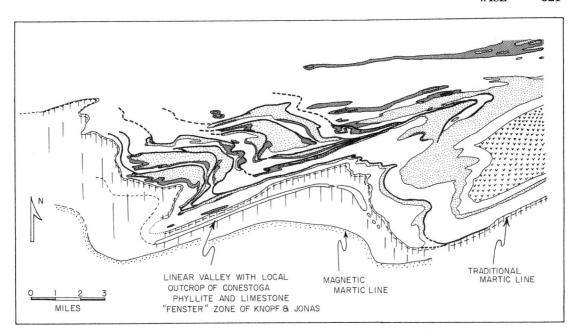

FIGURE 4. Geologic map of folded imbricate thrusts of the Martic Region. Geology modified from Cloos and Hietanen (1941). Symbols same as Figure 5.

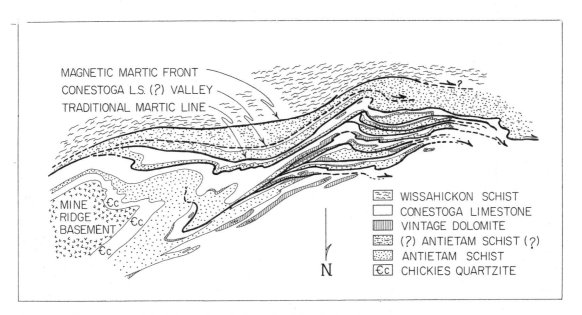

FIGURE 5. Cross section of the Martic thrust belt produced by viewing the region down the plunge of F_1 axes. Terminology is given in Figures 8 and 10. The figure is constructed by viewing Figure 4 along a line bearing S. 60° W. The pattern of several master thrusts splaying into lesser thrusts along their distal ends is clear. The apparent severe overturning of Mine Ridge basement is unreal, the effect being produced by F_2 folding still present in this projection. If this Figure 5 is now viewed at low angle from the lower left parallel with the F_2 axis of the Mine Ridge uplift, the effect of the second folding is removed and an approximation of a true geologic section appears.

The present configuration of the S_1 flow planes is illustrated in Figure 9 (top) on a southeast to northwest cross section along the Susquehanna River. The flow planes have steeply south-dipping roots in the lower Susquehanna region, flattening past recumbency over the Mine Ridge– Tucquan uplift and then passing back through recumbency in the Lancaster Valley. The axes of this fold system strike approximately N. 60 E. ($F_{0 \times 1}$ axes on Fig. 10) and the axial planes are parallel with the dominant schistosity of the region.

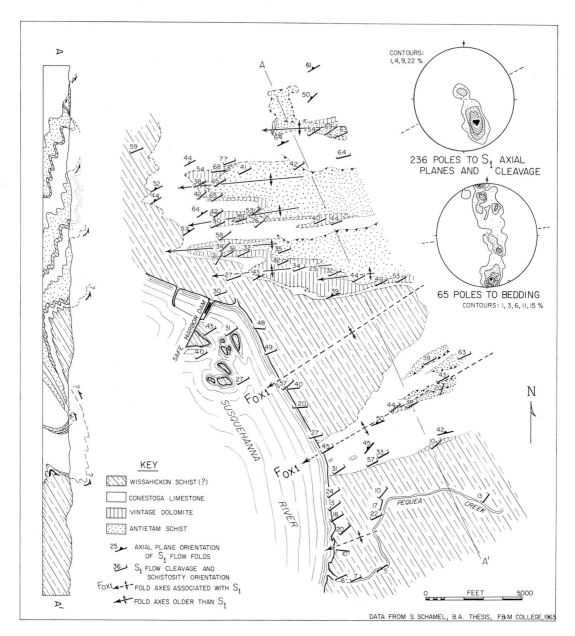

FIGURE 6. Superposition of S_1 planes and F_1 folds onto an older fold system, Safe Harbor, Pennsylvania. The older folds, apparently the distal ends of the thrust sheets, trend N. 80–85° E. as indicated by the map pattern and the π diagram of poles to least distorted beds. The S_1 planes have a uniform pattern of steepening dip across the area and a strike 20–30 degrees more northeasterly than the fold axes as indicated in the upper equal area plot and the dashed $F_{0 \times 1}$ fold axes. All data after S. Schamel, Franklin and Marshall College unpublished Bachelors thesis, 1963.

F_2 Fold System

Most outcrops in the region show evidence of a second set of S-surfaces (S_2) and associated folds (F_2) superimposed on the earlier structures. Most of these S-surfaces resemble strain-slip cleavage and remain relatively constant in orientation with a steep south dip (Fig. 9, center) but vary in degree of development from faint in some areas to being the most readily visible axial planes of the Lancaster Valley. The strikes of the F_2 system control the most obvious tectonic grain of the region and are subparallel with the axis of the Mine Ridge basement uplift.

In plunging westward beneath the metamorphics, the Mine Ridge basement mass bows S_1 schistosity into the Tucquan Antiform. The details of this transition down plunge (Figs. 9–11) of an arch in bedding ($F_{0 \times 2}$) changing into an arch in first cleavage ($F_{1 \times 2}$) illustrate the value of separating the two kinds of fold axes. In spite of being caused by the same uplift, the axis of the Tucquan Antiform does not join directly with the axis of the Mine Ridge Anticline but rather is offset $1\frac{1}{2}$ mi (2.4 km) to the south (Fig. 11). If bedding and first cleavage were not parallel, then when arched by a second folding, the crests of the $F_{0 \times 2}$ and $F_{1 \times 2}$ structures must be offset from each other (Fig. 12). To explain the direction of offset at Mine Ridge, the S_1 cleavage originally had to cut down section to the north, passing into basement at this location.

The pattern of dips of S_1 cleavage reflects the disruption of an older pattern by the F_2 Mine Ridge uplift and gives indications of a railroad tie-shaped piece of relatively rigid basement forming that uplift (Fig. 13). Contours of S_1 dip are irregular and shallow in the southeast in the Doe Run–Avondale area on the "railroad bed" adjacent to the uplifted "railroad tie." Northward the dips steepen sharply to 50 and 70 degrees along the junction of the Mine Ridge structure, being upturned sharply along the side of the "tie." Further evidence of the Mine Ridge D_2 disruption appears in the petrofabric orientation of isoclinal F_1 fold axes striking 30–40 degrees more northeasterly, diagonally into the south flank of Mine Ridge (Fig. 14A). In addition, the S_1 schistosity, as flow planes for the F_1 folds, is not in simple parallelism with Mine Ridge. Instead, orientation of the planes varies from location to location along the Martic Line with a general strike more northeasterly than Mine Ridge itself (Fig. 14B). On the nose of Mine Ridge (Fig. 13) the dip contours are closely bunched at either side of the Ridge as a result of sharp bending and draping over the edges of the "rail-road tie." The flat dips of S_1 on the nose reflect earlier S_1 structures simply uplifted on the top of the "tie." Westward along plunge, the "tie" is more deeply buried, the edge effects progressively subdued and dip contours spaced more evenly.

Uplift of the rigid Mine Ridge "railroad tie" in D_2 time distorted older surfaces including the regional garnet isograd, upwarped along the Tucquan Antiform. The Martic contact also was draped over the edges, top and sides of the "tie" to produce the several segments of Figure 2. Greatest complications to the Martic structures were produced by the D_2 tilting between the Mine Ridge nose and the Susquehanna River. Here the nearly recumbent S_1 flow planes originally cut stratigraphically deeper in passing northward through the imbricate thrust pile (Fig. 7B), a requirement similar to that needed to explain the offset of $F_{0 \times 2}$ and $F_{1 \times 2}$ axes at the Mine Ridge nose (Fig. 12). Consequently, advancing anticlinal noses of F_1 folds had the stratigraphy of synclines (Fig. 7C). With later D_2 tilting past recumbency along the north flank of the "railroad tie," the original antiforms with their synclinal stratigraphy became synforms with synclinal stratigraphy (Fig. 7D). Were it not for the continuous pattern of S_1 axial plane schistosity passing across the Tucquan Antiform into the Peach Bottom Syncline, these folds would appear as ordinary folds with an anomalous north dip of axial planes.

BASEMENT DEFORMATIONS

Basement deformation of the region involved much more than simple D_2 uplift of the Mine Ridge "railroad tie." Some D_2 thrusting occurs along the north flank of Mine Ridge and very complicated D_2 or younger faulting involving basement and Cambrian quartzites occurs in the Honeybrook Upland to the northeast. The possibility of major basement allochthany of the entire region cannot be discounted at present.

Mine Ridge itself suffered relatively minor basement folding during D_1 time as indicated by the open S_1 folds of the sub-Cambrian unconformity preserved on its plunging nose (segment C, Fig. 2, and $F_{0 \times 1}$ on Fig. 10). On the Mine Ridge nose, flow planes of these S_1 folds cut down section into basement in passing northward whereas eastward in the Doe Run–Avondale area, the same flow planes cut down section southward into the basement masses (Fig. 13) with reversal in original S_1 pattern occurring 5–10 mi (8–16 km) east of the plunging nose of Mine Ridge. These interrelationships of dip of S_1 in

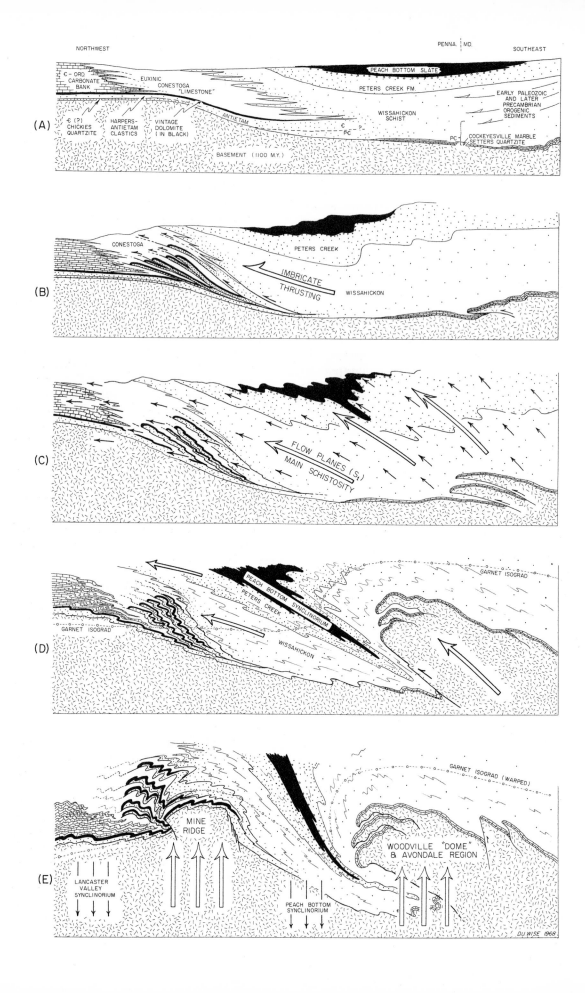

passing into basement are illustrated in Fig. 7B,C as the effects of a basement lip within a continuous flow pattern.

The case for basement thrusting in the Doe Run–Avondale Area based on plunging fold relationships was strongly made by Mackin (1962). The basement sheet of the Woodville Dome (Fig. 15) carries a northeast fold axis trend parallel with the $F_{0 \times 1}$ axes of the surrounding metasediments (Fig. 14A). Onto this structure is printed the present dominant D_2 tectonic grain of folds parallel with Mine Ridge complicated by the $F_{0 \times 2}$ folds plunging southwestward through the basement sheets. Thus, the Woodville Dome results from the reinforcement of an $F_{0 \times 1}$ anticline by an $F_{0 \times 2}$ anticline. As discussed by Anderson et al. (1964) the small folds at the northwest corner of the Woodville Dome (Fig. 15) are late structures which could have been produced by strike-slip displacements along the Cream Valley fault zone (Fig. 15) during late D_2 time or as a younger event.

These Doe Run–Avondale basement thrusts may have been initiated at the same time as the imbricate thrusting of sediments along the type Martic Area, but probably had their major basement movement during D_1 time as part of a more mobile basement deeper within the geosyncline. They are interpreted on Figure 7 as the complexly thrusted, rolled and greatly uplifted south flank of the Peach Bottom Synclinorium further modified by D_2 events.

A line of serpentine bodies occurs along the south flank of the Peach Bottom Synclinorium near the Avondale–Doe Run basement masses. Emplacement of the serpentines prior to the end of D_1 time is suggested by the presence of S_1 structures in one of them (Lapham and McKague, 1964) and by a 455–465 m.y. K–Ar age date from a contact zone of one of them (Lapham and Bassett, 1964). The relationships add some support to the interpretation of the Peach Bottom Synclinorium as marking a major tectonic boundary on the edge of intense basement flowage and deformation prior to and during D_1 time. Southwestward in Maryland, the same relationship holds with the Synclinorium marking the western edge of serpentine bodies and of basement uplifted by mantled gneiss domes. It may bound a zone of major crustal shortening and overriding.

YOUNGER STRUCTURAL EVENTS

A number of younger structural events have been recognized in the Martic Region by means of fabric studies but none seem to have had significant effect on map patterns. Intense local folds have been documented by Freedman et al. (1964) as a D_3 event in the lower Susquehanna region and in the area along the Susquehanna north of the Martic Line. Hanscom (1965) records a number of younger S-surfaces in the Tucquan Antiform west of the Susquehanna. Very late kink folding is common in the region and is transitional into many major joint sets (Newell and Wise, 1964). These kinks seem to be associated with a system of joints maintaining a constant symmetry plane orientation across the region (Wise and Grauch, 1966). The final event was the injection of basaltic dikes of probable Triassic age with north to northeast trend across the region.

FIGURE 7. Sequence of structural development in the Martic Region. Section extends approximately 25 mi from SE to NW. (A) Predeformation configuration of stratigraphic units. Deeper in the geosyncline to the SE, the Glenarm Series may be Precambrian; here it is shown interfingering with the Cambrian clastics and also providing a SE boundary for the euxinic Conestoga "Limestone" which progressively overlaps Cambrian shelf carbonates to the NW. The sedimentary geometry suggests NW migration of the shelf lip and slope with time. (B) Imbricate thrusting above the lip and slope of the basement shelf. The chief décollement plane is the Antietam–Harpers schistose quartzite, first lubricating horizon above basement. (C) Superposition of S_1 flow pattern on the imbricate thrust belt. The pattern cuts into basement at its roots but is forced to cut deeper in the section to the NW in passing through the imbricate thrusts and into the lip of the basement shelf. Deeper in the geosyncline some basement thrusting is taking place. (D) Intensification of the same S_1 flow pattern as (C) with uprolling of the basement thrusts on a large basement anticlinorium having the Peach Bottom Synclinorium ahead of it. Because of the flow pattern cutting down section through the imbricate thrusts, the forward advancing anticlinal noses will have the stratigraphy of synclines and vice versa. (E) F_2 basement disruption of the region by the Mine Ridge Uplift, draping the flowed imbricate thrusts across it. On the north flank, the anticlines with synclinal stratigraphy have been rotated past recumbency to appear locally as synclines with synclinal stratigraphy. Deeper in the geosyncline the Woodville Dome Region with its already folded thrusts is further folded along new F_2 axes. The garnet isograd (Figs. D and E) is further warped over the F_2 structures.

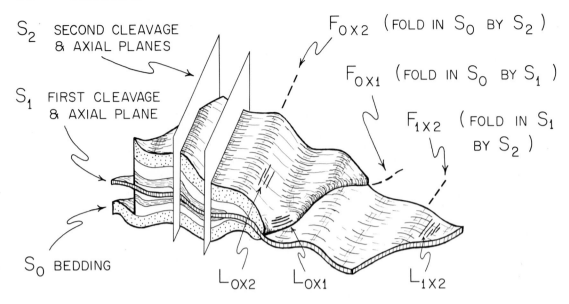

S_2 SECOND CLEAVAGE & AXIAL PLANES

S_1 FIRST CLEAVAGE & AXIAL PLANE

S_0 BEDDING

$F_{0 \times 2}$ (FOLD IN S_0 BY S_2)

$F_{0 \times 1}$ (FOLD IN S_0 BY S_1)

$F_{1 \times 2}$ (FOLD IN S_1 BY S_2)

$L_{0 \times 2}$ $L_{0 \times 1}$ $L_{1 \times 2}$

FIGURE 8. Multiple Fold Terminology. Bedding is designated S_0 with first cleavage and axial planes as S_1. These are refolded along a second set of cleavages and axial planes, S_2. The successive deformations creating S-surfaces are indicated by D_1, D_2, etc. The fold axes of the several systems are designated by an **F** with subscripts indicating the S-surfaces intersecting to form the folds. Note that none of the three axes indicated ($F_{0 \times 1}$, $F_{0 \times 2}$, $F_{1 \times 2}$) need be parallel to each other. Lineations are numbered in similar fashion.

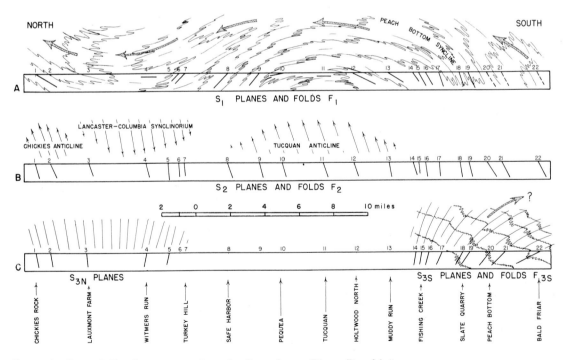

FIGURE 9. Dips of S-surface systems along the Susquehanna River. Petrofabric studies at 22 stations show the pattern of constant dip of S_2 surfaces (center section) across the region. The S_2 surfaces influence the isoclinal fold system of S_1 planes (top section) across the region and are in turn disrupted locally by younger deformations (bottom section). From Freedman, Wise, and Bentley (1964). Reprinted by Permission, Geological Society of America.

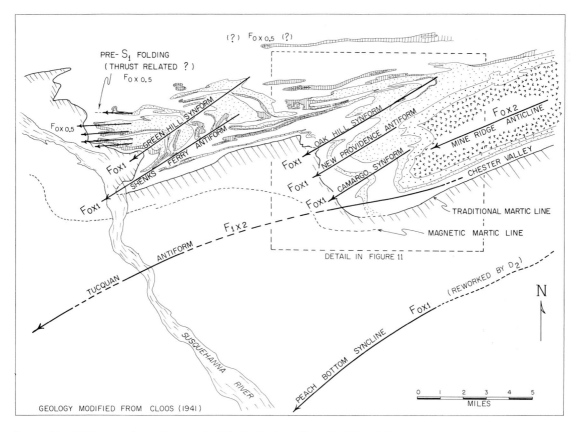

FIGURE 10. Fold axis relationships in the Martic Region. Note the Mine Ridge $F_{0 \times 2}$ anticline does not project directly into the $F_{1 \times 2}$ Tucquan Antiform.

MARTIC PROBLEM AND APPALACHIAN CURVATURE

An associated problem of the type Martic area involves the so-called Susquehanna Orocline, one of the sharpest bends of the Appalachian system. Fabric evidence in the area offers some control on hypotheses for the cause of the curvature. The earliest tectonic lines in the sub-Conestoga contacts reflect the present orocline trends (Figs. 3 and 16). The folds of the regional D_1 system are more northeasterly in trend, superimposed on the still older facies trends and apparently independent of the present curvature. The D_2 basement uplifts are again parallel to the older sub-Conestoga trends and reflect the present sharp curvature of the orocline. Thus, models of simple linear development followed by late-stage bending of the entire mountain system seem unlikely from the fabric evidence. Strike-slip displacements along large faults such as the Cream Valley zone are possible but no strong evidence for major distortions of the orocline by this method appear in the present fabric data. The fabric data instead suggest a relatively early

curvature of unknown cause with overprinting of the several deformation systems each with its distinctive strike, the D_2 being the most obvious on tectonic maps.

CONCLUSIONS

The Martic problem concerns the nature of transition of the inner to the outer Piedmont in the Central Appalachians. As posed in the 1930's the problem had two facets: the age of the Glenarm Series of the inner (eastern) Piedmont and the structural relationships of this Series with the Cambro-Ordovician limestones of the outer Piedmont to the west. Neither facet has been completely solved. The close parallelism of the Martic Line in Pennsylvania with facies overlap or unconformity boundaries beneath the Cambrian and/or Ordovician Conestoga "Limestone" suggests that the line may be largely a facies boundary marking the edge of a deeply subsided basin to the southeast. Along this basin edge the effects of several stages of deformation have been concentrated; (1) early thin-skinned imbricate thrusting of sediments; (2) regional metamorphism and flowage toward the northwest with probable major basement folding and thrusting deeper within the geosyncline; (3) brittle movement of large basement blocks near the line, folding of earlier basement thrusts deeper in the geosyncline, and overprinting of strain-slip cleavage and slip folds in the overlying metasediments; the final stages involved more brittle behavior, local intense folding (4) and widespread development of kink bands and joints (5).

The regional fold picture implies that the iso-

clinal flow planes of stage (2) in passing to the north cut down section through the imbricate sheets of stage (1), a pattern also dictated by offset of the Tucquan cleavage antiform from the Mine Ridge compositional anticline. The imbricate thrust system of stage (1) is most clearly discernible when the area is viewed looking S 60° W. along the plunge of stage (2) fold axes. Three major sheets, splaying into additional sheets near their distal northwest edges, involve the Antietam, Vintage, and Conestoga

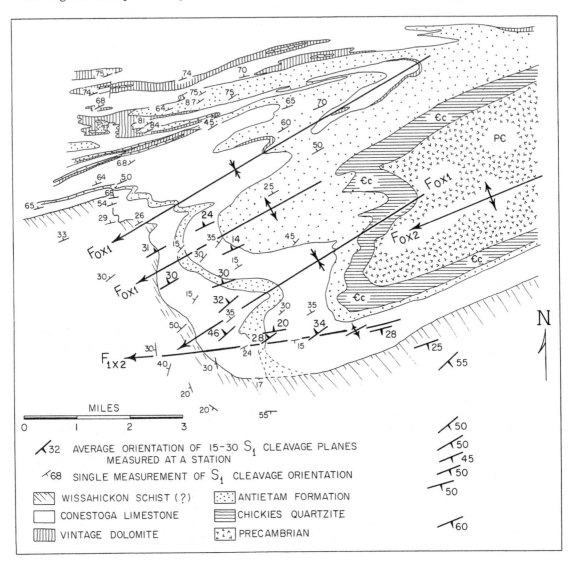

FIGURE 11. Detailed relationships of fold axes and S_1 planes on the plunging nose of Mine Ridge. Note the older $F_{0 \times 1}$ folds trend more northeasterly than Mine Ridge and that the S_1 cleavage has not folded them. On the other hand cleavage dips change sharply on either edge of the Mine Ridge basement uplift ($F_{0 \times 2}$). The Tucquan anticlinal axis in the cleavage ($F_{1 \times 2}$) projects into the south edge of the basement uplift, the two axes being offset by 1½ mi (2.4 km). Data compiled from Cloos and Hietanen (1941) and unpublished Franklin and Marshall College Theses of Forth and Halpin (1963), Saylor and Bowman (1963), and Rios (1966).

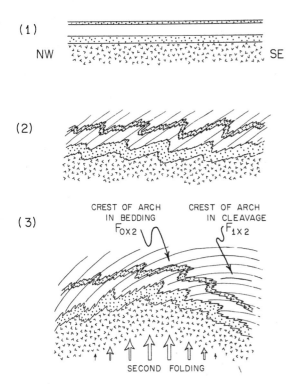

(1)

NW   SE

(2)

(3)   CREST OF ARCH   CREST OF ARCH
        IN BEDDING   IN CLEAVAGE
        F_{0x2} F_{1x2}

SECOND FOLDING

FIGURE 12. Offset of crest of a cleavage arch (F_{1x2}) from the crest of a bedding arch (F_{0x2}). The requirements are nonparallelism of bedding and first cleavage (center) followed by a second period of arching. Note that application of this geometry to Mine Ridge as shown in Figure 11, necessitates that the cleavage cut down section through bedding in passing northward.

formations as presently mapped. A higher sheet, traditionally regarded as part of the Wissahickon Schist of the Martic front, is interpreted as a fourth major thrust sheet of southerly facies of Antietam Schist, carrying Conestoga on its back. In this interpretation the real Martic contact, in all likelihood a thrust in this area, would be one ridge south of its traditional location, coinciding with a prominent aeromagnetic anomaly.

The sequence of structural events still has quite uncertain correlation with absolute age dates. Dates deep in the Maryland Piedmont suggest a late Precambrian age of the Glenarm Series with intense deformations and intrusions in earliest Paleozoic time (Hopson, 1964). As suggested in the present paper the uppermost and western-most of the Glenarm rocks interfinger with facies of the Cambrian clastics much as proposed by Hopson (1964) for Maryland.

The age of the Conestoga "Limestone" as Ordovician (Stose and Jonas, 1923) and possibly

Cambrian (Rodgers, in press) marks a maximum age for beginning the thrust and deformation sequence. The micas in the metamorphics along the Martic front yield Devonian K–Ar age dates (Lapham and Bassett, 1964) but at present it is uncertain which of the deformations is being dated by this method. The final stages of jointing and kinking may well have been related to Triassic block faulting to the northwest.

Among broader Appalachian problems, the structural sequence of the Martic region argues against a late-stage simple bending of the Pennsylvania Salient or Susquehanna Orocline. In addition the data do not place severe limits on the amount of crustal shortening or basement telescoping of the region: 5–10 mi (8–16 km) of telescoping along the Peach Bottom Synclinorium is a conservative estimate; major nappes to the northwest may involve basement sheets, raising the possibility that the entire Martic region is riding piggyback on much deeper major thrusts.

The intervening quarter century since Cloos' work on the Martic Problem has seen some advancements, largely as detailing of the structural sequence. However, many of the original problems still remain and Cloos' basic mapping and statement of them still stand as landmarks for future exploration.

ACKNOWLEDGMENTS

This work incorporates data and ideas from a number of field theses done under the author's direction at Franklin and Marshall College. These include: Senior Honors theses by S. Schamel, R. Hanscom; Senior field theses by D. Barry, J. Bowman, M. Dawson, M. Forth, R. Getz, M. Gibbons-Neff, D. Halpin, R. McEldowney; a M.S. thesis by J. Rios; and National Science Foundation supported Undergraduate Research Training studies by T. Anderson, D. Drake, W. Newell, R. Grauch. Some direct support of the author was provided by the Geological Survey of Pennsylvania and considerable indirect support by Franklin and Marshall College.

The debt to John Rodgers' idea of facies changes is acknowledged along with many helpful conversations with him, C. Hopson, J. Freedman, R. Bentley, D. Lapham, H. Mackin, and E. Watson.

Critical readings by K. Weaver, G. Fisher, J. Freedman, J. Rodgers and J. Warner improved the manuscript.

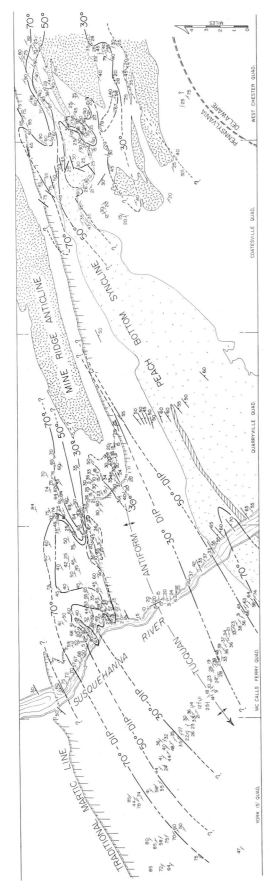

FIGURE 13. Dip of S_1 primary schistosity in southeastern Pennsylvania. Contours are on average dips of S_1; northward dip on the north side of Mine Ridge—Tucquan Axis, south dips in all other places.

A.

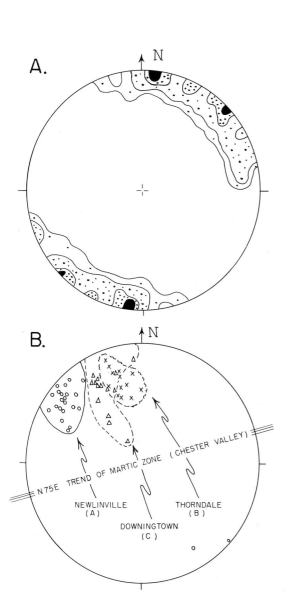

B.

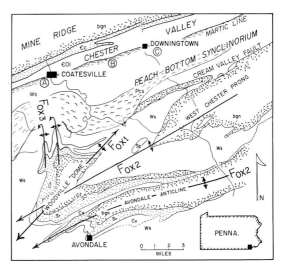

FIGURE 15. Multiple fold axes in basement sheets near the Woodville "Dome," Pennsylvania. The oldest $F_{0 \times 1}$ axis is parallel with the same axes of Figure 14A, just to the north, and involves folding of thrust sheets as suggested in Figure 7. This complex was refolded along $F_{0 \times 2}$ axes parallel with Mine Ridge and the Chester Valley–Martic Line. Younger $F_{0 \times 3}$ axes could be related to right lateral displacement on the Cream Valley Fault Zone. Symbols: Bgn, Baltimore Gneiss; Sr, Setters Quartzite; Cv, Cockeysville Marble; Ws, Wissahickon Schist; Pcs-Peters Creek Schist; €c, Cambrian clastics; €ol, Cambro-Ordovician(?) limestones. From Anderson, Drake, and Wise (1964).

FIGURE 14. Relationship of F_1 fold axes and S_1 planes to the trend of the Mine Ridge–Chester Valley Martic Line segment. (A) Equal area plot of 470 $F_{0 \times 1}$ isoclinal fold axes in the Peach Bottom Synclinorium just north of the Woodville Dome. Contours: 1,4,5,6 percent. Note the axes are much more NE trending than the Martic Line– Mine Ridge N75E trend. (B) Equal area plot of 53 poles of S_1 schistosity from 3 different areas on the Martic Line between Coatesville and Downingtown. Areas are indicated by letters on Figure 15. Note that each area has its characteristic strike which need not be parallel to the Martic Line. From Anderson, Drake, and Wise (1964).

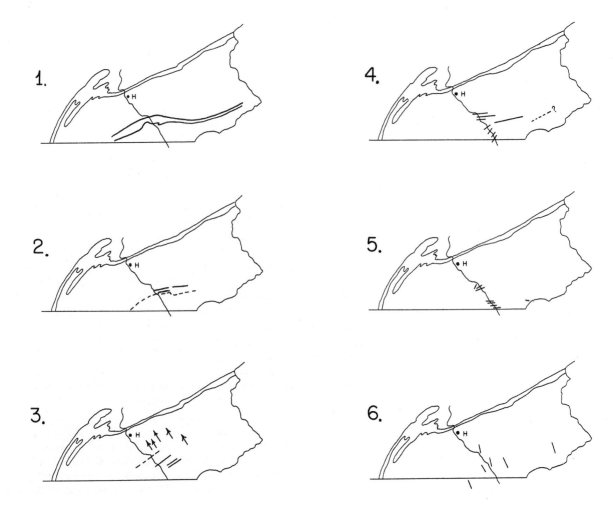

1.

2.

3.

4.

5.

6.

SEQUENCE OF TECTONIC TRENDS IN THE
LOWER SUSQUEHANNA REGION

FIGURE 16. Sequence of tectonic trends in the Pennsylvania Piedmont as related to Appalachian curvature: (1) Cambrian or Ordovician—lines of outcrop in the sub-Conestoga "unconformity." (2) Thrusting in the Martic imbricate zone with associated or slightly younger folds in the Safe Harbor Region. (3) Dominant flowage period creating regional schistosity (S_1) and F_1 fold axes. Flow pattern extends northward to become axial planes of nappe structures with smear lineations showing direction of tectonic transport. (4) Mine Ridge uplift and establishment of S_2 fracture cleavage and local tight folds. Lancaster Valley folding, present trend of Chickies Anticline established. Possible associated movements of basement in Doe Run Area and Cream Valley Fault. (5) Local centers of folding or fracture cleavage development in relatively brittle rocks. State Line folding zone is best example. (6) Late stage kink zones transitional into small faults and into master joint sets. Symmetry planes of joint plots indicated. Possibly associated with creation of Triassic Basins.

REFERENCES

Anderson, T. H., Drake, D. E., and Wise, D. U., 1964, Megapetrofabric of the Coatesville–Doe Run Area, Pennsylvania: Proc. Penna. Acad. Sci., v. 38, p. 174–182.

Bromery, R. W., Zandle, G. L., and others, 1959, Aeromagnetic Maps of the Quarryville and Conestoga Quadrangles, Lancaster County, Pennsylvania: U. S. Geol. Survey, Geophysical Investigations Maps GP-218, 219.

Cloos, E., and Hietanen, A., 1941, Geology of the "Martic Overthrust" and the Glenarm Series in Pennsylvania and Maryland: Geol. Soc. Am. Special Paper 35, 207 p.

Freedman, J., Wise, D., and Bentley, R., 1964, Pattern of Folded Folds in the Appalachian Piedmont along Susquehanna River: Geol. Soc. Am. Bull., v. 75, p. 621–638.

Hanscom, R., 1965, Structure of the Wissahickon Schist just west of the Susquehanna River, York County, Penna.: Proc. Penna. Acad. Sci., v. 39, p. 331–336.

Hopson, C. A., 1964, The crystalline rocks of Howard and Montgomery Counties, Maryland: in The Geology of Howard and Montgomery Counties, Maryland Geol. Survey, p. 27–215.

Knopf, E. B., and Jonas, A. I., 1929, Geology of the McCalls Ferry–Quarryville District, Pennsylvania: U. S. Geol. Survey Bull. 799, 156 p.

Lapham, D. M., and McKague, L., 1964, Structural Patterns associated with the serpentinites of southeastern Pennsylvania: Geol. Soc. Am. Bull., v. 75, p. 639–660.

————, and Bassett, W. A., 1964, K–Ar Dating of rocks and tectonic events in the Piedmont of southeastern Pennsylvania: Geol. Soc. Am. Bull., v. 75, p. 661–668.

Mackin, J. H., 1962, Structure of the Glenarm Series in Chester County, Pennsylvania: Geol. Soc. Am. Bull., v. 73, p. 403–409.

McKinstry, H., 1961, Structure of the Glenarm Series, Chester County, Pennsylvania: Geol. Soc. Am. Bull., v. 72, p. 557–578.

Miller, B. L., 1935, Age of the schists of the South Valley Hills: Geol. Soc. Am. Bull., v. 46, p. 715–756, p. 2021–2040.

Newell, W. L., and Wise, D. U., 1964, Independent joint system superimposed on metamorphic fabric of Glenarm Series near Coatesville, Penna.: Proc. Penna. Acad. Sci., v. 38, p. 150–153.

Rodgers, J., in press, The eastern edge of the North American continent during the Cambrian and early Ordovician: in Zen et al., Eds., Studies of Appalachian Geology, Northern and Maritime: Interscience, New York, p. 141–150.

Stose, G. W., and Jonas, A. I., 1923, Ordovician overlap in the Piedmont Province of Pennsylvania and Maryland: Geol. Soc. Am. Bull., v. 34, p. 507–524.

————, 1935, Highlands near Reading, Pennsylvania, an erosion remnant of a great overthrust sheet: Geol. Soc. Am. Bull., v. 46, p. 757–780.

————, and Stose, A., 1939, Geology and mineral resources of York County, Pennsylvania: Geol. Survey of Penna. Bull. C–67.

Weaver, K. N., 1954, The geology of the Hanover Area, York County, Pennsylvania: unpublished Ph.D. thesis, Johns Hopkins University.

Wise, D. U., 1966, Structural sequence near the type Martic Area of Pennsylvania (abstract): Program NE Section Geol. Soc. America Meetings, p. 47.

————, and Grauch, R. I., 1967, Regional joint pattern superimposed on metamorphic rocks of southeastern Pennsylvania: Proc. Penna. Acad. Sci., v. 15, p. 104–110.

————, and Kauffman, M. E., Editors, 1960, Some tectonic and structural problems of the Appalachian Piedmont along the Susquehanna River: Guidebook 25th Annual Field Conference of Pennsylvania Geologists, 99 pp.

————, and Werner, M. L., 1969, Tectonic transport domains and the Pennsylvania Elbow of the Appalachians (abstract): Program Am. Geophysical Union Meetings.

Chapter 23

Investigations of the Sedimentary Record in the Piedmont and Blue Ridge of Virginia*

WILLIAM RANDALL BROWN

INTRODUCTION

GEOLOGISTS HAVE LEARNED much about the landward, or miogeosynclinal, portions of geosynclines from which great fold mountains marginal to continents have grown; but numerous uncertainties remain regarding the nature of the seaward, or eugeosynclinal, portions. King (1959, p. 62), in discussing the Appalachian eugeosyncline, emphasizes some of these uncertainties when he says that the eugeosynclinal deposits might have been spread in part upon simatic ocean bottom, and asks "How do eugeosynclinal deposits connect with the very different yet contemporaneous miogeosynclinal deposits? Were eugeosyncline and miogeosyncline each synclinal in fact—that is, separate troughs?"

In 1959, Drake, Ewing, and Sutton, in a summary report based on about twenty years of extensive geophysical study along the continental margin of eastern North America by these men and numerous associates, pointed out the presence of two sedimentary troughs paralleling the coast. One trough is under the shelf, the other is under the continental slope and rise; the two are separated by a ridge in the basement near the edge of the shelf. Sediment in the shelf trough attains thicknesses as great as 17,000 ft (5,200 m) and is of shallow-water origin; that under the slope and rise is locally over 20,000 ft (6,100 m) thick and contains features suggestive of a deep-water

origin (Drake, Ewing, and Sutton, 1959, p. 179–183 and Fig. 29). Can it be, as these writers suggest, that the Appalachian geosyncline during its early stages of development had a configuration much like that of the present eastern margin of North America—that is, the miogeosyncline was essentially a shallow shelf trough, the eugeosyncline was a deep-water trough under the continental slope and rise, and the two were separated by a basement ridge which later became the present Blue Ridge–Green Mountains linear zone of basement uplift? This exciting suggestion has been considerably elaborated upon by Dietz (1963, 1963a) and Dietz and Holden (1966).

This hypothesis of Drake, Ewing, and Sutton can be tested, in part, in terms of the sedimentary record in the Appalachian geosyncline. It is well known that most sediments in the miogeosynclinal portion of this geosyncline are of relatively shallow-water origin, like those of the present shelf trough. Much less is known, however, concerning the environments of deposition of the eugeosynclinal sediments. Hurst (1955), Mellen (1956), Hurst and Schlee (1962), and Hadley and Goldsmith (1963), working with the Ocoee of the southern Appalachians, and Fisher (1963) and Hopson (1964), working with the Wissahickon of Maryland, have presented evidence to show that large parts of these great eugeosynclinal sedimentary masses were deposited in deep water by turbidity currents.

The purpose of this paper is threefold: (1) to point out and interpret some features of the eugeosynclinal metasediments of the Virginia

* The writer is grateful to the State Geologist of Virginia and to Shell Oil Company for permission to publish this paper and to the University of Kentucky Research Committee for partial financial support.

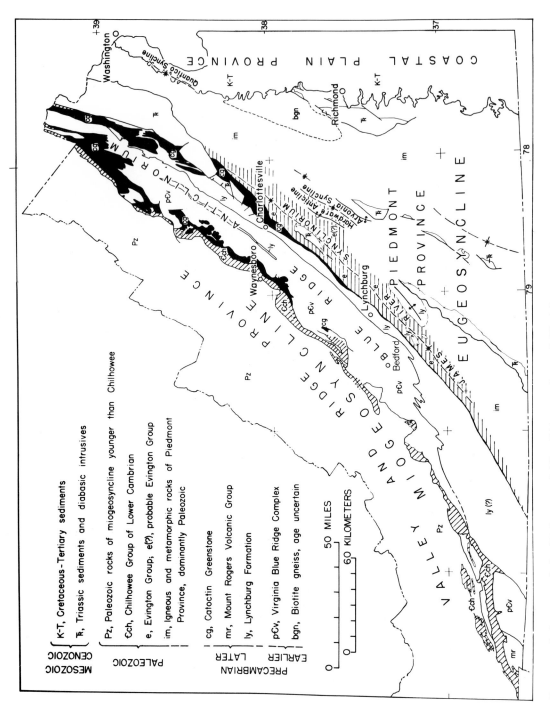

FIGURE 1. Index map of the Piedmont and Blue Ridge provinces of Virginia (after the Geologic Map of Virginia, 1963).

Piedmont, including features indicative of a deep-water origin for parts of the Lynchburg Formation and Evington Group; (2) to summarize the results of a paleocurrent study made in the Lower Cambrian Chilhowee Group of the Appalachian miogeosyncline; and (3) to present an attempt at reconstruction of stages in the early development of the Appalachian geosyncline at the latitude of central Virginia.

VIRGINIA BLUE RIDGE COMPLEX

The oldest known rocks exposed in Virginia are the earlier Precambrian basement rocks of the Blue Ridge region, known collectively as the Virginia Blue Ridge Complex (Table 1, and Brown, 1958, p. 7; Allen, 1963, p. 10). This complex comprises the core of the Blue Ridge anticlinorium, the great structure which dominates the Blue Ridge structural province from the vicinity of Bedford northeastward; it also comprises an, as yet, incompletely determined part of the rocks of the southwestern plateau portion of the Blue Ridge Province of Virginia (see papers by Espenshade (Chapter 14) and Rankin (Chapter 16), this volume). Rocks of comparable age possibly may be found in the eastern Piedmont between Richmond and Washington, D.C., if the basement doming which characterizes the Baltimore area of Maryland extends into Virginia

as interpreted by Jonas (1928) many years ago (Fig. 1).

In the eastern part of the Blue Ridge Province the Virginia Blue Ridge Complex is unconformable beneath the Lynchburg paragneiss of late Precambrian age; farther west it is overlain in successive onlap by the late Precambrian Swift Run Formation and Catoctin greenstone, and by the Lower Cambrian Unicoi Formation of the Chilhowee Group. The Complex is composed chiefly of more or less gneissic granitic rocks but it also includes some of the still older material of surficial origin, named Moneta Gneiss, which was intruded by and/or granitized into the granitic rocks. Radiometric determinations on the latter show these events to have taken place about one billion years ago (Marble, 1935, p. 351; Tilton et al. 1958, p. 1473).

LYNCHBURG FORMATION

In late Precambrian time seas advanced westward across the eroded surface of the Virginia Blue Ridge Complex, depositing, on the east, the great thickness of pelitic sediment and graywacke known as the Lynchburg Formation and, farther west, a greatly thinned edge of similar material which has been named Swift Run Formation (Stose and Stose, 1946, p. 18–19). Both the Lynchburg and Swift Run are overlain by

TABLE 1. Principal Stratigraphic Units in the Region of This Paper and Some Possible Approximate Correlations between Rocks of the Eugeosyncline and Miogeosyncline

		Valley and Ridge–Blue Ridge (Miogeosyncline)		Piedmont (Eugeosyncline)
Paleozoic	Silurian	Clinch Sandstone	?	Buffards Formation
	Ordovician	Juniata Formation Martinsburg Formation		Arvonia Formation
		Moccasin and Bays Formations		
	Cambrian	Cambro-Ordovician carbonates	Evington Group	Slippery Creek Greenstone Mount Athos Formation Archer Creek Formation
		Rome Formation Shady Dolomite		
		Chilhowee Group	Erwin Formation Hampton Formation Unicoi Formation	Candler Formation
Precambrian	Late Precambrian	Catoctin Greenstone Swift Run Formation		Catoctin Greenstone Lynchburg Formation
	Earlier Precambrian	Virginia Blue Ridge Complex		

Catoctin volcanics except locally in the west-central and southwestern Piedmont where the Lynchburg is overlain by the Candler Formation of the Evington Group (Table 1 and Fig. 1). The Lynchburg crops out along the southeast flank of the Blue Ridge anticlinorium from near the Maryland line southwestward beyond James River, and is probably continuous thence into North Carolina, a total distance of about 240 mi (385 km). Near the city of Lynchburg it has a thickness of from 10,000 ft (3,050 m) to 14,000 ft (4,260 m). Lynchburg Formation has been mapped as occupying most of the area of the southwestern plateau portion of the Blue Ridge province (Jonas, 1928; Stose and Stose, 1957; Dietrich, 1959), but more extensive and detailed mapping may show that appreciable parts of this area are underlain by older metasediments. The Swift Run crops out mainly on the northwestern flank of the anticlinorium in the northern part of the Blue Ridge Province where it is mostly 10–50 ft (3–15 m) thick but according to Bloomer and Werner (1955, p. 587) it reaches a maximum thickness of 400 ft (130 m).

The Lynchburg contains numerous bodies of amphibolite and hornblende gneiss, many of which are concordant with bedding but some of which are discordant. The concordant bodies may be mostly ophiolitic lavas and tuffs of the sort which characterize the initial stage of magmatism in many geosynclines. Discordant bodies, like most of those in the vicinity of Charlottesville, were possibly feeders to Catoctin fissure eruptions. Extensive concordant sheetlike, and shorter thick pod-shaped or irregularly shaped bodies of ultramafic rocks, especially chlorite–actinolite schist, serpentine, soapstone, and, more rarely, peridotite, are also common in the Lynchburg Formation, especially in its upper part. These are associated with the hornblendic rocks but appear generally to be distinct from them.

The basal few hundred to 1,000 ft (600 m) of the Lynchburg is generally massive, thick-bedded, and arkosic; locally it includes cobbles of granitic gneiss, obviously derived from the underlying basement. This basal conglomeratic material generally appears to grade into the granitic basement rock so that it is difficult to determine exactly where to place the contact. Bloomer and Werner (1955, p. 583–585) interpret this to mean that the conglomerate was deposited, not upon a clean erosion surface, but upon a deep granitic saprolite. The abundance of comparatively fresh, angular, obviously detrital plagioclase feldspar in the rock, however, gives one cause to doubt this interpretation. Perhaps the

situation is not so much a matter of gradation as it is of resemblance of overlying material to the basement rock from which it was derived nearby. On the basis of the lithology of the basal conglomeratic unit and its lack of current features it seems most likely that the unit was deposited rapidly in fairly deep water, possibly off a comparatively rugged, steeply inclined area undergoing rapid subsidence. The theory of deep-water deposition is greatly strengthened by plentiful deep-water features only a little higher in the section.

In most areas, the main body of the Lynchburg Formation consists predominantly of a monotonous sequence of alternating granulose metagraywacke beds and schistose pelitic beds. Many granulose beds are graded, and fragments or zones of slumped material are fairly common (Figs. 2 and 3; see also Allen, 1963, Figs. 14–17). The metagraywacke is mostly medium gray, "salt and peppery" feldspathic biotite–quartz gneiss—or, more accurately, granulite, because light and dark minerals are not segregated into distinct bands. Pelitic layers are mostly fine to medium grained biotite–muscovite schist. In some places the granulite is predominant, elsewhere the formation is largely schist. Graphitic schist typified by thin parallel laminations is common. The Fauquier Formation of the northern Piedmont of Virginia (Furcron, 1939, p. 37–38), a probable correlative of the Lynchburg, contains blue and white marble near its top where it dips beneath Catoctin greenstone.

The Swift Run Formation, which because of its stratigraphic position above the basement and below Catoctin greenstone is presumed to be the northwestern edge of the Lynchburg Formation, is described by Allen (1963, p. 38–39) as a heterogenous unit composed of coarse clastic material derived from erosion of the Virginia Blue Ridge Complex and the volcanic debris of the early stages of Catoctin vulcanism. It has not been studied by the present writer.

In the main body of the Lynchburg Formation the characteristic thin-bedded rhythmic alternation of granulose beds of graywacke composition with schistose beds of pelitic composition, the graded bedding in the graywacke, the structures suggestive of submarine slumping, and the prevailing lack of current-induced features are strongly suggestive of a flysch facies. Few characteristic sole markings have been noted (Fig. 3A); but well-exposed bedding surfaces are scarce, and bedding-plane slip accompanying deformation would tend to mar or destroy such markings.

FIGURE 2.   (A) Lynchburg Formation half a mile east of Schuyler, 20 mi (32 km) southwest of Charlottesville, Virginia. Granulose beds of graywacke, some of which show grading, alternate with beds of mica schist.   (B) Close-up of Lynchburg Formation near the above locality. Note sharp bottoms and gradational tops of graded beds at and below the hammer head and at the end of the hammer handle.

FIGURE 3.   (A) Irregular bottom markings (flame structure) at base of graded bed of metagraywacke in Lynchburg Formation, north side of James River opposite Lynchburg, Virginia. 6 in. (15 cm) rule gives scale.   (B) Lynchburg Formation containing clasts of slumped shaly and sandy material. Highway 6, 2 mi (3.2 km) north of Schuyler and 18 mi (29 km) southwest of Charlottesville, Virginia.

Hopson (1964, p. 70–117) and Fisher (1963, and Chapter 21 this volume) on the basis of similar features and compositional analysis have shown the Wissahickon Formation of Maryland likewise to be of the flysch facies. Using radiometric dating, Hopson (1964) and Steiger, Hopson, and Fisher (manuscript in preparation) also show the Glenarm Series, of which the Wissahickon is a part, almost certainly to be between 1 b.y. and early Cambrian in age. Furthermore, much of the Great Smoky Group of the Ocoee Series of southeastern Tennessee and northern Georgia, considered to be of late Precambrian age, shows similar features and appears to be of the flysch facies (King, 1949, p. 632–634; Hurst, 1955; Hurst and Schlee, 1962; and Mellen, 1956). As Hopson points out, it appears likely that in late Precambrian time an appreciable part of the sediment deposited in the eugeosynclinal areas of the southern Appalachians was deposited by turbidity currents in deep water.

CATOCTIN GREENSTONE VOLCANICS

Beginning in late Precambrian time and probably continuing into the lowermost Cambrian, lavas and tuffs of the great Catoctin period of vulcanism were spread widely in an area extending from southern Pennsylvania, across Maryland, and virtually across Virginia. From Maryland southwestward, Catoctin volcanics crop out in two main belts, one on each limb of the Blue Ridge anticlinorium. The northwestern belt, essentially along the crest of the Blue Ridge Mountains, extends nearly to James River; the southeastern belt along the western edge of the Piedmont is prominent in ridges and small mountains to just east of Lynchburg where the flows become much thinner (Fig. 1). Southwestward from Lynchburg almost to North Carolina, however, some greenstone is present almost everywhere at the normal stratigraphic position for the Catoctin.

In the Blue Ridge, the main mass of volcanics, which commonly includes some sedimentary members, occurs between the basement or the top of the Swift Run Formation, where present, and the bottom of the Lower Cambrian Chilhowee Group; individual flows occur well down within the Swift Run and well up into the Unicoi Formation, the lowest unit in the Chilhowee (Bloomer and Werner, 1955, p. 587). In the Piedmont the Catoctin is essentially at the top of the Lynchburg Formation but locally occurs either just below or just above the change from Lynchburg to Candler lithology; flows also occur

interspersed through thousands of feet of Candler Formation. Further east in the Piedmont, greenstones are extremely common and some are probably of the Catoctin period of vulcanism. Greenstones occur beneath the Ordovician slates in both the Arvonia and Quantico Synclines of Virginia.

In the Blue Ridge, Reed (1955, p. 893–894) interpreted the Catoctin to be plateau basalt, poured out subaerially, and eroded in places before deposition of the Chilhowee (King, 1949, p. 523). Bloomer and Werner (1955, p. 592) and Brown (1958, p. 24–25), on the other hand, working with the Catoctin farther south and east where much sediment is interspersed with the flows and the flows were more likely subaqueous, interpreted it to be spilitic.

CHILHOWEE GROUP

The Chilhowee Group of clastics constitutes the initial deposit in the miogeosynclinal portion of the southern Appalachians (King, 1949, p. 521). No attempt will be made to give a detailed description of the Chilhowee or of its stratigraphic relations; for these the reader is referred to Butts (1940), King (1949, 1950), King and Ferguson (1960), Bloomer and Werner (1955), Allen (1963), and others. The primary intent of the discussion here is to summarize and attempt to interpret some of the results of an environmental and paleocurrent study of the Chilhowee Group in central and southwestern Virginia and northeastern Tennessee, made in 1964 for Shell Oil Company. First, however, some general remarks are in order.

The Chilhowee lies between the Precambrian and the Lower Cambrian Shady or Tomstown dolomite and is considered to be the lowest unit in the Lower Cambrian (Table 1 and Fig. 1). It is extremely extensive and crops out chiefly along the northwest flank of the Blue Ridge uplift from Alabama into Pennsylvania; its probable somewhat younger extension is the "basal sand" encountered in drill holes in Kentucky and other states to the west (McGuire and Howell, Table 2-1, 1963). It ranges in thickness from about 2,500 ft (760 m) in northern Virginia to 7,500 ft (2,280 m) in northeast Tennessee (King and Ferguson, 1960, p. 33).

The Chilhowee is unconformable upon earlier Precambrian basement rocks in southwest-central Virginia, upon late Precambrian Catoctin greenstone in northern Virginia, upon Mount Rogers volcanics in southwest Virginia, and upon the Ocoee Series in Tennessee (King, 1949, p. 529–530, 632–634; 1950, p. 13–14; Stose and Stose, 1949, p. 314–315). Bloomer and Werner (1955,

p. 594–599) contend that in central Virginia no unconformity exists between Cambrian and late Precambrian clastic and volcanic rocks and that volcanics which occur in the lower Chilhowee "represent a final surge of Catoctin vulcanism." These volcanics in the lower Chilhowee are of particular significance because they offer possibilities of correlation, not only within the Chilhowee Group itself (King, 1949, p. 521–523), but also between the Chilhowee and certain eugeosynclinal rocks of the Piedmont to the east. This last possibility will be considered later.

In the area of the Shell study the Chilhowee is the product of one continuous sedimentary cycle. The sediments are predominantly of shallow-water marine and coastal origin, the latter including bars, beaches, barrier islands, and deltaic-fringe deposits. Many large orthoquartzite masses are coarse-grained at the top and become progressively finer grained downward, grading below into laminated and organically bored quartz wacke and siltstone (Fig. 4A).

Broadly speaking, the sediments from old to young become increasingly mature and reflect increasing stability of environment and a gradual lowering of the lands supplying sediment. There is a general tendency, particularly in the northern part of the area, for the sediments to show a predominance of quartz wacke, feldspathic sandstones, and arkosic conglomerates in their lower part; siltstones and shales in their medial part; and orthoquartzites, kaolinitic sandstones, and thinly laminated organically bored quartz wackes in their upper part. It is upon this basis that the subdivisions, Unicoi (oldest), Hampton, and Erwin (youngest) formations, generally used in this area and used in this account, have been separated (Table 1). Throughout much of the area and especially in northeastern Tennessee, however, it is impossible to tell one formation from another by lithology alone.

Current Directions

The area of paleocurrent study is the belt of Chilhowee outcrop, 3–15 mi (5–24 km) wide, extending from near Waynesboro, 22 mi (35 km) west of Charlottesville, Virginia, southwestward to mideastern Tennessee, a distance of nearly 300 mi (480 km) (Figs. 1 and 5–7). Crossbedding was the predominant current-direction indicator used, but ripple crests and rare groove casts and parting lineations were used locally. The most common and useful type of cross-bedding is medium-scale (sets 3 in. to 3 ft or 7.6 cm to 1 m high), planar or nearly planar, with dips of 15–25 degrees. Festoon cross-bedding is also extremely common. One of the few places that large-scale cross-bedding (sets over 3 ft or 1 m high) occurs is in Erwin (Hesse) quartzite at Backbone Rock (Station 115b) near the northeastern tip of Tennessee. Throughout the area, most bedding is steeply inclined or overturned and virtually all readings had to be rotated back

FIGURE 4. (A) Rippled surface of Erwin quartzite in quarry 4 mi (6.4 km) east of Natural Bridge Station and 22 mi (35 km) northwest of Lynchburg, Virginia. The surface of the quartzite is pocked with *Scolithus* borings. (B) Rhythmic graded beds in sandy calcareous Archer Creek Formation of the Evington Group 6 mi (9 km) east–northeast of Lynchburg, Virginia.

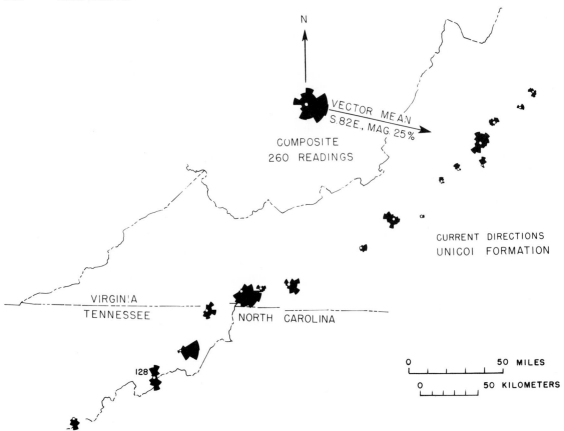

FIGURE 5. Paleocurrent directions in the Unicoi Formation of the Chilhowee Group at various localities in western Virginia and northeastern Tennessee.

to flat by means of stereonet. This procedure may involve some error as explained by Whitaker (1955).

In general, cross-bedding is appreciably more common and current trends are much stronger in the southwestern half of the area than in the northeastern. The seemingly more erratic current trends in areas of few cross-beds suggest that few readings may be inconclusive. In some places, the scarcity of readings was due to poor exposure or to the obscure nature of cross-bedding in some very clean sandstones.

Figures 5–7 show the plots of current roses for localities in the Unicoi and Erwin formations and for the entire Chilhowee Group, as well as composite roses and vector means for all readings in each of these units. Some roses include all readings from a single stratigraphic section; others include readings from a group of closely spaced partial sections. Current roses at individual localities give a much more complete picture of the current situation than would vector means, particularly where the results are strongly bidirectional. The composite roses for the Unicoi, Er-

win, and Chilhowee as a whole are quite similar, but vector means are S. 82° E., N. 50° E., and N. 68½° E., respectively. Without the abundant divergent northwesterly readings at Backbone Rock in northeastern Tennessee (Station 115, Fig. 6), the three vector means would be appreciably closer together.

Probably more striking than the resemblance between composite roses for all measurements in the Unicoi and Erwin is the resemblance between roses for the Unicoi and Erwin at specific localities (Figs. 5 and 6). In general, the changes in directions from Unicoi to Erwin sedimentation were less than the differences in directions from locality to locality within either sedimentation stage. This similarity lends support to the conclusion previously stated that the Chilhowee Group is the product of one sedimentary cycle.

The dominantly eastward direction of sediment transport indicated for the Chilhowee in the region of this study agrees well with that of the Lower Cambrian Weverton Formation in Maryland (Whitaker, 1955). It is in direct contrast, however, with the dominantly westward transport

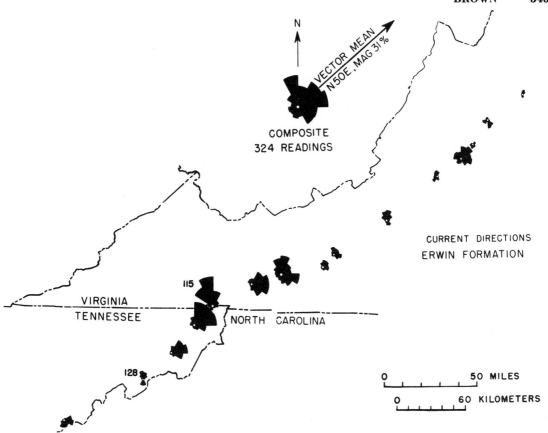

FIGURE 6. Paleocurrent directions in the Erwin Formation of the Chilhowee Group at various localities in western Virginia and northeastern Tennessee.

higher in the Paleozoic section (Meckel, Chapter 3 this volume).

Paleogeography

The meaning of these current directions in terms of paleogeography is not completely clear; partly because of the diverse nature of marine currents and partly, perhaps, because measurements are limited to such a narrow belt of outcrop along the regional strike. Considering the composite roses and vector means, the dominant east–southeast direction for the Unicoi may be an expression of a strong movement of sediment from relatively rugged lands to the west–northwest. The shift northeastward during Erwin deposition, along with cleaner, more mature sands, might be expected under conditions of lower borderlands, shallower water, and perhaps longshore drift to the northeast.

It may be noteworthy that the strongest and most consistent easterly to southeasterly current trends measured are in northeast Tennessee and extreme southwest Virginia where the Chilhowee is thickest. Most readings in this part of the area were from the Shady Valley thrust sheet; only Station 128 (Fig. 7), which shows both strong northward and southward current movement, is in the Cove Mountain window (King and Ferguson, 1960, Section 17 and Pl. 1 and 16). Rocks in this window, according to King's and Ferguson's (1960, p. 33–35) reconstruction, originally lay west of, and presumably were closer to shore than the rocks of the Shady Valley thrust block. The reversals of current direction, probably resulting from the back and forth movement of tides and waves near shore, appear to agree well with this reconstruction. There is no obvious explanation for the local strong north–northwestward movement of sediment at station 115 in the Erwin. The sands at this place are clean and show large round-crested ripples and large-scale (over 3 ft or 1 m high) cross-bedding. Possibly they were deposited as bars at a large tidal inlet.

In the northeast-central part of the area of study, westward to northward current directions are locally dominant. This is largely the area where late Precambrian rocks are absent beneath

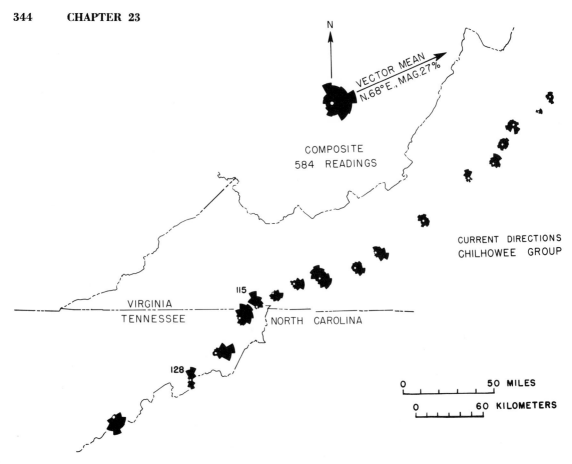

FIGURE 7. Composite of paleocurrent directions in the entire Chilhowee Group at various localities in western Virginia and northeastern Tennessee.

the Chilhowee, and the Lower Cambrian rests unconformably upon earlier Precambrian plutonics. Do these current directions indicate that there was a local land area within what is now the Blue Ridge Plateau during Lower Cambrian, and perhaps also in late Precambrian time? Only with further data can this question be answered.

EVINGTON GROUP

This group includes all rocks, except Catoctin Greenstone, which lie in the James River synclinorium of the central western Piedmont of Virginia and are younger than Lynchburg Formation and older than Triassic (Fig. 1; Espenshade, 1954, p. 14–23; Brown, 1953; 1958, p. 28–38). It has a maximum thickness of 10,000 ft (3,050 m) or more and includes, from old to young: Candler phyllite and schist; Archer Creek graphitic schist and gray marble (Fig. 4B); Mount Athos quartzite, schist, and white marble; and Slippery Creek greenstone (Table 1). Although this chronology appears certain north of Roanoke River, Redden (1963), working farther south has found evidence for the reverse sequence.

The Candler Formation overlies the main mass of Catoctin Greenstone and, in places, is interlayered with numerous lesser bodies of greenstone. The Unicoi Formation of the Chilhowee Group also overlies the Catoctin, although at least in part unconformably, and includes greenstones which King (1949, p. 522–523) considers widely separated in time from the Catoctin but which Bloomer and Werner (1955, p. 594) think "represent a final surge of Catoctin vulcanism." In either case, the relation of the Candler and Unicoi formations to the main mass of Catoctin and to later greenstones seems definitely to make them at least partly correlative. Also, if the above sequence is correct, the whole Evington Group must be at least as young as Cambrian, and it almost certainly is not correlative with the Glenarm Series, which is probably Precambrian, judging from available radiometric dates (Hopson, 1964, p. 203–207).

Without help from radiometric dating of the Slippery Creek greenstone or the discovery of fossils, it does not presently appear possible to make more than a guess at correlations of other formations in the Evington Group with rocks in the miogeosyncline to the west.

Rocks of the Evington Group appear to be dominantly deep-water marine. In those in the Lynchburg quadrangle, current bedding is extremely rare; whereas thin laminations, graded bedding, and, locally, rhythmic bedding are common (Fig. 4B). The formations change much more rapidly across the strike than along it, suggesting that the area of deposition was long and narrow (a trough?). Mount Athos quartzite is thickest and coarsest to the west, and, very locally (4 mi or 6½ km east of Lynchburg), it has a basal conglomerate containing well-rounded quartz pebbles as much as 1 in. (2.5 cm) long. The Mount Athos quartzite is commonly calcareous and feldspathic, but it is also frequently clean and vitreous. Could this be clean Chilhowee or younger miogeosynclinal sand washed beyond the shelf?

Just southeast of Charlottesville, very thick Catoctin greenstone dips steeply southeastward beneath Candler Formation of the James River synclinorium and seems to reappear (at least sizable masses of greenstone appear) on both flanks of the large overturned Hardware anticline 14 mi (22 km) further southeast (Fig. 1). Between the synclinorium and this anticline the Candler changes (or grades) from dense phyllite into sandy schist (feldspathic metagraywacke). This schist, which has been mapped as Evington(?) Group, continues with little change across the anticline and for several miles beyond to the Ordovician Arvonia slate in the Arvonia syncline. In vicinity of the slate it is locally finely conglomeratic, is interlayered with numerous greenstones and metadacites, and is intruded by batholithic Hatcher granodiorite (Smith, Milici, and Greenberg, 1964; Brown, 1965, 1966). If this Evington(?) schist, overlying and interlayered with greenstones, is indeed Candler Formation, or its equivalent, the formation coarsens eastward! If the greenstones on the flanks of the Hardware anticline are Catoctin, the Evington(?) schist in the core of this anticline is equivalent timewise to the Lynchburg Formation of the far western Piedmont which also underlies Catoctin but which coarsens to the west. These relations suggest movement of sediment from both the east and the west during Lynchburg sedimentation and a dominant, or at least more proximal, source area to the east during Candler

sedimentation. It is interesting to note that in Maryland about 120 mi (193 km) northeast of this region, Hopson (1964) has shown that the Sykesville Formation, which may be partly correlative with the Lynchburg Formation, was deposited by westward moving currents (see also Fisher, Chapter 21 this volume).

ROCKS IN THE ARVONIA SYNCLINE

The Arvonia syncline in the north-central Piedmont of Virginia is a most noteworthy feature because in it lies the fossiliferous Arvonia slate of late Ordovician age (Fig. 1). A similar feature, and perhaps a former continuation of the Arvonia trough, is the Quantico syncline, partly covered by Coastal Plain sediments, 14 mi (22 km) south–southwest of Washington, D.C. This also contains fossil-bearing slates, but their exposure is limited and they have been less studied than the Arvonia (Watson and Powell, 1911).

Arvonia slate overlies, probably with large unconformity, the Evington(?) schists and volcanics just mentioned, and lies with distinctive basal conglomerate upon Hatcher granodiorite (the "granite near Carys Brook" of Stose and Stose, 1948, p. 406), which intrudes these Evington(?) schists and volcanics. Apparently, Hatcher granodiorite was intruded into at least fairly deeply buried Evington(?) sediments of possible Lower Cambrian age, the region was elevated, and the granodiorite was exposed by erosion before deposition of the Arvonia slate in upper Ordovician time. Perhaps this intrusion and uplifting were related to the granitic magma series of Maryland, described by Hopson (1964, p. 201) as possibly having begun as early as 570 million years ago (Cambrian) but no later than about 490 million years ago (Early Ordovician).

Arvonia slate is mostly dark gray, lustrous, siliceous, and hard. It lacks the rhythmic bedding characteristic of parts of the Martinsburg slate of Pennsylvania (McBride, 1962, p. 44–45). The brachiopods, echinoderms, bryozoans, trilobites, and possible pelecypods which have been found are certainly much more characteristic of a shallow-shelf environment than of deep water. The fossil occurrences are spotty, and the possibility that the fossils were washed or slumped in must be considered. The presence of long unbroken tubes resembling crinoid stems argues against this possibility, however.

In the northeastern part of the Arvonia syncline, the Arvonia Formation includes or has at its top the Bremo Quartzite Member, which consists of about 1,000 ft (300 m) of from very

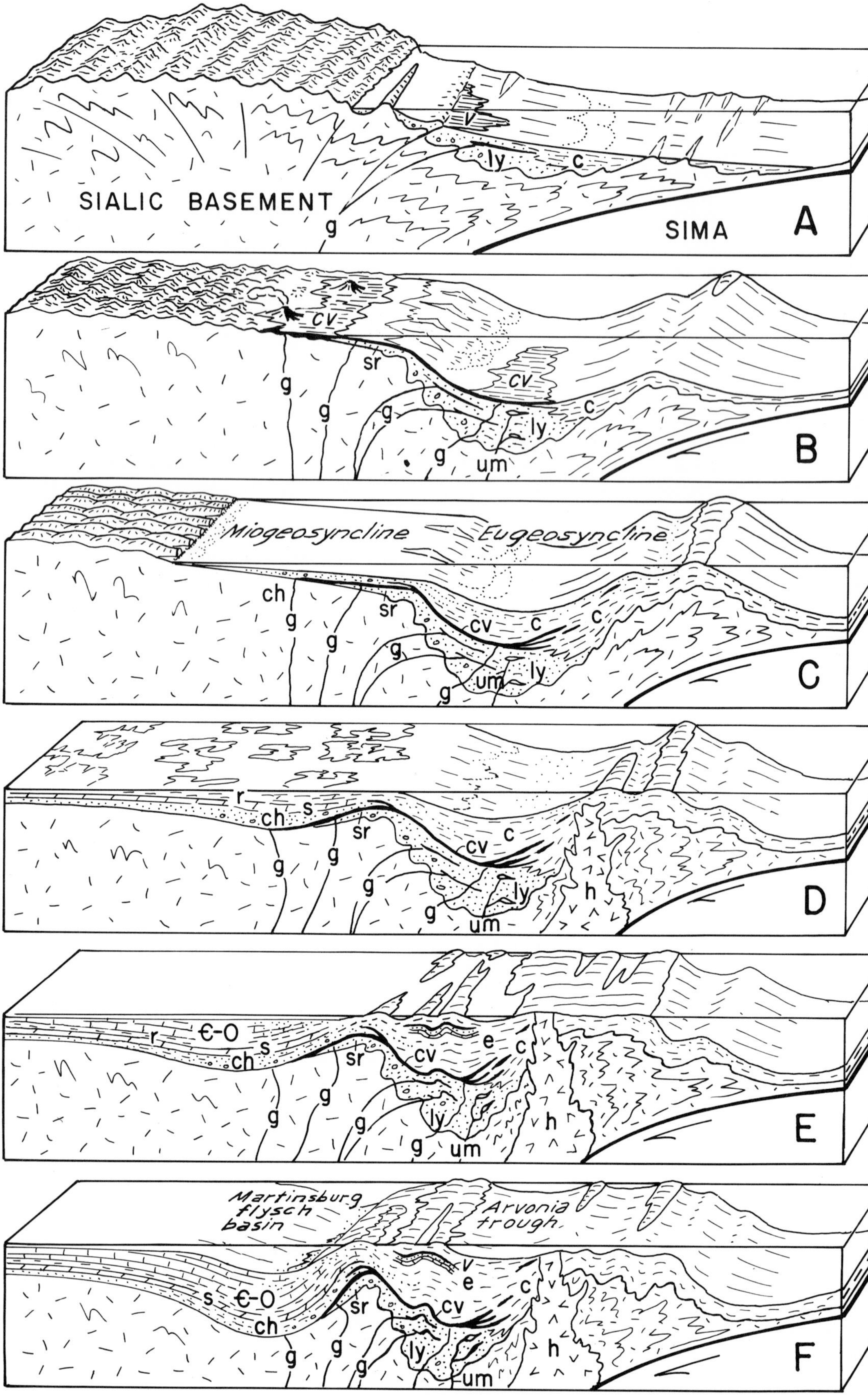
v
ly
c
SIALIC BASEMENT
g
SIMA
A
cv
sr
g
g
g
cv
c
ly
g
um
B
Miogeosyncline
Eugeosyncline
ch
g
sr
g
cv
c
c
g
um
ly
g
C
r
ch
s
sr
g
g
cv
c
g
ly
h
g
um
D
Є-O
r
s
ch
sr
cv
e
c
g
g
g
ly
h
g
um
E
Martinsburg flysch basin
Arvonia trough
v
e
c
s
Є-O
sr
cv
ch
g
g
ly
h
g
um
F

thick to thin bedded quartzite (metaquartz wacke) and interbedded gray slate and schist. The quartzite is locally conglomeratic and is sparsely cross-bedded. Current directions from the cross-beds are inconclusive, but the position of the quartzite mass and of conglomeratic zones suggests that it was introduced into the Arvonia basin from the northeast.

Conglomeratic schist, called Buffards Formation (Brown, 1965), in a mass perhaps 1,500 ft (460 m) thick overlies the Arvonia Formation, apparently unconformably, and extends along the medial part of the Arvonia syncline from near James River 17 mi (27 km) southwestward to the limit of mapping and beyond. The Buffards is somewhat conglomeratic nearly everywhere, but its coarsest part, by far, is to the north–northwest, where it contains flattened pebbles and cobbles of vein quartz as much as 8 in. (20 cm) long. In this portion, it is also coarsely pyroclastic. Because the Buffards overlies Arvonia slate, it must be as young as upper Ordovician, and it may well be lower Silurian—perhaps the product of rapid erosion promoted by Taconic uplift of some sort to the northwest.

SUMMARY OF CONCLUSIONS

The inferred development of the miogeosynclinal and eugeosynclinal portions of the southern Appalachians at the latitude of central Virginia from late Precambrian to Late Ordovician time is illustrated in Figure 8. It is assumed that the crumpling of the continental margin is due to sea-floor thrusting.

Block A. Near the start of late Precambrian time: Mountains completed about 1 billion years ago are being eroded, and coarse sediment (earliest Lynchburg Formation) is being deposited in fairly deep water relatively near the land. Further from land, shaly, silty Candler (c) facies is being deposited. It is inferred that these sediments were deposited upon thin sialic crust because sialic rocks over 1 b.y. old (i.e., Baltimore Gneiss and possibly gneiss of northeastern Virginia Piedmont) later become exposed in this region. There is some submarine volcanic activity (v), especially the extrusion of basaltic tuffs.

Block B. Near the end of Precambrian time: Lynchburg graywacke and pelitic sediment (ly) has accumulated to great thicknesses in deep water, partly at the base of and against the continental slope; and, later, partly in a deep trough separated from the open sea by a ridge which has begun to rise. Sediment moves into this trough from both the east and the west, transported in considerable part by turbidity currents. To the north there is submarine sliding of material westward into the trough. This slumped material becomes the Sykesville Formation of Maryland (Hopson, 1964). To the west, there is some subsidence and waves and currents have begun to cut a shelf into the continental margin. Swift Run sediments (sr) are deposited upon this shelf. At a late stage, Catoctin basaltic flows and pyroclastics (cv) are spread upon land to the west and upon the sea floor to the east; volcanics and sediments become interstratified. Feeder dikes and sills of gabbro (g) penetrate the underlying rocks, especially Lynchburg Formation. Numerous ultramafic bodies are also intruded into the Lynchburg at about this time.

Block C. Lower Cambrian time: The load of sediment in the eugeosyncline promotes subsidence here and to a lesser extent in the miogeosyncline. The shelf is widened westward. Chilhowee clastics are deposited upon the shelf, partly upon Catoctin greenstone and partly upon basement; some sand is washed beyond the edge of the shelf to be incorporated with other finer, less cleanly washed sediments of the Candler Formation of the Evington Group. Further east, sediment is brought in chiefly from the east.

Block D. Rome tidal-flat stage near the end of Lower Cambrian time: The shelf has been cut far westward, and greater subsidence near the edge of the shelf has produced a shallow trough. Filling has caught up with subsidence, however, and widespread tidal-flat conditions prevail in the miogeosyncline. Sediments of the Evington Group continue to accumulate in the eugeosynclinal trough. About this time or somewhat later,

---

FIGURE 8. Blocks illustrating diagrammatically the inferred development of the eugeosynclinal and miogeosynclinal portions of the southern Appalachians at the latitude of Virginia from late Precambrian to Late Ordovician time. (A) near start of late Precambrian time; (B) near the end of Precambrian time; (C) Lower Cambrian time; (D) Rome tidal-flat stage near end of Lower Cambrian time; (E) Middle Ordovician time; (F) Late Ordovician time. Symbols are: ly, Lynchburg Formation; c, Candler facies; v, submarine volcanics; cv, Catoctin greenstone; g, gabbro; um, ultramafics; sr, Swift Run Formation; e, Evington Group; ch, Chilhowee Group; sv, Slippery Creek volcanics; s, Shady Dolomite; r, Rome Formation; h, Hatcher granodiorite; Є-O, Cambro-Ordovician carbonates.

Hatcher granodiorite intrudes Lynchburg Formation and rocks of the Evington(?) Group of the central eugeosyncline.

Block E. Middle Ordovician time: Considerable carbonate has accumulated in the miogeosyncline. Some carbonate and clean sand (Archer Creek and Mount Athos Formations) have been deposited in the eugeosyncline, and there has been at least local renewal of submarine volcanic activity. Some compression and the uplifting of low lands has just occurred in the eugeosyncline; from these lands muds and fine sands are being washed westward into the miogeosyncline to form rocks like the Moccasin mudrock and Bays sandstone. Hatcher granodiorite is exposed by erosion.

Block F. Late Ordovician time: The miogeosynclinal trough has deepened and the flysch facies of the Martinsburg is accumulating; the chief source of sediment is to the southeast (McBride, 1962). A more shallow trough also has developed in the eugeosyncline; in this, siliceous muds, later to become Arvonia slate, are being deposited at depths below wave base but not at depths characterized by turbidity flow.

REFERENCES

Allen, Rhesa M., Jr., 1963, Geology and mineral resources of Green and Madison counties: Virginia Div. Min. Resources, Bull. 78, 102 p.

Bloomer, R. O., and Werner, H. J., 1955, Geology of the Blue Ridge region in central Virginia: Geol. Soc. America Bull., v. 66, p. 579–606.

Brown, W. R., 1953, Structural framework and mineral resources of the Virginia Piedmont: Kentucky Geol. Survey Spec. Pub. no. 1, p. 88–111; Virginia Geol. Survey Reprint no. 16.

————, 1958, Geology and mineral resources of the Lynchburg quadrangle, Virginia: Virginia Division of Mineral Resources, Bull. 74, 99 p.

————, 1965, Geologic map of the Dillwyn quadrangle, Virginia: Virginia Div. Min. Resources.

————, 1966, Road log and guide to the Geology of the Arvonia slate district, Virginia: 1966 Pick and Hammer Club field trip, Virginia Div. Min. Resources.

Butts, C., 1940, Geology of the Appalachian Valley in Virginia: Virginia Geol. Survey Bull. 52, 568 p.

Dietrich, R. V., 1959, Geology and mineral resources of Floyd County of the Blue Ridge Upland, southwestern Virginia: Bull. Virginia Polytechnic Inst., v. 52, no. 12, 160 p.

Dietz, R. S., 1963, Collapsing continental rises: an actualistic concept of geosynclines and mountain building: Jour. Geology, v. 71, p. 314–333.

————, 1963a, Alpine serpentines as oceanic rind fragments: Geol. Soc. America Bull., v. 74, p. 947–952.

————, and Holden, J. C., 1966, Deep sea deposits in but not on the continents: American Assoc. Petroleum Geologists Bull., v. 50, p. 351–362.

Drake, C. L., Ewing, M., and Sutton, G. H., 1959, Continental margins and geosynclines: The East Coast of North America north of Cape Hatteras: Physics and Chemistry of the Earth, v. 3, p. 110–198.

Espenshade, G. H., 1954, Geology and mineral deposits of the James River–Roanoke River manganese district, Virginia: U. S. Geol. Survey Bull. 1008, 155 p.

Fisher, G. W., 1963, The petrology and structure of the crystalline rocks along the Potomac River, near Washington, D. C.: Unpublished Ph.D. dissertation, Johns Hopkins University.

Furcron, A. S., 1939, Geology and mineral resources of the Warrenton quadrangle, Virginia: Virginia Geol. Survey, Bull. 54, 94 p.

Hadley, J. B., and Goldsmith, R., 1963, Geology of the eastern Great Smoky Mountains, Tennessee and North Carolina: U. S. Geol. Survey Prof. Paper 349-B, 118 p.

Hopson, C. A., 1964, The crystalline rocks of Howard and Montgomery Counties, in The geology of Howard and Montgomery counties: Maryland Geol. Survey, p. 27–215, 270–337.

Hurst, V. J., 1955, Stratigraphy, structure, and mineral resources of the Mineral Bluff quadrangle, Georgia: Georgia Dept. Mines, Mining and Geology, Bull. 63, 137 p.

————, and Schlee, J. S., 1962, Ocoee metasediments, north central Georgia–southeast Tennessee: Georgia Dept. Mines, Mining and Geol., Geol. Soc. Amer. Southeastern Sec., Field Excursion Guidebook No. 3, 28 p.

Jonas, A. I., 1928, Geologic map of Virigina: Virginia Geol. Survey.

King, P. B., 1949, The base of the Cambrian in the southern Appalachians: Am. Jour. Sci., v. 247, p. 513–530, 622–645.

————, 1950, Geology of the Elkton area, Virginia: U. S. Geol. Survey Prof. Paper 230, 82 p.

————, 1959, The evolution of North America: Princeton Univ. Press, Princeton, N. J., 190 p.

————, Ferguson, H. W., and Hamilton, W., 1960, Geology of northeasternmost Tennessee: U. S. Geol. Survey, Prof. Pap. 311, 136 p.

Marble, J. P., 1935, Age of allanite from Amherst County, Virginia: Am. Jour. Sci., 5th Ser., v. 30, p. 349–352.

McBride, E. F., 1962, Flysch and associated beds of the Martinsburg Formation (Ordovician), Central Appalachians: Jour. Sed. Petrology, v. 32, p. 39–91.

McGuire, W. H., and Howell, Paul, 1963, Oil and gas possibilities of the Cambrian and lower Ordovician in Kentucky: Spindletop Research for Dept. of Commerce, Commonwealth of Kentucky, 216 p.

Mellen, J., 1956, Pre-Cambrian sedimentation in the northeast part of Cohutta Mountain quadrangle, Georgia: Georgia Geol. Survey, Georgia Mineral Newsletter, v. 9, p. 46–61.

Redden, J. A., 1963, Stratigraphy and metamorphism of the Altavista area, p. 77–99, in Geological excursions in southwestern Virginia: Virginia Polytechnic Institute Engrg. Ext. Series, Geol. Soc. America Southeastern Sect. Guidebook II, 99 p.

Reed, J. C., Jr., 1955, Catoctin formation near Luray, Virginia: Geol. Soc. America Bull., vol. 66, no. 7, p. 871–896.

Smith, J. W., Milici, R. C., and Greenberg, S. S., 1964, Geology and mineral resources of Fluvanna County: Virginia Div. Mineral Resources, 62 p.

Stose, A. J., and Stose, G. W., 1957, Geology and mineral resources of the Gossan Lead District and adjacent areas in Virginia: Virginia Div. Min. Resources, Bull. 72, 279 p.

Stose, G. W., and Stose, A. J., 1948, Stratigraphy of the Arvonia slate, Virginia; Am. Jour. Sci., v. 246, p. 393–412.

————, 1949, Ocoee Series of the southern Appalachians: Geol. Soc. Amer. Bull., v. 60, p. 267–320.

Stose, A. J., and Stose, G. W., 1946, Geology of Carroll and Frederick counties, Maryland: Maryland Geol. Survey, p. 11–131.

Tilton, G. R., Wetherill, G. W., Davis, G. L., and Hopson, C. A., 1958, Ages of minerals from the Baltimore Gneiss near Baltimore, Maryland: Geol. Soc. America Bull., v. 69, p. 1469–1474.

Watson, T. L., and Powell, S. L., 1911, Fossil evidence of the age of the Virginia Piedmont slates: Am. Jour. Sci., 4th ser., v. 31, p. 33–44.

Whitaker, J. C., 1955, Direction of current flow in some Lower Cambrian clastics in Maryland: Geol. Soc. America Bull., v. 66, p. 763–766.

Chapter 24

The Carolina Slate Belt*

HAROLD W. SUNDELIUS

INTRODUCTION

THE SOUTHERN APPALACHIAN PIEDMONT, an area of metamorphic and plutonic rocks lying between the Blue Ridge crystalline rocks on the west and the Cretaceous and younger Atlantic Coastal Plain sedimentary rocks on the east, is one of the least known geologic provinces of North America. Deep weathering, a generally flat to rolling terrain with few outcrops, a dearth of fossils, and a very complex stratigraphic, structural, and metamorphic framework, make the geology of the region difficult to decipher. Many geologists, therefore, look to the lightly metamorphosed rocks of the Carolina slate belt (hereafter referred to as slate belt) in the eastern Piedmont as a possible key to interpretation of the more highly metamorphosed rocks of the Charlotte and Inner Piedmont belts to the west. The underlying assumption is, of course, that rocks in the slate belt can be divided into rock stratigraphic units that can be identified in the Charlotte, Kings Mountain, and Inner Piedmont belts. Recent geologic quadrangle mapping within the slate belt has indeed demonstrated that distinctive stratigraphic units can be recognized and traced along strike for miles (Conley, 1962a,b; Conley and Bain, 1965; Stromquist, Choquette, and Sundelius, 1966; Stromquist and Sundelius, 1969; Sundelius and Taylor, 1968; H. W. Sundelius, unpublished data). Because the areas thus far mapped in detail represent only a small part of the entire slate belt, it is perhaps premature to extrapolate these units very far within the slate belt, much less across strike into the higher grade Piedmont rocks.

* Publication authorized by the Director, U. S. Geological Survey.

Recent investigations of slate belt lithologies have also led to a clearer understanding of the ancient tectonic and sedimentary environments of this part of the Piedmont. For example, the modern assumption that rocks of the Piedmont represent eugeosynclinal accumulations within the Appalachian mobile belt seems to be confirmed by the character of the rocks within the slate belt. In addition, limited radiometric age data (White and others, 1963) and a trilobite fossil discovery (St. Jean, 1965), seem to support the hypothesis that much of the Piedmont is of early Paleozoic age.

Finally, coincident with recent investigations in the slate belt has been a renewal of mineral exploration in this region. The slate belt was the locus of perhaps the earliest mineral exploration and mining in the United States. Gold was discovered in Cabarrus County, North Carolina, in 1799, and systematic mining began in 1802 or 1803 (Stuckey, 1965, p. 295). In addition to gold, some copper, lead, zinc, silver, and tungsten have been produced from mines within the slate belt (Stuckey, 1965, p. 102). Current interest in this area is due in part to recognition of the similarity of the slate belt rocks and those associated with the ore deposits in New Brunswick and near Timmins, Ontario. Mineral deposits in the slate belt have been described by Pardee and Park (1948) and Parker (1963).

It is the purpose of this paper to review briefly the results of recent geologic investigations within the slate belt and to summarize what is known at present. As the writer is acquainted with the slate belt only in south-central North Carolina, any new data presented will be limited to that area. Other information used in this review is a synthesis of data from the recent literature. Particular emphasis is given to descriptions of the lithologies encountered in the slate belt. Stratigraphic data are presented and summarized where available, but because of the limited amount of detailed mapping in the slate belt and in the rest of the Piedmont, no regional correlations will be attempted.

ACKNOWLEDGMENTS

The author is particularly grateful to his colleagues of the U. S. Geological Survey, especially Arvid A. Stromquist, Henry Bell III, Jarvis B. Hadley, and John C. Reed, Jr., who introduced him to the frustrations and rewards which are a part of Piedmont geology. Because of the large area involved and the author's limited regional experience, it was necessary to depend to a considerable extent on the published reports of many individuals. For the contributions of all these geologists, the author is indeed appreciative. The careful and thoughtful reviews of Lynn Glover III and David H. McIntyre of the U. S. Geological Survey were extremely helpful.

LOCATION

Carolina slate belt is a geologic term not wholly accurate but honored by tradition. The rocks to which the term refers are neither confined to the Carolinas, nor distributed neatly in a belt. Indeed, the rocks are not even composed largely of slate. As presently used, the term slate belt refers to greenschist grade, generally fine-grained volcanic and sedimentary rocks of probable early Paleozoic age. The belt extends southwest for more than 400 mi (640 km) from near Petersburg, Virginia (lat. 37°12′ N. long. 77°30′ W.), to south of Milledgeville in central Georgia (lat. 33°05′ N. long. 83°15′ W.) (see tectonic map at the end of this volume). In Virginia these rocks are known as the Virgilina Volcanic Group (Jonas, 1932), and in Georgia, they are referred to as the Little River Series (Crickmay, 1952). On the west the slate-belt rocks are in contact with medium-grade metamorphic rocks of probable Paleozoic age. To the east, rocks of the slate belt extend under Cretaceous and Tertiary deposits of the Atlantic Coastal Plain.

Slate-belt lithologies are recognized in cores from deep wells at considerable distances east of the Coastal-Plain overlap. In North Carolina, wells apparently penetrated slate-belt rocks in Camden County about 8 mi (12.8 km) north of Elizabeth City (lat. 36°15′ N. long. 76°15′ W.); in Bladen County a few miles southeast of Kelly (lat. 34°28′ N. long. 78°19.6′ W.); in Pender

County 4 mi (6.4 km) south of Atkinson (lat. 34°31.7′ N. long. 78°10.3′ W.); near Merrimon (lat. 34°58.6′ N. long. 76°38.2′ W.) in Carteret County; and perhaps also in Onslow County in the southeast part of the North Carolina Coastal Plain (Stuckey, 1965, p. 90, 91) (see Fig. 1). In South Carolina, H. S. Johnson, Jr. (written communication, 1967), reports that slate-belt rocks were found at depths of 900–1,500 ft (300–500 m) in several holes in the Savannah River Plant area (Fig. 1). Siple (1958, p. 67) notes that rhyolite breccia was found in a well at Dillon, South Carolina (lat. 34°24′ N. long. 79°28′ W.) (Fig. 1). Milton and Hurst (1965), in a report on the subsurface rocks of Georgia, describe volcanic and sedimentary rocks from several wells drilled on the Coastal Plain. In the opinion of the writer, some of these rocks resemble closely those of the Carolina slate belt.

Rocks of the slate belt compose much of the eastern Piedmont and crop out in two regions. A large western region includes the Virgilina synclinorium of Virginia and North Carolina, the Great Slate Formation of Olmsted (1824) in North Carolina, and its extension into South Carolina and Georgia. A less extensive eastern area is between the southern tip of the Richmond Triassic basin in Virginia, and Moore County in south-central North Carolina, where it passes under Coastal-Plain deposits. The eastern and western parts of the slate belt are separated to the north by a region of medium-grade metamorphic rocks, largely gneiss and schist, and to the south by the Deep River Triassic basin. Within both the eastern and western subdivisions of the slate belt, these rocks are intruded by granitic to gabbroic sills, stocks, and batholiths of probable Paleozoic age and by northwest- to north-trending Triassic diabase dikes.

HISTORY OF GEOLOGIC INVESTIGATIONS IN THE CAROLINA SLATE BELT

Early Investigations

Geologic investigations have been conducted in the Carolina slate belt since the 1820's. The early workers, Olmsted (1824), Mitchell (1827), Emmons (1856), and Kerr (1875), established the general distribution of these fine-grained, low-grade slaty rocks, speculated about their age, and held the opinion that they were all sedimentary rocks. The presence of volcanic rocks was noted by Lieber in 1860 (Overstreet and Bell, 1965a, p. 18) and was well established by the turn of the century (Williams, 1894; Diller,

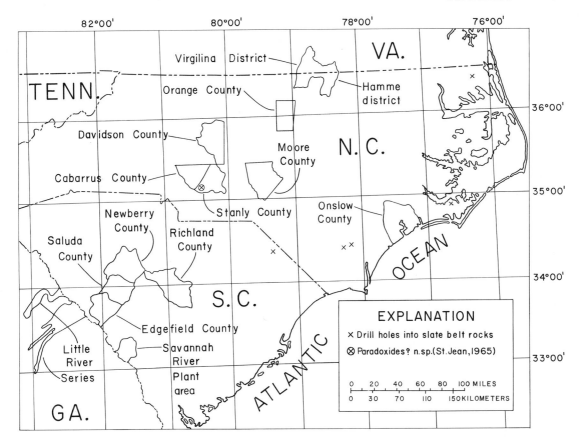

FIGURE 1. Generalized index map of part of Virginia, North Carolina, South Carolina, Tennessee, and Georgia.

1899). From 1900 to 1917, detailed studies were made of several of the mining districts by Weed and Watson (1906), Laney (1910, 1917), and Pogue (1910). Apart from Stuckey's study of the pyrophyllite deposits of Moore County, North Carolina (Fig. 1), in 1928, geologic investigations in the Carolina slate belt lagged until the 1940's when a cooperative program of water resources investigations was initiated by the State of North Carolina and the U. S. Geological Survey (Mundorff, 1948; LeGrand and Mundorff, 1952). More information about the early investigations are given by Stuckey (1965, p. 93, 94, 100, 101) and Overstreet and Bell (1965a, p. 18, 83–85).

Investigations Since 1955

About 1955, an upsurge of interest in the Carolina slate belt resulted in the initiation of detailed geologic quadrangle mapping in North Carolina by geologists of the North Carolina Division of Mineral Resources and the U. S. Geological Survey.

Conley's (1962a,b) maps of the Albemarle quadrangle, North Carolina (Fig. 2), and of Moore County, North Carolina (Fig. 1) are the first modern detailed general purpose geologic maps to be published dealing with areas within the Carolina slate belt. Conley's work led to the establishment of several mappable stratigraphic units which have been named and extended to the east and northeast in a report by Conley and Bain (1965). Mapping by Stromquist, Choquette, and Sundelius (1966) in the Denton quadrangle, by Sundelius and Taylor (1968) in the Gold Hill quadrangle, and by H. W. Sundelius (unpublished data) in the Mount Pleasant quadrangle, all in south-central North Carolina (Fig. 2), confirms in large part the stratigraphy established by Conley (1962a,b) and by Conley and Bain (1965). Recently, Stromquist and Sundelius (1969) have revised part of the stratigraphy of Conley and Bain (1965).

Members of the U. S. Geological Survey have participated in a cooperative reconnaissance mapping program with the State of North Carolina.

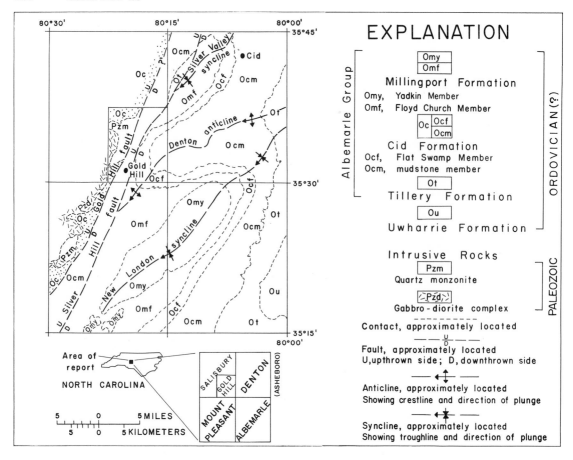

FIGURE 2. Generalized geologic map of the Albemarle, Denton, Gold Hill, and Mount Pleasant quadrangles in south-central North Carolina.

Some results have been published by Mundorff (1948), LeGrand and Mundorff (1952), Floyd, (1965), and Bain (1966); other reports are being prepared for publication. The geology and mineral deposits of the Hamme tungsten district of North Carolina and Virginia (Fig. 1) have been discussed by Parker (1963).

In South Carolina, Overstreet and Bell (1965b) have prepared a geologic map of the Piedmont and Blue Ridge provinces based on U. S. Department of Agriculture county soil survey maps and a limited amount of fieldwork. They also have written a report setting forth their interpretations of the geology of the crystalline rocks of South Carolina (Overstreet and Bell, 1965a, and Overstreet, Chapter 25 this volume).

Aeromagnetic and Bouguer gravity maps prepared by the U. S. Geological Survey (Henderson and Gilbert, 1966; Watkins and Yuval, 1966) cover a 1,000-sq mi (2,600 sq km) area of south-central North Carolina, including the Albemarle, Denton, Gold Hill, and Mount Pleasant quad-

rangles (Fig. 2), and are a valuable adjunct to geologic mapping.

AGE OF THE ROCKS OF THE CAROLINA SLATE BELT

The dearth of fossils in the southern Piedmont has caused considerable uncertainty as to the geologic age of rocks in the slate belt. Most of the early workers considered these rocks to be of Precambrian age (Kerr, 1875; Williams, 1894; Nitze and Hanna, 1896), and the Geologic map of the United States (Stose and Ljungstedt, 1932) so depicts them. More recently, most workers have considered them to be Paleozoic.

Direct evidence for the age of the slate-belt rocks is meager. In 1963, White and others reported Ordovician (440 to 470 ± 60 m.y.) lead-alpha ages for zircons collected from felsic crystal–lithic tuff of the Uwharrie Formation of Conley and Bain (1965). In 1965, St. Jean reported the first authentic fossils to be found within the slate belt. Fragments of the thorax and pygidium of two trilobites were found in stream rubble in Island Creek in Stanly County, North Carolina (lat. 35°13' N. long. 80°23.7' W.), about 2½ mi south of the south border of the Mount Pleasant quadrangle (Figs. 1 and 2). The loose pelitic rock in which the fossil fragments were embedded is described as being similar to the rock cropping out upstream and is probably derived from the Floyd Church Member of the Millingport Formation of Stromquist and Sundelius (1969). St. Jean believes these fragments represent a new species probably of the Middle Cambrian genus *Paradoxides*, and he suggests that they are not older than Early Cambrian, nor younger than Middle Cambrian. It should be noted that the rocks in which these fossils occur are stratigraphically above the Uwharrie Formation dated by White and others (1963) as being of Ordovician age. St. Jean notes that micropygous trilobites are common in eugeosynclinal belts and hence would be compatible with the slate-belt environment.

STRATIGRAPHY

Despite the general lack of detailed mapping within the slate belt, numerous regional correlations have been made, both along and across the strike of these rocks. Slate-belt rocks in North and South Carolina are considered correlative along strike with the Little River Series of Georgia to the south and with those of the Virgilina and Hamme districts to the north in North Carolina and Virginia (Fig. 1). In addition, the Virgilina rocks have long been correlated (Laney, 1917) with the fossiliferous Ordovician rocks of the Arvonia (centered on lat. 37°42' N. long. 78°20' W.) and Quantico (lat. 38°32' N. long. 77°22' W.) areas of Virginia (see tectonic map at the end of this volume). A general lithologic similarity has compelled numerous geologists (Emmons, 1856, p. 51; Keith and Sterrett, 1931; Kesler, 1936, p. 34; King, 1955, p. 343; Stromquist and Conley, 1959, p. 3; and Overstreet and Bell, 1965a, p. 19, 44, 46) to correlate slate-belt rocks across strike with those of the Kings Mountain belt to the west in North Carolina and South Carolina.

Generalized to detailed geologic mapping has been carried on in several separate areas within the slate belt. The area that has received the most detailed study is in south-central North Carolina and includes the old Gold Hill and Cid mining districts, the Albemarle, Denton, Gold Hill, and Mount Pleasant quadrangles (Fig. 2), and Moore County to the east (Fig. 1). In 1965,

Conley and Bain established a stratigraphic sequence based largely on Conley's work in the Albemarle quadrangle and in Moore County and extended the sequence through reconnaissance ground–water investigations by Bain (1966), Floyd (1965), and others to include most of the slate belt in North Carolina west of the Deep River Triassic(?) basin (see tectonic map at end of this volume). Conley and Bain (1965) recognized a lower sequence of volcanic and sedimentary rocks that is separated by an angular unconformity from an overlying volcanic sequence which they named the Tater Top Group. Stromquist and Sundelius (1969) confirm in general the stratigraphy of the lower sequence but do not recognize the postulated unconformity and believe that the volcanic rocks included in the Tater Top Group are actually interbedded with the formations of the Albemarle Group and Uwharrie Formation. A comparison of Conley and Bain's stratigraphy and the revisions suggested by Stromquist and Sundelius is shown in Figure 3. Stromquist and Sundelius subdivided the sequence of rocks primarily on the basis of the relative proportions of volcanic and sedimentary material. Thus, the Uwharrie Formation is a largely rhyolitic to rhyodacitic pyroclastic unit that is overlain by the distinctive finely laminated shales of the Tillery Formation. The Flat Swamp Member of the Cid Formation is a remarkably persistent volcanic-rich unit that separates the blocky mudstones of the lower part of the Cid Formation from the volcanic-rich shale, siltstone, and sandstone of the Millingport Formation. According to Conley and Bain (1965), the slate-belt rocks of North Carolina form a section that is at least 30,000 ft (9 km) thick.

In the Virgilina district along the Virginia and North Carolina State line, Laney (1917) subdivided the slate belt sequence into the following stratigraphic units arranged from oldest to youngest: Hyco Quartz Porphyry, Goshen Schist, Aaron Slate, and Virgilina Greenstone. Parker (1963, p. 21, 22) also recognized most of these lithologies in the Hamme district which lies just east of the Virgilina district (Fig. 1).

In 1965, Overstreet and Bell (1965a,b) published the first map of the crystalline rocks of South Carolina along with a bulletin in which they set forth their ideas concerning stratigraphic correlations within the Piedmont of South Carolina. As noted previously, their geologic map is based largely on the interpretation of bedrock geology from published county soil maps. Ac-

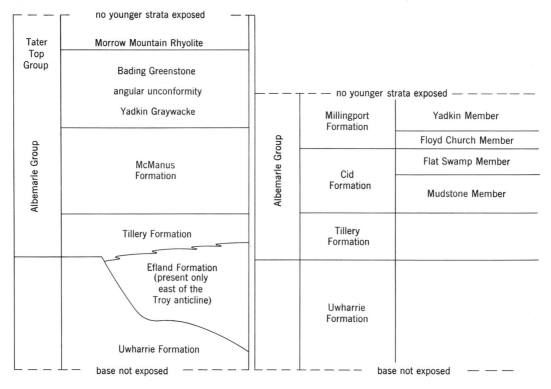

Conley and Bain (1965)   Stromquist and Sundelius (1969)

FIGURE 3.   Chart showing stratigraphic nomenclature of the lower Paleozoic slate-belt strata in south-central North Carolina. Note that Stromquist and Sundelius do not recognize the Tater Top Group of Conley and Bain. According to Stromquist and Sundelius, the volcanic units comprising the Tater Top Group are conformable units within the Uwharrie Formation and the Albemarle Group.

cordingly, they regard their map as provisional and the map units as lithologic rather than as time-stratigraphic units. In addition, they support the idea that the regional geologic belts within the Piedmont, i.e., Carolina slate, Charlotte, Kings Mountain, and Inner Piedmont, are metamorphic zones imposed on a regional stratigraphic sequence and that these metamorphic zones cut across the stratigraphic units in places. The metamorphism tends to decrease in intensity from the central part of the Inner Piedmont on the west to the slate belt on the east. For a summary of their conclusions, see the paper by Overstreet, Chapter 25 this volume.

Overstreet and Bell have had to base their stratigraphic interpretations largely on lithologic differences and structural divergences because of the lack of detailed geologic mapping, of fossils, and of any established stratigraphy. Accordingly, their interpretations should be regarded as a remarkable initial attempt, but one that will certainly be modified as more detailed investigations are conducted.

CAROLINA SLATE BELT LITHOLOGIES

Sedimentary Rocks

The Carolina slate belt includes rocks generally considered representative of the eugeosynclinal suite. In the slate belt these rocks are mainly lightly metamorphosed epiclastic (nonvolcanic) and volcaniclastic rocks including siltstone, claystone, mudstone, and wackes interbedded with rhyolitic, rhyodacitic, andesitic, and basaltic volcanic rocks, including tuff, lapillistone, pyroclastic breccia, and lava flows. Because of the intertonguing nature of these rocks, the proportions of volcaniclastic, epiclastic and flow rocks range both laterally and vertically. In addition, there is probably a considerable range in the proportion of primary or reworked pyroclastic material and of volcanic and epiclastic material within the various lithologic units. Subaqueous deposits, including both shallow- and deep-water varieties, and subaerial deposits appear to be present. Stratification ranging from fine, rhythmic laminations to thick graded beds, poor sorting, virtual absence of fossils, abrupt facies changes, and only minor amounts of quartzite and limestone are other eugeosynclinal attributes common to rocks of the slate belt. Bedded chert and pillow lavas apparently are lacking, however.

Mudstone, siltstone, and argillite. The dominant sedimentary rock in the slate belt is a generally well-bedded and commonly finely laminated metapelite that has been termed slate, volcanic slate, shale, mudstone, argillite, and siltstone. In many places this rock has been metamorphosed to phyllite and schist composed essentially of quartz, sericitic muscovite, chlorite and albite. The Aaron Slate of the Virgilina district (Laney, 1917), the Tillery Formation, the mudstone member of the Cid Formation, and the Floyd Church Member of the Millingport Formation (Conley and Bain, 1965; Stromquist and Sundelius, 1969) are representatives of this lithologic type in Virginia and North Carolina. Monroe Slates was the name formerly given to these rocks in south-central North Carolina (Nitze and Hanna, 1896), and argillite is the designation used for this rock type on the map of the crystalline rocks of South Carolina (Overstreet and Bell, 1965b). These rocks compose a part of the Little River Series of Georgia (Crickmay, 1952, p. 31–33).

These fine-grained rocks may be further divided into three major groups: (1) a medium to dark gray, blocky, massive mudstone with beds 6–24 in. (15–60 cm) thick; (2) an olive-gray to brown, obscurely bedded to well-bedded siltstone; and (3) a distinctive laminated argillite with individual laminae ranging from 0.04 to 0.3 in. (0.1–8 mm) in thickness. These laminations are regular, uniform, and persistent, and are graded, silt-sized particles at the base passing upward to clay-sized particles at the top. In appearance the laminations are similar to finely developed glacial varves, which, in fact, they have been called (Thiesmeyer, 1939).

All the pelitic rocks are composed essentially of quartz, feldspar, muscovite, and chlorite. Epidote, biotite, pyrite, sphene–leucoxene, and magnetite are also present in lesser amounts. Most of these minerals are considered to be secondary. Table 1A gives mean modes of argillites from Newberry County, South Carolina, and Table 1B

TABLE 1.   Mineralogy of Slate-Belt Argillite, Mudstone, and Siltstone

A.   Mean modes of six argillites, Newberry County, South Carolina (Fig.1) (McCauley, 1961a, p. 14)

	Volume percent
quartz	58.8
mica	34.7
epidote	0.3
plagioclase	1.2
opaques	4.7
	99.7

B.   Mean norms of argillites, mudstones, and siltstones from south-central North Carolina (Fig. 2). (Computed from unpublished chemical analyses by Paul Elmore, Samuel Botts, H. Smith, Gillison Chloe, and Lowell Artis, U. S. Geol. Survey)

	Mean of eight massive argillites and mudstones (from Mount Pleasant and Gold Hill quadrangles)	Mean of eight laminated argillites and siltstones (from Denton, Asheboro, and Salisbury quadrangles)
	Weight percent	Weight percent
quartz	29.4	35.7
albite	23.7	15.2
sericitic muscovite	27.7	22.0
Mg-chlorite	5.8	6.7
Fe-chlorite	7.4	3.5
epidote	2.7	2.9
ilmenite	1.8	1.6
magnetite	1.4	1.1
apatite	0.6	0.5
		Fe-chloritoid 10.3
	100.5	99.5

gives mean norms of argillites from south-central North Carolina. These rocks are commonly so fine grained that microscopic point counts are not feasible. For this reason, normative mineralogy was calculated using representative chemical analyses of these rocks and standard analyses of minerals known or suspected to occur in these rocks.

Column 1 of Table 2 represents the mean composition of eight chemical analyses of laminated argillite and siltstone from the Denton, Asheboro, and Salisbury quadrangles of North Carolina (Fig. 2), and column 2 represents the mean composition of 11 analyses of more massive and unlaminated argillite and mudstone from the Mount Pleasant and Gold Hill quadrangles, North Carolina (Fig. 2). Other analyses of slate-belt argillite, mudstone, and siltstone, that are, in general, similar to those presented in Table 2 may be found in Laney, 1910, p. 28; Pogue, 1910, p. 41; Stuckey, 1928; Conley, 1962b, p. 12; McCauley, 1961a, p. 7, 14, and 16; Butler, 1964, p. 104. A review of Table 2 shows that the mean compositions of slate-belt argillite, mudstone, and siltstone are, in general, chemically similar to the mean compositions of shale and slate cited. The slate-belt rocks do contain a higher proportion of ferrous to ferric iron and a lesser amount of combined water than does the average shale. In terms of these two constituents, these rocks are more comparable with the average slate, and the differences obviously reflect changes resulting from the low-grade metamorphism imposed on the slate-belt rocks. A more striking chemical difference, however, appears to be a somewhat lower K_2O/Na_2O ratio in the slate-belt argillite, mudstone, and siltstone as compared with the average shale and slate noted. The relatively large Na_2O content of the slate-belt metasedimentary rocks has been long recognized (Laney, 1910, p. 28; Pogue, 1910, p. 42; Stuckey, 1928; Crickmay, 1952, p. 33; Council, 1954; King, 1955, p. 344; and Conley, 1962a, p. 12) and has been the basis for the conclusion that these rocks are somewhat atypical. This difference can be seen readily in Figure 4 which compares K_2O/Na_2O and Al_2O_3/Na_2O ratios for mean slate-belt argillite, siltstone, and mudstone, and mean shale, slate, graywacke, dacite, and igneous rock. Both of these ratios emphasize major

TABLE 2. Mean Composition of Slate Belt Argillites (in weight percent)

	1†	2†	3	4	5	6	7	8
SiO_2	62.60	63.10	58.10	60.64	56.30	63.58	59.14	66.7
Al_2O_3	17.90	17.40	15.40	17.32	17.24	16.67	15.34	13.5
Fe_2O_3	1.10	1.50	4.02	2.25	3.83	2.24	3.08	1.6
FeO	5.80	4.80	2.45	3.66	5.09	3.00	3.80	3.5
MgO	2.40	2.20	2.44	2.60	2.54	2.12	3.49	2.1
CaO	1.00	0.99	3.11	1.54	1.00	5.35	5.08	2.5
Na_2O	1.80	2.50	1.30	1.19	1.23	3.98	3.84	2.9
K_2O	2.60	2.90	3.24	3.69	3.79	1.40	3.13	2.0
H_2O^-	0.25	0.18⎱	⎰5.00	0.62	0.38⎱	⎰0.56	⎰1.15	⎰0.6
H_2O^+	3.60	3.10⎰		3.51	3.31⎰			⎱2.4
TiO_2	0.84	0.93	0.65	0.73	0.77	0.64	1.05	0.6
P_2O_5	0.18	0.23	0.17	—	0.14	0.17	0.30	0.2
MnO	0.15	0.11	—	—	0.10	0.11	0.12	0.1
CO_2	<0.05	<0.05	2.63	1.47	0.84	—	—	1.2
Others*	—	—	1.44	0.38	3.44	—	—	0.5
	100.22	99.94	99.95	99.60	100.00	99.82	99.52	100.4

Column headings:

1. Mean composition of 8 analyses of laminated argillite and siltstone from Denton, Asheboro, and Salisbury quadrangles, N.C. (Fig. 2).
2. Mean composition of 11 analyses of massive argillite and mudstone from Mount Pleasant and Gold Hill quadrangles, N.C. (Fig. 2).
3. Mean composition of shale (Clarke, 1924, p. 24).
4. Mean composition of 36 analyses of slate (Eckel, 1904).
5. Mean composition of 33 analyses of Precambrian slate (Nanz, 1953).
6. Mean composition of dacite and dacite–obsidian (Nockolds, 1954, p. 1015).
7. Mean composition of igneous rock (Rankama and Sahama, 1955, p. 159).
8. Mean composition of graywacke (Pettijohn, 1963, p. 15).

* Includes SO_3, S, BaO, FeS_2, C.

† Unpublished chemical analyses by Paul Elmore, Samuel Botts, H. Smith, Gillison Chloe, Lowell Artis, U. S. Geol. Survey.

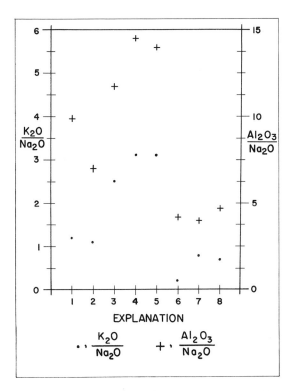

FIGURE 4. Comparison of K_2O/Na_2O and Al_2O_3/Na_2O ratios of slate-belt argillites and those of mean shale, slate, Precambrian slate, graywacke, dacite, and igneous rock. Numbers 1-8 refer to Column headings in Table 2.

changes that occur in chemical weathering and in the depositional environment. Because Al_2O_3 is the most inert and Na_2O the most mobile oxide, Pettijohn (1957, p. 103) considers the Al_2O_3/Na_2O ratio to be a good index of the maturity of pelitic rocks. Note that with respect to both of these ratios, the slate-belt argillites, siltstones, and mudstones lie somewhat between mean dacite, igneous rock, and graywacke, on the one hand, and the chemically more differentiated mean shale and slate on the other. It would appear then that these slate-belt rocks contain significant quantities of relatively unweathered pyroclastic material, probably rhyodacitic ash.

Volcanic wacke and siltstone. Volcanic wacke and siltstone, sandy siltstone, and fine sandstone are representative of perhaps the second most abundant clastic lithology in the slate belt. This lithology is recognized in Virginia, North Carolina, and South Carolina. In south-central North Carolina, Conley and Bain (1965) have called a lithology of this type the Yadkin Graywacke (the Yadkin Member of the Milling-port Formation of Stromquist and Sundelius, 1969).

These rocks are dark grayish green, generally massive, thick-bedded, and, in places, contain graded beds ranging from fine sand-sized clasts at the base to silt-sized particles at the top. Bifurcations, small-scale crossbedding, and slump structures occur locally. Sorting is poor, and in thin section the rock is characterized by a disrupted framework of quartz, plagioclase, and silt to sand-sized lithic fragments in a fine-grained matrix of sericitic muscovite, chlorite, quartz, and plagioclase. On the average, the framework composes about 40 percent of the rock by volume. Rock fragments consist of volcanic clasts with plagioclase microlites, pelite fragments, and polycrystalline quartz clasts. Secondary epidote and clinozoisite, magnetite, ilmenite, and apatite occur in accessory quantities. Secondary pyrite cubes as much as 2 in. (5 cm) on a side are characteristic of this lithology. Table 3 presents both a mean mode and a mean norm based on five thin sections and five chemical analyses from the Yadkin Member of the Millingport Formation, Mount Pleasant quadrangle, North Carolina (Fig. 2). Table 3 also shows the mean chemical composition of these rocks based on five chemical analyses from the same unit. This mean composition is comparable with that of the laminated and unlaminated argillite reported in Table 2. The somewhat higher SiO_2 and lower Al_2O_3 content in the volcanic wackes and siltstones may be attributed to their larger average grain size. The K_2O/Na_2O ratio of 1 and the Al_2O_3/Na_2O ratio of 5.8 for these slate-belt wackes and siltstones are close to those of the average graywacke, i.e., K_2O/Na_2O of 0.7 and Al_2O_3/Na_2O of 4.7 (see Fig. 4). These values confirm that the Yadkin Member is largely volcanic wacke and siltstone containing a significant proportion of pyroclastic debris.

Quartzite, conglomerate, and limestone. Other lithologic types less abundantly represented in the slate-belt suite include quartzite, polymictic and oligomictic conglomerate, and thin limestone beds. Several varieties of quartzite have been reported from slate-belt rocks in Virginia, North Carolina, South Carolina, and Georgia, including muscovite-, pyrophyllite-, and kyanite-bearing types. Some of these rocks described as quartzites may actually be recrystallized silicic tuff and chert.

Conglomerate beds and lenses are reported from Virginia, North Carolina, South Carolina, and Georgia. In the Hamme district of North Carolina and Virginia (Fig. 1), Parker (1963) found a conglomerate metamorphosed to sericite–chlorite phyllite that in places contains as much

TABLE 3. Mean Mineralogy and Chemical Composition of Volcanic Wackes and Siltstones from the Yadkin Member of the Millingport Formation, Mount Pleasant Quadrangle, North Carolina (Fig. 2)

1		2		3	4
	Volume percent		Weight percent	Weight percent	Weight percent
Framework	41.5	albite	23.8	SiO_2	65.20	66.75
quartz	13.9	muscovite	27.7	Al_2O_3	16.30	13.54
albite	14.6	quartz	31.6	Fe_2O_3	2.20	1.60
epidote	9.5	Mg-chlorite	5.2	FeO	3.90	3.54
opaque and		Fe-chlorite	2.5	MgO	1.90	2.15
heavy minerals	2.4	epidote	3.4	CaO	1.00	2.54
calcite	1.1	pyrite	1.7	Na_2O	2.80	2.93
		magnetite	2.5	K_2O	2.80	1.99
Matrix	58.5	ilmenite	1.6	H_2O^-	0.13	0.55
mainly sericitic		apatite	0.4	H_2O^+	2.70	2.42
muscovite, chlorite,			———	TiO_2	0.86	0.63
quartz, and plagioclase;			100.4	P_2O_5	0.17	0.16
lithic fragments less				MnO	0.10	0.12
than 5 percent				CO_2	<0.05	1.24
				*Others	—	0.45
					100.06	100.61

Column headings:
   1. Mean mode based on 5 thin sections.
   2. Mean norm based on 5 chemical analyses.
   3. Mean composition based on 5 chemical analyses (unpub. chemical analyses by Paul Elmore, Samuel Botts, H. Smith, Gillison Chloe, and Lowell Artis, U. S. Geol. Survey).
   4. Mean chemical composition of 61 graywackes (Pettijohn, 1963, p. 7).

* Includes SO_3, S, BaO, C.

as 50 percent lithic fragments. The fragments are of two types: rounded, gray, fine-grained quartzite pebbles that reach a maximum of $1\frac{1}{4}$ by 2 by 4 in. (3.2 by 5 by 10 cm) in size; and flat plates or chips of pelitic rock that are as much as 5 in. (13 cm) in long dimension. Volcanic lithic clasts were not observed by Parker. In the Efland Formation, Conley and Bain (1965, p. 124, 125) found argillaceous lithic conglomerate with rounded fragments as much as 2 in. (5 cm) in diameter. The fragments are volcanic flow rocks containing plagioclase microlites in a hematitic matrix. Conley and Bain (1965, p. 125) also described the Denny Conglomerate Member, a polymictic conglomerate containing well-rounded to subrounded pebbles as much as 6 in. (15 cm) in diameter of quartzite, felsic volcanic rock, feldspathic quartzite, quartz–hornblende gneiss, and jasper or red rhyolite. These pebbles are set in a wacke matrix, and, in fact, these rocks grade along strike into wackes. Oligomictic quartz–pebble conglomerates are reported from North Carolina, South Carolina, and Georgia.

Thin beds and lenticular bodies of calcareous mudstone and limestone have been recognized in a few places in the slate belt, mainly in drill core, fresh material on mine dumps, or from newly excavated bedrock exposures. Because of the intense chemical weathering prevalent in the southeast, these calcareous units are easily dissolved and commonly are marked only by a brown, sticky clay. Therefore, sedimentary carbonate rocks may be more abundant in the slate belt than has been recognized. Hatchell (1964) reports a calcareous phyllite with as much as 30 percent calcite, present mainly as quartz–calcite lenses. Conley and Bain (1965) recognized limestone beds 1 in. (2–3 cm) thick in the McManus Formation, and Stromquist and Sundelius (1969) have described calcareous siltstone in the Floyd Church Member of the Millingport Formation.

Volcanic Rocks in the Carolina Slate Belt

Interbedded with the previously described sedimentary rocks are a variety of volcanic rocks ranging in composition from rhyolite and rhyodacite to andesite and basalt and from flows to pyroclastic deposits. Tables 4 and 5 show mean chemical compositions of the mafic and felsic volcanic rocks, and Figure 5 shows variation diagrams based on 30 chemical analyses, for Al_2O_3, FeO, MgO, CaO, and $Na_2O + K_2O$. Figure 5 shows that although there is a considerable scatter of points, generalized variation trends can be drawn. The alkali–lime index derived from the curves shown on Figure 5 is between 58 and 59, placing these rocks in the calc-alkaline group. This value may be compared with the alkali–line index of 59 determined by Butler (1964, p. 109) for 20 analyses of slate-belt volcanic rocks. Calc-alkaline suites are characteristic of orogenic regions and are composed typically of the basalt–andesite–dacite–rhyolite association. Twenty-six of the analyses are from the Denton, Gold Hill, and Mount Pleasant quadrangles, North Carolina (Fig. 2); of these analyses, three are from Pogue, 1910, p. 54, 57, 67; 23 are from unpublished data of A. A. Stromquist and H. W. Sundelius. In addition, two analyses are from Butler, 1964, p. 104; two are from McCauley, 1961a, p. 16.

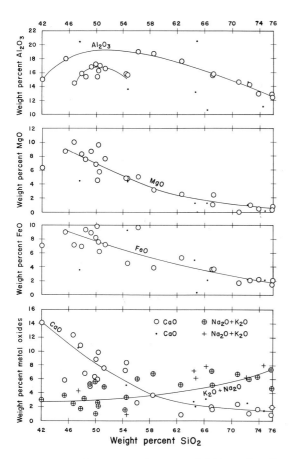

FIGURE 5. SiO_2 variation diagrams of felsic and mafic volcanic rocks from the Carolina slate belt; points within circles from unpublished analyses of A. A. Stromquist and H. W. Sundelius; other points from published analyses of Pogue (1910), Butler (1964), and McCauley (1961a).

Mafic volcanic rocks. It is apparent from Table 4 that the mafic volcanic rocks of the slate belt range from basalt to andesite in composition. These rocks commonly are lightly metamorphosed and altered and are now "greenstones." Both flow rock and pyroclastic material are found in bodies ranging from thin and local interbeds to lenses and large units whose thickness is measured in tens or hundreds of feet and which can be traced for several miles.

There are fine-grained, massive, and amygdular varieties of flow rocks. Amygdules are crudely spherical to ellipsoidal in shape and may be as much as 2 in. (5 cm) in diameter. They are generally filled with quartz, epidote, and chlorite, and less commonly with calcite or feldspar. Some of the flow rocks have a faintly porphyritic texture with feldspar or amphibole phenocrysts as much as 2 in. (5 cm) long. Pillow structures have not been recognized, but flow breccias have been noted by Stromquist and Sundelius (1969) in the Flat Swamp Member of the Cid Formation.

The mafic pyroclastic rocks are green, generally obscurely bedded, and range from fine tuff to rocks with various proportions and sizes of lithic fragments and crystal clasts, including lithic tuff, lithic–crystal tuff, crystal-lithic tuff, tuff breccia, and lapilli tuff. The fragments range from subrounded to angular and 0.08–19.5 in. (2 mm to 50 cm) in size. Most fragments are of mafic volcanic rock, including fragments of flow rocks with fluidal arrangement of plagioclase microlites, amygdules, and vesicular texture, and fragments of mafic pyroclastic rocks. Lenticular pieces that may represent flattened pumice fragments are also noted. Crystals and crystal clasts $\frac{1}{16}$–$\frac{3}{16}$ in. (1.6–4.8 mm) in size of intermediate to basic plagioclase also occur in these rocks. Sundelius (1964) reported the occurrence of accretionary lapilli in andesitic lithic tuff from the Yadkin Member of the Millingport Formation in the Mount Pleasant quadrangle, North Carolina (Fig. 2), which he interpreted as suggesting subaerial deposition.

Felsic volcanic rocks. Table 5, which includes representative analyses of the felsic volcanic rocks of the slate belt, shows that these rocks are mainly rhyolites and rhyodacites. Both extrusive and pyroclastic felsic volcanic rocks occur as interbeds and lenses associated with sedimentary and mafic volcanic rocks. These rocks are composed essentially of the secondary minerals, quartz, albite, microcline, sericitic muscovite, kaolinite, chlorite, and epidote and

TABLE 4.   Mean Chemical Composition of Mafic Volcanic Rocks from the Carolina Slate Belt (in weight percent)

	1	2	3	4	5	6
SiO_2	50.40	54.6	47.48	66.28	50.83	51.33
Al_2O_3	16.50	13.6	20.42	10.62	14.07	18.04
Fe_2O_3	2.20	7.1	6.41	6.41	2.88	3.40
FeO	7.40	9.3	3.61	2.11	9.06	5.70
MgO	6.80	4.3	4.53	1.15	6.34	6.01
CaO	7.60	5.1	10.52	3.17	10.42	10.07
Na_2O	2.70	0.3	2.00	6.09	2.23	2.76
K_2O	0.70	0.6	2.30	1.73	0.82	0.82
H_2O^-	0.15⎱	2.6	0.03⎱	0.61	—	—
H_2O^+	3.40⎰		1.65⎰		0.91	0.45
TiO_2	0.89	1.5	0.52	—	2.03	1.10
P_2O_5	0.20	—	—	—	0.23	0.16
MnO	0.18	0.2	0.22	—	0.18	0.16
CO_2	0.72	—	0.11	1.47	—	—
	99.84	99.2	99.80	99.64	100.00	100.00

Column headings:

1. Mean chemical composition of 14 greenstones from Denton, Gold Hill, and Mount Pleasant quadrangles, North Carolina (Fig. 2) (based on unpublished chemical analyses by Paul Elmore, Samuel Botts, H. Smith, Gillison Chloe, Lowell Artis, U. S. Geol. Survey).
2. Mean composition, slate belt mafic phase, Newberry County, South Carolina (Fig. 1) (McCauly, 1961a, p. 16).
3. Amygdaloidal basalt, Orange County, North Carolina (Fig. 1) (Butler, 1964, p. 104).
4. Andesite, Davidson County, North Carolina (Fig. 1) (Pogue, 1910, p. 67).
5. Mean composition normal tholeiitic basalt (Nockolds, 1954, p. 1021).
6. Mean composition of 56 "central basalts" (Nockolds, 1954, p. 1021).

accessory pyrite, pyrrhotite, magnetite, and sphene–leucoxene.

The felsic flow rocks include hard, aphanitic, black to dark-gray rhyolite and rhyolite porphyry. Dark, flinty rocks that occur as dikes cutting the flows may be devitrified obsidian or rhyolite. Pluglike bodies of felsite or vitrophyre, flow banding, and autobreccias are also noted. Spherulites of feldspar and quartz ranging from less than 1/4–3 in. (6.3–7.7 mm) in diameter have been described from several places in North Carolina (Butler, 1963; Conley, 1962a; Stromquist and Sundelius, 1969). The felsic porphyries contain phenocrysts of orthoclase, oligoclase, and quartz that range in size from 0.004 to 0.17 in. (0.1–3 mm).

The felsic pyroclastic rocks include varieties of fine tuff, lithic tuff, crystal tuff, crystal-rich welded ash-flow tuff, devitrified vitric tuff, and crystal–lithic tuff. These rocks are commonly blue-gray to light gray. The lithic fragments in the felsic pyroclastic rocks range from subrounded to angular and from less than 1 in. to 2 ft (2.5–50 cm) in diameter. These fragments are mainly of light-colored felsite, rhyolite porphyry, collapsed pumice, and felsic pyroclastic material that is internally fragmented. Fragments of basic volcanic rocks and of mudstone, argillite, and silt-

stone are less common. Euhedral crystals and crystal clasts of plagioclase and quartz that range in size from 0.04 to 0.08 in. (1–2 mm) occur in some of the tuffs. Vitroclastic textures are noted in thin section, and Conley (1962a) interprets wispy particles as devitrified shards.

Intrusive Rocks within the Carolina Slate Belt

Intrusive rocks in the Carolina slate belt range from granitic to gabbroic and perhaps lamprophyric in composition. In size the bodies range from narrow dikes and sills to plutons miles across, and in age, they range from early Paleozoic to Triassic(?). Some of the intrusive rocks are metamorphosed and deformed; others are apparently unmetamorphosed.

Large plutons are common in the northern half of the western slate belt in North Carolina and in the eastern slate belt. In South Carolina, large plutons are located near Kershaw (lat. 34°29′ N. long. 80°39.5′ W.), near Winnsboro (lat. 34°22.4′ N. long. 81°5.8′ W.), in the vicinity of Chester (lat. 34°42′ N. long. 81°11′ W.) and in Richland, Saluda, and Edgefield Counties in the west-central part of the State (Fig. 1) (Overstreet and Bell, 1965a, p. 29–32). In the Hamme–Virgilina district (Fig. 1), the Redoak Granite, Buffalo

TABLE 5.   Mean Chemical Composition of Felsic Volcanic Rocks from the Carolina Slate Belt (in weight percent)

	1	2	3	4	5	6
SiO$_2$	69.40	64.8	64.54	74.67	72.33	66.26
Al$_2$O$_3$	14.90	20.5	13.13	10.78	14.56	15.39
Fe$_2$O$_3$	0.85	5.4	2.25	1.25	0.15	2.14
FeO	2.90	0.3	5.01	2.11	2.22	2.23
MgO	1.10	—	1.18	—	0.91	1.57
CaO	1.90	2.1	2.81	1.47	2.55	3.68
Na$_2$O	4.10	2.7	4.18	5.31	3.40	4.13
K$_2$O	2.10	3.5	2.99	2.68	2.82	3.01
H$_2$O$^-$	0.12 }	} 0.5	0.02 }	} 0.59	} 0.30	—
H$_2$O$^+$	1.60 }		1.12 }			0.68
TiO$_2$	0.55	0.7	0.67	—	—	0.66
P$_2$O$_5$	0.12	—	—	—	—	0.17
MnO	0.12	—	0.13	—	—	0.07
CO$_2$	0.30	—	0.69	1.30	—	—
S	—	—	0.22	—	—	—
	100.06	100.5	98.94	100.16	99.24	99.99

Column headings:
1. Mean chemical composition of 9 felsic tuffs and flows from Denton, Gold Hill, and Mount Pleasant quadrangles, North Carolina (Fig. 2) (based on unpublished chemical analyses by Paul Elmore, Samuel Botts, H. Smith, Gillison Chloe, and Lowell Artis, U. S. Geol. Survey).
2. Mean composition slate belt felsic phase, Newberry County, South Carolina (Fig. 1) (McCauley, 1961a, p. 16).
3. Vitric–crystal tuff, Orange County, North Carolina (Fig. 1) (Butler, 1964, p. 106).
4. Rhyolite, Davidson County, North Carolina (Fig. 1) (Pogue, 1910, p. 54).
5. Dacite, Davidson County, North Carolina (Fig. 1) (Pogue, 1910, p. 57).
6. Mean chemical composition of 115 rhyodacite and rhyodacite–obsidian analyses (Nockolds, 1954, p. 1014).

Granite, and an albite granodiorite have been described (Parker, 1963). These plutons generally range from granodiorite to quartz monzonite in composition and have zones of contact alteration that range from narrow to hundreds of feet across. Overstreet and Bell (1965a, p. 30) observe that in vertical section the plutons of the slate belt in South Carolina have an inverted teardrop shape. In addition to these plutons, there are numerous lenses, pods, and irregular bodies of granitic composition. In the Hamme district, Parker (1963) noted microcline–quartz and quartz–muscovite pegmatites.

Dioritic to gabbroic bodies ranging from small dikes, sills, and irregularly shaped bodies to plutons several miles across also intrude the slate-belt rocks (see tectonic map at end of this volume). In North Carolina, diorite and gabbro plutons have been mapped in the northern part of the western slate belt. These are medium- to coarse-grained, generally massive, hornblende–plagioclase–pyroxene rocks. Numerous smaller saussuritized gabbroic dikes, sills, and irregular bodies are found within the slate belt. These rocks are essentially altered aggregates of plagioclase, actinolite (tremolite), epidote (clinozoisite), and chlorite containing accessory sphene–leucoxene,

magnetite, and ilmenite. Relatively fresh, unmetamorphosed, very fine grained diabase dikes of Triassic(?) age are common. These Triassic dikes are generally steeply dipping, and most are 3–10 ft (0.9–3 m) wide, but a few are as much as 100 ft (30 m) wide. In the southern part of the slate belt most dikes trend northwest, but in northern North Carolina, their strike changes to north and northeast.

METAMORPHISM WITHIN THE CAROLINA SLATE BELT

The degree of metamorphism is not uniform throughout the Carolina slate belt, and there is a gradation from relatively unfoliated sedimentary and volcanic rocks to slate, phyllite, and schist. In general, the amount of shearing and recrystallization is higher as the Charlotte belt is approached; Conley (1962a) also noted an increase in metamorphism toward the east in south-central North Carolina, i.e., toward the Deep River–Wadesboro Triassic(?) basin. Despite this variation, the metamorphic grade nowhere exceeds the greenschist facies except in local contact metamorphic aureoles.

The felsic volcanic rocks are commonly light-colored schists and phyllites with the mineral assemblage quartz – albite – (microcline) – sericitic muscovite – chlorite – (biotite) – epidote – (clinozoisite) and probably belong to the quartz–albite–microcline–epidote–biotite subfacies of the greenschist facies (Fyfe, Turner, and Verhoogen, 1958, p. 223). The mafic volcanic rocks are green phyllites and schists with the mineral assemblage albite – (oligoclase) – actinolite – (tremolite) – epidote – (clinozoisite) – chlorite – sphene – (calcite) and seem to belong to the albite–epidote–actinolite–chlorite–sphene subfacies of the greenschist facies (Fyfe, Turner, and Verhoogen, 1958, p. 223). Plagioclase generally is altered to chlorite, epidote, and calcite, and K-feldspar commonly is sericitized and cloudy. The original ferromagnesian minerals are changed to chlorite, clinozoisite–epidote, actinolite, and calcite.

At places within the slate belt, stocks of adamellite, granodiorite, and diorite cut the slate-belt rocks, and adjacent to these bodies, narrow contact aureoles containing garnetiferous–mica schist, mica–hornblende schist, and biotite schist have developed (Stuckey, 1965, p. 97). In South Carolina, Heron and Johnson (1958) report the development of garnet, kyanite, and staurolite within a narrow zone of argillite in contact with gneiss. The rocks are hydrothermally altered, principally by silicification, at numerous places, particularly near mineralized zones.

STRUCTURES WITHIN THE CAROLINA SLATE BELT

Rocks of the Carolina slate belt seem to be deformed into a series of northeast-trending folds that range from broad and open to tightly compressed. Some of the rocks have a closely spaced slaty cleavage, which has led to the misconception that nearly all the rocks of the slate belt are, in fact, slates. Some rocks are essentially massive and have little or no cleavage. In places, bedding and slaty cleavage are nearly parallel, whereas in other locations, there is a considerable divergence between them. Folds range from symmetrical to asymmetric and overturned, and bedding ranges from nearly horizontal to vertical.

Stuckey (1965, p. 100) observed that the direction of dip of slaty cleavage changes from a northwest dip in the southern part of North Carolina, i.e., south of 36° N. latitude, to a southeast dip north of that general line (see tectonic map at end of this volume). King (1950, 1964) has speculated that the Appalachian system may indeed be two-sided with an axis within the Inner Piedmont. In that case, slaty cleavage and the axial planes of folds might well be expected to dip northwest in the slate belt.

In the Virgilina district (Fig. 1), Laney (1917) considered the rocks to be distributed in closely compressed folds in a syncline with both bedding and slaty cleavage dipping steeply to the southeast. After mapping in the Hamme district to the east (Fig. 1), Parker (1963) proposed that the axis of the Virgilina synclinorium be moved about 12 mi (19.2 km) to the east in order to include both the Virgilina and Hamme areas as part of a larger synclinorium.

Stuckey (1965, p. 91–93) interprets the region of schist and gneiss that divides the slate belt in the northern part of North Carolina into an eastern and a western area as a pre-slate-belt metamorphic terrane perhaps equivalent in age to the rocks of the Charlotte belt. According to Stuckey, slate-belt rocks in both areas dip away from this intervening gneiss and schist terrane, suggesting that this is an anticline, accentuated, perhaps, by emplacement of the Rolesville granitic pluton.

In south-central North Carolina, the author and others (Conley 1962a; Conley and Bain, 1965); Stromquist and Sundelius, 1969; Stromquist, Choquette, and Sundelius, 1966; Sundelius and Taylor, 1968 have recognized and mapped a series of major folds, including the New London syncline, the Denton anticline, and the Silver Valley syncline (Fig. 2). In Newberry County, South Carolina, McCauley (1961b) interprets the slate-belt terrane as a large syncline adjacent to an anticline that exposes the Charlotte belt rocks immediately to the west.

The relationship of the Carolina slate belt rocks to the medium-grade metamorphic rocks to the west is difficult to assess. As indicated previously, Overstreet and Bell (1965a) believe that in South Carolina, these belts within the Piedmont represent metamorphic zones that, in places, cut across stratigraphic units.

Beginning at the South Carolina–North Carolina line and continuing northeast for approximately 85 mi (136 km), there is a rather sharp break between the low-grade slate-belt rocks on the east and the plutonic rocks of the Charlotte belt on the west. This sharp break is shown quite clearly on the aeromagnetic map of several quadrangles within this area (Henderson and Gilbert, 1966). Laney (1910) recognized this feature and named it the Gold Hill fault. Mapping in the Denton, Gold Hill, and Mount Pleasant quadrangles by A. A. Stromquist, H. W. Sundelius, and A. R. Taylor has confirmed the presence of a major fault zone which has been named the Gold Hill–Silver Hill fault zone. Within this zone occur steeply dipping, highly fractured, and closely cleaved or foliated slate, phyllite, and schist. To the east of this fault zone, the slate-belt rocks are, in general, characterized by gentle dips, are not foliated, and do not have closely spaced slaty cleavage, except locally. In the Denton quadrangle (Stromquist, Choquette, and Sundelius, 1966), the distinctive finely laminated Tillery Formation has been brought up within the fault zone against the mudstone member of the Cid Formation on the east, and the Silver Valley syncline has been cut off by the fault zone. In the Gold Hill quadrangle (Sundelius and Taylor, 1968) and in the Mount Pleasant quadrangle, the mudstone member of the Cid Formation is upthrown against the Floyd Church Member of the Millingport Formation on the east, and the Denton anticline is truncated by the fault zone. Immediately west of the fault zone are basalt flows, tuffs, and tuff breccia along with interbeds of rhyodacitic tuff and crystal tuff, a sequence lithologically similar to rocks of the slate belt. To the west, these volcanic rocks appear to be intruded by a complex ranging in composition from diorite–gabbro to quartz monzonite. Farther west is an intimately mixed assemblage of biotite and hornblende schist, amphibolite, and coarse-grained, almost pegmatitic granitic rock. Much of the mineralization in the Gold Hill and Cid districts shown on the geologic maps of the Denton, Gold Hill, and Mount Pleasant quadrangles is within or adjacent to this fault zone. North of the northernmost extension of this fault zone, the slate belt–Charlotte belt boundary seems much less definite, and Mundorff (1948) describes the border near Greensboro, North Carolina, as an interfingering of altered volcanic rocks and sheared granite.

SUMMARY

This short review attempts to collect and synthesize results of geologic investigations in the slate belt since 1955. The pace of these investigations has accelerated each year.

An important finding has been that detailed geologic mapping is possible in the slate belt. Stratigraphic units have been established and traced for considerable distances, and provide a firm basis for structural interpretation. Geologic mapping can be successful if, in addition to standard geologic techniques, geophysical methods, including aeromagnetics, aeroradiometrics, and gravity, are utilized.

It has been shown that the association of lithologies is eugeosynclinal in character. Calc-alkaline volcanic rocks such as basaltic, andesitic, rhyodacitic and rhyolitic flows and pyroclastics are interbedded with claystone, mudstone, argillite, siltstone, and volcanic wacke. These rocks are predominantly unfossiliferous, contain numerous graded beds, undergo abrupt facies changes, and appear to include both subaerial and subaqueous deposits. They were probably deposited in basins between tectonic ridges and volcanic islands, which acted as source areas. The general low grade of metamorphism suggests that rocks of the slate belt were laterally and, perhaps, vertically marginal to the Inner Piedmont zone of maximum deformation and metamorphism. A single fossil discovery suggesting a Middle Cambrian age coupled with lead-alpha dating of one pyroclastic unit that indicates an Ordovician age suggests that slate-belt rocks are early Paleozoic in age. Based on lithologic differences, structural divergences, and isotopic dating of intrusive rocks, Overstreet and Bell (1965a) postulate geologic ages as young as Mississippian for certain slate-belt rocks.

The highly unstable sedimentary environment in which these rocks were deposited suggests that facies changes may preclude large-scale regional correlations within the slate belt based solely on the results of small areas of widely scattered detailed geologic mapping. Thorough understanding of the slate belt may require a considerable amount of detailed geologic quadrangle mapping.

REFERENCES

American Geophysical Union, Special Committee for the Geophysical and Geological Study of the Continents, 1964, Bouguer gravity anomaly map of the United States (exclusive of Alaska and Hawaii): U. S. Geol. Spec. Map, 2 sheet, scale 1:2,500,000.

Bain, G. L., 1964, Metavolcanic and metasedimentary rocks in Chatham and Randolph counties, North Carolina: Carolina Geol. Soc. Fieldtrip Guidebook, p. 1–6, map.

———, 1966, Geology and ground–water in the Durham area, North Carolina: North Carolina Dept. Water Resources Groundwater Bull. 7, 147 p.

Becker, F. G., 1895, Reconnaissance of the gold fields of the southern Appalachians: U. S. Geol. Survey 16th Ann. Rept., 1894, Pt. 3, p. 251–319.

Butler, J. R., 1963, Rocks of the Carolina slate belt in Orange County, North Carolina: Southeastern Geology, v. 4, no. 3, p. 167–185.

———, 1964, Chemical analyses of rocks of the Carolina slate belt: Southeastern Geology, v. 5, no. 2, p. 101–112.

———, 1965, Guide to the geology of York County, South Carolina: South Carolina State Devel. Board, Div. Geology, Geol. Notes, v. 9, no. 2, p. 27–36, map.

Clarke, F. W., 1924, Data of geochemistry: 5th ed., U. S. Geol. Survey Bull. 770, 841 p.

Conley, J. F., 1962a, Geology of the Albemarle quadrangle, North Carolina: North Carolina Div. Mineral Resources Bull. 75, 26 p., map.

—————, 1962b, Geology and mineral resources of Moore County, North Carolina: North Carolina Div. Mineral Resources Bull. 76, 40 p., map.

—————, and Bain, G. L., 1965, Geology of the Carolina slate belt west of the Deep River–Wadesboro Triassic basin, North Carolina: Southeastern Geology, v. 6, no. 3, p. 117–138, map.

Council, R. J., 1954, A preliminary geologic report on the commercial rocks of the Volcanic Slate series, North Carolina: North Carolina Div. Mineral Resources Inf. Circ. 12, 30 p.

Crickmay, G. W., 1952, Geology of the crystalline rocks of Georgia: Georgia Dept. Mines, Mining and Geology Bull. 58, 54 p.

Diller, J. S., 1899, Origin of *Paleotrochis*: Am. Jour. Sci., 4th ser., v. 7, p. 337–342.

Eckel, E. C., 1904, On the chemical composition of American shales and roofing slates: Jour. Geology, v. 12, p. 25–29.

Emmons, E., 1856, Geological report of the midland counties of North Carolina: North Carolina Geol. Survey, 351 p.

Floyd, E. O., 1965, Geology and ground-water resources of the Monroe area, North Carolina: North Carolina Dept. Water Resources Ground-water Bull. 5, 109 p.

Fyfe, W. S., Turner, F. J., and Verhoogen, J., 1958, Metamorphic reactions and metamorphic facies: Geol. Soc. America Mem. 73, 259 p.

Georgia Div. Mines, Mining, and Geology, 1939, Geologic map of Georgia: scale 1:500,000, prepared in cooperation with U. S. Geol. Survey.

Hatchell, W. O., 1964, Petrology of metasedimentary and volcanic rocks along Harmon Creek in the Irmo N.E. quadrangle, South Carolina: South Carolina State Devel. Board, Div. Geology, Geol. Notes, v. 8, no. 3-4, p. 35–45.

Henderson, J. R., and Gilbert, F. P., 1966, Aeromagnetic map of the Mount Pleasant, Albemarle, Denton, and Salisbury quadrangles, west-central North Carolina: U. S. Geol. Survey Geophys. Inv. Map GP-581.

Heron, S. D., Jr., and Johnson, H. S., Jr., 1958, Geology of the Irmo Quadrangle, South Carolina: South Carolina Div. Geology, scale 1:24,000.

Jonas, A. I., 1932, Structure of the metamorphic belt of the southern Appalachians: Am. Jour. Sci., 5th Ser., v. 24, p. 228–243.

Keith, Arthur, and Sterrett, D. B., 1931, Description of the Gaffney and Kings Mountain quadrangles [South Carolina–North Carolina]: U. S. Geol. Survey Geol. Atlas, Folio 222, 13 p.

Kerr, W. C., 1875, Report of the geological survey of North Carolina. Volume 1, Physical geography, résumé, economical geology: Raleigh, N.C., 325, 120 p., map.

Kesler, T. L., 1936, Granitic injection processes in the Columbia quadrangle, South Carolina: Jour. Geology, v. 44, no. 1, p. 32–42.

King, P. B., 1950, Tectonic framework of southeastern United States: Am. Assoc. Petrol. Geologists Bull., v. 34, no. 4, p. 635–671.

—————, 1955, A geologic section across the southern Appalachians—An outline of the geology in the segment in Tennessee, North Carolina, and South Carolina, *in* Russell, R. J., ed., Guides to southeastern geology: New York, Geol. Soc. America, p. 332–373.

—————, 1964, Further thoughts on tectonic framework of the southeastern United States: Virginia Polytech. Inst. Dept. Geol. Sci. Mem. 1, p. 5–31.

Laney, F. B., 1910, The Gold Hill mining district of North Carolina: North Carolina Geol. and Econ. Survey Bull. 21, 137 p., map.

—————, 1917, The geology and ore deposits of the Virgilina district of Virginia and North Carolina: North Carolina Geol. and Econ. Survey Bull. 26, 176 p., map.; also pub. as Virginia Geol. Survey, Bull. 14.

LeGrand, H. E., and Mundorff, M. J., 1952, Geology and ground water in the Charlotte area, North Carolina: North Carolina Div. Mineral Resources Bull. 63, 88 p.

McCauley, J. F., 1961a, Rock analyses in the Carolina slate belt and the Charlotte belt of Newberry County, South Carolina: Southeastern Geology, v. 3, no. 1, p. 1–20.

—————, 1961b, Relationships between the Carolina slate belt and the Charlotte belt in Newberry County, South Carolina: South Carolina State Devel. Board, Div. Geology, Geol. Notes, v. 5, no. 5, p. 59–66.

Milton, C., and Hurst, V. J., 1965, Subsurface "basement" rocks of Georgia: Georgia Dept. Mines, Mining and Geology Bull. 76, 56 p.

Mitchell, E., 1827, Report on the Geology of North Carolina, Part III: Raleigh. N.C., [North Carolina] Board Agriculture p. 1–27.

Mundorff, M. J., 1948, Geology and ground water in the Greensboro area, North Carolina: North Carolina Div. Mineral Resources Bull. 55, 108 p.

Nanz, R. H., 1953, Chemical composition of pre-Cambrian slates with notes on the geochemical evolution of lutites: Jour. Geology, v. 61, p. 51–64.

Nitze, H. B. C., and Hanna, G. B., 1896, Gold deposits of North Carolina: North Carolina Geol. Survey Bull. 3, 200 p.

Nockolds, S. R., 1954, Average chemical compositions of some igneous rocks: Geol. Soc. America Bull., v. 65, no. 10, p. 1007–1032.

North American Geologic Map Committee, 1965, Geologic map of North America: Washington, D. C., U. S. Geol. Survey, 2 sheets, scale 1:5,000,000.

Olmsted, D., 1824, Report on the geology of North Carolina, conducted under the direction of the Board of Agriculture: North Carolina Board Agriculture, Papers on Agricultural Subjects, pt. 1, 44 p.

Overstreet, W. C., and Bell, H., III, 1965a, The crystalline rocks of South Carolina: U. S. Geol. Survey Bull. 1183, 126 p.

—————, 1965b, Geologic map of the crystalline rocks of South Carolina: U. S. Geol. Survey Misc. Geol. Inv. Map I-413, scale 1:250,000.

Pardee, J. T., and Park, C. F., Jr., 1948, Gold deposits of the southern Piedmont: U. S. Geol. Survey Prof. Paper 213, 156 p.

Parker, J. M., 1963, Geologic setting of the Hamme tungsten district, North Carolina and Virginia: U. S. Geol. Survey Bull. 1122-G, 69 p.

Pettijohn, F. J., 1957, Sedimentary rocks: 2d ed., New York, Harper & Bros., 718 p.

—————, 1963, Chemical composition of sandstones—excluding carbonate and volcanic sands: U. S. Geol. Survey Prof. Paper 440-S, 21 p.

Pogue, J. E., Jr., 1910, Cid mining district of Davidson County, North Carolina: North Carolina Geol. and Econ. Survey Bull. 22, 144 p.

Rankama, K. K., and Sahama, Th. G., 1955, Geochemistry: Chicago, Univ. Chicago Press, 912 p.

St. Jean, J., Jr., 1965, New Cambrian trilobite from the Piedmont of North Carolina [abs.]: Geol. Soc. America Spec. Paper 82, p. 307–308.

Siple, G. E., 1958, Stratigraphic data from selected oil tests and water wells in the South Carolina Coastal Plain: South Carolina State Devel. Board, Div. Geology, Mineral Industries Lab. Monthly Bull., v. 2, no. 9, p. 62–68.

Stose, G. W., and Ljungstedt, O. A., comps., 1932, Geologic map of the United States: Washington, D. C., U. S. Geol. Survey, 1:2,500,000.

Stromquist, A. A., and Conley, J. F., 1959, Geology of the Albemarle and Denton quadrangles, North Carolina: Carolina Geol. Soc. Fieldtrip Guidebook, 36 p.

————, and Sundelius, H. W., 1969, Stratigraphy of the Albemarle Group of the Carolina slate belt in central North Carolina: U. S. Geol. Survey Bull. 1274-B, 22 p.

————, Choquette, P. W., and Sundelius, H. W., 1966, Bedrock geologic map of the Denton quadrangle, North Carolina: U. S. Geol. Survey open-file report.

Stuckey, J. L., 1928, The pyrophyllite deposits of North Carolina: North Carolina Dept. Conserv. and Devel. Bull. 37, 62 p.

————, 1965, North Carolina: Its geology and mineral resources: Raleigh, N.C., North Carolina Dept. Conserv. and Devel., 550 p.

————, and Conrad, S. G., 1958, Explanatory text for geologic map of North Carolina: North Carolina Div. Mineral Resources Bull. 71, 51 p.

Sundelius, H. W., 1964, Accretionary lapilli in rocks of the Carolina slate belt, Stanly County, North Carolina: U. S. Geol. Survey Prof. Paper 475-B, p. 42–44.

————, and Taylor, A. R., 1968, Geologic map of the Gold Hill quadrangle, North Carolina: U. S. Geol. Survey open-file report.

Thiesmeyer, L. R., 1939, Varved slates in Fauquier County, Virginia: Virginia Geol. Survey Bull. 51-D, p. 105–118.

U. S. Geological Survey and American Association of Petroleum Geologists, 1961, Tectonic map of the United States, exclusive of Alaska and Hawaii: 2 sheets, scale 1:2,500,000 [1962].

Virginia Div. Mineral Resources, 1963, Geologic map of Virginia: Charlottesville, scale 1:500,000.

Watkins, J. S., and Yuval, Z., 1966, Simple Bouguer gravity map of the Mount Pleasant, Albemarle, Denton, and Salisbury quadrangles, west-central North Carolina: U. S. Geol. Survey Geophys. Inv. Map GP-582.

Weed, W. H., and Watson, T. L., 1906, The Virgilina copper deposits: Econ. Geology, v. 1, p. 309–330.

White, A. M., Stromquist, A. A., Stern, T. W., and Westley, Harold, 1963, Ordovician age for some rocks of the Carolina slate belt in North Carolina: U. S. Geol. Survey Prof. Paper 475-C, p. 107–109.

Williams, G. H., 1894, The distribution of ancient volcanic rocks along the eastern border of North America: Jour. Geology, v. 2, p. 1–31.

Chapter 25

The Piedmont in South Carolina[*]

WILLIAM C. OVERSTREET

INTRODUCTION

IN SOUTH CAROLINA, 95 percent of the area of crystalline rocks is in the Piedmont province. Crystalline rocks occupy a small area in the extreme western part of the State in the Blue Ridge physiographic province, but these rocks are separated from the rocks of the Piedmont by the Brevard fault zone, which passes northeastward across the State (Reed and Bryant, 1964). In central South Carolina the crystalline rocks are unconformably overlain by sedimentary rocks of the Atlantic Coastal Plain. For a few miles east of the west edge of the overlap, small exposures of crystalline rocks are in the beds of streams that cut through the sedimentary rocks. The area of Piedmont crystalline rocks thus defined between the east edge of the Brevard fault zone and the west edge of the Coastal Plain in South Carolina is about 11,000 sq mi. (17,700 sq km).

This review of the Piedmont in South Carolina is based on a synthesis of the geology of the crystalline rocks of the State completed by the writer and Henry Bell III in 1960 (Overstreet and Bell, 1965a,b). That synthesis derives from a literature extending back to 1802 (Drayton, 1802), but is supported mainly by the work of Tuomey (1844, 1848), Lieber (1858a,b, 1859, 1860), Hammond (1883), Sloan (1908), King (1955), and the important contributions in the soil maps of the U. S. Department of Agriculture.

Detailed areal studies published recently by Butler (1966), Brown and Cazeau (1964), Paradeses and others (1966), Ridgeway and others (1966), Cazeau (1966), and regional syntheses by Bain (1964), King (1964), Hadley (1964),

and Dietrich (1964) have contributed to this summary, as have lead-alpha age data given by White and others (1963), the economic geology of McCauley and McCauley (1964), McCauley and Butler (1966), Conley (1962), and ground–water studies of Floyd (1965), Bain and Thomas (1966), and Marsh and Laney (1966).

Acknowledgment is gratefully made of the encouragement and help given by Henry Bell III in the preparation of this summary, and of the constant interest in this work shown by Henry S. Johnson, Jr., State Geologist of South Carolina.

GEOLOGIC BELTS

The complex geology of the crystalline rocks in the Piedmont of South Carolina is poorly known owing to deep weathering of the rocks and to the scarcity of detailed geologic maps. However, the main types of rocks can be inferred from the distribution of residual soils depicted on county soils maps prepared by the U. S. Department of Agriculture. The residual soils and the rocks from which they formed have a northeast-trending beltlike distribution that has long been known in South Carolina and is one of the striking geologic aspects of the southeastern States. It is thought that these belts are zones of different response to variations in regional metamorphism and tectonic activity by originally similar sedimentary and volcanic rocks (Overstreet and Bell, 1965a, Pl. 2 and 3).

The beltlike distribution of soils and rocks in South Carolina was originally noticed in 1802 (Drayton, 1802, p. 10–11). Over the years, the belts he identified have been variously redefined by Dickson (1819), Sloan (1908, Pl. 1), and Jonas (1932, p. 230–231), until in 1955, P. B. King proposed names for geologic belts in the

[*] Publication authorized by the Director, U. S. Geological Survey.

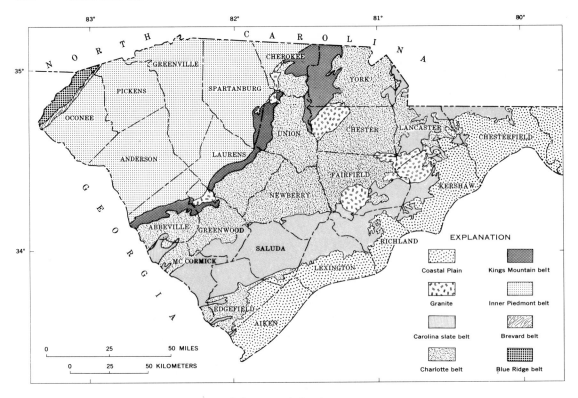

FIGURE 1. Map of the geologic belts in South Carolina.

southern Appalachians (King, 1955, p. 337–338). Seven of these geologic belts are recognized in South Carolina, four in the Piedmont physiographic province. From southeast to northwest the geologic belts in the Piedmont are known as the Carolina slate belt, the Charlotte belt, the Kings Mountain belt, and the Inner Piedmont belt (Fig. 1). The westernmost part of the Inner Piedmont belt in South Carolina is in the Blue Ridge physiographic province. Flanking the Inner Piedmont belt on the west, and separating the rocks of the Inner Piedmont from the rocks of the Blue Ridge belt, is the Brevard belt, a prominent structural feature marked by aligned valleys in the Brevard fault zone.

Some major rock units in the geologic belts in the Piedmont have stratigraphic features in common which persist across the belts. The stratigraphic succession inferred from these pertinent features indicates that the total thickness of rocks must be very great, but the thickness is unknown. As presently exposed, these rocks are much modified by folding, faulting, regional and contact metamorphism, and igneous intrusion (Fig. 2).

A long and complex succession of intrusive events is disclosed by the shape, structure, composition, and relations of the bodies of plutonic igneous rocks exposed in the geologic belts (Fig.

3). The order of events observed in the plutonic intrusive rocks is matched by successions of dikes.

Apparent ages of minerals in the intrusive rocks give the only reference for the probable ages of the metamorphosed sedimentary and volcanic rocks in the geologic belts in the Piedmont of South Carolina, because fossils have not yet been found there. A few fossils are known, however, in similar and possibly correlative rocks in North Carolina and Virginia.

The geologic belts in the Piedmont differ from each other mainly in the changes effected in the original sedimentary and volcanic rocks since they were deposited. The belts are primarily zones of different grades of regional metamorphism (Fig. 4). They parallel the metamorphic isograds, which trend east in the southern part of the state and northeast in the central and northern part (Overstreet and Bell, 1965a, Pl. 3; 1965b), and which cut across the bedding of stratigraphic units and across major unconformities (Fig. 2). Therefore, the grade of metamorphism is thought to be the principal factor defining the geologic belts.

Carolina Slate Belt

The easternmost belt of low-rank metamorphic rocks is called the Carolina slate belt (Figs. 1 and

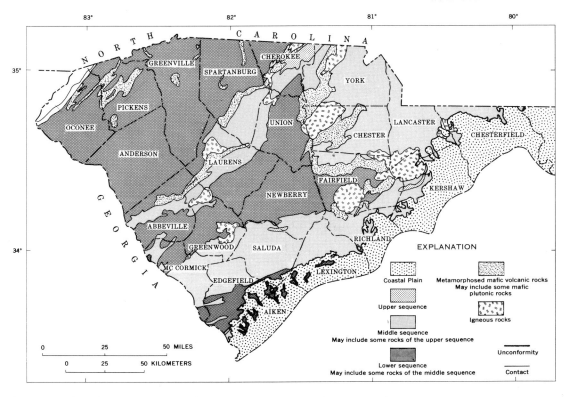

FIGURE 2. Map of the main stratigraphic units and major unconformities in the Piedmont of South Carolina.

4). Its southeastern edge is covered by sedimentary rocks of the Atlantic Coastal Plain, and the northwestern edge merges with the gneisses, schists, and granitoid rocks of the Charlotte belt (Overstreet and Bell, 1965a, p. 19). The principal rocks in the belt are amphibolite (including chlorite schist), argillite, and muscovite schist (Table 1); few rocks in the belt are true slates.

Charlotte Belt

The broad central part of the Piedmont in South Carolina between the slate belt on the southeast and the Kings Mountain belt on the northwest (Fig. 1) was called the Charlotte belt by King (1955, p. 346–350). King noted that the belt contains more granite than the other belts. Indeed, granitoid textures are common, and intrusive plutons are a conspicuous aspect of the Charlotte belt, but much of the granitoid rock has strong compositional layering apparently inherited from bedded sequences of original sedimentary and volcanic rocks. This granitoid paragneiss is commonly a fine-grained, epidote-bearing gneiss and migmatite of the albite–epidote amphibolite facies. Locally, the grade of regional metamorphism rises to the staurolite–kyanite subfacies (Fig. 4), and adjacent to parts of the large

plutons, the grade rises to the sillimanite–almandine subfacies. The Charlotte belt is, therefore, a conspicuous granitoid zone of moderate metamorphic grade between two belts of lower grade rocks. The belt is also notable for the swarms of mafic dikes, which appear to be feeders for volcanic flows in the middle and upper sequences (Fig. 2) in the Piedmont.

Kings Mountain Belt

The Kings Mountain belt consists of sericite schist, hornblende schist, and sparse quartzite and marble. The belt is well defined in the central part of the Piedmont of South Carolina (Fig. 1). In Cherokee and York Counties, the Kings Mountain belt includes a variety of formations named by Keith and Sterrett (1931, p. 4–6), but these formations have not been traced individually farther to the south and southwest. Their probable extensions are in units identified as hornblende schist and sericite schist (Overstreet and Bell, 1965b). A well-defined unconformity separating an upper sequence of rocks from a middle sequence (Fig. 2) in the Kings Mountain belt can be correlated with an unconformity having the same stratigraphic position in the slate belt. Correlation of formations defined by Keith and

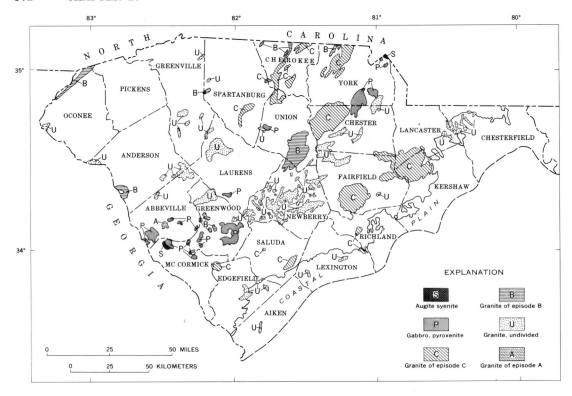

FIGURE 3. Map of the larger masses of intrusive rocks in South Carolina.

Sterrett with individual units in the slate belt, however, has not been made.

Inner Piedmont Belt

The widest belt in the Piedmont of South Carolina is the Inner Piedmont belt west of the Kings Mountain belt (Fig. 1). The boundary between these two belts appears to represent a change in metamorphic grade instead of the juxtaposition of rocks of different ages. On its northwest side the Inner Piedmont belt terminates in the Brevard fault zone. The prevailing high metamorphic grade makes the origin, sequence, and geologic history of the rocks in the Inner Piedmont difficult to define. The belt occupies the zone of regional metamorphic climax in the Piedmont of South Carolina.

STRATIGRAPHIC SEQUENCES

The metamorphosed sedimentary and volcanic rocks in the Carolina slate belt, Charlotte belt, Kings Mountain belt, and Inner Piedmont belt in South Carolina are thought to consist of three stratigraphic sequences (Fig. 2 and Table 1) separated by two unconformities (Overstreet and Bell, 1965a, p. 10–11). These sequences each contained graywacke, shale, felsic and mafic tuffaceous shale, tuff, and lava. Sparse conglom-

erate, sandstone, and limestone were interbedded with the main components. Each sequence was deposited in a subsiding basin. In South Carolina, the western parts of the successive basins are more or less superimposed on each other. Along strike to the northeast and southwest the basins extend into Virginia and Alabama. Eastward the basins are covered by unconsolidated sedimentary rocks of the Atlantic Coastal Plain. Well cuttings disclose characteristic basin rocks under the Coastal Plain sediments (Siple, 1958, p. 67), and data from airborne magnetometer surveys (Reed and Owens, 1967) have been interpreted to show that these rocks extend eastward to the edge of the continental shelf. Westward the rocks extend at least as far as the Brevard belt. Thus, the three stratigraphic sequences in the Piedmont of South Carolina are inferred to be the exposed western third of an immense volume of eugeosynclinal sediments whose depositional basins reach eastward to the continental margin. How far eastward and how late in geologic time volcanic activity persisted in the covered parts of these basins is uncertain. Volcanic ash (Rooney and Kerr, 1967, p. 740; Heron and Johnson, 1966, p. 61; Sandy and others, 1966, p. 17, 23–25) and possible rhyolitic flows (Crawford and others, 1966, p. 1, 34–40)

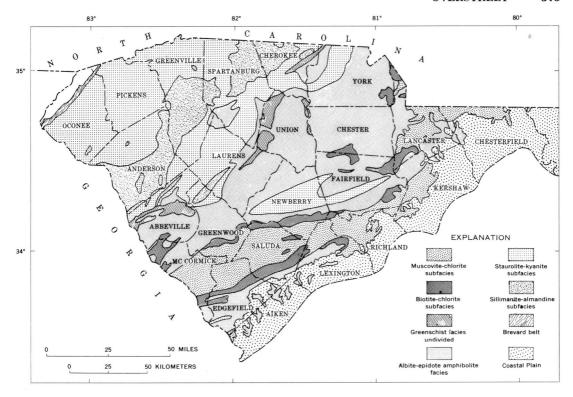

FIGURE 4.   Map of the major metamorphic facies in South Carolina.

have been found in Cretaceous and Tertiary rocks in the Coastal Plain of the Carolinas and Georgia, but the sources have not been identified in the Piedmont.

The three inferred stratigraphic sequences (Fig. 2) have been called the lower, middle, and upper sequences (Overstreet and Bell, 1965a, p. 12–13). Formal stratigraphic names have not been applied to the sequences, but several units in them are known by formation names in the area of the Gaffney and Kings Mountain quadrangles (Keith and Sterrett, 1931; Espenshade and Potter, 1960, p. 64–85). Formation names have also been used for the rocks near the Brevard belt (Keith, 1905, 1907).

The strongest evidence that the stratigraphic sequences extend across the geologic belts is the distribution of mafic metamorphic rocks derived from andesitic lavas and tuffs in the central part of the Piedmont (Fig. 2). These rocks are near the base of the middle sequence, and they define a series of north-plunging folds. In the southern part of the area they are in the slate belt in Fairfield and Kershaw Counties (Fig. 1). They cross from the slate belt into the Charlotte belt in Fairfield County. In Chester County, they are intruded by a granite pluton, which separates them from like rocks in the Kings Mountain belt in

Chester and York Counties. What are probably the same rocks reappear on the west flank of the Kings Mountain belt in Cherokee County and extend southwestward into the Inner Piedmont belt.

Distinctive manganese-rich sedimentary beds in the upper sequence are present in the Inner Piedmont, Kings Mountain, Charlotte, and Carolina slate belts. Similar rocks are unknown in the middle and lower sequences.

Other evidence that the stratigraphic sequences extend across the geologic belts is found in the inferred traces of the main unconformities (Fig. 2). The existence of these unconformities is not proven, and the positions shown on Figure 2 are tentative. Of these, the best defined is the unconformity between the lower and middle sequences, which is in part indicated by the distribution of metaandesite in the fan of folds between Fairfield and York Counties. The unconformity between the middle and upper sequences has been seen in the slate belt and Kings Mountain belt. None of the unconformities has been recognized in the Inner Piedmont belt in South Carolina, but the probable position of the one between the lower and middle sequences was indicated in an adjacent area in North Carolina (Overstreet, Yates, and Griffitts, 1963).

TABLE 1. Inferred Stratigraphic Succession and Sequence of Igneous Episodes in the Piedmont of South Carolina (Adapted from Overstreet and Bell, 1965a, Table 2)

Inferred geologic age	Sequences of sedimentary and pyroclastic rocks and intrusive episodes	Inner Piedmont belt	Kings Mountain belt	Charlotte belt	Carolina slate belt
Permian		Syenite pegmatite Gabbro, pyroxenite, and norite Yorkville Quartz Monzonite	Syenite Gabbro and pyroxenite Yorkville Quartz Monzonite	Minette and syenite pegmatite Syenite Gabbro, pyroxenite, and norite Yorkville Quartz Monzonite	Minette and syenite pegmatite Syenite Gabbro, pyroxenite, and norite
Permian and Carboniferous	Intrusive episode C	Muscovite pegmatite Cherryville Quartz Monzonite	Cherryville Quartz Monzonite	Granite in circular plutons at Chester, Winnsboro, and Liberty Hill	Granite in circular plutons at Winnsboro, Liberty Hill, and Cayce
Carboniferous	Upper sequence	Marble (in part) Biotite schist (in part) Quartzite (in part)	Marble Sericite schist (in part) Quartzite Sericite schist (in part)	Mica gneiss (in part)	Argillite (in part) Muscovite schist (in part)
			——— Unconformity ———		
	Middle sequence	Biotite granite gneiss			

		Toluca Quartz Monzonite	Metamorphosed gabbro	Metamorphosed mafic dikes	Metamorphosed mafic dikes
Devonian through Ordovician	Intrusive episode B	Toluca Quartz Monzonite	Oligoclase tonalite		
	Middle sequence	Metamorphosed gabbro and soapstone Biotite schist (in part) Marble (in part) Quartzite (in part) Henderson Gneiss Biotite gneiss and migmatite (in part) Hornblende gneiss (in part)	Sericite schist (in part) Hornblende schist	Mica gneiss (in part)	Argillite (in part) Muscovite schist (in part) Quartzite Quartz–microcline gneiss Amphibolite
			Unconformity		
Cambrian and Late Precambrian	Intrusive episode A	Biotite gneiss at Iva, Anderson County	Not recognized	Porphyritic granite in Abbeville County and gneissic granodiorite in York County	Not recognized
	Lower sequence	Hornblende gneiss (in part) Biotite schist (in part) Biotite gneiss and migmatite (in part)	Granitoid gneiss	Granitoid gneiss	Granitoid gneiss
			Unconformity (not observed in Piedmont)		
Basement			Unobserved in Piedmont		

The sedimentary and volcanic rocks of all sequences are intruded by igneous plutons and dikes of different ages and composition. Specific intrusive and unconformable relations between the igneous plutons and dikes, on the one hand, and the sedimentary and volcanic units in the three stratigraphic sequences, on the other hand, have been used as criteria to identify individual unconformities or stratigraphic sequences in the different belts (Overstreet and Bell, 1965a, p. 45–48, 52–54).

Lower Sequence

The lower sequence of sedimentary and volcanic rocks (Fig. 2, Table 1) is exposed in the Charlotte and Inner Piedmont belts and east of the Carolina slate belt. In the Charlotte belt and in the exposures east of the slate belt, the lower sequence consists of graywacke, arkose, shale, and pyroclastic rocks typically metamorphosed to the albite–epidote amphibolite facies. Much of the resulting rock is gneiss and migmatite to which the name granitoid gneiss was given (Table 1). In the Inner Piedmont belt the lower sequence is typically metamorphosed to staurolite–kyanite subfacies and sillimanite–almandine subfacies biotite schist, biotite gneiss, and migmatite (Table 1). Specific units of granitoid gneiss in the Charlotte belt and the area east of the slate belt have not yet been correlated with specific units in the Inner Piedmont.

Mafic lava and tuff are much less common in the lower sequence than they are in the middle and upper sequences. Thus, most of the central part of the Inner Piedmont belt, the central and southwestern part of the Charlotte belt, and the area east of the slate belt, have sparse mafic schist and gneiss. The mafic metamorphic rocks on the flanks of the Inner Piedmont belt (Fig. 2) may be part of the middle sequence.

The basement on which the lower sequence is presumed to rest has not been identified in the Piedmont of South Carolina. The basement is assumed to consist of old plutonic gneisses and schists like the pre-Ocoee rocks in the Blue Ridge west of the Brevard belt (King, 1955, p. 359–360).

Middle Sequence

The middle sequence consists of felsic and mafic lava and tuff, tuffaceous argillite, graywacke, and minor carbonate rocks. These rocks extend in broad folds from the Carolina slate belt across the Charlotte belt into the Kings Mountain belt, and appear in tight folds at higher metamorphic grade on the flanks of the Inner Pied-

mont belt. Rocks here included in the middle sequence were thought by Keith and Sterrett (1931, maps) to be part of the Precambrian basement on which later rocks of the Kings Mountain belt were deposited. Southeast of the area studied by Keith and Sterrett, however, the rocks they assigned to a Precambrian basement seem to overlie older rocks of the lower sequence.

The metamorphosed mafic lavas at and near the base of the middle sequence could be identified readily from soils maps; hence, they were given a dominant role in stratigraphic interpretations of the South Carolina Piedmont (Overstreet and Bell, 1965a, p. 21–22, 44–45; 1965b). Argillaceous rocks such as argillite, sericite schist, muscovite schist, mica gneiss, and sillimanite schist, however, are more common components of the middle sequence than the mafic rocks. The mafic lavas are generally the oldest rocks of the middle sequence, and they rest unconformably on the lower sequence, which is pierced by swarms of metamorphosed mafic dikes near the base of the middle sequence. Locally, as in Fairfield County, the mafic lavas are underlain by thin felsic flows and sedimentary strata which overlie the lower sequence. The upper part of the mafic lava becomes increasingly interlayered with argillaceous sedimentary rocks in interbeds that range in thickness from a few inches to hundreds of feet. Thus, the lavas are supplanted upward in the middle sequence by sedimentary rocks.

The sedimentary parts of the middle sequence originally included fine-grained, poorly bedded to laminated argillite, tuffaceous argillite, graywacke, felsic and mafic agglomerate, felsic flows and ash, and sparse conglomerate, sandstone, and carbonate-rich layers.

Some small masses of quartzite, muscovite quartzite, pyrophyllite quartzite, and kyanite quartzite in the middle sequence may not be of sedimentary origin. If they were originally sedimentary, they have been much modified by hydrothermal processes or ancient cycles of weathering.

Upper Sequence

The upper sequence was originally composed of shale, pyroclastic rocks, local thin carbonate-rich beds, and rare manganese-rich beds. In the slate belt (Fig. 2) these rocks unconformably overlie felsic and mafic volcanic rocks of the middle sequence that are intruded by dikes of metagabbro which do not extend into the upper sequence. In the Kings Mountain belt the rocks of the upper sequence rest unconformably on units of the lower and middle sequences consisting of hornblende schist, sericite schist, and biotite gneiss intruded by metagabbro which does not reach into the upper sequence (Overstreet and Bell, 1965a, p. 44–51). Rocks of the upper sequence in the Carolina slate belt are interpreted to be part of the same stratigraphic unit represented by the upper sequence in the Kings Mountain belt because of probable similar original lithology of the upper sequence in the slate belt and Kings Mountain belt, because the upper sequence in both belts unconformably overlies similar rocks, and because metagabbro is lacking in the upper sequence in both belts.

Areas of rocks of the upper sequence large enough to show on Figure 2 are unknown in the Charlotte and Inner Piedmont belts, but several units in these two belts are almost certainly part of the upper sequence. The most likely of these units are manganese-rich schist, marble, and quartzite in the Inner Piedmont belt and manganese-rich schist in the Charlotte belt (Overstreet and Bell, 1965a, p. 48). Parts of the mica gneiss unit (Table 1) in the northern part of the Charlotte belt in South Carolina are very likely part of the upper sequence, particularly in eastern York and Chester Counties and Lancaster County.

INTRUSIVE EPISODES

The relations among the igneous rocks in the Piedmont of South Carolina disclose a long and complex history of intrusive episodes. The plutonic intrusive rocks and dike rocks older than the diabase of Late Triassic(?) age have crosscutting relations that define three major episodes of igneous activity (Overstreet and Bell, 1965a, p. 14–16) which are called, from oldest to youngest, episode A, episode B, and episode C in Figure 3. Each of the intrusive episodes is thought to be connected with one of the sequences of sedimentation and volcanic activity and to have occurred late in the evolution of the sequence (Table 1).

Little is known of the distribution of rocks belonging to episode A. Episode B is thought to be the most important in the central and western parts of the Piedmont in South Carolina, and episode C is thought to be the most important in the eastern part of the Piedmont. Some of the rocks emplaced in episode B, like the Toluca Quartz Monzonite in the Inner Piedmont belt, were later deformed and partly recrystallized about the time the Yorkville Quartz Monzonite and Cherryville Quartz Monzonite of episode C were intruded.

All these intrusive rocks are intersected by diabase dikes of Late Triassic(?) age; therefore, they are older than Late Triassic(?). Although evidence for Cretaceous and Tertiary felsic volcanic activity has been observed in the Coastal Plain of the Carolinas and Georgia (Rooney and Kerr, 1967, p. 740; Heron and Johnson, 1966, p. 61; Sandy and others, 1966, p. 17, 23–25; Crawford and others, 1966, p. 1, 34–40), no rocks younger than Late Triassic(?) have been recognized in the Piedmont of South Carolina.

Age determinations by the lead-alpha method on zircon and monazite from granitic rocks in the Piedmont in North and South Carolina have given three groups of apparent ages in general agreement with observed field relations (Overstreet and Bell, 1965a, p. 90–113). The agreement of the lead-alpha ages with field data increases with decreasing age of the rocks. The three groups of ages are about 550 m.y., 450 m.y., and 260 m.y., corresponding to rocks in intrusive episodes A, B, and C.

Episode A

Intrusive rocks of episode A have been difficult to identify with certainty. All the intrusive rocks observed in South Carolina cut the metamorphic rocks of the lower sequence, but it has been difficult to find igneous rocks truncated by the unconformity between the lower and the middle sequences. At one place, a locality in Anderson County, a coarse-grained biotite orthogneiss containing coarse euhedral crystals of zircon intrudes biotite schist of the lower sequence in the Inner Piedmont belt and is unconformably overlain(?) by rocks of the middle sequence in the Kings Mountain belt. Similar gneiss was not found in the Kings Mountain belt; therefore, this orthogneiss may have been emplaced during intrusive episode A. Gneissic granodiorite closely intruded by swarms of metamorphosed mafic dikes in York County may also have been emplaced during intrusive episode A, because the dikes seem to be feeders for lava flows that are at the base of the middle sequence, but the age of the rock into which the gneissic granodiorite is intruded is uncertain.

Episode B

Most of the known intrusive rocks in the Piedmont of South Carolina are emplaced in metasedimentary and metavolcanic rocks of the middle sequence (Overstreet and Bell, 1965a, p. 15). A few of these intrusive rocks are clearly overlain unconformably by rocks of the upper sequence and can be related to episode B.

Felsic intrusives. Felsic rocks intrusive into the middle sequence and unconformably overlain by the upper sequence can be divided into rocks intruded before or after a unit of gabbro that is also unconformably overlain by the upper sequence (Table 1).

The older felsic rocks of episode B are represented by oligoclase tonalite (Espenshade and Potter, 1960, p. 70) in the Kings Mountain belt in York County. Equivalent rocks in the other belts are as yet unknown. The oligoclase tonalite is fractured and intruded by metagabbro of episode B (Table 1).

Younger felsic rocks of episode B, like the Toluca Quartz Monzonite of the Inner Piedmont belt (Table 1) in Cherokee and Spartanburg Counties, intrude the gabbro and contain inclusions of it. The Toluca Quartz Monzonite tends to form migmatitic complexes that are cut by the massive intrusive rocks of episode C.

Probably much of the granite in Newberry, Laurens, Greenville, Anderson, and Abbeville Counties, shown as "Granite, undivided" on Figure 3, is granite of episode B, but insufficient data are available to permit assigning a position to these rocks.

Mafic rocks. The mafic plutonic rocks of episode B include metagabbro in the Kings Mountain belt and Inner Piedmont belt, soapstone in the Inner Piedmont, and metamorphosed dikes in the Carolina slate belt and Charlotte belt (Table 1). In the Kings Mountain belt a unit of oligoclase tonalite is intruded by metagabbro of episode B, and both rocks are unconformably overlain by schistose pyroclastic rocks of the upper sequence (Espenshade and Potter, 1960; Overstreet and Bell, 1965a, p. 52–54; Butler, 1966). Metamorphosed mafic and ultramafic rocks ranging in composition from gabbro to soapstone crop out as small, irregularly shaped masses and short dikes in the Inner Piedmont. Inclusions of these mafic rocks in Toluca Quartz Monzonite of episode B in the Inner Piedmont have thick reaction rims of biotite. The metamorphosed dikes of episode B in the Carolina Slate belt and Charlotte belt are commonly gabbro and pyroxenite which are chloritized and epidotized.

Episode C

Rocks of intrusive episode C (Fig. 3 and Table 1) consist of discordant granitic plutons and associated pegmatite dikes, and discordant bodies of gabbro, norite, syenite, syenite pegmatite, and minette (Overstreet and Bell, 1965a, p. 15–16). Discordant plutons of granite in the Carolina slate belt and Charlotte belt form distinctive circular to elliptical bodies. The discordant granites of episode C in the Kings Mountain belt and Inner Piedmont belt cut across the unconformity between the middle and upper sequences and tend to form highly elongate bodies. Pegmatite dikes from which commercial muscovite was mined formed about the same time as the granites of episode C, because the pegmatites are rarely intruded by any later rock. Discordant bodies of gabbro, syenite, syenite pegmatite, and minette of episode C intrude the youngest granitic rocks in the Piedmont of South Carolina but are not intruded by the granites.

Granite. The most distinctive intrusive rocks of episode C are felsic plutons, circular to oval in plan, in the Carolina slate belt and Charlotte belt. They are also possibly the most potassic granites in South Carolina. The circular pluton near Winnsboro, Fairfield County, is the best example (Fig. 3). The rock is coarse-grained biotite granite with fine-grained marginal selvages and local hornblendic phases rich in sphene. Inclusions of wall rocks as much as 100 ft in length oriented parallel to the contacts are common near the rim of the pluton. Longest axes of many inclusions plunge steeply. The inclusions define a flow banding which indicates that the walls of the pluton dip steeply inward, and that flowage was upward and outward. Thus, the pluton is interpreted to have a carrotlike shape. The wall rock is locally brecciated. In the brecciated areas even small fragments of the wall rocks are intricately ruptured and threaded with granite, but there is little or no rotation or turbulent transport of blocks. The granite has produced very little contact alteration of its wall rocks, a feature characteristic of the late felsic plutons in the slate belt.

Granite plutons of episode C in the Charlotte belt are structurally controlled and located in folds near the contacts of the granitoid gneiss unit of the Charlotte belt and amphibolite and argillite of the slate belt (Table 1). The distinctive oval pluton of coarse-grained porphyritic biotite granite with fine-grained marginal phases in Chester County resembles the circular plutons in the slate belt in composition, structure, and age. Although the area of this oval pluton reaches batholithic dimensions (Fig. 3), its walls converge downward like the circular plutons in the slate belt.

The Cherryville Quartz Monzonite in Cherokee County and the Yorkville Quartz Monzonite in York and Cherokee Counties (Fig. 3 and Table 1) are examples of elongate discordant felsic plutons of episode C in the Kings Mountain and Inner Piedmont belts. The contacts of these plutons are steeply dipping, and the plutons tend to be many times longer than they are wide. The wall rocks may be notably metamorphosed at the contacts: sillimanite schist, corundum gneiss, pyroxene granulite, and cordierite gneiss were formed adjacent to the Yorkville (Potter, 1954, p. 149–157). In southwestern Cherokee County the effect of contact metamorphism from the Cherryville and Yorkville Quartz Monzonites has been to convert typical low-grade schists of the Kings Mountain belt into high-grade schist and gneiss identical in appearance with rocks of the Inner Piedmont belt.

Muscovite pegmatite. Commercial muscovite pegmatite dikes in the Piedmont of South Carolina were emplaced during intrusive episode C and are confined to the Inner Piedmont belt despite the fact that the major bodies of granite in South Carolina are in the Charlotte belt. The mica pegmatites are common in biotite schist in the Inner Piedmont belt, but they are uncommon in biotite gneiss, migmatite, Henderson Gneiss, biotite granite gneiss, and hornblende gneiss (Table 1). Efforts to relate the distribution of commercial muscovite pegmatite to bodies of granite fail to explain why these pegmatite dikes are restricted to zones of high-grade metasedimentary rocks and are absent from the largest areas of exposed granite in South Carolina. A more fundamental control than mere contiguity to granite seems to be necessary. Such control was suggested by Griffitts (1958, p. 83–97). He found that swarms of muscovite pegmatite dikes in the Carolinas are rarely associated with large bodies of granite. They are near small bodies of granite where the pegmatite magma entered rocks in which the pressure and temperature of the staurolite–kyanite subfacies of regional metamorphism prevailed. Apparently commercial muscovite tended not to crystallize in lower or higher temperature environments. Griffitts also postulated the need for some as-yet-undeciphered structural control for the formation of the muscovite pegmatites. To these criteria must be added a lithologic control, because regionally the swarms of muscovite pegmatite dikes are mainly in schistose rocks instead of gneissic or massive rocks (Overstreet and Bell, 1965a, p. 70).

Mafic rocks and syenite. Distinctive masses of gabbro, pyroxenite, norite, syenite, syenite pegmatite, and minette of episode C (Table 1) are known in the Piedmont of South Carolina (Fig. 3). The most characteristic examples of gabbro are intrusive into granite and granitoid gneiss of the Charlotte belt, but similar rocks occur in the other belts as well. Most of these gabbro masses are circular to kidney shaped in

plan. Their weathered outcrops usually form topographic depressions, the largest of which are as much as 100 ft below the surrounding land surface. Immediately adjacent rocks tend to form low ridges which follow the outline of the gabbro body. Coarse-grained to extremely coarse-grained hornblende gabbro with relict olivine and hypersthene is the principal variety of rock. Norite is present but rare. Commonly the gabbro grades into or is intruded by coarse hornblendite. Most of the hornblende gabbro and hornblendite appears to have been originally pyroxene gabbro and pyroxenite.

Syenite is associated with, and presumably a differentiate from, the gabbro. The syenite usually forms bold outcrops. The largest body of syenite in South Carolina is in the southwestern part of the Charlotte belt (Fig. 3), but smaller masses are known in the Kings Mountain belt and Carolina slate belt, and important syenite pegmatites are in the Inner Piedmont belt (Table 1).

Syenite pegmatite composed of microcline and biotite or vermiculite with little or no quartz is also a differentiate from the gabbro of episode C. Such pegmatite forms dikes within or some distance from the parent gabbro. They are known in all belts except the Kings Mountain belt (Table 1). The syenite pegmatites characteristically have thick wall zones of vermiculite, rich in coarse-grained zircon. Some have been mined for vermiculite in South Carolina, and others in North Carolina have been mined for zircon. The vermiculite wall zone may be present where the pegmatite intrudes felsic rocks as well as where it is in mafic rocks.

Lamprophyre dikes consisting of dominant fine-grained biotite with potassium feldspar, sparse coarse-grained biotite, and local, equant peppercorn-size grains of quartz are associated with the largest body of syenite in the Charlotte belt and are found as isolated dikes in the Carolina slate belt. These minette and quartz minette dikes are differentiated from the syenite and resemble in mineral composition the wall zones of the syenite pegmatites, except that coarse and copious zircon is lacking.

The syenite, syenite pegmatite, and minette are the youngest pre-Triassic intrusive rocks in the Piedmont of South Carolina. Only the diabase of Late Triassic(?) age is known to intrude them.

METAMORPHISM

Three definable episodes of regional metamorphism are thought to have affected the Piedmont in South Carolina, and an older event, known in the Blue Ridge belt, is inferred to have affected rocks of the unexposed basement in the Piedmont (Overstreet and Bell, 1965a, p. 114–115). Variable, local contact metamorphism was associated with the intrusion of igneous rocks in the three intrusive episodes. Superposition of two or more episodes of regional metamorphism, and local diverse effects of contact metamorphism, make many of the rocks polymetamorphic. Distribution of the resultant major metamorphic facies in the Piedmont of South Carolina is shown on Figure 4.

If Figure 4 is compared with Figure 1, it can be seen that the Carolina slate belt is essentially defined by rocks of the greenschist facies, including the muscovite–chlorite subfacies and the biotite–chlorite subfacies. Likewise the Charlotte belt, except for the common occurrence of granitic plutons (Fig. 3), is underlain dominantly by rocks of the albite–epidote amphibolite facies with a local rise in Newberry and Fairfield Counties to the staurolite–kyanite subfacies. The Inner Piedmont belt is marked by the highest grade metamorphic rocks in the Piedmont. The rocks rise in grade from the staurolite–kyanite subfacies on the flanks of the belt to the sillimanite–almandine subfacies in the core of the belt. The cumulative effect of regional metamorphism in three episodes created the geologic belts in the Piedmont of South Carolina.

Effects of the first episode of regional metamorphism in the Piedmont are scarcely identified as yet. This metamorphic episode may be represented by biotitic orthogneiss in the Inner Piedmont belt at Iva, Anderson County (Table 1). The few lead-alpha ages available for zircon from igneous rocks of episode A indicate that the first episode of regional metamorphism to affect the lower sequence of sedimentary and volcanic rocks in the Piedmont might have taken place in late Precambrian(?) or Cambrian time.

The second and strongest episode of regional metamorphism in the Piedmont of South Carolina seems to have taken place near the close of deposition of the middle sequence, possibly in Ordovician time when the Toluca Quartz Monzonite and equivalent rocks were emplaced. It brought parts of the Inner Piedmont belt to the sillimanite–almandine subfacies and staurolite–kyanite subfacies. Probably at the same time the Charlotte belt was in part brought to the albite–epidote amphibolite facies, and the older rocks of the slate belt, at least locally, were brought to the greenschist facies. The rocks of the upper sequence and igneous episode C were not affected by the second episode of regional metamorphism.

The third episode of regional metamorphism may be Carboniferous to Permian in age (Overstreet and Bell, 1965a, p. 114). It is the only episode to affect the sedimentary and volcanic rocks of the upper sequence, but it also affected the older rocks, largely in a retrogressive fashion.

Response of the rocks in the slate belt to this late episode of metamorphism was complex, and local geologic factors seem to have exerted considerable control. In areas where the rocks of the Carolina slate belt were already metamorphosed, particularly along the present margins of the belt, this episode resulted in some retrogressive effects. A description of this retrogressive response has been given for a part of the slate belt in North Carolina where chloritoid is reported to be pervasively altered to biotite and sericite (Stromquist and Conley, 1959, p. 9). Where slate belt rocks were unmetamorphosed prior to this event, as in the northern part of Saluda County and western Kershaw County, they were raised to the greenschist facies. Possibly much of the slate belt was metamorphosed at this time.

In the Charlotte belt, the third episode of regional metamorphism seems to have been a period of extensive fracturing and epidotization associated with the intrusion of felsic and mafic plutons during intrusive episode C.

This episode of metamorphism in the Kings Mountain belt affected the upper sequence of sedimentary and volcanic rocks, as well as older units. Folding was much tighter than in the slate belt, and some very high-grade contact-metamorphic rocks were formed adjacent to elongate discordant plutons of quartz monzonite of episode C.

In the Inner Piedmont belt the third episode of metamorphism was accompanied by fracture deformation, cross folding, and the intrusion of discordant plutons of episode C. Retrogressive metamorphic effects were the most common response in the Inner Piedmont. Sillimanite was widely replaced by white mica, hornblende was locally converted to biotite and chlorite, earlier biotite was extensively recrystallized into fine-grained aggregates of new biotite, and feldspar was partly but widely altered to muscovite.

The third episode of metmorphism closed with extensive faulting and mylonitization in the Brevard belt. A persistent west–southwest-trending fault of similar or somewhat younger age was formed between Lake Murray in Lexington County and the Georgia border of McCormack County (Overstreet and Bell, 1965a, Pl. 1). It may extend northeastward, in part under sedimentary rocks of the Coastal Plain, into North

Carolina (Henry Bell, III, and A. A. Stromquist, oral communication, 1967), and it continues southwestward in Georgia (Henry Bell, III, oral communication, 1967).

Further faulting, some producing ultramylonite younger than the mylonite of the Brevard belt, has been recognized in the western Piedmont and Blue Ridge of North Carolina by Conley and Drummond (1965a,b). It can be assumed that similar faults are in the Piedmont of South Carolina, but they have not been identified.

ORE DEPOSITS

The geologic synthesis presented here holds that the rocks exposed in the Piedmont of South Carolina formed under a wide variety of conditions of temperature and pressure so that the Piedmont cannot consist only of the deep and barren roots of old mountain systems. The distribution of known ore deposits has long been recognized as conforming to regional zones in the Piedmont (Sloan, 1908): sheet muscovite, sillimanite, kyanite, and monazite are found in the high-grade rocks and gold in the low-grade rocks (Overstreet and Bell, 1965a, Pl. 3). It is now thought that metals, minerals, and ores heretofore recognized as curiosities, or not recognized in the State, may exist in exploitable abundance. Deep weathering may have destroyed or modified the surface expression of base-metal deposits, for example, to the point where customary methods of surface exploration fail to disclose potential deposits. New methods must be introduced and widely used in South Carolina to study the regional zoning and distribution of metals and minerals in the weathered rocks of the Piedmont. New exploration techniques perfected in this part of South Carolina should have wide application in many other parts of the world where the rocks are deeply weathered.

REFERENCES

Bain, G. L., 1964, Road log of the Chatham, Randolph and Orange County areas, North Carolina: Carolina Geol. Soc. Ann. Mtg. Field Trip Guidebook, Raleigh, 23 p.

_____, and Thomas, J. D., 1966, Geology and ground-water in the Durham area, North Carolina: North Carolina Dept. Water Resources Ground-Water Bull. 7, 147 p.

Brown, C. Q., and Cazeau, C. J., 1964, Geology of the Clemson quadrangle, South Carolina: South Carolina State Devel. Board, Div. Geology Map Ser. 9.

Butler, J. R., 1966, Geology and mineral resources of York County, South Carolina: South Carolina State Devel. Board, Div. Geology Bull. 33, 65 p.

Cazeau, C. J., 1966, Geology of the La France quadrangle, South Carolina: South Carolina State Devel. Board, Div. Geology Map Ser. 10.

Conley, J. F., 1962, Geology and mineral resources of Moore County, North Carolina: North Carolina Div. Mineral Resources Bull. 76, 40 p.

_____, and Drummond, K. M., 1965a, Ultramylonite zones in the western Carolinas: Southeastern Geology, v. 6, no. 4, p. 201–211.

_____, 1965b, Faulted alluvial and colluvial deposits along the Blue Ridge front near Saluda, North Carolina: Southeastern Geology, v. 7, no. 1, p. 35–59.

Crawford, T. J., Hurst, V. J., and Ramspott, L. D., 1966, Extrusive volcanic rocks and associated dike swarms in central-east Georgia: Geol. Soc. America, Southeastern Sec., Field Trip Guidebook 2, 1966, Athens, Ga., Univ. Georgia, Dept. Geology, 53 p.

Dickson, J., 1819, Notices on the mineralogy and geology of parts of South and North Carolina: Am. Jour. Sci., 1st ser., v. 3, no. 1, p. 1–4.

Dietrich, R. V., 1964, Igneous activity in the southern Appalachians, in Lowry, W. D., ed., Tectonics of the southern Appalachians: Virginia Polytech. Inst. Dept. Geol. Sci. Mem. 1, p. 47–61.

Drayton, J., 1802, A view of South Carolina, as respects her natural and civil concerns: Charleston, W. P. Young, 252 p.

Espenshade, G. H., and Potter, D. B., 1960, Kyanite, sillimanite, and andalusite deposits of the southeastern States: U. S. Geol. Survey Prof. Paper 336, 121 p.

Floyd, E. O., 1965, Geology and ground-water resources of the Monroe area, North Carolina: North Carolina Dept. Water Resources Ground-Water Bull. 5, 109 p.

Griffitts, W. R., 1958, Pegmatite geology of the Shelby district, North Carolina: U. S. Geol. Survey open-file rept., 123 p.

Hadley, J. B., 1964, Correlation of isotopic ages, crustal heating and sedimentation in the Appalachian region, in Lowry, W. D., ed., Tectonics of the southern Appalachians: Virginia Polytech. Inst. Dept. Geol. Sci. Mem. 1, p. 33–45.

Hammond, H., 1883, South Carolina. Resources and population, institutions and industries: Charleston, S.C., South Carolina State Board Agriculture, 726 p.

Heron, S. D., and Johnson, H. S., Jr., 1966, Clay mineralogy, stratigraphy, and structural setting of the Hawthorn Formation, Coosawhatchie district, North Carolina: Southeastern Geology, v. 7, no. 2, p. 51–63.

Jonas, A. I., 1932, Structure of the metamorphic belt of the southern Appalachians: Am. Jour. Sci., 5th ser., v. 24, p. 228–243.

Keith, Arthur, 1905, Description of the Mount Mitchell quadrangle [North Carolina–Tennessee]: U. S. Geol. Survey Geol. Atlas, Folio 124, 10 p.

_____, 1907, Description of the Pisgah quadrangle [North Carolina–South Carolina]: U. S. Geol. Survey Geol. Atlas, Folio 147, 8 p.

_____, and Sterrett, D. B., 1931, Description of the Gaffney and Kings Mountain quadrangles [South Carolina–North Carolina]: U. S. Geol. Survey Geol. Atlas, Folio 222, 13 p.

King, P. B., 1955, A geologic section across the southern Appalachians—an outline of the geology in the segment in Tennessee, North Carolina, and South Carolina, in Russell, R. J., ed., Guides to southeastern geology: New York, Geol. Soc. America, p. 332–373.

King, P. B., 1964, Further thoughts on tectonic framework of southeastern United States, in Lowry,

W. D., ed., Tectonics of the southern Appalachians: Virginia Polytech. Inst. Dept. Geol. Sci. Mem. 1, p. 5–31.

Lieber, O. M., 1858a, Report on the survey of South Carolina; being the first annual report ...: 2d ed., Columbia, S.C., 133 p.

————, 1858b, Report on the survey of South Carolina, being the second annual report ...: Columbia, S.C., 145 p.

————, 1859, Report on the survey of South Carolina; being the third annual report ...: Columbia, S.C., 223 p.

————, 1860, Report on the survey of South Carolina, being the fourth annual report ...: Columbia, S.C., 194 p.

Marsh, O. T., and Laney, R. L., 1966, Reconnaissance of the ground-water resources in the Waynesville area, North Carolina: North Carolina Dept. Water Resources Ground-water Bull. 8, 131 p.

McCauley, C. K., and Butler, J. R., 1966, Gold resources of South Carolina: South Carolina State Devel. Board, Div. Geology Bull. 32, 78 p.

————, and McCauley, J. F., 1964, Corundum resources of South Carolina: South Carolina State Devel. Board, Div. Geology Bull. 29, 9 p.

Overstreet, W. C., and Bell, H., III, 1965a, The crystalline rocks of South Carolina: U. S. Geol. Survey Bull, 1183, 126 p.

————, 1965b, Geologic map of the crystalline rocks of South Carolina: U. S. Geol. Survey Misc. Geol. Inv. Map I-413.

————, Yates R. G., and Griffitts, W. R., 1963, Geology of the Shelby quadrangle, North Carolina: U. S. Geol. Survey Misc. Geol. Inv. Map I-384.

Paradeses, W., McCauley, J. F., and Colquhoun, D. J., 1966, The geology of the Blythewood quadrangle: South Carolina State Devel. Board, Div. Geology Map Ser. 13.

Potter, D. B., 1954, High alumina metamorphic rocks of the Kings Mountain district, North Carolina and South Carolina: U. S. Geol. Survey open-file report, 204 p.

Reed, J. C., Jr., and Bryant, B., 1964, Evidence for strike-slip faulting along the Brevard zone in North

Carolina: Geol. Soc. America Bull., v. 75, p. 1177–1196.

————, and Owens, J. P., 1967, Interpretation of basement rocks beneath the Atlantic Coastal Plain from reconnaissance aeromagnetic data [abs.]: Geol. Soc. America Ann. Mtg., New Orleans, Program, p. 182–183.

Ridgeway, D. C., McCauley, J. F., and Colquhoun, D. C., 1966, Geology of the Blaney quadrangle, South Carolina: South Carolina State Devel. Board, Div. Geology Map Ser. 11.

Rooney, T. P., and Kerr, P. F., 1967, Mineralogic nature and origin of phosphorite, Beaufort County, North Carolina: Geol. Soc. America Bull., v. 78, no. 6, p. 731–748.

Sandy, J., Carver, R. E., and Crawford, T. J., 1966, Stratigraphy and economic geology of the Coastal Plain of the central Savannah River area, Georgia: Geol. Soc. America, Southeastern Sec., Field Trip 3, Guidebook, Athens, Ga., Univ. Georgia, Dept. Geology, 30 p.

Siple, G. E., 1958, Stratigraphic data from selected oil tests and water wells in the South Carolina Coastal Plain: South Carolina State Devel. Board, Div. Geology, Mineral Industries Lab. Monthly Bull., v. 2, no. 9, p. 62–68.

Sloan, E., 1908, Catalogue of the mineral localities of South Carolina: South Carolina State Devel. Board, Div. Geology, reprint, 505 p. [1958].

Stromquist, A. A., and Conley, J. F., 1959, Field trip guidebook, geology of the Albemarle and Denton quadrangles, North Carolina: Carolina Geol. Soc., 36 p.

Tuomey, M., 1844, Report on the geological and agricultural survey of the State of South Carolina: Columbia, S.C., A. S. Johnston, 63 p.

————, 1848, Report on the geology of South Carolina: Columbia, S.C., A. S. Johnston, 293 p., appendix.

White, A. M., Stromquist, A. A., Stern, T. W., and Westley, H., 1963, Ordovician age for some rocks in the Carolina slate belt in North Carolina: U. S. Geol. Survey Prof. Paper 475-C, p. C107-C109.

Chapter 26

The Piedmont in Georgia

VERNON J. HURST

INTRODUCTION

SINCE THE GEOLOGIC MAP of Georgia (Stose and Smith, 1939) was published, more than 5,000 sq mi of the Georgia Piedmont have been mapped geologically by reconnaissance standards or better (Fig. 1). The new work allows major revision of concepts previously held. This review is presented as a brief working summary of major stratigraphic, tectonic, metamorphic, and chronologic concepts derived from the recent work, most of which is unpublished.

Several terms used commonly in the past are not perpetuated in this review, as Carolina Gneiss, Talladega Series, Amicalola Belt, Ashland Schist, and Wedowee Formation. The term Carolina Gneiss, defined by Keith in 1901–1903, has been so broadly applied to unrelated rock groups as to nullify its usefulness, now that detailed stratigraphic relations are beginning to emerge. Talladega Series, coined by Smith in the late 1800's, likewise encompasses unrelated rocks. The Ashland Schist and Wedowee Formation, too, are catch-all terms. To remain useful these terms would have to be redefined and very much restricted. Confusion can be avoided by discontinuing their use.

Several lithologic belts are designated by numbers in Figure 2. Some of the boundaries are similar to the ones previously drawn by Crickmay (1952, p. 6). Where available information is sufficient the belts are delineated so as to group related units, but the main purpose of the division of the area into belts is to facilitate rapid review.

GENERAL GEOLOGIC RELATIONS

The belt of metamorphic and igneous rocks in Georgia generally is called the Piedmont or Crystalline Piedmont, although in addition to the Piedmont physiographic province it also includes the Upland and Highland physiographic provinces which are extensions of the Blue Ridge.

The crystalline rocks are bounded on the west by the Cartersville Fault, which is generally marked by, or near to, a prominent scarp. Shales and limestones of Paleozoic age lie immediately west of the fault throughout most of its course in Georgia; the prominence of the scarp relates to the type of rock east of the fault. In northern Georgia where quartzite and metagraywacke dominate east of the fault, the scarp is pronounced. In Bartow County where quartzites are west of the fault and where granite and schist lie immediately east of it, the erosional scarp does not correspond closely to the fault. In Polk County the erosional scarp again corresponds closely with the fault because metagraywacke again is the prominent rock type east of the fault, and Paleozoic formations to the west intersect the fault at a high angle.

The crystalline rocks are bounded on the east by an erosional unconformity at the base of the Cretaceous deposits of the Coastal Plain. The surface expression of the unconformity is the Fall Line. The general dip of the unconformity is southeast at 8–15 ft/mi. (3.8–7.2 m/km). In detail the surface of the unconformity is gently undulating with local relief of as much as 150 ft (46 m). The Fall Line is clearly visible at most places in east Georgia as a low scarp. In middle Georgia and on to the southwest the scarp commonly is lacking.

Metamorphic grade is low on the west side of the Piedmont, but rises rapidly toward the southeast, and remains high southeastward all the way to the Fall Line.

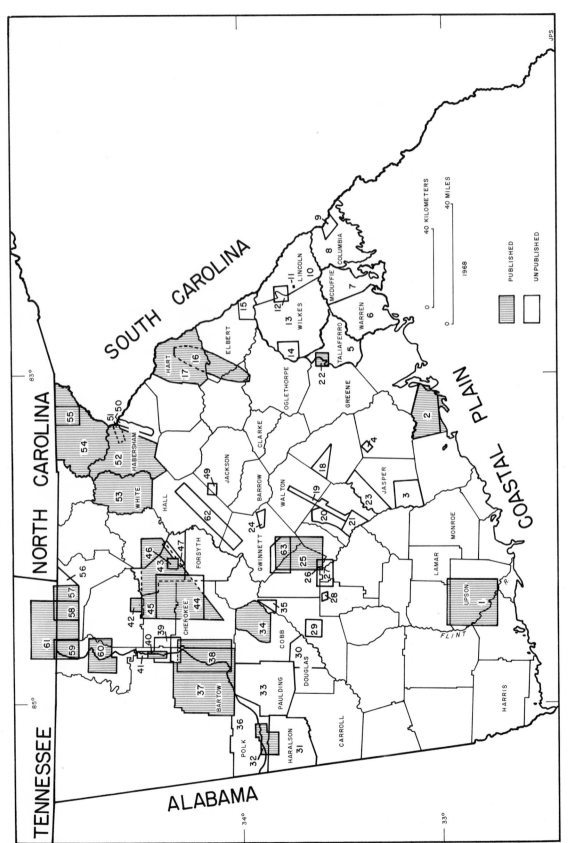

FIGURE 1. Geologic Mapping in Georgia since 1939. Compilation by Jim W. Smith and Samuel M. Pickering

KEY

1. Clarke, James W. (1952)
2. La Moreaux, Philip E. (1946)
3. Matthews, Vincent III (1967)
4. Meyers, Carl Weston (1967)
5. Crawford, Thomas J., Taliaferro County, Unpublished Geologic Map.
6. Crawford, Thomas J., Warren County, Unpublished Geologic Map.
7. Crawford, Thomas J., McDuffie County, Unpublished Geologic Map.
8. Crawford, Thomas J., Columbia County, Unpublished Geologic Map.
9. McLemore, William H. (1965)
10. Crawford, Thomas J., Lincoln County, Unpublished Geologic Map.
11. Hurst, Vernon J. (1959)
12. Fouts, James A. (1966)
13. Crawford, Thomas J., Wilkes County, Unpublished Geologic Map.
14. Cook, Robert B., Jr. (1967)
15. Austin, Roger S. (1965)
16. Furcron, A. S., and Teague, Kefton H. (1945)
17. Grant, Willard H. (1958)
18. Lawton, David E. (1967)
19. Reade, Ernest H., Jr. (1960)
20. Schulz, Roger S. (1961)
21. Gardner, Charles Harwood (1961)
22. Medlin, Jack H., and Hurst, Vernon J. (1967)
23. Fountain, Richard C. (1961)
24. Grant, Willard H. (1949)
25. Herrmann, Leo A. (1954)
26. Cofer, Harland E. (1948)
27. Holland, Willis A. (1954)
28. King, James A. (1957)
29. Cofer, Harland E. (1958)
30. Schepis, Eugene L. (1952)
31. Crawford, Thomas J., Haralson County, Unpublished Geologic Map.
32. Webb, James E. (1958)
33. Hurst, Vernon J., and Crawford, Thomas J., Paulding County, Unpublished Geologic Map.
34. Hurst, Vernon J. (1956)
35. Higgins, Mike W. (1965)
36. Pinson, William H., Jr. (1949)
37. Croft, M. G. (1963)
38. Kesler, Thomas L. (1950)
39. Smith, James W. (1959)
40. Stewart, J. W. (1958)
41. Smith, William L. (1958)
42. Power, W. Robert, and Reade, Ernest H. (1962)
43. Sever, Charles W. (1946)
44. Fairley, William M. (1965)
45. Furcron, A. S., and Teague, Kefton H. (1945)
46. Cofer, Harland E., Jr. in Stewart and others (1964).
47. Bowen, Boone M., Jr. (1961)
48. Holland, Willis, Forsyth County, Unpublished Geologic Map.
49. Klett, William Y., Jr. (1967)
50. Brent, William B. (1952)
51. Pruitt, Robert G., Jr. (1952)
52. Crawford, Thomas J., Habersham County, Unpublished Geologic Map.
53. Otwell, Larry, White County, Unpublished Geologic Map.
54. Furcron, A. S., and Teague, Kefton H. (1951)
55. Giles, Robert T. (1966)
56. Nuttall, Brandon D. (1950)
57. Hurst, Vernon J. (1955)
58. Hurst, Vernon J. (1956)
59. Salisbury, John W. (1961)
60. Furcron, A. S., Teague, Kefton H., and Calver, J. L. (1947)
61. Hurst, Vernon J., and Schlee, John S. (1962)
62. Holland, Willis, and Hurst, Vernon J., Geologic Maps along the Brevard Belt, Unpublished.
63. Mohr, David W. (1965)

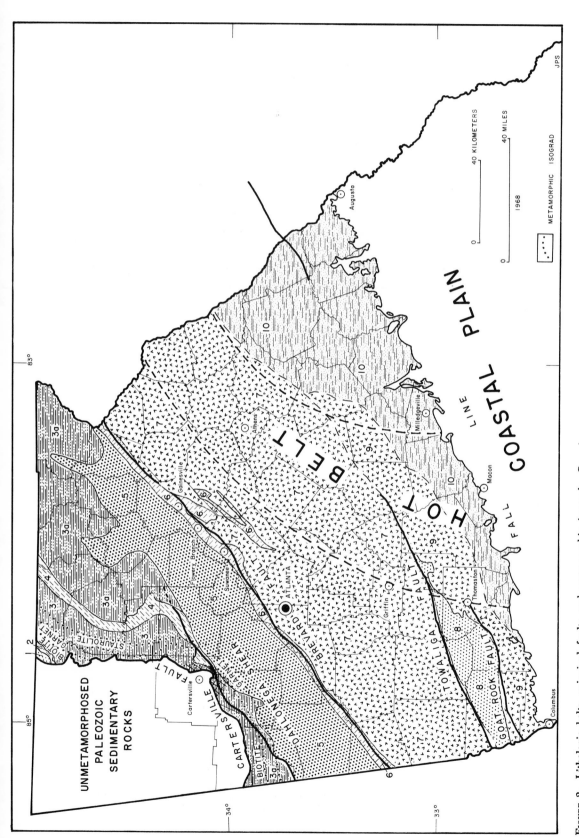

FIGURE 2. Lithologic belts, principal faults and metamorphic isograds, Georgia. See text for description of numbered belts.

Belt 1

The rocks in the belt immediately east of the Cartersville Fault in northernmost Georgia (No. 1 of Fig. 2) are chiefly quartzite, arkosic quartzite, siltstone, shale, or phyllite, but also include minor limestone. Rodgers (1953) correlates the westernmost quartzite with the Nebo Quartzite of the Chilhowee group of Cambrian age and the other rocks with the Sandsuck Shale of the Ocoee Series of late Precambrian age. The Nebo Quartzite is intensely faulted, though none of the faults appear to have caused any great displacement except the Cartersville Fault, which is the western boundary of the unit. Probably the disruption of the sequence is more apparent than real. Original textures are obscured by the intense deformation and by silicification, but original grain shapes, conglomeratic texture and cross-bedding are recognizable at places. *Scolithus* tubes have been found in the road cut at Parksville dam (lat. 35.1° N. long. 84.7° W.).

The quartzites are interbedded with thin phyllites or shales. The quartzite beds range in thickness from a few inches to massive beds as much as 30 ft (10 m) thick.

The limestone in this belt is dark colored, laminated, and cut by numerous thin calcite veinlets. Thin sections show fine granular carbonate with very fine pigmenting matter. Some of the limestone is dolomitic. The limestone crops out along U. S. Highway 65 about 1.5 mi east of the dam at Parksville, Tennessee (lat. 35.1° N. long. 84.7° W.), as reported by Hurst and Schlee (1962, p. 8); a similar limestone crops out in the same stratigraphic position to the south in the Cohutta Mountain quadrangle (lat. 34.9° N. long. 84.8° W.) where it has been described by Salisbury (1961, p. 30–32).

The rocks in belt 1 are mostly right-side up, as shown by cross-bedding and graded bedding. Though bounded by the Cartersville Fault on the west and the Sylco Creek Fault(?) on the east, their attitude and position indicate that they are younger than the phyllitic rocks to the east and are probably of lower Cambrian age.

Belt 2

The rocks in belt 2 are mainly gray-green laminated phyllites with a total thickness of at least 1500 ft (450 m). The individual phyllite layers are ½–2 in. (1–5 cm) thick, occasionally as much as 6 in. (15 cm) thick. Fine-grained, carbonate-rich beds are interlayered with finer grained, less competent, mica-rich or quartz–mica-rich beds. About 900 ft (275 m) stratigraphically above the base of the phyllite unit is a horizon of thinly interbedded quartzites and calcareous quartzites. The quartzite-rich zone is 30–50 ft (9–15 m) thick. Individual beds of quartzite are generally 1–4 in. (2.5–10 cm) thick; a few exceed 1 ft (0.3 m). The quartzite and calcareous quartzite beds constitute a third to a half of the zone, the other part being interbedded phyllite similar to that above and below the zone.

Stratigraphically beneath the gray-green phyllite and conformable with it is a band of dark laminated pyritic phyllite or slate. Layers within the dark slate are mostly less than ¾ in. (1.9 cm) thick and are commonly marked by concentrations of pyrite cubes, many of them up to 1 in. (2.5 cm) across. The slate is about 600 ft (183 m) thick, except where thinned by faulting. There are excellent exposures a few miles north of the Georgia–Tennessee line along the Ocoee River gorge (lat. 35.1° N.; long. 84.5° W.). This unit can be correlated confidently with the Nantahala Slate in the trough of the Murphy synclinorium, 17 mi (27.3 km) to the southeast, because the sequence and structure of intervening formations have been worked out in detail (Hurst, 1955; Hurst and Schlee, 1962).

Belt 3

The rocks of belt 3 are mainly metagraywacke (commonly conglomeratic) and phyllite or mica schist. They belong to the Great Smoky Group of late Precambrian age (Hurst, 1963, Pl. 2).

A study of current structures in the northern part of belt 3 (Mellen, 1956) shows that the sediment source for at least a part of the Great Smoky Group lay to the north or northeast. Consistent with Mellen's conclusion, the proportion of conglomeratic units decreases to the southwest, though there are still conglomeratic layers on the west side of the State in Haralson County. The stratigraphy of the northern part of the belt has been described in detail (Hurst, 1955, p. 27–45). The Great Smoky Group in Haralson County has been described by Webb (1958, p. 19–24).

Where belt 3 is crosshatched in Figure 2, the Great Smoky Group is cut out by the convergence of the Cartersville Fault and another unnamed fault to the east. Geologic relations are particularly complex in this area: slightly metamorphosed lower Cambrian formations are intricately folded, highly faulted and intruded by the Corbin Granite. This is an area of widespread mineralization and hydrothermal alteration.

The rocks in belt 3A are correlative with belt 3. The thicknesses and sequences of beds are very similar, as well as zonations of conglomeratic versus fine-grained formations and the patterns made by the presence or absence of current structures, but lithologies are different due to the higher grade of metamorphism in belt 3A. The calcareous quartzites in the western part of belt 3 correspond to pseudodiorites in the eastern part of belt 3 and in belt 3A; iron-rich, aluminous schists to the west correspond to staurolite schists to the east. The metagraywacke units are similar in both belts, though textures have been modified more toward the east, where the metagraywacke is generally called biotite gneiss.

Belt 4

Belt 4 is the Murphy Marble belt. The oldest unit in the belt is the Nantahala Slate. Above it are quartzites, other slates, phyllites, a band of marble up to 250 ft (76.2 m) thick, and sericite schist. These rocks have been described in detail by Hurst (1955, p. 45–46) and Fairley (1965, p. 16–41). Detailed stratigraphic study of the marble in the Tate area (lat. 34.4° N. long. 84.3° W.) has been carried on by geologists with the Georgia Marble Company in the last few years but none of their work has been published.

Belt 4 is the trough of a major synclinorium. Though there is much local faulting due to upward-outward crowding of the rocks in the central part of the syncline, none of the units are cut out for any great distance as far south as Cherokee County. There the belt is apparently cut off by a major unnamed fault which extends along what previously has been called the Dahlonega shear zone. Marble is exposed at the surface at two places farther to the southwest in Haralson County.

The structures between belts 1 and 4 are known well enough to confidently correlate the Murphy Marble with the carbonates in belt 1. The formations below the marble in belt 4 correlate very well with those in belt 2. Probably the marble is of Cambrian age.

Belt 5

Belt 5 contains a stratigraphic sequence of various schists, metavolcanics and lesser quartzites. Though intruded by a variety of rocks ranging from granite to gabbro, and very commonly migmatized, this sequence is well-defined from western Georgia at least as far to the northeast as Forsyth County. Good geologic maps are available for much of the belt (Fig. 1).

The stratigraphic sequence in the middle portion of belt 5 is represented in Figure 3 (Hurst, 1956). Prominent quartzites about a hundred feet thick are overlain by several hundred feet

of fine- to medium-grained layered rocks which range from fine mica schist to paragneiss. Underneath the quartzite is a sequence of mica schists 1000–2000 ft (305–610 m) thick characterized by abundant garnets. Below the garnetiferous mica schists is another mica schist unit up to 3,000 ft thick in which garnets characteristically are scarce and in which there are thin amphibolite bands. Below the unit are amphibolite rocks and other metavolcanics. The contact relations of the mica schist unit and the metavolcanics are well revealed at several outcrops as gradational by interbedding.

Kyanite–quartz schists are found in this stratigraphic sequence above the metavolcanics, often in the mica schist unit, but also bounding the garnet–mica schists. It is not clear from the work done so far whether the kyanite–quartz schists are a single stratigraphic unit.

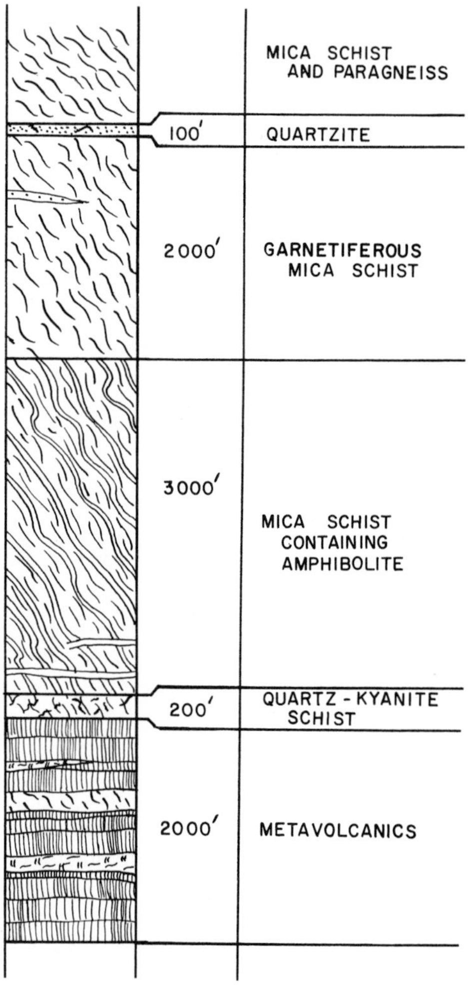

FIGURE 3. Generalized section of metamorphic rocks in belt 5 of Figure 2.

The metavolcanics are well developed in Paulding County where metatuffs, metamorphosed amygdaloidal basalts, and other metabasalts can be recognized. Apparently, only the lower part of the Cobb County sequence crops out in Paulding County and on to the southwest. Graphitic schists, which are not conspicuous in Cobb County, are prominent in Paulding, Haralson, and Carroll Counties. The Cobb County quartzite extends for several miles to the southwest in Douglas County but becomes inconspicuous in the stratigraphy farther to the southwest.

Northeast of Cobb County, the quartzite becomes even more prominent, and there are at least two quartzite horizons. The second quartzite begins as a quartz-rich zone in the garnet– mica schists in the middle of Cobb County and develops into a continuous quartzite layer more than a hundred feet thick in Forsyth County. On to the northeast in Hall County, quartzites diminish and it appears that the outcropping units belong mainly to the lower part of the Cobb County sequence.

Various gneisses and migmatites locally obscure the stratigraphic relations in belt 5, but the broad outline of the stratigraphy is still clear: a horizon of quartzites is overlain by paragneiss and underlain by garnet–mica schists. The garnet– mica schists become less garnetiferous downward, become interbedded with amphibolites, and grade into a thick unit of metavolcanics, principally metabasalts.

The age relation between this stratigraphic sequence and that in belt 3 is unknown. Major faulting separates the two belts from the Alabama line northeastward at least as far as Cherokee County.

The relative age of the rocks in belt 5 and the Brevard zone, belt 6, likewise is unclear. Attempts have been made to relate the two by detailed mapping in southeastern Cobb County (Jim Smith), Flowery Branch to Suwanee in Gwinnett County (Willis A. Holland and Vernon J. Hurst), the White Sulfur Springs area in Hall County (Vernon J. Hurst) and Habersham County (Thomas J. Crawford). Various gneisses obscure the stratigraphic relations of the two belts in the areas so far mapped.

Belt 6

Belt 6 is the Brevard zone. The principal rock types in it are mica schist and biotite gneiss. Less abundant rock types are marble, quartzite, and amphibolite.

The schists are fine- to medium-grained, gray quartz–muscovite schists containing minor feldspar, and commonly graphitic. Biotite and garnet are common accessories. Staurolite schists and sillimanitic schists occur locally. The schists are interbedded with thin fine- to medium-grained quartzites in which current structures are preserved. The quartzites and most, if not all, of the schists are metasedimentary units.

The marble typically is a thin band of bluish gray, fine-grained, magnesian marble. Its thickness and quality vary along strike due to facies changes. Locally it is doubled by folding and interrupted by faulting and flowage. Still, the marble crops out almost continuously from the vicinity of Suwanee (lat. 34.2° N. long. 84.1° W.) to the North Carolina–Georgia line.

Brevard-type rocks crop out extensively in middle Georgia east of the main zone. Though they are yet to be mapped in detail, they have been traced by reconnaissance methods in southeastern Hall County and portions of Jackson, Barrow, and Gwinnett Counties, where they are folded.

Belt 7

Belt 7 has been studied in Hart County (Grant, 1958) and in the Stone Mountain–Lithonia district (Herrmann, 1954). Gneisses predominate, though sillimanitic schists, staurolitic schists, garnet–mica schists, quartzites and amphibolites locally are well developed.

Good exposures made along major highways during the last ten years show that many of the gneisses are metasedimentary units. Field work is hampered in this belt by the difficulty of distinguishing paragneisses from orthogneisses.

Prominent quartzites which never have been mapped are in Walton and Clarke Counties. They might be equivalent to the quartzites in belt 6 in Gwinnett, Barrow, and Jackson Counties. It is likewise possible that they are equivalent to the Hollis Quartzite in belt 8.

Belt 8

The Woodland Gneiss (Clarke, 1952, p. 7) and the Pine Mountain Series (Galpin, 1915, p. 74) consisting of the Sparks Schist, overlain by the Hollis Quartzite and the Manchester Schist occupies belt 8. To the west, in Alabama, the Chewacla Marble is found in the Pine Mountain Series between the Hollis Quartzite and the Manchester Schist.

The Woodland Gneiss is a biotite granite gneiss. It was described by Clarke (1952, p. 6–7) as comprising the base of the stratigraphic column between the Towaliga Fault and the Goat Rock Fault. Clarke described the Hollis Quartzite as unconformably overlying the Woodland Gneiss and conformably underlying the Manchester Formation (Manchester Schist). The Hollis Quartzite is up to 800 ft (244 m); it is bedded, and locally conglomeratic. The Manchester Formation consists of mica schist, graphitic mica schist, and kyanite–muscovite schist.

Willard H. Grant recently has studied the belt along the Towaliga Fault.

Belt 9

Belt 9 is characterized by a variety of granites, gneisses, and basic rocks. Clarke (1952) described charnockitic hypersthene–quartz monzonite and hypersthene gabbro, biotite–oligoclase gneiss, and epidote amphibolite gneiss in the Thomaston area of Upson County. Hurst and Holland (1959, p. 74) have briefly described coronites near Culloden in Monroe County. Matthews (1967) has described norite in Jasper County. A variety of granitic and basic rocks within the belt have never been described.

Belt 10

The principal and probably the oldest unit in belt 10 is a series of gneisses which are referred to as the Kiokee Series, prominent in Columbia County (McLemore, 1965). These gneisses extend westward from Columbia County across central McDuffie County and into central Warren County, with little change.

The Little River Series underlies most of Taliaferro, Wilkes, and Lincoln Counties and parts of northern Columbia, McDuffie, and Warren Counties, except where there are granites. Similar rocks extend onto the southwest in belt 10, though they have not been studied in detail.

The Little River Series was named by Crickmay (1952) from exposures along the Little River in Wilkes, Lincoln, and McDuffie Counties. He equated the series with the rocks of the Carolina slate belt in North and South Carolina.

Probably the oldest unit in the Little River Series is the extensive metadacite in Lincoln and Wilkes Counties. Beginning in east-central Wilkes County near Metasville, the metadacite outcrop broadens to a width of 4 mi near Lincolnton and maintains this width eastward to the Savannah River. A smaller area underlain by metadacite is near Broad and Anthony shoals in northeastern Wilkes County. The metadacite in adjacent parts of Elbert County has been described by Austin (1965). The metadacites and related rocks are laced by thousands of basic dikes, and generally are deeply weathered. In the past they have been mistakenly identified as granitic rocks.

Other prominent rock types in belt 10 are epidote–hornblende gneiss, hornblende gneiss, biotite gneiss, amphibolite, quartz–muscovite schist, kyanite–quartz–sericite schist, sillimanite–quartz–sericite schist, staurolite–muscovite schist, quartzite, and phyllite.

Subsurface Basement Rocks of the Coastal Plain

Igneous and metamorphic rocks extend southeast of the Fall Line under the coastal plain sediments at least 150 mi (240 km) and probably farther. Test borings that have penetrated the basement within about 35 mi (56 km) of the Fall Line typically have encountered diorite, schist, gneiss, and other crystalline rocks similar to those encountered at the surface in belt 10. Deep borings within a belt 35–85 mi (56–136 km) southeast of the Fall Line typically have encountered "red beds" and diabase. Farther to the southeast, deep wells have encountered volcanic rocks, as basalt, rhyolite, and tuff, feldspathic quartz sandstone, dark shale, and occasionally granite. The general distribution of igneous and metamorphic rocks beneath the coastal plain sediments has been shown by Hurst (1960, p. 13). Detailed descriptions of the rocks have been given by Milton and Hurst (1965).

LARGE SCALE FAULTING

Movements along major faults in the Piedmont (Fig. 2) partly predate and partly postdate the Appalachian orogeny. Considerable movement postdates the last period of metamorphism which was about 250 million years ago, but no major movement has been recorded in historic time.

Cartersville Fault

The best known of the Piedmont faults is the Cartersville overthrust which is along or near the western boundary of the igneous and metamorphic rocks. Its existence was reported first in 1890 and 1895 by Hayes, Campbell, and Brooks (manuscript report in files of U. S. Geological Survey) and amplified by subsequent work (Hayes, 1901). Recent work by Kesler (1950) cast doubt on the existence of a large-scale overthrust in the Cartersville district. Since Kesler's work, four special studies have been made for clarification: (1) Webb (1958) mapped 60 mi astride the Cartersville Fault in Polk and Haralson Counties; (2) James W. Smith (1959) prepared a detailed map of 50 sq mi in the Fairmount area (lat. 34.4° N. long. 84.6° W.); (3) Salisbury (1961) mapped the northwest corner of the Cohutta Mountain quadrangle (lat. 34.9° N. long. 84.8° W.) ; (4) Hurst and Schlee (1962) mapped the Ocoee River gorge in southeast Tennessee (lat. 35.1° N. long. 84.5° W.).

The general relationship observed in Fairmount (lat. 34.4° N. long. 84.6° W.) and in Haralson County is shown schematically in Figure 4. Meta-sediments which are right-side up and dipping gently to the southeast are on top of highly folded, slightly metamorphosed to unmetamorphosed lower Paleozoic rocks. The metasediments are lithologically dissimilar to any rocks in the Paleozoic section from Cambrian to Pennsylvanian. Either the metasediments are Precambrian and have been thrust into contact with much younger Paleozoic sediments (a displacement of several thousand feet would be necessary in either case) or there is a remarkable facies change along the fault. As detailed below, there is strong evidence for the fault aside from the lithologic evidence.

Cambro-Ordovician rocks are along the west side of the Cartersville Fault everywhere in Georgia except in the Rockmart area (lat. 34° N. long. 85.1° W.), where there are Mississippian rocks. Tightly folded lower Cambrian quartzites lie east of the fault from Alaculsey (lat. 34.9° N. long. 84.7° W.) to the Tennessee line, along the fault in the vicinity of Cartersville, and west of the fault in western Georgia. The fault truncates stratigraphic units and also disrupts the regional pattern of metamorphic grade. The metamorphic isograds in north Georgia (Hurst and Schlee,

1962, p. 10) continue to approach the fault as far south as Fairmount, where the biotite isograd is truncated, and diverge from the fault farther south.

What has been called the Cartersville Fault actually is a system of faults, particularly in the Cartersville district. Movements partly preceded and partly followed the last period of metamorphism. Total displacement varies from not very great in northern Georgia to great in the Cartersville District, and probably great on to the west side of Georgia.

Dahlonega Shear Zone

The Murphy Marble belt is cut off in Cherokee County by a zone of distributive movement which has been called the Dahlonega shear zone. The zone appears to continue southwest from Cherokee County at least as far as the state line. Marble appears along it in Haralson County.

The strike faults shown by LaForge and Phalen (1913) in north Georgia and extended southward by Bayley (1928) as the Whitestone fault now are known to be small trough faults that do not warrant representation as a major continuous fault.

Recent mapping by Thomas J. Crawford shows a portion of the shear zone in Haralson County, but no detailed study has yet been undertaken.

Brevard Zone

One of the largest postulated faults is the Brevard zone which extends northeast across

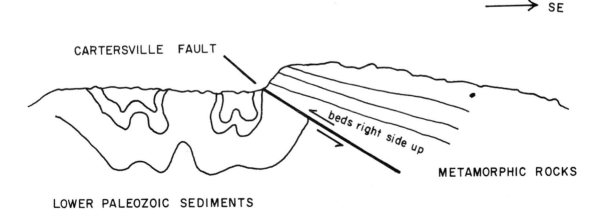

FIGURE 4. Geologic relations along the Cartersville Fault; somewhat schematic.

the State from Heard County through north Atlanta, Gainesville, and on into North Carolina. The North Carolina portion of the fault recently was studied by Reed and Bryant (1964, p. 1177–1196) who concluded that it is a strike-slip fault of great magnitude comparable to the San Andreas Fault in California. They postulated right-lateral displacement of at least 135 mi occurring during late Paleozoic or early Triassic time or both. Conspicuous evidence of faulting is everywhere along the zone, particularly in North Carolina.

Recent tectonic activity along the Brevard zone has been postulated by Husted and Strahley (1960). They report the coincidence of epicenters of several recent intermediate and minor quakes with the trace of the Brevard zone. The epicenters, however, were estimated from verbal reports of tremors, so their locations are only approximate.

If the Brevard zone is a major strike-slip fault, it is remarkable how a thin band of marble persistently crops out along the zone for several hundred miles. Until more field data are available and until it can be shown conclusively that the rocks on the two sides of the zone are dissimilar, the possibility of the Brevard zone being a major fold complicated by trough faulting rather than a major strike-slip fault should not be ruled out.

There is considerable but moderate earthquake activity in Georgia. Unfortunately the epicenters of most of the shocks have been fixed by seismic stations outside the region or estimated from shock effects observed at the surface. Inaccuracies in fixing the foci of the quakes so far has precluded the possibility of accurately correlating recent movements with specific tectonic elements.

Towaliga Fault

The Towaliga Fault, a northwest-dipping zone of distributive movement marked by mylonites and extending at least 125 mi across western Georgia and eastern Alabama, was named by Crickmay (1933, p. 171). A portion of it was described by Clarke (1952, p. 72–73). More recently Willard H. Grant has completely mapped the fault from the eastern Lamar County line to the Flint River, and has done reconnaissance work along the fault westward into Harris County.

Goat Rock Fault

The Goat Rock Fault was named by Crickmay (1933). Subsequently a portion of it was studied by Clarke (1952, p. 73–80) who described it as a "southeast-dipping zone of distributive move-

ment which extends from the Thomaston quadrangle westward for sixty-five miles into Alabama where it is covered by sediments of the Atlantic Coastal Plain." According to Clarke, the fault does not turn northward, as shown on the 1939 Geologic Map of Georgia, but continues eastward across the Thomaston quadrangle with no sign of dying out. Willard H. Grant has traced the fault eastward across the Yatesville quadrangle.

Other Faults

The unnamed fault shown in Figure 2 in the northeast part of the Piedmont was mapped in South Carolina by Overstreet and Bell (1965) as a "lengthy high angle normal fault that trends E–NE, is associated with the sandstones and shales of Late Triassic age near Pageland, South Carolina, and cuts across phyllites of the slate belt SW of Columbia." The extension of the fault into Georgia was mapped by Thomas J. Crawford in 1965 (unpublished).

REGIONAL METAMORPHISM

The pattern of regional metamorphism on the west side of the Piedmont might be called classic. Metamorphic grade is low on the west, rises rapidly toward the east, and continues high most of the way to the Fall Line, though retrograde metamorphism is conspicuous at several places, as along the Brevard zone (belt 6). Near the Fall Line, the metamorphic rocks appear to be lower grade; however, staurolite and garnet which are characteristic of moderately high regional metamorphism occur throughout and sillimanite is common. The appearance of lower metamorphic grade is due, at least partly, to extensive hydrothermal alteration.

The metamorphic isograds have been traced in detail on the west side of the Piedmont and are remarkably regular (Fig. 2). The biotite isograd has been truncated by the Cartersville Fault—evidence of post metamorphic movement. Low grade metamorphism commonly extends west of the fault. In Gordon County and parts of Bartow County, for example, metamorphism can be traced for several miles into the Cambrian rocks by observing chlorite porphyroblasts, spotting of the shales and the recrystallization of micas.

At least two periods of metamorphism have been recognized, though their effects have not been clearly separated. Whether the present distribution of isograds is mainly a consequence of one metamorphic event or the net effect of at least two is yet to be worked out.

The highest metamorphic temperatures were obtained along a broad belt extending northeast–

southwest through Athens and Milledgeville and making a small angle with the regional trend. This "hot belt" characteristically contains the rare earth-bearing gneisses—gneisses with accessory xenotime and monazite—whose micas give K/A ages of 250 m.y.

Superimposed on the pattern of regional metamorphism are local anomalies. Between Columbus and Macon are charnockitic rocks and coronites (dry metamorphism?). Andalusite is abundant near Milledgeville and is common to the northeast in Wilkes and Lincoln Counties. In contrast, andalusite has not been reported on the west side of the Piedmont. Could its presence on the east side relate to the clustering there of granitic intrusives?

In the Cartersville district and in mineralized areas within belt 5 sericitization has resulted from hydrothermal alteration. In other areas, as along the Brevard zone, sericitization has been interpreted as retrograde metamorphism. The Brevard rocks have been subject to conditions of high grade metamorphism, but subsequent sericitization and chloritization have imparted a low-grade appearance. The graphitic schists, interbedded with the other Brevard rocks, through their retarding effect on recrystallization probably have contributed to the low-grade look.

Widespread muscovitization has been observed in eastern Cobb County. In the same area, xenotime and monazite which are characteristic accessory minerals in the "hot belt" gneisses show up.

Fractures in gneisses and granites throughout the Piedmont generally are coated with zeolites. Zeolitization is very prominent on the east side of the Piedmont where this type of alteration is widespread and intense, particularly in the extreme southwest.

Age of Metamorphism

The age of the Piedmont rocks has been investigated by Gruenenfelder and Silver (1958), Pinson, Fairbairn, Hurley, Herzog, and Cormier (1957), Long, Kulp, and Ecklemann (1959), and Kulp and Eckelmann (1961).

The micas in the "hot belt" give ages of 250 m.y. Apparent age increases both to the northwest and to the southeast (Fig. 5). Apparently the 250 m.y. event dates the last main period of regional metamorphism. On the flanks of the "hot belt," where recrystallization did not reset the K/A clock, the rocks show older ages. A few sericites on the east side of the Piedmont in South Carolina also give a 250 m.y. age. Possibly

the sericitic alteration along the cooler flanks of the "hot belt" can be attributed to water vapor escaping from the hot belt during metamorphism.

Most, or nearly all, of the granites in Georgia have a metamorphic texture. None of the granites yet dated are younger than the 250 m.y. event (late Paleozoic).

PIEDMONT CARBONATES

In reviewing the lithology of the crystalline area one is struck by the dearth of carbonates. While limestones and the dolomites are major rock types in northwest Georgia and on the Atlantic Coastal Plain, the only carbonates known in the Piedmont are along two narrow belts, the Murphy Marble belt and the Brevard belt. A third belt, the Pine Mountain Series, extended to the southwest into Alabama contains some carbonate.

In the Murphy Marble belt, marble is found as far south as Canton (lat. 34.25° N. long. 84.5° W.) and shows again in Haralson County 2–3 mi (3–5 km) northeast of Buchanan (lat. 33.75° N. long. 85.2° W.). The belt appears to have been cut out by faulting in Cherokee County. The isolated occurrence in Haralson County appears to be along this fault.

In the Brevard zone, marble crops out as far south as Suwanee. No marble has been found farther south along the Brevard belt as now mapped. About 20 ft of marble and 60 ft of other calcareous rocks have been encountered in a drill hole 2 mi north of Griffin in Spalding County.

Carbonates are abundant in Paleozoic rocks of Georgia. The dearth of carbonates in the Piedmont, considered in connection with the time of metamorphism, strongly suggests a preponderance of Precambrian rocks.

VOLCANISM

At least three different periods of volcanism are distinguishable: (1) pre-Mesozoic, (2) Triassic or Jurassic, and (3) upper-Cretaceous or lower-Tertiary.

The pre-Mesozoic metavolcanics consist mainly of metabasalts, commonly amygdaloidal, but include metatuffs and a variety of other types. These are extensively developed in Carroll, Paulding, and Cobb Counties and somewhat less well developed on to the northeast as far as the North Carolina line. Two different ages of pre-Mesozoic metavolcanics can be distinguished: those which are a part of the Cobb County stratigraphic sequence (belt 5 in Fig. 2) and those which are in the Great Smoky Group (belts 3 and 3A).

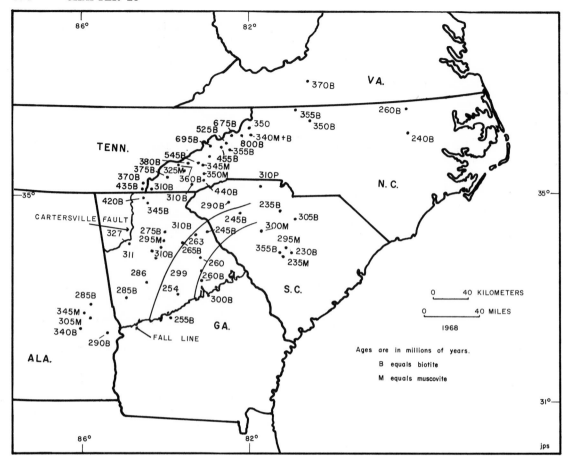

FIGURE 5. Distribution of radiometric dates in southeastern United States. The position of the "Hot Belt" is shown for Georgia and part of South Carolina.

The metavolcanics in belts 7 and 9 are most extensively developed in Jasper and Putnam Counties. Little detailed information is available.

In the southeastern part of the Piedmont, pre-Mesozoic metavolcanics include extensive metadacites and a variety of other rocks, including dikes that range in composition from rhyolite to gabbro. All were regionally metamorphosed toward the close of the Paleozoic, about 250 million years ago.

The Triassic or Jurassic rocks are diabase dikes and possibly granophyric porphyries and hornblende andesites. These have not been regionally metamorphosed. A few of the diabase dikes are amygdaloidal.

Volcanism is evident in late Cretaceous or early Tertiary time in the southeastern part of the Piedmont: at the top of the Tuscaloosa Formation of Cretaceous age is a layer 12–28 ft thick of tuff and possibly rhyolite except where the layer has been removed by pre-Barnwell erosion (the Barnwell Formation is upper Eocene).

DIKES

Unmetamorphosed diabase dikes, commonly believed to be of Triassic age but possibly of Jurassic age (de Boer, 1967), are fairly common throughout the Georgia Piedmont along and east of belt 6. They are notably scarce west of belt 6, where there are older, metamorphosed basic dikes.

In the southeastern part of the Piedmont, notably in Elbert, Oglethorpe, Greene, Wilkes, Lincoln, McDuffie, and Columbia Counties unmetamorphosed rhyolite dikes are common. They are generally less than 10 ft thick, though some exceed 100 ft in thickness. The rhyolite dikes, diabase dikes, and the dikes of intermediate composition are so numerous in some areas as to constitute swarms. Possibly some of these are feeder dikes for Triassic (Jurassic?) volcanism. One can speculate that some of the rhyolites might be feeder dikes for the tuffs found at the top of Cretaceous strata in the upper coastal plain.

REFERENCES

Austin, R. S., 1965, The geology of southeast Elbert County, Georgia: unpublished Master's Thesis, Univ. Georgia.

Bayley, W. S., 1928, Geology of the Tate quadrangle, Georgia: Georgia Geol. Survey Bull. 43.

Bowen, B. M., Jr., 1961, The structural geology of a portion of southeastern Dawson County, Georgia: unpublished Master's Thesis, Emory Univ.

Brent, W. B., 1952, The Taccoa Quartzite and adjacent rocks in Stevens County, Georgia: unpublished Master's Thesis, Cornell Univ.

Clarke, J. W., 1952, Geology and mineral resources of the Thomaston quadrangle, Georgia: Georgia Geol. Survey Bull. 59.

Cofer, H. E., 1948, Petrology, petrography, mineralogy and structure of the Arabia Mountain Gneiss, DeKalb County, Georgia: unpublished Master's Thesis, Emory Univ.

————, 1958, Structural relations of the granites and the associated rocks of South Fulton County, Georgia: unpublished Ph.D. dissertation, Univ. Illinois.

Cook, R. B., Jr., 1967, A geologic study of the west-central part of Wilkes County, Georgia: unpublished Master's Thesis, Univ. Georgia.

Crawford, T. J., 1957, Geology of part of Indian Mountain, Polk County, Georgia, and Cherokee County, Alabama: Georgia Mineral Newsletter, v. X, no. 2, p. 39–51.

Crickmay, G. W., 1933, The occurrence of mylonites in the crystalline rocks of Georgia: Am. Jour. Sci., 5th Ser., v. 26, p. 161–177.

————, 1952, Geology of the crystalline rocks of Georgia: Georgia Geol. Survey, Bull. 58.

Croft, M. G., 1963, Geology and ground water resources of Bartow County Georgia: U. S. Geol. Survey Water Supply Paper 1619-FF.

de Boer, J., 1967, Paleomagnetic–tectonic study of Mesozoic dike swarms in the Appalachians: Jour. Geophys. Research, v. 72, p. 2237–2250.

Fairley, W. M., 1965, The Murphy syncline in the Tate quadrangle, Georgia: Georgia Geol. Survey Bull. 75.

Fountain, R. C., 1961, The geology of the northwestern portion of Jasper County, Georgia: unpublished Master's Thesis, Emory Univ.

Fouts, J. A., 1966, The geology of the Metasville area, Wilkes and Lincoln Counties, Georgia: unpublished Master's Thesis, Univ. Georgia.

Furcron, A. S., and Teague, K. H., 1945, Sillimanite and massive kyanite in Georgia: Georgia Geol. Survey Bull. 51, 76 p.

————, 1951, Mineral resources of Union, Towns, Lumpkin and White Counties, Georgia: Georgia Dept. Mines, Mining and Geology, map 6.

————, Teague, K. H., and Calver, J. L., 1947, Talc deposits of Murray County, Georgia: Georgia Geol. Survey Bull. 53, 75 p.

Galpin, S. L., 1915, The feldspar and mica deposits of Georgia: Georgia Geol. Survey Bull. 30.

Gardner, C. H., 1961, The geology of central Newton County, Georgia: unpublished Master's Thesis, Emory Univ.

Giles, R. T., 1966, Petrography and petrology of the Rabun Bald area, Georgia–North Carolina: unpublished Master's Thesis, Univ. Georgia.

Grant, W. H., 1958, The geology of Hart County, Georgia: Georgia Geol. Survey Bull. 67.

————, 1949, The lithology and structure of the Brevard Schist and Hornblende Gneiss in the Lawrenceville, Georgia area: Hornblende Gneiss in the Lawrenceville, Georgia area: unpublished Master's Thesis, Emory Univ.

Gruenenfelder, M., and Silver, L. T., 1958, Radioactive age dating and its petrologic implications for some Georgia granites (abs.): Geol. Soc. America Bull., v. 69, p. 1574.

Hayes, C. W., 1901, Geologic relations of the iron ore deposits in the Cartersville district, Georgia: Am. Inst. Min. Eng. Trans., v. 30, p. 403–419.

Herrmann, L. A., 1954, Geology of the Stone Mountain–Lithonia District, Georgia: Georgia Geol. Survey Bull. 61.

Higgins, M. W., 1965, Geology of a portion of Sandy Springs quadrangle, Georgia: unpublished Master's Thesis, Emory Univ.

Holland, W. A., 1954, The geology of the Penola Shoals area, DeKalb County, Georgia: unpublished Master's Thesis, Emory Univ.

Hurst, V. J., 1955, Stratigraphy, structure, and mineral resources of the Mineral Bluff quadrangle, Georgia: Georgia Geol. Survey Bull. 63.

————, 1956, Geologic map of the Kennesaw Mountain–Sweat Mountain area, Cobb County, Georgia: Georgia Dept. Mines, Mining and Geology.

————, 1956, Geology of the Epworth quadrangle: unpublished Geol. map.

————, 1959, The geology and mineralogy of Graves Mountain, Georgia: Georgia Geol. Survey Bull. 68.

————, 1960, Oil tests in Georgia: Georgia Dept. Mines, Mining, and Geology Inf. Circ. 19.

————, and Holland, W. A., 1959, Coronites near Culloden, Georgia: Georgia Acad. Sci. Bull. v. 17, p. 74.

————, and Schlee, J. S., 1962, Ocoee Metasediments of North Central Georgia–Southeast Tennessee: Geol. Soc. America, Southeastern Sec., Ann. Mtg., 1962, Guidebook 3.

Husted, J., and Strahley, H. W., 1960, The Blue Ridge Fault Zone and the distribution of Southern Appalachian Earthquakes (abs.): Georgia Acad. Sci. Bull., v. 18, p. 14.

Kesler, T. L., 1950, Geology and mineral deposits of the Cartersville district, Georgia: U. S. Geol. Survey Prof. Paper 224.

King, J. A., 1957, The petrography and structure of a portion of Soapstone Ridge, DeKalb and Clayton Counties, Georgia: unpublished Master's Thesis, Emory, Univ.

Klett, W. Y., Jr., 1967, The geology of North Central Jackson County and South Central Hall County, Georgia: unpublished Master's Thesis, Univ. Georgia.

Kulp, J. L., and Eckelmann, F. D., 1961, Potassium–Argon Isotopic ages on micas from the Southern Appalachians: New York Acad. Sci., v. 91, art. 2, p. 408-419.

LaForge, L., and Phalen, W. C., 1913, Description of the Ellijay quadrangle: U. S. Geol. Survey Geol. Atlas, folio 187.

La Moreaux, P. E., 1946, Geology of the Coastal Plain of East-Central Georgia: Georgia Geol. Survey Bull. 50.

Lawton, D. E., 1967, Geology of the Hard Labor Creek area, Morgan County, Georgia: unpublished Master's Thesis, Univ. Georgia.

Long, L. E., Kulp, J. L., and Eckelmann, F. D., 1959, Chronology of major metamorphic events in the Southeastern United States: Am. Jour. Sci., v. 257, p. 585–603.

McLemore, W. H., 1965, The geology of the Pollard's Corner area, Columbia County, Georgia: unpublished Master's Thesis, Univ. Georgia.

Matthews, V. III, 1967, Geology and petrology of the pegmatite district in southeastern Jasper County, Georgia: unpublished Master's Thesis, Univ. Georgia.

Medlin, J. H., and Hurst, V. J., 1967, Geology and mineral resources of the Bethesda Church area, Greene County, Georgia: Georgia Geol. Survey Inf. Circ. 35.

Mellen, J., 1956, Pre-Cambrian sedimentation in the northeast part of the Cohutta Mountain quadrangle, Georgia: Georgia Mineral Newsletter, v. IX, p. 46–61.

Meyers, C. W., 1967, The geology of the Presley's Mill area, northwest Putnam County, Georgia: unpublished Master's Thesis, Univ. Georgia.

Milton, C., and Hurst, V. J., 1965, Subsurface "Basement" Rocks of Georgia: Georgia Dept. Mines, Mining and Geology Bull. 76.

Mohr, D. W., 1965, Regional setting and intrusion mechanics of the Stone Mountain Pluton: unpublished Master's Thesis, Emory Univ.

Nuttall, B. D., 1950, The Hartwell–Oconee Contact in North Georgia: unpublished Master's Thesis, Cincinnati Univ.

Overstreet, W. C., and Bell, H., 1965, The crystalline rocks of South Carolina: U. S. Geol. Survey Bull. 1183.

Pinson, W. H., Jr., 1949, Geology of Polk County, Georgia: unpublished Master's Thesis, Emory Univ.

————, Fairbairn, H. W., Hurley, P. M., Herzog, L. F., and Cormier, R. F., 1957, Age study of some crystalline rocks of the Georgia Piedmont (abs.): Geol. Soc. America Bull., v. 68, p. 1781.

Power, W. R., and Reade, E. H., 1962, The Georgia Marble district (Field excursion): Georgia Dept. Mines, Mining and Geology, Guidebook 1.

Pruitt, R. G., Jr., 1952, The Brevard Zone of north-easternmost Georgia: unpublished Master's Thesis, Emory Univ.

Reade, E. H., Jr., 1960, The geology of a portion of Newton and Walton Counties, Georgia: unpublished Master's Thesis, Emory Univ.

Reed, J. C., Jr., and Bryant, B., 1964, Evidence for strike-slip faulting along the Brevard Zone in North Carolina: Geol. Soc. America Bull., v. 75, p. 1177–1196.

Rodgers, J., 1953, Geologic map of East Tennessee: Tenn. Dept. Conserv., Div-Geol. Bull. 58.

Salisbury, J. W., 1961, Geology and mineral resources of the northwest quarter of the Cohutta Mountain quadrangle: Georgia Geol. Survey Bull. 71.

Schepis, E. L., 1952, Geology of eastern Douglas County, Georgia: unpublished Master's Thesis, Emory Nniv.

Schultz, R. S., 1961, The geology of northwestern Newton and southwestern Walton Counties, Georgia: unpublished Master's Thesis, Emory Univ.

Sever, C. W., 1964, Geology of ground-water resources of crystalline rocks, Dawson County, Georgia: Georgia Geol. Survey Inf. Circ. 30.

Smith, J. W., 1959, Geology of an area along the Cartersville Fault near Fairmount, Georgia: unpublished Master's Thesis, Emory Univ.

Smith, W. L., 1958, The geology of the Conasauga Formation in the vicinity of Ranger, Georgia: unpublished Master's Thesis, Emory Univ.

Stewart, J. W., 1958, Earthquake history of Georgia: Georgia Mineral Newsletter, v. XI, no. 1-4, p. 127–128.

————, and others, 1964, Geologic and hydrologic investigation at the site of the Georgia Nuclear Laboratory, Dawson County, Georgia; U. S. Geol. Survey Bull. 1133-F, 90 p.

Stose, G. W. and Smith, R. W., 1939, Geologic Map of Georgia: Georgia Dept. Mines, Mining and Geology.

Webb, J. E., 1958, Reconnaissance geologic survey of part of Polk and Haralson Counties, Georgia: Georgia Mineral Newsletter, v. XI, no. 1, p. 19–24.

Chapter 27

Structure and Petrology of the Harford County Part of the Baltimore–State Line Gabbro–Peridotite Complex*

DAVID L. SOUTHWICK

INTRODUCTION

THE BALTIMORE–STATE LINE peridotite–gabbro complex, consisting of variably metamorphosed gabbro, ultramafic rock, quartz diorite, and albite granite, extends from southernmost Pennsylvania to near Laurel, Md., where it is overlapped by sedimentary rocks of the Coastal Plain (Fig. 1). It is the largest mafic pluton in the northeast Piedmont (Larrabee, 1966, sheet 3). For years this mass has been called the Baltimore Gabbro or Baltimore Gabbro complex (Cloos and Hershey, 1936; Hopson, 1964); recently it has been referred to informally as the Baltimore gabbro–State Line complex by Thayer (1967) in order to emphasize the fundamental relationship between the State Line chromite district (Pearre and Heyl, 1960) and the rocks near Baltimore. In this paper the informal name Baltimore–State Line complex is used.

These rocks are the subject of classic papers by G. H. Williams on the description and classification of mafic and ultramafic rocks (1884, 1890) and the uralitization process (1886). Recently they have been studied by C. A. Hopson (1964, p. 132–154) who presented an integrated structural and petrologic interpretation of the whole complex.

According to Hopson (1964), the Baltimore–State Line complex is a stratiform sheet that was

emplaced into essentially flat-lying strata, deformed contemporaneously, and later folded and faulted during gneiss doming and regional metamorphism. The principal points offered in support of this hypothesis are (Hopson, 1964, p. 149): (1) concordant country rock contacts; (2) rhythmic layering parallel to contacts, with cumulate textures well developed in the layered rocks; (3) symmetrical distribution around mantled gneiss domes, except where there was faulting; and (4) the concentration of ultramafic rocks near contacts that would correspond to the base of the sheet. Thayer (1967), on the other hand, contends that the Baltimore–State Line complex is fundamentally alpine and lists the following diagnostic characteristics: (1) the occurrence of olivine-rich peridotite containing large podiform chromite deposits; (2) the relative scarcity of pyroxenite; (3) the random intermingling of ultramafic and gabbroic rocks, as opposed to regular grouping of them in typical stratiform units; (4) the association of quartz-rich diorites and albite granites; and (5) the occurrence of gabbro dikes and irregular masses in metaperidotite at Hunting Hill quarry, west of Rockville, Md.

Most of the modern, detailed petrologic work on which this controversy is based has been done in the vicinity of Baltimore City (Cohen, 1937; Herz, 1951; Hopson, 1964) or in the serpentinites at the northeast end of the complex (Lapham and McKague, 1964). Relatively little de-

* Publication authorized by the Director, U. S. Geological Survey.

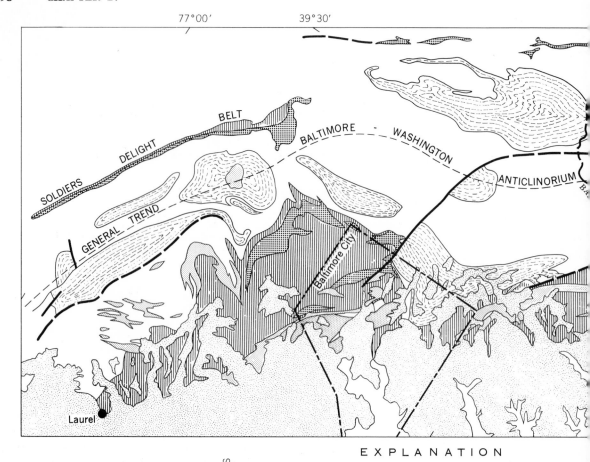

77°00'   39°30'

SOLDIERS DELIGHT BELT

BALTIMORE - WASHINGTON

GENERAL TREND

ANTICLINORIUM

Baltimore City

Laurel

EXPLANATION

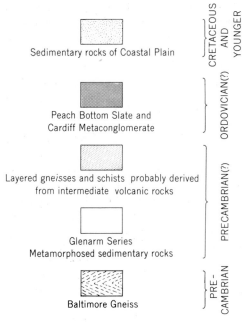

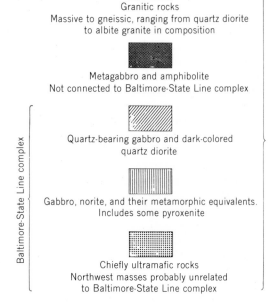

Sedimentary rocks of Coastal Plain — CRETACEOUS AND YOUNGER

Peach Bottom Slate and Cardiff Metaconglomerate — ORDOVICIAN(?)

Layered gneisses and schists probably derived from intermediate volcanic rocks — PRECAMBRIAN(?)

Glenarm Series
Metamorphosed sedimentary rocks

Baltimore Gneiss — PRE-CAMBRIAN

Granitic rocks
Massive to gneissic, ranging from quartz diorite to albite granite in composition

Metagabbro and amphibolite
Not connected to Baltimore-State Line complex

Quartz-bearing gabbro and dark-colored quartz diorite

Gabbro, norite, and their metamorphic equivalents. Includes some pyroxenite

Chiefly ultramafic rocks
Northwest masses probably unrelated to Baltimore-State Line complex

Baltimore-State Line complex

LOWER TO MIDDLE PALEOZOIC

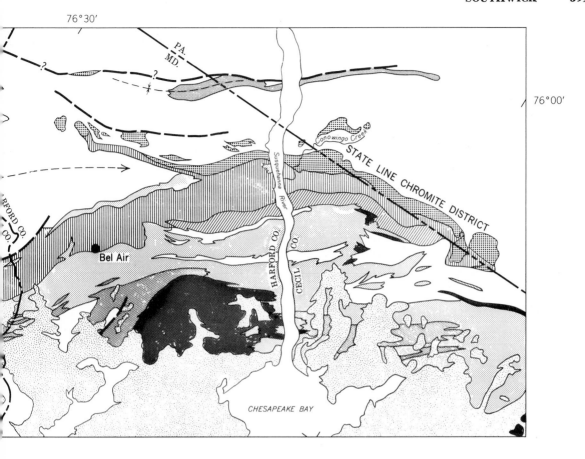

76°30'

76°00'

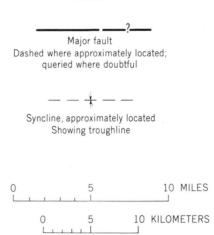

Contact

Major fault
Dashed where approximately located;
queried where doubtful

Syncline, approximately located
Showing troughline

FIGURE 1. Generalized geologic map showing the Baltimore–State Line complex
and its geologic setting. Compiled from many sources; Harford County, Md., part
after Southwick and Owens (1968).

tailed information is published about the central part of the complex in Harford County, Md., northeast of Baltimore. This paper deals primarily with that segment and is the result of a cooperative mapping project of the U. S. Geological Survey and the Maryland Geological Survey.

DISTRIBUTION OF ROCKS

At the Susquehanna River, the Baltimore–State Line complex consists of a mile wide (1.6 km) zone of ultramafic rocks on the northwest, a zone of gabbro about 3 mi (4.8 km) wide in the middle, and a zone of quartz gabbro and quartz diorite about 1.5 mi (2.4 km) wide on the southeast (Fig. 1). Contacts are broadly concordant

with bedding schistosity in wall rocks of the Wissahickon Formation, and this, plus the distribution of lithologies, led Knopf (1921, p. 89) to postulate that the mass was a huge, upturned sheet with its gravity-differentiated ultramafic basal phase along the northwest edge.

This simple pattern does not extend far southwest of the Susquehanna, however (Fig. 2). About 4.5 mi (7.2 km) from the river, the ultramafic rocks on the northwest side of the complex veer westward away from the main mass of gabbro, forming a dike-like offshoot into the country rock. Serpentinized olivine-rich rocks predominate in the northern part of this offshoot, whereas the southern part is a confused tangle of talc- and amphibole-rich rocks derived from

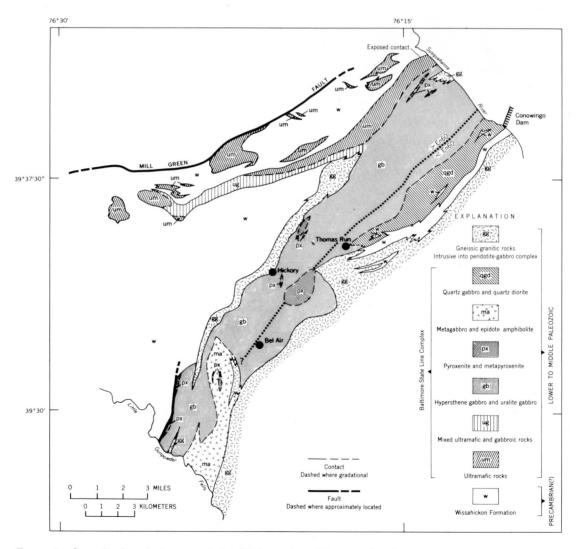

FIGURE 2. Generalized geologic map of the Baltimore–State Line complex in Harford County, Md. The dotted line through gb approximately separates gabbro containing orthopyroxene more magnesian than En_{60} (on northwest) from gabbro containing orthopyroxene less magnesian than En_{60}.

intimately mixed pyroxenite and gabbro. Pod-like bodies lying north of the offshoot, though chiefly serpentinite, also contain significant volumes of metamorphosed gabbro and pyroxenite.

The central part of the complex is chiefly gabbroic rock that has been variably metamorphosed. Between Bel Air and the Susquehanna, most of the rock is hypersthene gabbro that is slightly to completely uralitized, and massive except for local shear zones. Virtually unaltered and uralitized gabbro are patchily intermixed in the manner described by Williams (1886, Pl. 4, p. 72), and it is common to find all stages in the conversion of gabbro to uralite over a distance of a few feet.

Pyroxenite and metapyroxenite form several small lenslike bodies near the northwest margin of the gabbro belt and an oval area about 1.5 mi (2.4 km) long near its center, southeast of Hickory. Most of these rocks are composed chiefly of pale green, fibrous actinolite; relict pyroxene is rare. They grade into gabbro on all sides (Fig. 2).

Between Bel Air and Little Gunpowder Falls (and further to the southwest) the dominant rock is thoroughly recrystallized lineated epidote amphibolite. It is riven by numerous small masses of gneissic quartz diorite and is intruded by a large apophysis south of Bel Air. The amphibolite of this area was interpreted by Insley (1928, p. 321) as Baltimore Gneiss rather than gabbro because it is lineated and at least partly quartz-bearing. It is directly on strike with undoubted gabbro, however, and probably is regionally metamorphosed gabbro that has been locally modified by additions of water and perhaps silica from closely subjacent intrusive quartz diorite. The rock plainly is not inside a mantled dome and only superficially resembles the amphibolite in true Baltimore Gneiss.

A belt of quartz gabbro and quartz diorite borders the hypersthene gabbro on the southeast roughly from Conowingo Dam to Thomas Run. It continues northeastward into Cecil County, but its extent there is imperfectly known. The variable appearance and extremely poor exposure of these rocks have led to conflicting interpretations, but earlier workers all agree that they are somehow related to the gabbro complex. They were mapped as "meta-gabbro, quartz–gabbro, and hornblende–gabbro" by Mathews and Johannsen (1904) and were considered as some sort of "contact phase" of the gabbro by Insley (1928, p. 310–315). Similar rocks in Cecil County were termed "meta-gabbro or quartz–hornblende–gabbro and hornblende–gabbro" by

Bascom (1902, p. 121–124) who pointed out the difficulty of distinguishing them from nearby hornblende–quartz diorite and hornblende–quartz monzonite. Some rocks in this belt are dark colored, hornblende-rich, medium to coarse grained uralite gabbros with a few percent of quartz, and others are rather light colored biotite–hornblende–quartz diorite. Commonly the leucocratic rocks contain abundant dark inclusions, and the map pattern suggests that some larger areas of gabbroic rock are unusually large inclusions in quartz diorite.

Net veins, generally richer in plagioclase and somewhat coarser grained than the enclosing rock, are common in the gabbro. Some contain primary hornblende and are close to diorite in composition. Most veins are sharply bounded and show no chilling at the margins. Commonly they wander across all earlier structures, but where layering is present in the host rock the veins have some tendency to follow it. The veins rarely exceed a foot (30 cm) in thickness; they pinch and swell, and commonly branch. In many ways they are similar to the subpegmatitic gabbro dikelets described by Baragar (1960), and are identical in appearance to veins in the Canyon Mountain complex of Oregon, figured by Thayer (1963b, p. 59, Fig. 10).

STRUCTURE

Contact Relations

Contacts of the Baltimore–State Line complex are very poorly exposed in Harford County. The only exposed contact is in a highly sheared, badly weathered, subvertical concordant zone between talcose serpentinite and rocks of the Wissahickon Formation in a roadcut on Maryland route 623, about 0.4 mi (0.64 km) southeast of Broad Creek (Fig. 2). A steep to vertical attitude is indicated for most of the northwest contact of the complex by its straight trace and the steep magnetic gradient across it (Bromery and others, 1964). Generally the contact can be mapped rather closely from the distribution of float, and it appears to be essentially parallel to bedding schistosity in the Wissahickon Formation.

Along part of its northwest boundary and almost all of its southeast boundary the complex is bordered by younger granitic plutons (Fig. 1). Because the younger intrusives have plainly invaded the gabbroic rocks, these contacts may not be related to the original shape of the mafic pluton. Southwest of Bel Air, uralite gabbro, and amphibolite are shot through with small masses of gneissic quartz diorite. Along a segment of

Gunpowder Falls in the White Marsh quadrangle, Baltimore County, metamorphosed gabbro occurs toward the top of the valley slopes, whereas the riverbed is chiefly gneissic quartz diorite. The gabbro seems to form a relatively thin roof over quartz diorite between Bel Air and eastern Baltimore County, as interpreted from the field relations just described and the lower magnetic and gravity anomalies over this part of the complex (Bromery and others, 1964; R. W. Bromery, unpublished gravity map).

Internal Structural Elements

Structural elements of both igneous and metamorphic origin occur in the Baltimore–State Line complex (Cohen, 1937; Hopson, 1964, p. 135–141). Some structures formed by crystal settling and related magmatic processes early in the history of the complex, others formed during deformation and flowage while much of the complex was incompletely crystallized, and still others formed in response to metamorphic events after complete solidification. Structural elements recognized by Hopson (1964, p. 135–144) are listed in Table 1. Hopson emphasized that foliation that formed during the flowage and protoclastic stage grades into that formed after complete solidification, and suggested that foliation developed continuously throughout a long period of deformation that began while the gabbro was still partly liquid.

In Harford County no structural features that can be ascribed unequivocally to quiescent crystal settling have been found in the complex. This is not to dispute the cumulate magmatic origin of layering in certain ultramafic and gabbroic rocks of the Baltimore area that have settled textures and other related magmatic features (Hopson, 1964), but merely to point out that such features have not yet been found in Harford County.

Compositional layering on the scale of a few inches or a few feet was noted at 10 localities, in fresh and metamorphosed gabbroic and ultramafic rocks. In all these places the layering appears to be locally developed in the midst of massive rocks. Generally the layering is parallel to contacts between major rock units, but near the mouth of Conowingo Creek in Cecil County, layering in gabbro strikes N. 20° W., nearly perpendicular to the nearby gabbro–ultramafite contact. Data are insufficient to determine whether this divergence is due to folding or to semisolid flowage as the rocks were emplaced. Cumulate texture is lacking, and there is no evidence to determine whether or not the layering is primary igneous stratification.

Subvertical dikelike layers of ultramafic rock occur in gabbro on the east side of the Susquehanna River about 0.4 mi (0.64 km) south of Conowingo Creek (Bascom, 1902, p. 132–134). They consist of pyroxenite, harzburgite, and serpentinite, and are in contact with thoroughly uralitized feldspar-rich gabbro which displays a prominent foliation parallel to the layering. Contacts of the ultramafic layers are abruptly gradational to razor sharp, and some are lineated. One 7-ft thick layer (2.1 m) consists chiefly of coarse harzburgite at the edges and grades to bronzitite in the center. Bronzite crystals (En_{77-82}) in the peridotite are as long as 1 in. (2 cm). Streaky, broadly lenticular flow layers that vary in pyroxene content are parallel to the contacts, and tabular orthopyroxene crystals within the layers display a marked preferred orientation parallel to the contacts. Roughly half of a 23-ft layer (7 m) consists of harzburgite and the other half is tough, massive serpentinite. Lensoid layers up to 6 in. thick of serpentinized dunite occur in the harzburgite, and the serpentinite contains vague streaks of coarse, partly serpentinized pyroxene crystals. The olivine in these units is Fo_{88-90}, precisely the same composition as that in the main ultramafic zone some 1.5 mi (2.4 km) away. No olivine occurs in the gabbroic wall rocks.

These ultramafic units are interpreted as thick flow layers resulting from mixing and shearing together of semisolid peridotite and more fluid gabbro mush during syntectonic emplacement.

TABLE 1.

Structures formed by crystal settling and related magmatic processes. Represent quiescent crystallization.	1. rhythmic layering 2. inch-scale layering 3. graded layering 4. tabular plagioclase foliation (or igneous lamination of Hess, 1960, p. 51)
Structures formed during deformation and flowage of crystal mush. Represent plastic deformation.	1. irregular, bulbous layering 2. schlieren 3. boudinaged layers 4. feldspathic net veins and related amphibolite dikes 5. foliation and lineation that antedate the net veins
Structures formed during metamorphism; rocks completely solid.	1. foliation and lineation that crosscut the net veins and younger granitic intrusions into the gabbro.

The sharp contacts of the layers, their gneissic structure, and the strong compositional contrast between adjacent plagioclase-rich, olivine-free gabbro and olivine-rich harzburgite argue that these are not deformed primary layers in a stratiform complex. Moreover, these layers are almost 1.5 mi (2.4 km) from the main ultramafic belt, and would be that far above it if the complex were considered to be stratiform. Inasmuch as the olivine in the layers is the same composition as that in the main ultramafic belt, it would appear that the layers are out of place so high in the sequence. Regarding the layering problem, Thayer (1960, p. 251) states:

> The layers in alpine-type intrusions are much less persistent . . . but may be more sharply defined than in the stratiform complexes. The layers in some of the Philippine rocks [Zambales Complex] have knife-sharp contacts and appear uniform across outcrops a meter or so long, but on larger exposures the thinner ones pinch out in a few meters or tens of meters. Many of the layers are defined by a combination of strong foliation and marked contrast in mineral composition. Centimeter-thick layers of anorthosite may alternate with peridotite layers of similar thickness in sequences consisting mainly of various kinds of gabbro in layers up to several meters thick. In other places, as near the Acoje mine in Luzon, layers of dunite tens of meters across alternate with noritic rocks in the boundary zone between peridotite and norite. Such abrupt compositional changes as from anorthosite to olivine-rich peridotite across a width of a millimeter, or from dunite to norite in a few centimeters, is [sic] not duplicated in the Stillwater complex. In other words, adjoining layers in alpine-type complexes may show much greater compositional contrast than those in stratiform complexes. On the other hand, most alpine-type peridotite and much gabbro are massive and seemingly structureless away from peripheral zones.

Structural elements developed during postconsolidation metamorphism of the complex include lineation of hornblende in recrystallized amphibolite and various cleavages developed in serpentinite (Lapham and McKague, 1964). Metamorphic foliation is absent or very poorly developed in most of the massive hypersthene gabbro and uralite gabbro but is recognizable in parts of the quartz gabbro–quartz diorite belt. Narrow shear zones and small faults, generally unmappable at the scale of this study, are numerous.

Regional Structure

Near Baltimore, the gabbro–peridotite complex has been interpreted by Hopson (1964, p. 132–135) as an essentially stratiform sheet that differentiated in place, was caught up in compressional folding and gneiss doming during later stages of its crystallization, and continued to be deformed and metamorphosed after solidification. Hopson concluded that a central belt of gabbro in and near Baltimore City had been folded into a syncline overturned to the southeast by the rise of gneiss domes on the west and north; that the Soldiers Delight belt to the west was the greatly thinned, olivine-rich margin of the sheet that had been folded over a row of gneiss domes; and that an eastern belt, which crosses Harford County, was the steeply dipping east limb of an arch over other gneiss domes lying just northeast of Baltimore (Hopson, 1964, p. 136, Fig. 33). The serpentinized pods of peridotite that have intruded the Wissahickon Formation northwest of the main complex are interpreted by Hopson (1964, p. 149) as bodies of partly serpentinized dunite and peridotite that were squeezed sideways from their locus of crystallization in the complex by the upward pressure of gneiss domes rising from below.

Although the succession of rocks from peridotite on the northwest to pyroxene-bearing quartz diorite on the southeast along the Susquehanna River has been explained as a stratiform sheet tilted southeastward, relations a few miles southwest along strike cannot be explained by this simple hypothesis. In central Harford County, a long, thin arm of metamorphosed peridotite and gabbro diverges from the main complex, and between it and the main complex is the northeast-plunging nose of the Baltimore–Washington anticlinorium (Fig. 1). If this arm is the thin edge of a stratiform sheet folded over an anticlinal axis, as Hopson (1964, p. 135–137; Fig. 33) interpreted the similar Soldiers Delight belt west of Baltimore, one has to explain the scarcity of olivine-bearing rocks along the west side of the main complex, which should be the floor, and the concentration of olivine-rich rocks on the north side, or hypothetical top, of the arm. In other words, the ultramafites do not close around the nose of the anticlinorium as they should if they are the basal part of a simply folded sheet (Fig. 3a,b). The observed map pattern (Fig. 3C) can be explained by thrusting and wrench-faulting a folded sheet in various ways, as the reader can quickly verify for himself. The absence of mapped faults within the complex or at its contacts does not invalidate this line of reasoning, for faults are extremely hard to demonstrate in this terrane. In my opinion, however, the map pattern in no way requires that the complex ever was a continuous, solid sheet that has been folded and faulted in place. The distribution of rocks may have been imposed as solid or nearly solid peridotite and more fluid crystal-rich gabbroic magma were deformed during emplacement. The mechanical differences between

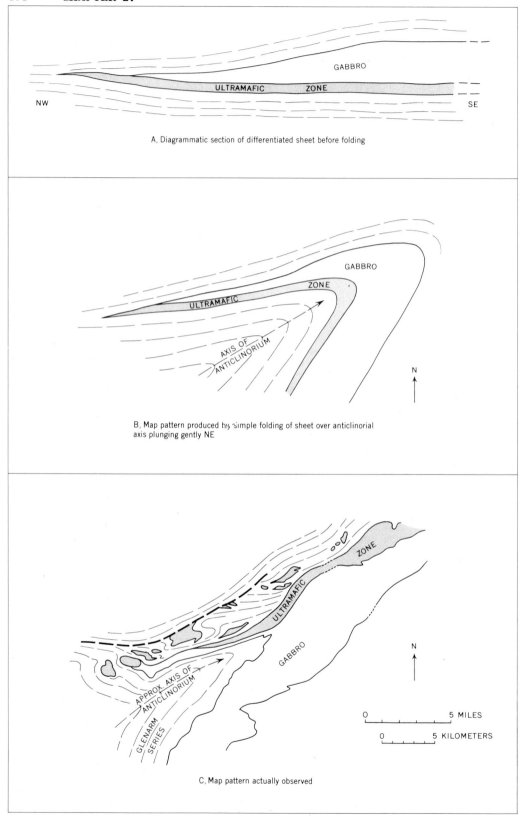

A, Diagrammatic section of differentiated sheet before folding

B, Map pattern produced by simple folding of sheet over anticlinorial axis plunging gently NE

C, Map pattern actually observed

FIGURE 3. Diagrams comparing observed map pattern in part of Harford County (C) with hypothetical map pattern (B) produced by simple folding of a stratiform complex (A) over an anticlinal axis plunging gently northeast. See text for details.

the two components of such an intrusion might lead to independent rates of movement, and almost certainly would cause extensive internal disruption and structural discontinuities. In other words, much of the deformation may have been pre- rather than postconsolidation, and may have involved flowage rather than faulting in the usual sense.

PETROGRAPHY AND MINERALOGY

Ultramafic Rocks

Rocks that originally were dunite, peridotite, olivine pyroxenite, and pyroxenite occur in the serpentinite belt on the northwest side of the complex, in the ultramafic offshoot, and in lenticular bodies that lie northwest of the main complex. Serpentinization and steatization have been very extensive in Harford County, and relict features of the original ultramafic rocks are scarcer than in the west Baltimore–Soldiers Delight area (Hopson, 1964) or the State Line district (Bascom, 1902, p. 94; Pearre and Heyl, 1960, p. 719; D. M. Lapham, oral communication, 1966).

Relict olivine in the dunitic rocks is highly magnesian, having compositions of Fo_{86-92} as determined by optical (Poldervaart, 1950) and X-ray (Yoder and Sahama, 1957) methods. Extensive to complete serpentinization has obscured whatever original fabric may have been present (Battey, 1960).

To a great extent the mesh texture which typically results from serpentinization of olivine-rich rocks has been modified by subsequent shearing and recrystallization. Much of the serpentinite is composed of a felted mat of serpentine minerals that is crosscut by shear zones containing platy serpentine and thin veinlets of cross-fiber asbestos. Variable amounts of talc, carbonate, chlorite, tremolite, and magnetite are common. Talc-rich rocks that completely lack relict textures are associated with serpentinite, especially in the ultramafic offshoot and the lenses northwest of it.

Most of the larger ultramafic lenses have cores of massive or talcose serpentinite that are jacketed by various kinds of talc-rich schist. Between the talc-rich rocks and the country rock there commonly is a zone of rocks rich in chlorite, actinolite, or both. The talc-bearing rocks and the chloritic "blackwall" zone probably formed during regional metamorphism, by reaction between serpentinite and aluminous wall rocks, as Chidester (1962) has so convincingly demonstrated in Vermont.

Gabbro and its metamorphic equivalents form 1–10 percent of most serpentinite lenses and about 50 percent of one. In most places the gabbro and ultramafic rock are intimately mixed and difficult to map separately with the available exposure.

Unaltered pyroxenite is nearly as rare as unaltered peridotite in Harford County. Most of the original pyroxene has been converted to a tangle of fibrous amphiboles, talc, chlorite, and magnetite; relict pyroxene is relatively scarce. Optical properties indicate a composition near En_{80} for orthopyroxene in pyroxenites associated with olivine-bearing rocks on the northwest side of the complex, in dike-like pyroxenite bodies near the center of the complex, and in pyroxenite masses not associated with olivine-bearing rocks toward the southeast margin. None of the orthopyroxene is zoned. Clinopyroxene, some of it zoned, forms a significant part of some pyroxenites and locally may be dominant. An outward increase in optic angle from 52 degrees at the core to 58 degrees at the rims of zoned crystals may indicate progressive Ca-enrichment.

Chromite has been mined from the ultramafic rocks of the complex in northeastern Maryland and nearby Pennsylvania. Both massive and disseminated deposits occur; the massive ore bodies are podiform or sacklike and commonly are lineated (Pearre and Heyl, 1960; T. P. Thayer, oral commun., 1966); the main ore mass at the Wood mine in Cecil County had a cross section as large as 35×300 ft (10×90m) and was mined 720 ft (216 m) down the plunge (Pearre and Heyl, 1960, p. 777). The chromite is typically granulated and some shows "pull-apart" texture (Thayer, 1964, p. 1505). Most of the ore has a gangue of serpentine, talc, or the chromian chlorite kämmererite; olivine has been reported with chromite from the Wood mine (T. P. Thayer, oral communication, 1966). A modern analysis of cleaned chromite from the Reed mine near Jarrettsville, Harford County, shows 58.0 percent Cr_2O_3, 23.4 percent FeO, and 9.1 percent Al_2O_3 (Pearre and Heyl, 1960, p. 748, Table 6, col. 17). The Cr: Fe ratio of 2.2:1 seems to be fairly typical of chromite from the Maryland–Pennsylvania area, judging from other data in the table just cited.

Hypersthene Gabbro and Uralitized Gabbro

Hypersthene, augite (characteristically "diallage"), and calcic plagioclase are the principal minerals of the gabbro. The ratio of hypersthene to augite is variable, and rocks properly classed as norite or augite gabbro are not rare; however, the amounts of hypersthene and augite are subequal in many rocks, and the term hypersthene gabbro, used in a loose, descriptive sense, properly describes much of the mass. Olivine is lacking and primary hornblende is relatively rare.

Most of the unaltered gabbro has allotriomorphic or hypidiomorphic granular texture. Plagioclase and hypersthene appear to have crystallized almost simultaneously and were followed closely by augite, which in some rocks is interstitial to them. Poikilitic pyroxene was not observed, but primary brownish green hornblende, where present, tends to form crystals as long as 2 cm which are notably poikilitic. The hornblende also forms angular interstitial fillings between plagioclase crystals; it commonly is in smooth contact with rounded islands of pyroxene, from which it appears to have formed by magmatic reaction. Primary hornblende is not common in the complex but occurs sporadically near the southeastern edge in gabbros that contain iron-rich orthopyroxene and abundant magnetite. Texturally it is quite distinct from secondary

colorless to blue-green amphiboles, which generally have a fuzzy, acicular habit and replace both pyroxene and plagioclase.

Relatively fresh gabbro commonly is complexly intermingled with completely uralitized rock. Almost any large outcrop in the area northeast of Bel Air will yield a suite of samples showing the step-by-step transformation of hypersthene gabbro to a rock rich in fibrous amphibole, epidote, and recrystallized plagioclase. Modal analyses of some typical fresh and partly uralitized rocks are given in Table 2.

Optical data indicate that the hypersthene in most gabbro is in the composition range En_{65}–En_{72} (Table 3), but within a poorly defined zone along the southeast edge of the complex it is more iron rich, in the range En_{47}–En_{57} (Fig. 2). Zoned orthopyroxene crystals are uncommon, and the few observed had iron-rich rims. Exsolved clinopyroxene in orthopyroxene is almost universal and takes two forms: planar lamellae in some rocks and blebs or droplets in others.

The gabbros containing iron-rich orthopyroxene generally also contain more opaque min-

TABLE 2.   Modal Analyses of Relatively Unaltered Hypersthene Gabbro from the Baltimore–State Line Complex in Harford and Cecil Counties, Maryland

	1	2	3	4	5	6	7	8
Plagioclase	47.1	58.2	62.3	65.6	48.4	42.8	51.8	52.4
Hypersthene	31.9	19.3	19.7	10.5	38.1	23.9	16.8	9.1
Augite	18.1	14.2	14.2	1.3	6.9	19.8	12.3	21.8
Uralitic amphibole	1.8	7.7	2.5	14.5	4.7	10.9	16.8	7.9
Biotite	—	—	—	0.6	—	—	—	—
Opaques	0.3	0.5	1.3	6.0	1.4	1.4	2.2	8.1
Epidote	—	0.1	—	0.6	—	—	—	0.7
Sphene	—	—	—	—	—	tr	—	—
Apatite	0.2	—	—	0.8	0.1	tr	tr	tr
Talc	0.4	—	—	—	0.4	0.6	—	—
Chlorite	—	—	—	—	—	0.6	0.1	tr
Quartz	—	—	—	0.1	—	—	—	—
Carbonate	0.2	—	—	—	—	—	—	—
Total (vol %)	100.0	100.0	100.0	100.0	100.0	100.0	100.0	100.0
Plagioclase An content	90	91	92	81	85	93	94	88
Hypersthene En content	71	72	69	47	70	72	65	66
No. points	1576	1494	1462	1460	1489	1436	1371	1488

Column headings:
1. Hypersthene gabbro, road along Susquehanna River just above Conowingo Creek, Cecil County (C-79a).
2. Hypersthene gabbro, Bel Air bypass (U.S. Route 1) approximately 1.3 mi north of Winters Run, Harford County (B-458).
3. Hypersthene gabbro, railroad cut along Susquehanna River 500 ft south of Conowingo Creek, Cecil County (C-30).
4. Iron-rich hypersthene gabbro, railroad cut along Susquehanna River about 1.3 mi south of Conowingo Creek, Cecil County (C-32).
5. Hypersthene gabbro, railroad cut along Susquehanna River about 0.9 mi south of Conowingo Creek, Cecil County (C-33b).
6. Hypersthene gabbro, Susquehanna River 3,000 ft north of Glen Cove, Harford County (C-125).
7. Hypersthene gabbro, just east of Maryland Route 136, 0.6 mi south of Poplar Grove, Harford County (D-320).
8. Hypersthene gabbro, 1,800 ft southwest of bridge over Deer Creek on U.S. Route 1, Harford County (B-350).

erals and apatite, and some that contain a little biotite and traces of quartz appear to be transitional to quartz-bearing mafic diorites on the southeast.

The plagioclase of least altered gabbro is typically calcic bytownite or sodic anorthite. Compositions, determined optically by the Turner (1947) and Köhler (1941) universal stage methods using unpublished correlation curves compiled by T. L. Wright, range from An_{77} to An_{94}, the largest grouping being between An_{88} and An_{92} (Table 3). Most of the plagioclase is unzoned, but a few thin marginal rinds of more sodic composition were seen.

The clinopyroxene of the hypersthene gabbro was not studied in detail. Optic angles in the range 51 degrees to 54 degrees indicate that the pyroxene probably is a normal Ca-bearing variety. By optical methods, Herz (1951, p. 991)

found composition range of $En_{56}Fs_5Wo_{39}$ to $En_{40}Fs_{21}Wo_{39}$ for clinopyroxenes in the Baltimore area.

Quartz-Bearing Gabbro and Diorite

The primary minerals of these rocks are strongly zoned plagioclase, quartz, blue-green or olive-green hornblende, reddish brown biotite, augite, and hypersthene. Most rocks contain both hornblende and biotite, some contain only hornblende, and still others contain hornblende, biotite, and two pyroxenes. Secondary minerals include pale blue-green hornblende, a colorless clinoamphibole tentatively identified as cummingtonite, epidote, clinozoisite, chlorite, and recrystallized sodic plagioclase. Dark-colored, gabbrolike rocks contain abundant pyroxene and hornblende with only a few percent of quartz and biotite, whereas light-colored, dioritic rocks con-

TABLE 3.   Optical Properties and Compositions of Some Hypersthenes from the Harford County belt of the Baltimore–State Line Complex, Together with the Compositions of Coexisting Plagioclase. (N_γ measured in Na light by standard immersion methods; 2H measured in white light on Leitz 4-axis universal stage. Each determination is the average of three or more measurements.)

		N_γ	Corrected $2H_x$	Comp.* from N_γ	Comp.* from $2H_x$	Comp. coexisting plagioclase
Pyroxenite	C-522	1.693	$(-)75°$	En 78	En 78	An_{91}†
	B-370	zoned	$(-)82.5°$ (core)		En 83 (core)	—
			$(-)75.5°$ (rim)		En 78 (rim)	
	WD-1	1.697	$(-)73°$	En 74	En 77	—
	WD-2	1.691	$(-)81°$	En 79	En 82	—
	ED-6	—	$(-)71°$	—	En 75	—
	C-31	—	$(-)79°$	—	En 81	—
Hypersthene Gabbro Central and Northwest	C-30	1.704	—	En 69	—	An_{92}
	C-33G	1.703	$(-)64°$	En 70	En 71	An_{85}
	C-79a	1.701	$(-)65°$	En 71	En 72	An_{90}
	C-125	1.700	—	En 72	—	An_{93}
	D-320	1.709	—	En 65	—	An_{94}
	B-350	1.707	—	En 66	—	An_{88}
	B-353	—	$(-)65°$	—	En 72	An_{93}
	C-356	1.701	$(-)63°$	En 71	En 70	An_{91}
	B-371a	—	$(-)57°$	—	En 65	An_{90}
	B-458	1.700	—	En 72	—	An_{91}
	B-459	1.701	—	En 71	—	An_{80}
	C-513	1.703	—	En 70	—	An_{88}
	C-514	1.699	—	En 72	—	An_{77}
	C-516	1.700	—	En 72	—	An_{90}
Hypersthene Gabbro Southeast	C-32	1.731	—	En 47	—	An_{81}
	C-33a	1.721	$(-)47.5°$	En 55	En 53	An_{81}
	C-92	—	$(-)47°$	—	En 53	An_{68}
	B-373	1.719	$(-)50°$	En 57	En 57	An_{82}
	B-518	1.718	$(-)48°$	En 57	En 55	An_{90}
Dark Quartz Diorite	C-51	1.721	$(-)48°$	En 55	En 55	An_{85} (core)
						An_{45} (rim)

* Using correlation curves of Hess, 1960, p. 27.
† 2% of plagioclase in pyroxenite.

tain abundant quartz, biotite, and hornblende with only traces of pyroxene (Table 4). Strongly zoned plagioclase occurs in all the rocks as subhedral to euhedral crystals that have bytownite cores (An_{85}) and sodic labradorite or andesine rims (An_{45-60}) and that commonly are enclosed by hornblende, biotite, or quartz. Pyroxene, where present, forms rounded grains jacketed by olive-green or brown hornblende; apparently it crystallized early and was converted to hornblende by magmatic reaction. Biotite followed hornblende in the crystallization sequence and was followed by quartz, which fills wedge-shaped interstices between earlier minerals.

Iron-oxide minerals appear to have formed both early and late in the crystallization sequence. Primary pyroxenes contain numerous angular opaque inclusions reaching a maximum of about 0.05 mm in diameter which contrast in shape and size with rounded, bloblike opaque grains

TABLE 4. Modal Analyses of Quartz Gabbro and Related Quartz Diorite in Harford County Southwest of Conowingo Dam

	1	2	3	4
Plagioclase*				
(strongly zoned)	34.0	24.6	48.5	47.3
Hypersthene	0.8	0.2	1.5	—
Augite	11.0	4.6	5.0	tr
Amphibole	39.1†	51.0†	33.8†	18.7
Biotite	2.1	1.0	1.4	9.5
Quartz	1.5	9.1	7.6	20.2
Opaques	5.4	2.0	1.4	2.8
Epidote and				
clinozoisite	6.0	6.8	0.3	0.5
Apatite	0.1	0.3	0.5	0.4
Tourmaline	tr	tr	—	—
Sphene	tr	tr	tr	—
Chlorite	tr	0.4	tr	0.6
Muscovite	tr	tr	tr	tr
Talc	—	tr	—	—
Total (vol %)	100.0	100.0	100.0	100.0
Number of points	1364	1570	1554	1554

* Includes minor sericitic alteration.

† Consists of about 50 percent poikilitic primary olive-geeen hornblende, 25 percent secondary blue-green calc–amphibole, and 25 percent colorless clino-amphibole, tentatively identified as cummingtonite.

Column headings:

1. Quartz-bearing hornblende gabbro, 0.5 mi west of Darlington, Harford County (C-104).
2. Mafic hornblende quartz diorite, 0.5 mi southwest of Hopkins Cove, Harford County (C-505a).
3. Quartz-bearing gabbro, north flank of Sugar Hill, just north of Hopkins Branch, Harford County (C-51).
4. Biotite–hornblende quartz diorite, Hollands Branch, 500 ft north of Trappe Church Road (C-90).

as large as 0.5 mm that invariably are associated with biotite and seem to have crystallized late.

Secondary minerals and textures imposed by low-grade regional metamorphism (and perhaps to some extent by deuteric processes) are ubiquitous; they obscure primary features to various degrees. The calcic cores of zoned plagioclase have been partly replaced by clinozoisite or epidote and secondary plagioclase of more sodic composition. Fibrous amphiboles, blue-green and colorless, have partly replaced pyroxene and, to a lesser extent, primary hornblende. Biotite, plagioclase, and primary hornblende are bent, and quartz exhibits pronounced mosaic extinction.

Dark inclusions are abundant in the dioritic rocks, especially near the gradational contact with gabbro. The most common inclusions consist of confusedly intergrown, fine to medium grained green hornblende, intermediate plagioclase, epidote, and a little quartz. They show all degrees of digestion and assimilation by quartz diorite; in places so many inclusions have been digested that the composition of the host has been made noticeably more mafic. The composition and distribution of inclusions indicate that they probably are pieces of gabbro, but textural proof of this has been obliterated.

Although the inclusions show that some of the quartz diorite is younger than the gabbro, it probably is not much younger, and there is good evidence that the two are genetically related:

(1) The cores of zoned plagioclase in the quartz diorite are as calcic as the plagioclase in hypersthene gabbro.

(2) The quartz diorite contains small amounts of iron-rich hypersthene.

(3) Normal hypersthene gabbro grades into quartz diorite through a poorly defined intermediate zone of gabbro that contains iron-rich hypersthene and small amounts of primary hornblende, biotite, and quartz.

The quartz diorite may be a late differentiate that has reinjected and reacted with earlier formed rocks of the same magma suite. Extremely poor exposure and widespread low-grade metamorphic effects hamper a more rigorous investigation of this rather speculative hypothesis.

In field relations, textures, and composition the quartz diorite is remarkably similar to dioritic rocks associated with gabbro in the alpine-type Canyon Mountain Complex in Oregon (Thayer, 1963a, 1967). Albite granites, which form an important part of the Canyon Mountain Complex, have not been found in Harford County, but are associated with gabbro of the Baltimore–State Line complex at Relay, Md., south of Baltimore (Hopson, 1964, p. 155–160).

Epidote Amphibolite

The epidote amphibolites typically are medium to coarse grained and consist of blue-green hornblende, intermediate plagioclase, epidote, and quartz; magnetite and sphene are the commonest accessories. Hornblende forms weakly aligned prismatic crystals that are sieved with quartz inclusions. Plagioclase and epidote are intergrown and formed by the degradation of a more calcic plagioclase which is preserved in a few rocks. Quartz, as recrystallized patches associated with plagioclase and epidote and as inclusions within hornblende, generally constitutes less than 10 percent of the rock. Most of the quartz probably was generated by metamorphic reactions (Kretz, 1963), but some may have been added from subjacent intrusive quartz diorite.

These rocks are more completely recrystallized than uralitized gabbro, for relict igneous textures are lacking. The amphibole is in well-formed poikilitic prisms rather than the aggregate of fine needles so typical of uralitized gabbro; the plagioclase, epidote, and quartz tend to form a typical crystalloblastic mosaic. Lineation of hornblende is apparent, though weak.

Contact Metamorphic Effects

No contact metamorphic effects have been found near the Baltimore–State Line complex in Harford County, or in the Baltimore area (Hopson, 1964, p. 141–142). Hopson suggests that failure to find contact effects may be due to the absence of wall-rock exposures suitably close to the gabbro, or to later regional metamorphism and tectonism that may have erased or obscured them; it is also possible that the gabbro had crystallized sufficiently prior to emplacement that it was incapable of causing extensive thermal metamorphism.

Rocks of the Wissahickon Formation near the gabbro are in the garnet zone of regional metamorphism in southern Harford County and in the biotite or chlorite zones near the Susquehanna River. In the southern area the metamorphic grade increases away from the gabbro and is clearly related to the Baltimore Gneiss domes (Hopson, 1964, p. 142; Southwick and Owens, 1968). The gabbro has exerted no mappable effect on the regional metamorphic pattern and appears to have been in place during the climax of metamorphism.

CHEMICAL PETROLOGY

Because the Baltimore–State Line complex has had a complicated history of crystallization, injection, and regional metamorphism, it is difficult to determine or even approximate its bulk composition or that of the original magma. Chilled margins that might represent the bulk composition of original magma have not been found. There is ample evidence that parts of the mass are crystal accumulates (the layered sequences in west Baltimore, for example; Hopson, 1964, p. 135–144), and these rocks obviously do not represent primary magmas. Moreover, much of the complex has been uralitized or converted to amphibolite, and it is likely that these altered rocks are not chemically identical to the gabbros from which they were derived.

New and selected older analyses of rocks from the complex are given in Table 5. An effort was made to select unweathered and unaltered samples, but such material is decidedly scarce. The analyses are grouped according to their degree of alteration, which is indicated roughly by the content of H_2O^+. Gabbro containing about 0.4 percent H_2O^+ is incipiently altered and has about 5 percent secondary amphibole, whereas specimens with more than about 1.6 percent H_2O^+ generally have no relict pyroxene.

A striking feature of the gabbro analyses is the low content of Na_2O and the consequently high ratio of CaO to Na_2O; the four analyses of least altered gabbro average 0.81 percent Na_2O. For comparison, the average Na_2O content of analyzed gabbros from the upper gabbro zone of the Stillwater Complex is 1.88 percent (Hess, 1960, p. 92–94), and from the Astralabe–DeLangle stock, Alaska, is 2.5 percent (Rossman, 1963). Border facies of the Stillwater, Bushveld, and Skaergaard Complexes respectively contain 1.87, 1.58, and 2.44 percent Na_2O (Hess, 1960, p. 152). Available data suggest that alpine gabbros may be poorer in Na_2O than the stratiform examples just cited, but there are too few analyses of unaltered alpine rocks to permit a general conclusion. Three analyses of gabbro from the Canyon Mountain Complex, Oregon, average 0.86 percent Na_2O (T. P. Thayer, unpublished data); two analyses from the Troodos Plutonic Complex, Cyprus, average 0.53 percent (Wilson and Ingham, 1959); and one analysis from New Caledonia contains 0.75 percent (Lacroix, 1943, p. 18). The paucity of Na_2O in the Baltimore rocks is, of course, reflected in the high anorthite content of the plagioclase (see norms and Table 2).

The analyses in Table 5 and others selected from the literature, as plotted in Figure 4, indicate that rocks of the complex followed a differentiation trend toward iron and alkali enrichment. Iron enrichment does not seem so pronounced as in typical stratiform complexes (Hess, 1960, Pl. 11, A-D), but the curve must be interpreted with caution because sample distribution is poor, several analyses were made before 1900, and some analyzed samples are uralitized.

The question of uralitization is especially important. The analyzed uralite gabbros in Table 5 all contain significantly less MgO than fresh gabbros, to which they are otherwise chemically

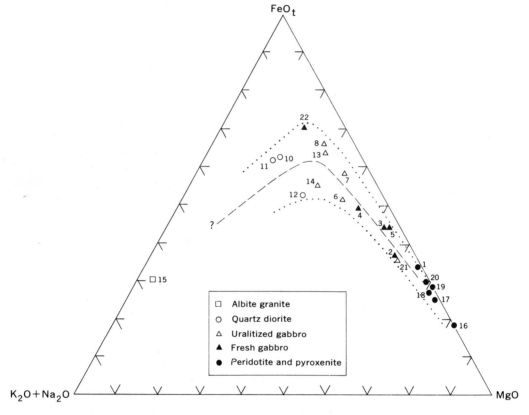

FIGURE 4. Triangular diagram showing possible differentiation trend (dashed line) of rocks in the Baltimore–State Line complex. FeO_t is total iron calculated as FeO. The dotted lines merely enclose all the data points.
Analyses used:

1-12. Correspond to analyses (column headings) in Table 5; analysis 9, which obviously is of metasomatic rock, is not plotted.

13. Hypersthene–augite gabbro, partly uralitized, Patapsco River at Ilchester, Baltimore Co., Md. Analyst: O. von Knorring *in* Hopson, 1964, p. 146.

14. Metagabbro, Ilchester, Howard Co., Md. Analyst: W. F. Hillebrand *in* Williams, 1895, p. 673.

15. Albite granite facies of Relay Quartz Diorite (of Knopf and Jonas, 1929), Rockburn Branch at the Patapsco River, Howard Co., Md. Analyst: O. von Knorring *in* Hopson, 1964, p. 159.

16 and 17. Websterite, Hebbville, Baltimore Co., Md. Analyst: J. E. Whitfield *in* Williams, 1890, p. 41.

18. Porphyritic lherzolite, Johnny Cake Road, Baltimore Co., Md. Analyst: T. M. Chatard *in* Williams, 1890, p. 39.

19. Websterite, Johnny Cake Road, Baltimore Co., Md. Analyst: J. E. Whitfield *in* Williams, 1890, p. 41.

20. Feldspathic lherzolite, dike on Western Maryland Railroad near Pikesville, Baltimore Co., Md. Analyst: Leroy McKay *in* Williams, 1890, p. 39.

21. Gabbro (contains 2.53 percent of H_2O), Wetheredsville, Baltimore Co., Md. Analyst: W. F. Hillebrand *in* Williams, 1895, p. 673.

22. Magnetite-rich hypersthene gabbro, Mt. Hope Station, Baltimore Co., Md. Analyst: Leroy McKay *in* Williams, 1886, p. 37.

similar. Whether or not this means that the Mg/Fe ratio is lowered during uralitization remains to be demonstrated by detailed chemical and mineralogic work on selected sample suites, but the possibility exists. Therefore, the use of uralitized samples in a plot like Figure 4 may give a false impression of magmatic iron enrichment.

Perhaps the best evidence for iron enrichment in this complex is mineralogic. Iron enrichment is indicated by the iron-rich hypersthene in gabbro transitional to quartz diorite (Fig. 2), by late magnetite in quartz diorite, and by the weak zoning to more iron-rich rims in some pyroxene crystals.

ORIGIN OF THE BALTIMORE–STATE LINE COMPLEX

It has already been mentioned that Hopson (1964) interpreted this complex as a contemporaneously deformed stratiform sheet, whereas Thayer (1967) considered it to be a typical alpine intrusion. Although Hopson interpreted the mass as fundamentally stratiform, he pointed out (1964, p. 148–150) that it has some alpine characteristics, and is neither an *ideal* stratiform nor an *ideal* alpine intrusion.

The fundamental characteristics of alpine complexes and the critical differences between alpine and stratiform complexes have been outlined by Thayer (1960, 1963a,b). According to Thayer (1967; personal discussion), the most important and diagnostic features of alpine complexes are:

1. Close areal and structural association of ultramafites, gabbroic, dioritic, and granophyric rocks.
2. Predominance of highly magnesian olivine over pyroxene in ultramafic parts of complexes.
3. Podiform chromite deposits.
4. Flow-layering and related structures and textures that are characteristic of high-grade metamorphic rocks.
5. Complicated structural relations between gabbroic and ultramafic rocks, such as intertonguing of major units along flow-layering; intrusive relations between various facies; and dikes of gabbro in peridotite or vice versa, without chilled margins.
6. Soda-rich dioritic and granophyric rocks within or near gabbroic rocks commonly are hybrid and accompanied by much albitization and brecciation.
7. Relict cumulate layering and textures may be preserved, but where relationships are clear, they antedate the formation of tectonic fabrics and related features.

If one accepts these features as diagnostic, the Baltimore–State Line complex must be classified as alpine. But does this necessarily mean that it was formed and emplaced according to the Bowen and Tuttle crystal mush hypothesis (Bowen and Tuttle, 1949; Thayer, 1960, 1963b, 1964)? If

"alpine" is a descriptive, nongenetic term, as Thayer agrees it should be (1967, personal discussion), then it should not carry this implication. Unfortunately, however, it does. The use of the word "alpine" to describe an *intrusive mechanism* as well as a *class of intrusions* leads to much confusion in the case of the Baltimore–State Line complex. Hopson (1964) logically explains many alpine (descriptive) features by a mechanism that is not alpine (genetic); Thayer (1967), on the other hand, contends that the alpine (descriptive) features discussed by Hopson require an alpine (genetic) interpretation. If alpine (genetic) complexes started out as essentially stratiform complexes deep in the crust or mantle and were reemplaced to higher levels during tectonism, as Thayer believes (1960, 1963b, 1964, 1967), many of the features developed should be similar to those produced by contemporaneously deforming a stratiform complex at higher crustal levels. In a sense, alpine (genetic) complexes are a special kind of deformed stratiform complex, and the interpretations of Hopson (1964) and Thayer (1967) are not as divergent as they might appear. The main differences have to do with the depth at which differentiation and deformation took place, and the origin of the layering.

It seems to me that the classification controversy interferes with the main purpose for studying this complex, which is to try to determine the geologic conditions under which it formed. Hopson (1964), Thayer (1967), and I all agree that the complex has had a complicated magmatic and tectonic history; whether it fits in the alpine or stratiform pigeonhole seems beside the point.

In Harford County, as in the Baltimore area (Hopson, 1964), country-rock contacts of the Baltimore–State Line complex are essentially concordant and sharp; chilled contacts and recognizable contact metamorphic effects are lacking. The overall shape of the complex is more or less sheetlike but is complicated on the northwest side by a long arm that diverges into the country rock (Fig. 2). Between this arm and the main complex is the axis of the Baltimore–Washington anticlinorium. Ultramafic rocks do not close around the anticlinorium as they should if they are the basal part of a simply folded stratiform sheet; this indicates either that there has been thrusting and strike-slip faulting of the sheet (for which there is no direct evidence) or that the stratiform character of the complex was completely disrupted during syntectonic emplacement, as parts of it moved independently of others. Rhythmic layering is uncommon, and layered

TABLE 5. Chemical Analyses of Pyroxenite, Gabbro, Metagabbro, and Quartz Diorite from the Baltimore–State Line Complex in Harford and Cecil Counties, Maryland

(New analyses indicated by asterisk)

	Pyrox-enite	Essentially fresh (unuralitized) gabbro				Uralitized gabbro			Quartz-bearing meta-gabbro	Dark quartz diorite		
	(1) WD-2*	(2) C-79a*	(3) B-458*	(4) WD-2R*	(5) Ba 128	(6) 320a*	(7) BG-5	(8) Ba 124	(9) C-11(Q)*	(10) A-49b*	(11) Ba 124a	(12) Ba 124b
Major oxides												
SiO_2	55.3	50.1	48.3	45.2	48.02	46.2	45.41	44.04	53.1	59.7	58.57	55.16
Al_2O_3	2.7	17.0	19.5	20.4	20.01	23.8	23.05	20.01	17.6	15.0	16.10	17.51
Fe_2O_3	1.0	0.33	1.3	3.8	1.13	1.4	1.52	4.22	3.2	2.3	2.89	2.62
FeO	11.2	6.7	6.2	6.2	7.29	4.8	8.35	8.61	7.3	7.0	6.12	5.83
MgO	24.8	11.1	8.8	8.3	10.05	4.5	5.89	5.01	3.6	2.6	2.33	4.35
CaO	2.8	12.5	14.2	13.5	11.42	14.8	12.52	11.68	9.8	6.7	7.39	8.50
Na_2O	0.20	0.93	0.60	1.2	0.51	1.0	0.76	1.24	1.3	2.4	2.11	1.83
K_2O	0.32	0.12	0.10	0.20	0.05	0.16	0.32	0.15	0.14	0.64	1.01	1.08
H_2O^+	0.36	0.33	0.44	0.50	0.67	1.7	1.52	2.01	1.6	1.2	1.27	2.01
H_2O^-	0.07	0.12	0.07	0.04	n.d.	0.09	n.d.	n.d.	0.06	0.07	0.21	0.21
TiO_2	0.26	0.27	0.05	0.60	0.23	0.50	0.62	2.24	1.8	1.7	1.41	0.64
P_2O_5	0.00	0.07	0.04	0.00	tr	0.35	0.13	0.52	0.50	0.53	0.37	0.21
MnO	0.15	0.15	0.20	0.00	0.18	0.14	0.09	0.28	0.18	0.24	0.18	0.15
CO_2	0.09	0.15	0.11	0.05	0.25	0.09	n.d.	n.d.	0.12	<0.05	n.d.	n.d.
SO_3	n.d.	0.21	0.13	n.d.	n.d.	0.19	n.d.	n.d.	n.d.	n.d.	n.d.	n.d.
Others	—	—	—	—	0.16	—	—	0.41	—	—	0.11	0.10
Total	99.25	99.63	100.04	99.99	99.98	99.72	100.18	100.42	100.30	100.08	100.07	100.17
Trace elements†												
Ba	0.001	0.002	0.001	0.001	—	0.003	—	—	0.005	0.015	r	r
Be	—	—	—	—	—	—	—	—	0.0001	0.0015	—	—
Co	0.005	0.005	0.005	0.003	—	0.002	—	—	0.003	0.0015	—	—
Cr	0.15	0.05	0.002	0.015	r	0.015	—	—	0.002	0.0003	—	—
Cu	0.007	0.07	0.005	0.05	—	0.002	—	—	0.005	0.002	—	—
Ga	—	0.001	0.001	0.0015	—	0.001	—	—	0.002	0.0015	—	—
Mo	0.0005	0.0003	—	—	—	0.0003	—	—	—	—	—	—
Nb	—	—	—	—	—	—	—	—	—	0.0015	—	—
Ni	0.07	0.02	0.005	0.03	r	0.005	—	r	0.007	—	—	r
Pb	0.003	—	—	0.003	—	—	—	—	—	—	—	—
Sc	0.007	0.005	0.005	0.005	—	0.003	—	—	0.005	0.005	—	—
Sr	0.0003	0.02	0.015	0.02	r	0.03	—	—	0.03	0.02	r	—
V	0.015	0.02	0.02	0.07	r	0.01	—	r	0.02	0.01	r	r
Y	—	0.0007	—	—	—	0.007	—	—	0.0015	0.003	—	—
Yb	—	—	—	—	—	—	—	—	0.0001	0.0003	—	—
Zr	—	0.001	—	—	—	0.001	—	r	0.10	0.10	r	r
Norms												
Q	2.94	0.97	1.14	—	0.80	2.37	—	3.93	17.30	22.98	21.62	13.72
C	—	—	—	—	—	—	—	—	—	—	—	—
Z	—	—	—	—	—	—	—	—	—	—	0.13	0.03
Or	1.67	0.71	0.59	1.11	0.30	0.94	1.89	0.87	0.83	3.78	5.97	6.38
Ab	1.57	6.49	4.22	9.96	4.31	7.21	6.42	10.49	10.99	20.30	17.84	15.48
An	5.56	42.59	50.67	49.80	52.16	60.64	58.57	48.59	41.78	28.27	31.48	36.38
Th	—	0.37	0.23	—	—	0.34	—	—	—	—	—	—
Wo	3.25	7.52	7.85	7.09	1.87	4.14	1.13	2.48	1.17	0.63	1.15	1.84
En	62.25	27.63	21.91	12.15	25.02	11.20	12.90	12.47	8.96	6.47	5.80	10.83
Fs	19.66	11.86	10.60	4.35	12.30	7.09	11.63	5.59	8.12	8.59	6.86	7.72
Fo	—	—	—	6.05	—	—	1.24	—	—	—	—	—
Fa	—	—	—	2.24	—	—	1.23	—	—	—	—	—
Mt	1.39	0.48	1.88	5.56	1.64	2.03	2.20	6.12	4.64	3.34	4.19	3.80
Il	0.46	0.51	0.10	1.06	0.44	0.95	1.18	4.25	3.42	3.23	2.68	1.22
Ap	—	0.17	0.10	—	—	0.83	0.31	1.23	1.18	1.26	0.88	0.50
Pr	—	—	—	—	0.11	—	—	0.24	—	—	—	0.06
Cc	0.20	0.34	0.25	0.10	—	0.20	—	—	0.27	—	—	—
H_2O	0.43	0.45	0.51	0.54	0.67	1.79	1.52	2.01	1.66	1.27	1.48	2.22
Total	99.38	100.09	100.05	100.03	99.62	99.73	100.22	100.27	100.22	100.12	100.08	100.18
Di	6.22	14.52	15.21	13.63	3.63	8.08	2.24	4.80	2.31	1.25	2.29	3.61
Hy	78.94	32.50	25.15	9.97	35.57	14.35	23.43	15.75	15.94	14.44	11.51	16.78
Ol	—	—	—	8.29	—	—	2.47	—	—	—	—	—

sections seem to be local phenomena in the midst of essentially massive rock. Some layers are typified by great compositional contrast with enclosing rock, and are best explained as flowage features. The chemical and mineralogic trend from peridotite through gabbro to mafic quartz diorite is gradational, for the most part, and might be termed broad-scale cryptic zoning. Only the contacts between olivine-bearing and olivine-free rocks seem to be sharp. Massive, lineated, podiform chromite deposits occur in dunite and peridotite in the State Line district and in the peridotite lenses north of the divergent offshoot in Harford County. Late-stage gabbroic pegmatites and net veins are common in the gabbro zone.

All these features are compatible with synkinematic emplacement but do not require that the rocks had a long history of quiescent differentiation prior to deformation. Perhaps this part of the complex began to crystallize at a deeper level in the crust, where olivine-rich rocks with cumulate structures and textures accumulated. Before differentiation had progressed much beyond the ultramafic stage, however, the magma chamber

was squeezed tectonically, and its contents were injected into higher levels of the crust. Solid or semisolid peridotite, carrying chromite with it, moved upward along with a more fluid mixture of gabbroic magma and crystals. Because of tectonic agitation of the cooling crystallizing gabbroic magma, and possibly also because of its high crystal content, convective circulation was not established, and rhythmic layering was not produced on a large scale. Only the crudest sort of large-scale differentiation could take place as cooling and flowage progressed. The fact that large-scale differentiation (cryptic zoning) did occur may indicate that viscosity differences were not large in the mobilized gabbro mush and further suggests that cooling was slow enough to permit free diffusion.

Tectonic mixing of diverse rock types took place, but not as extensively as in many of the alpine complexes described by Thayer (1963b). The dike-like layers of foliated peridotite and pyroxenite within foliated olivine-free gabbro probably are the result of tectonic mixing, and some of the streakily interbedded gabbroic and pyroxenitic rocks in the ultramafic offshoot may

Column headings:
  (1) WD-2*: Bronzite pyroxenite from center of 7-foot pyroxenite layer, railroad cut on east side of Susquehanna River, about 0.3 mi south of Conowingo Creek, Cecil County, Md. Analyst:‡ U.S.G.S. Rapid, 1966.
  (2) C-79a*: Hypersthene gabbro, sharp bend on road along Susquehanna River about 200 ft above mouth of Conowingo Creek, Cecil County, Md. Analyst: U.S.G.S. Rapid, 1965. (Mode, col. 1, table 1)
  (3) B-458*: Hypersthene gabbro, Bel Air bypass (U.S. Route 1) about 1.3 mi north of Winters Run, Harford County, Md. Analyst: U.S.G.S. Rapid, 1965. (Mode, col. 2, table 1)
  (4) WD-2R*: Hypersthene gabbro between ultramafic layers, cut below railroad along Susquehanna River, about 0.3 mi south of Conowingo Creek, Cecil County, Md. Analyst: U.S.G.S. Rapid, 1966.
  (5) Ba 128: Norite, Oak Grove, Cecil County, Md. Analyst: W. F. Hillebrand in Leonard, 1901, p. 151.
  (6) 320a*: Coarse uralitized gabbro, pipeline excavation near Maryland Route 136, 0.9 mi south of Poplar Grove, Harford County, Md. Analyst: U.S.G.S. Rapid, 1965. Chiefly fibrous green amphibole, partly recrystallized calcic plagioclase, clinozoisite, and epidote with traces of relict pyroxene.
  (7) BG-5: Gabbro, Dublin, Harford County, Md. (probably uralitized because it contains 1.52 percent H_2O). Analyst: Penniman and Browne in Insley, 1919, suppl.
  (8) Ba 124: Metagabbro (hornblende gabbro; bojite of Herz, 1951), Rising Sun, Cecil County, Md. Analyst: W. F. Hillebrand in Leonard, 1901, p. 146.
  (9) C-11(Q)*: Quartz-bearing metagabbro composed chiefly of quartz, epidote, and fibrous actinolitic amphibole; hydrothermal addition of quartz likely. Small quarry on east side of Susquehanna River, 0.6 mi south of Conowingo Creek, Cecil County, Md. Analyst: U.S.G.S. Rapid, 1963.
  (10) A-49b*: Dark-colored biotite-hornblende quartz diorite, north end of Mountain Hill, Harford County, Md. Modally very similar to C-90, Table 3, col. 4. Analyst: U.S.G.S. Rapid, 1963.
  (11) Ba 124a: Quartz-biotite-hornblende gabbro, near foundry on Stone Run, Cecil County, Md. Analyst: W. F. Hillebrand in Leonard, A. G., 1901, p. 146.
  (12) Ba 124b: Quartz-biotite-hornblende gabbro, near Porter Bridge on Octoraro Creek, Cecil County, Md. Analyst: W. F. Hillebrand in Leonard, A. G., 1901, p. 146.

† Semiquantitative spectrographic analyses by W. B. Crandell and J. L. Harris, U.S. Geological Survey. Results are reported in percent to the nearest number in the series 1, 0.7, 0.5, 0.3, 0.2, 0.15, and 0.1, etc., which represent approximate midpoints of group data on a geometric scale. The assigned group for semiquantitative results will include the quantitative value about 30 percent of the time. In addition to those listed, the following elements were sought but not detected: Ag, As, Au, B, Bi, Cd, Ce, Ge, Hf, Hg, In, La, Li, Pd, Pt, Re, Sb, Sn, W, and Zn.
r: reported as oxide in old analysis; totaled under "other."

‡ Analyses noted as "U.S.G.S. Rapid" were done in the laboratories of the U.S. Geological Survey by methods similar to those described in Shapiro and others (1962). P. Elmore, S. Botts, H. Taylor, L. Artis, H. Smith, and G. Chloe, analysts.

also be. The last feldspathic residuum to be squeezed from the gabbro probably formed the gabbroic pegmatites and net veins that wander across all earlier structures.

As crystallization and tectonism continued, parts of the complex were separated from the main mass and injected into country rock. They may have been serpentinized and uralitized as they were emplaced; later they were partly converted to talcose rock by exchange reactions with the enclosing aluminous schists.

The emplacement history just outlined differs from that proposed by Thayer (1960, 1963a,b, 1964) for typical alpine complexes in that it does not require initial crystallization in the mantle, or remobilization of an already crystallized rock. Rather, reemplacement of the complex began before its less refractory parts had completely solidified at depth, and final crystallization was synkinematic. This interpretation differs from Hopson's (1964) in that the Harford County part of the complex, at least, is thought to have moved from the site where it began to crystallize. The depth of initial crystallization and the distance moved from the original magma chamber are unknown, but need not have been great.

Other points to be considered are the presence of very calcic plagioclase (about An_{90}) in the gabbro, and the prevalence of orthopyroxene. N. L. Bowen, in his classic chapter on assimilation (1928, p. 201–214), has shown that assimilation of aluminous sedimentary rock by basaltic magma should inhibit olivine formation and favor formation of orthopyroxene and anorthite. Thus, the assimilation of pelitic rock from the Wissahickon Formation may have influenced the mineralogy of the complex, as Bowen himself suggested (1928, p. 212). If the assimilation theory applies, it would mean that magma was injected into pelitic wall rocks while it was still hot, mostly liquid, and capable of magmatic reaction. It would require crystallization within the sialic crust and eliminate the possibility of extensive primary differentiation in the mantle. There is no direct evidence, such as partly digested schist inclusions or contact reaction zones, that large-scale assimilation took place. It is interesting to note, however, that only two chemical analyses of gabbro in Table 5 lack quartz in the norm.

ACKNOWLEDGMENTS

I am indebted to Davis M. Lapham, C. E. Brown, G. Malcom Brown, George W. Fisher, and A. J. Naldrett for their interest in and ideas on the problems of this complex, and I am especially grateful to T. P. Thayer and G. H. Espenshade

for their penetrating reviews of an early draft of this paper. All these men have visited the complex in my company, and it is safe to say that none of them agrees completely with my interpretation.

REFERENCES

Baragar, W. R. A., 1960, Petrology of basaltic rocks in part of the Labrador Trough: Geol. Soc. America Bull., v. 71, p. 1589–1644.

Bascom, Florence, 1902, The geology of the crystalline rocks of Cecil County: Maryland Geol. Survey, Cecil County, p. 83–148.

Battey, M. H., 1960, The relationship between preferred orientation of olivine in dunite and the tectonic environment: Am. Jour. Sci., v. 258, p. 716–727.

Bowen, N. L., 1928, The evolution of the igneous rocks: Princeton, N.J., Princeton Univ. Press, 334 p., *Reprinted* by Dover Publications, 1956.

———, and Tuttle, O. F., 1949, The system MgO–SiO_2–H_2O: Geol. Soc. America, Bull., v. 60, p. 439–460.

Bromery, R. W., Petty, A. J., and Smith, C. W., 1964, Aeromagnetic map of Bel Air and vicinity, Harford, Baltimore, and Cecil Counties, Md.: U. S. Geol. Survey, Geophys. Inv. Map GP-482.

Chidester, A. H., 1962, Petrology and geochemistry of selected talc-bearing ultramafic rocks and adjacent country rocks in north-central Vermont: U. S. Geol. Survey, Prof. Paper 345, 207 p.

Cloos, E., and Hershey, H. G., 1936, Structural age determination of Piedmont intrusives in Maryland: Natl. Acad. Sci., Proc., v. 22, p. 71–80.

Cohen, C. J., 1937, Structure of the metamorphosed gabbro complex at Baltimore, Md.: Maryland Geol. Survey, v. 13, p. 215–236.

Herz, N., 1951, Petrology of the Baltimore Gabbro, Maryland: Geol. Soc. America Bull., v. 62, p. 979–1016.

Hess, H. H., 1960, Stillwater igneous complex, Montana: Geol. Soc. America, Mem. 80, 225 p.

Hopson, C. A., 1964, The crystalline rocks of Howard and Montgomery Counties, *in* The geology of Howard and Montgomery Counties: Baltimore, Maryland Geol. Survey, p. 27–215.

Insley, H., 1919, The gabbros and associated intrusive rocks of Harford County, Md.: Ph.D. dissertation, Johns Hopkins Univ.

———, 1928, The gabbros and associated intrusive rocks of Harford County, Md.: Maryland Geol. Survey, v. 12, p. 289–332.

Knopf, E. B., 1921, Chrome ores of southeastern Pennsylvania and Maryland: U. S. Geol. Survey, Bull. 725-B, p. 85–99.

———, and Jonas, A. E., 1929, The geology of the crystalline rocks [of Baltimore County]: Maryland Geol. Survey, Baltimore County, p. 97–199.

Köhler, A., 1941, Drehtischmessungen an Plagioklaszewillingen von Tief- und Hoch-temperaturoptik: Miner. U. Petrog. Mitt., Bd. 53, p. 159–179.

Kretz, R., 1963, Note on some equilibria in which plagioclase and epidote participate: Am. Jour. Sci., v. 261, p. 973–982.

Lacroix, A., 1943, Les péridotites de la Nouvelle-Calédonie, leurs serpentines et leurs gîtes de nickel et de cobalt; les gabbros qui les accompagnet: Acad. Sci. Paris, mém., t. 66, 143 p.

Lapham, D. M., and McKague, H. L., 1964, Structural pattern associated with the serpentinites of southeastern Pennsylvania: Geol. Soc. America, Bull., v. 75, no. 7, p. 639–659.

Larrabee, D. M., 1966, Map showing distribution of ultramafic and intrusive mafic rocks from northern New Jersey to eastern Alabama: U. S. Geol. Survey, Misc. Geol. Inv. Map I-476.

Leonard, A. G., 1901, The basic rocks of northeastern Maryland and their relation to the granite: Am. Geologist, v. 28, p. 135–176.

Mathews, E. B., and Johannsen, A., 1904, Geologic map of Harford County: Baltimore, Maryland Geol. Survey, scale 1:62,500.

Pearre, N. C., and Heyl, A. V., Jr., 1960, Chromite and other mineral deposits in serpentine rocks of the Piedmont upland, Maryland, Pennsylvania, and Delaware: U. S. Geol. Survey, Bull. 1082-K, p. 707–833.

Poldervaart, Arie, 1950, Correlation of physical properties and chemical composition in the plagioclase, olivine, and orthopyroxene series: Am. Mineralogist, v. 35, p. 1067–1079.

Rossman, D. L., 1963, Geology and petrology of two stocks of layered gabbro in the Fairweather Range, Alaska, U. S. Geol. Survey, Bull. 1121-F, 50 p.

Shapiro, L., and Brannock, W. W., 1962, Rapid analysis of silicate, carbonate, and phosphate rocks: U. S. Geol. Survey, Bull. 1144-A, 56 p.

Southwick, D. L., and Owens, J. P., 1968, Geologic map of Harford County, Md.: Baltimore, Maryland Geol. Survey, County Geol. Map CGM-1, scale 1:62,500.

Thayer, T. P., 1960, Some critical differences between alpine-type and stratiform peridotite–gabbro complexes: Internat. Geol. Cong., 21st, Copenhagen, 1960, Rept., pt. 13, p. 247–259.

————, 1963a, The Canyon Mountain complex, Oregon, and the alpine mafic magma stem: U. S. Geol. Survey, Prof. Paper 475-C, p. C82–C85.

————, 1963b, Flow layering in alpine peridotite-gabbro complexes: Mineralog. Soc. America, Spec. Paper no. 1, p. 55–61.

————, 1964, Principal features and origin of podiform chromite deposits, and some observations on the Guleman–Soridağ district, Turkey: Econ. Geology, v. 59, p. 1497–1524.

————, 1967, Chemical and structural relations of ultramafic and feldspathic rocks in alpine intrusive complexes, in P. J. Wyllie, ed., Ultramafic and related rocks: New York, N.Y., J. Wiley & Sons, Inc., p. 222–239.

Turner, F. J., 1947, Determination of plagioclase with the four-axis universal stage: Am. Mineralogist, v. 32, p. 389–410.

Williams, G. H., 1884, On the paramorphosis of pyroxene to hornblende in rocks: Am. Jour. Sci., 3d ser., v. 28, p.259–268.

————, 1886, The gabbros and associated hornblende rocks occurring in the neighborhood of Baltimore, Md.: U. S. Geol. Survey Bull. 28, 78 p.

————, 1890, The non-feldspathic intrusive rocks of Maryland and the course of their alteration: Am. Geologist, v. 6, p. 35–49.

————, 1895, General relations of the granitic rocks in the middle Atlantic Piedmont Plateau: U. S. Geol. Survey, 15th Ann. Rept., p. 651–684.

Wilson, R. A. M., and Ingham, F. T., 1959, The geology and mineral resources of the Xeros–Troodos area: Cyprus Geol. Survey Dept., Mem. 1, 184 p.

Yoder, H. S., and Sahama, Th. G., 1957, Olivine X-ray determinative curve: Am. Mineralogist, v. 42, p. 475–491.

Chapter 28

Post-Triassic Tectonic Movements in the Central and Southern Appalachians as Recorded by Sediments of the Atlantic Coastal Plain*

JAMES P. OWENS

INTRODUCTION

LATE TECTONIC EVENTS, defined herein as those of post-Triassic age, have received little attention in any structural analyses of the Appalachian Mountains. The absence of post-Triassic formations within this mountain system has made structural studies embracing this period impossible.

A detailed examination of the adjacent Atlantic Coastal Plain, particularly the northern part (Virginia through New Jersey) has yielded a large amount of data relating to probable late tectonic movements within the central Appalachians. A survey of the literature was used to extrapolate these data into the southern Appalachians. The validity of the conclusions in this latter region is, therefore, limited to the quality of the published stratigraphic studies.

Although much of the evidence presented is circumstantial, an analysis of the Coastal Plain sediments and their distribution offers the best opportunity to unravel these late diastrophic events for the central and southern Appalachians. A basic assumption in this sedimentary tectonic analysis is that uplift in the source land (Appalachians) is accompanied by erosion, which will then supply clastics to the depositional basin

* Publication authorized by the Director, U. S. Geological Survey.

(coastal plain) in quantities that depend upon the intensity of the uplift and basin warping. Where uplift is intense, a clastic wedge will prograde into the basin. During periods of quiescence, biogenic remains become the dominant sediment type. The distribution of clastic and carbonate rocks within the basin, therefore, is a crude measure of tectonic activity in the source land.

REGIONAL SETTING

The Atlantic Coastal Plain is an irregularly thick, extensively dissected eastward-facing wedge of unconsolidated to semiconsolidated sedimentary rocks which borders the central and southern Appalachians. If the whole Coastal Plain province (Gulf, Mississippi embayment, and Atlantic) is considered as a single entity, the central and southern Appalachians are bounded on three sides by these unconsolidated sedimentary rocks. The Atlantic Coastal Plain lies athwart a relatively low-lying craton, the Piedmont physiographic province. Most of the basement upon which it rests consists of Piedmont-type rocks (largely low- to high-grade metamorphic, igneous, and older sedimentary rocks). Some authors have drawn attention to the similarity of the Piedmont rocks to those in the New England metamorphic belt. However, New England is much more mountainous than the low-lying southern Piedmont. It is possible that the physiography of the two regions was once similar and that the mass wasting of the southern Piedmont produced a platform, now the basement beneath the Coastal Plain, in addition to supplying large quantities of clastic material to the depositional basin itself.

Although the Atlantic Coastal Plain was long considered a classic example of a stable shelf, detailed geophysical studies supplemented by drill-hole data revealed that the basement underlying this plain was a much more irregular surface than had long been thought (U. S. Geol. Survey and Am. Assoc. Petroleum Geologists, 1961). The surface pattern in this part of the basin consists of a series of arches and troughs aligned for the most part normal to the Appalachian structures. As much as half the Coastal Plain, however, is inundated by the Atlantic Ocean. In recent years, this submerged part of the shelf has been investigated in varying detail by a number of different geophysical techniques, and our knowledge of this part of the shelf has been greatly enlarged. The seismic study of Drake and others (1959) of the submerged shelf off New Jersey was particularly enlightening in regard to the overall shape of the continental terrace (Fig. 1). As can be seen, the terrace has a deep double trough at the edge of the continental shelf. Drake and others (1959) noted that the configuration of the terrace in cross section was similar to that proposed for the Appalachian geosyncline by Kay (1951). Subsequent seismic studies off the southern Atlantic Coastal Plain (Hersey and others, 1959) revealed a somewhat similar outline to the shelf, although the two troughs bifurcated and were widely separated.

SEDIMENT DISTRIBUTION

Some recent authors (LeGrand, 1961, for example) note that the Atlantic Coastal Plain appears to be divided into two large, rather ill-defined basins, northern North Carolina roughly serving as the dividing zone. The northern province is characteristically a glauconite-rich area, whereas the southern basin is a carbonate province. The report by Dryden and Bryden (1956), shows that the heavy minerals in the southern Atlantic Coastal Plain are characterized by a stable suite (high zircon, tourmaline, and rutile) compared with those of the northern area, which has a "full" suite containing many less stable minerals (such as epidote, garnet, and chloritoid) particularly within the marine facies. These observations suggest that clastic sedimentation was more rapid in the north, whereas carbonate deposition and recycled clastic material indicate less rapid sedimentation (hence, greater structural stability) in the southern Atlantic Coastal Plain region.

This simple view, however, may only apply to particular time intervals between the two areas (early Tertiary for example); if the whole Coastal Plain is examined, its sedimentary history is much more complex.

The distribution of the arches, troughs, and deep furrows or trenches in the Atlantic seaboard basement was found to be remarkably similar to that in the basement of the Gulf Coast and Missis-

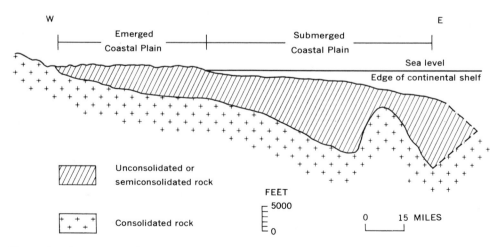

FIGURE 1.   Cross section of continental terrace in northern Atlantic Coastal Plain (modified from Drake and others, 1959). Section extends across New Jersey to the edge of the continental shelf in the Atlantic Ocean. Note double trough separated by a basement high. Only part of the deeper outer trough is shown.

sippi embayment (Fig. 2), a fact noted earlier by Murray (1961). The typical pattern consists of a series of troughs and arches oriented normal to the existing structural trends in the craton. Farther downdip, these troughs and arches are flanked by a deeper trench whose main axis is oriented parallel to the older structural trends. In cross section, the two areas are also remarkably similar (Fig. 3). Although the terrace is much larger in the Gulf Coast, the terrace off New Jersey (Fig. 1) has the same general form. If all these features are tectonic in origin, as is likely, a similar structural history is implied for both regions. The hypothesis that the Atlantic Coastal Plain is a stable shelf appears unwarranted if the Gulf Coast is considered to be an unstable shelf or platform.

This paper examines the distribution of the clastic rocks and associated carbonate rocks in relation to these structural elements, and attempts to date the activity of the positive and negative elements by a study of the type, age, thickness, and areal distribution of the sedimentary rocks. Because of the large area to be discussed, many generalizations have been made.

The Coastal-Plain deposits can be readily separated into four major sequences, discussed in Table 1 from youngest to oldest.

Sedimentation Sequence I

The deepest troughs adjacent to the foreland, the Salisbury and Albemarle embayments and the Suwannee straits or Apalachiola embayment, contain thick accumulations of clastic rocks most of which have been dated as Early Cretaceous, although Swain (1947, 1952) suggests a possible Jurassic(?) age for the basal beds in the deepest part of the Albemarle embayment.

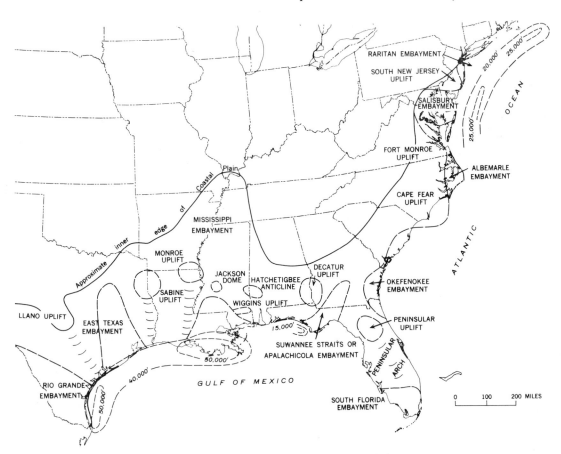

FIGURE 2. Major structural elements in the Gulf Coast and Atlantic Coastal Plain (modified mainly from Tipsword, 1962). Dashed lines, chiefly in offshore area, indicate approximate thickness of sediments within troughs. Off Atlantic Coast, these data represent total thickness of post-Triassic(?) sediments; those off Louisiana and Texas presumably represent only sediments of Cenozoic age (Hardin, 1962); and those off Florida represent sediments of post-Paleozoic age (Puri and Vernon, 1964). Note that most of the structural features in the Gulf region are larger scale than those along the Atlantic seaboard.

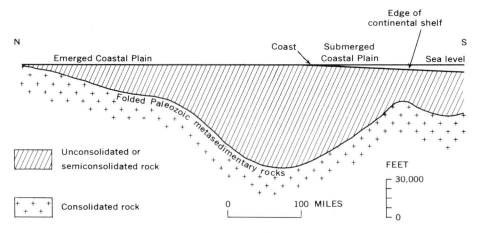

FIGURE 3. Cross section of continental terrace in Gulf Coast region (modified from Hardin, 1962). Section extends from Arkansas across Louisiana out into the Gulf of Mexico. Although this terrace is considerably larger, it has the same general form as the one off New Jersey.

In the Salisbury embayment, more than 4,000 ft (1,200 m) of Lower Cretaceous clastic rocks were penetrated in a deep well (Anderson, 1948). Updip in northeastern Maryland, the equivalents of these beds, rocks of the Potomac Group, crop out at the head of Chesapeake Bay. Interestingly enough, this is the only area on the Atlantic Coastal Plain where beds of this age appear at the surface. This may be a reflection on the magnitude of this particular downwarp. The continental character of the exposed beds is indicated by cut-and-fill structures, extensive large-scale cross stratification, and rapid lensing of the sandy and clayey beds. On the north side of the embayment, the Potomac Group consists largely of extensively crossbedded quartz sands and lesser gravels interstratified with black to variegated highly aluminous clays (kaolinite–illite mixtures). On the south side, the rocks are mainly arkosic sands interstratified with brown to olive-green, low aluminous clays (mainly montmorillonite–illite mixtures). The northerly rocks apparently were derived from the Triassic basin (multicycle), whereas the arkoses were shed from the granitic Piedmont terrain of Virginia.

A relatively thin dark-gray clay, the Arundel (the middle member of the Potomac Group), crops out in a more or less continuous arcuate belt along the head of Chesapeake Bay. The persistency of this clay, small-scale sedimentary bedding structures, and extensive beds of banded siderite suggest an estuarine to brackish-water origin. Thus, this unit records the farthest marine transgression onto the older craton during this sedimentation sequence.

In the deep subsurface, the Potomac units contain an extensive brackish-water fauna, suggesting more marine conditions basinward. Continental facies, however, still appear to be more abundant.

A somewhat similar stratigraphic distribution is suggested in the Albemarle embayment (Swain and Brown, 1964) where the marine facies (characterized by brackish-water fauna) overlies the

TABLE 1.

	Sedimentation sequences	Basal contact
IV	Fluviatile, marine in low-lying areas near coast. Pliocene to Quaternary in age.	Unconformable
III	Marine–continental, more continental facies than below. Locally very fossiliferous; middle through upper Miocene and perhaps lower Pliocene(?).	Unconformable
II	Marine–continental interbeds. Mostly shallow-water marine. Very fossiliferous and locally extremely glauconitic. Upper Cretaceous (Cenomanian)-lower Tertiary (Claiborne) in age.	Generally unconformable, but locally conformable
I	Fluviatile mainly, locally brackish in downdip areas. Lower Cretaceous mainly (Neocomian–Albian: some possible Jurassic).	Unconformable

nonmarine facies in the updip areas. Downdip, this entire sequence thickens to more than 3,000 ft (900 m), and the marine facies is more abundant (Swain, 1952). These rocks are arkoses and montmorillonitic shales similar to the Potomac Group in Virginia.

The clastic facies in the Suwannee straits–Apalachicola embayment is approximately 3,500 ft (1,000 m) thick in southwestern Georgia (Herrick and Vorhis, 1963) and more than 5,000 ft (1,500 m) thick in the western Florida panhandle (Applin and Applin, *in* Puri and Vernon 1964, p. 34–41). It is commonly referred to as the red-bed facies and consists in large part of arkosic sand interbedded with red to green clay. Nodular limestones are present downdip. Except for the limy facies the general lithology of the clastic facies resembles the Potomac Group of Virginia.

The carbonate facies is restricted to the central and southern Florida embayment (Fig. 4A). There, a southward-thickening sequence of interbedded carbonate rocks and evaporite of Jurassic(?) and Early Cretaceous age attains thicknesses of more than 8,000 ft (2,400 m). Northward, toward the peninsular arch, these carbonate rocks apparently interfinger with nearer shore marly sands and clays.

These Early Cretaceous deposits along the Atlantic seaboard, therefore, indicate large-scale warping and uplift in the central and southern Appalachians. Large volumes of clastic material filled the downwarped basins and prograded seaward, where it interfingered with marginal marine facies. The carbonate facies was deposited far out on the shelf at this time.

Sedimentation Sequence II

In order to evaluate the northern and southern Coastal Plain sediment distribution, this sequence will be discussed as three parts (Upper Cretaceous, early Tertiary, and middle Tertiary). In the north, the sedimentary rocks of this sequence form a single lithologic entity, whereas to the south there is a distinct change in sedimentation pattern through time.

Upper Cretaceous. The entire continental block along the eastern seaboard was depressed in the Late Cretaceous (beginning in Cenomanian time), and all the positive and negative structural elements were eventually overlapped by a sequence of intercalated marine-continental clastic facies (Fig. 4B). The marine facies compose most of these deposits, although continental deposits are more common in the lower part of the Upper Cretaceous (Cenomanian–Santonian). The older structural troughs no longer were the sites

of maximum downwarp and sediment accumulation, but new and smaller troughs (depocenters of Murray, 1961) were formed. In New Jersey, one of the best studied, younger downwarps is the Raritan embayment (Fig. 2). Nearly 1,500 ft (450 m) of interstratified clastic-rich and glauconite-rich beds accumulated in this trough. The section thins to 250 ft (75 m) on the south New Jersey uplift (Fig. 2) to the southwest; to the east it lies beneath the Atlantic Ocean or is covered by thick morainal deposits of Pleistocene age. No dominantly carbonate beds occur in the trough or are known from dredgings on the shelf in the northern Atlantic Coastal Plain.

One interesting feature of the Late Cretaceous beds is their cyclic nature (most commonly interpreted as a product of epeirogenesis) which indicates general basin-foreland instability during this period. A typical cycle of sedimentation in the marine–transitional environment of the northern clastic province is shown in Figure 5. As can be seen, a complete cyclic sequence includes three major lithologies: a basal transgressive glauconite sand (interpreted as the most seaward deposit), which is overlain by regressive silt, which is in turn overlain by a regressive nearshore sand. This well-developed cyclic sequence is repeated seven times in the Late Cretaceous through early Tertiary (middle Eocene).

The Late Cretaceous beds of North Carolina are similar to those of New Jersey, although there is a marked decline in the glauconite sand content. Although the rocks are calcareous, clastic material predominates throughout. Pure limestone beds are unknown even in the farthest seaward part of the embayment (Swain, 1947, 1952). There is no published information on the thickness and distribution of the Late Cretaceous in North Carolina or South Carolina.

In Georgia a second center of deposition, similar to the one in New Jersey, is present in the central coastal plain (Herrick and Vorhis, 1963). More than 2,300 ft (700 m) of Late Cretaecous rocks are present in the east–west oriented trough (Herrick and Vorhis, 1963, p. 44, 47) that lies generally to the north of the Suwanne straits. Again, the continental beds are most abundant in the lower part. Except in southeastern Georgia, these Late Cretaceous beds are clastic rocks (Fig. 4B). Limy beds of Late Cretaceous age are for the most part restricted to peninsular Florida. The distribution of Late Cretaceous deposits (Fig. 4B) is highly schematized because the clastic–carbonate boundary migrated back and forth across northern Florida during this time. The

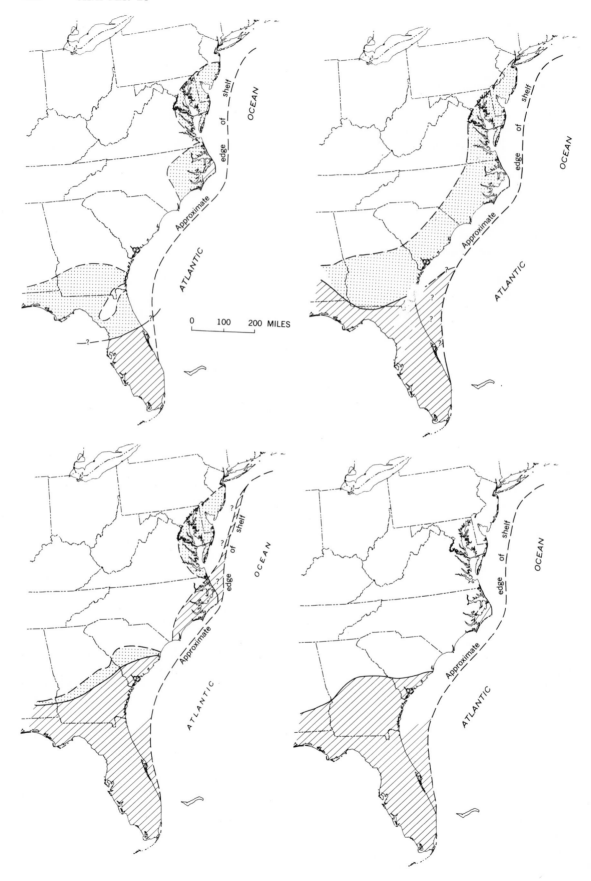

0   100   200 MILES

boundary shown represents the maximum northern position.

During Late Cretaceous time, there was a general transgression of the marine sequence onto the craton. A large volume of clastic material derived from the craton, however, precluded nearshore deposits of the carbonates, and indicates a mild degree of diastrophism in the central and southern Appalachians during Late Cretaceous time.

Early Tertiary. In the northernmost Atlantic Coastal Plain, clastic deposition on the emerged shelf area continued during the early Tertiary (Fig. 4C). Dredging in two of the canyons at the edge of the shelf, however, encountered a foraminiferal limestone of Eocene age (Stetson, 1949, p. 11). Thus, the carbonate wedge is present on the shelf in this area, but its shoreward distribution is not known. The early Tertiary beds are mainly intercalated sands of clastic and authigenic (glauconite) origin. A minor interruption in the dominant pattern, however, is indicated by the occurrence of a calcarenite facies in the Vincentown Formation of Paleocene age.

The early Tertiary beds have a known maximum thickness of about 500 ft (150 m) in the center of the Salisbury embayment (Anderson, 1948). The section thins updip and east–west along strike. As in the underlying Late Cretaceous beds, the early Tertiary formations exhibit well-developed cyclicity (two cycles can be outlined).

In North Carolina the carbonate facies transgresses well onto the emerged shelf (Fig. 4C) and overlaps most of the older Coastal Plain deposits in South Carolina and Georgia. In Florida it blankets the entire State. The wedge of clastic rocks is thin throughout this region, although it is thicker and areally more widespread in the lower Eocene than in the upper Eocene.

In Georgia, where they are best studied, the Eocene beds aggregate about 2,400 ft (700 m). They are thickest in the Suwannee straits and Okefenokee embayment (Herrick and Vorhis, 1963). In the Albemarle embayment area, these beds are about 1,000 ft (300 m) thick (Swain, 1952).

The widespread advance of the carbonate facies across the shelf in early Tertiary time suggests a gradual lessening of tectonic activity in this region. The major tectonic troughs, however, still were the sites of maximum sedimentation, particularly those in the Georgia–Florida area (Okefenokee embayment and Suwannee straits). In these latter areas, carbonate rocks form the bulk of the deposits.

Middle Tertiary. In the northernmost Atlantic Coastal Plain, beds of Oligocene age are not known to be present. In North Carolina, Swain (1952, p. 10) reports the presence of about 200 ft (60 m) of intercalated limy and sandy strata of Oligocene(?) age. This is shown on the map as part of the carbonate province (Fig. 4D).

To the south in Florida, Georgia, and South Carolina, thick limestones of Oligocene age crop out widely, indicating that shoreward movement of the carbonate facies reached its maximum in this region (Fig. 4D). The clastic facies of this age is nearly absent in this general region, but it apparently thickens westward in the Gulf Coastal Plain.

In summary, this second sedimentation sequence was deposited during a period in which there was a gradual encroachment of the carbonate facies onto the platform flanking the southern Appalachians, culminating in the Oligocene carbonate overlap. The platform flanking the central Appalachians remained largely a clastic province throughout, except for the appearance of the carbonate facies in North Carolina and at the shelf edge off New Jersey. In the Eocene and early in the Oligocene, tectonic activity in the central Appalachians was more intense and lasted longer than to the south.

Sedimentation Sequence III

This sedimentation sequence was deposited largely during the Miocene, although the upper part may include some strata of lower Pliocene age.

FIGURE 4. Distribution of clastic (dotted) and carbonate (hachured) facies along part of the eastern seaboard of the United States. (A) Jurassic(?) to Early Cretaceous time; clastic rocks of marine and nonmarine origin have been lumped because of insufficient data as to their areal distribution. The carbonate facies is thickest in Florida and much thinner in North Carolina where calcareous beds are interstratified with abundant clastic rocks. (B) Late Cretaceous time; the carbonate facies has moved updip, although compared with the Early Cretaceous they are much thinner. The clastic facies is again of marine and nonmarine origin. (C) Eocene time. There is maximum carbonate overlap during this epoch. (D) Middle Tertiary time; clastic facies has not been reported. Some uncertainty exists as to the inner margin of the carbonate rocks in North Carolina.

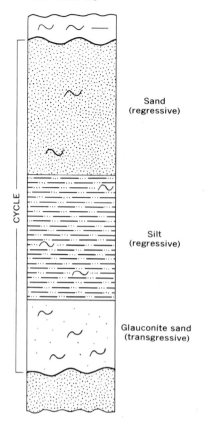

CYCLE

Sand
(regressive)

Silt
(regressive)

Glauconite sand
(transgressive)

FIGURE 5.   Typical cycle of sedimentation in the marine–transitional formations of Late Cretaceous through early Tertiary age. In all, seven cycles can be outlined during this time interval. The typical cycle as shown is asymmetrical.

A large-scale submergence of the northern part of the Atlantic Coastal Plain and to a lesser degree the southern part in the middle Miocene is indicated. The older structural elements were apparently reactivated, and thick sequences of sediments accumulated near the axes of these embayments.

In southern Maryland, in the Salisbury embayment, approximately 1,500 ft (450 m) of Miocene clastic rocks were encountered in a well (Anderson, 1948). These strata thin northward toward Atlantic City, N.J. (Richards, 1945), and southward toward Norfolk, Va. (Cederstrom, 1945). They are assigned to the Chesapeake Group and are chiefly composed of quartz, mica, and clay minerals but contain abundant shaly material. Glauconite, which is abundantly developed in the preceding sedimentation sequence, is much less abundant. The four formations of the Chesapeake Group apparently form two cycles of sedimentation, each characterized by a basal transgressive silt or clay which is overlain by a sandy regressive unit. The basal transgressive

glauconite sands shown in Figure 5 are not present in these beds. This pattern indicates crustal instability at this time throughout the Virginia–Maryland part of the Coastal Plain.

The Miocene section in the Albemarle embayment is only 800 ft (240 m) or less thick if the lower part is Oligocene in age as was suggested by Swain (1952). The southern limit of the Miocene in this region appears to be the Cape Fear arch, which probably was a physical barrier throughout the entire Miocene. The lithology of the sediments of Miocene and Pliocene? age in the Albemarle embayment are very similar to those in the Virginia–Maryland area.

In the South Carolina–Florida region, the Miocene has two components, clastic and carbonate rocks. Compared with the Oligocene, the clastic wedge of the Miocene has greatly broadened, and contact between the two facies is roughly at the northern boundary of Florida (Fig. 6A).

As in the northern Coastal Plain, the Miocene sedimentary rocks are thickest along the axes of the old troughs. They are as much as 700 ft (200 m) thick in the Okefenokee embayment and 500 ft (150 m) in the Suwannee straits–Apalachicola embayment (Herrick and Vorhis, 1963). Apparently these older structural troughs were reactivated at this time. The Miocene rocks, particularly in the lower part, are largely thinbedded limestones which interfinger with and are overlapped by clastic rocks (very micaceous clays for the most part).

Downdip, in peninsular Florida, the Miocene is represented by a thick series of very calcareous, locally dolomitic beds. A sand blanket which covers the central highland of Florida is considered to be of Miocene age by Puri and Vernon (1964), whereas others interpret it as a much younger deposit. Because of the uncertainty, these sands will be discussed with the Pliocene to Quaternary deposits.

In summary, the Miocene was a period of renewed uplift in the central and southern Appalachians. Within the basin, the older structural troughs were rewarped, and a large quantity of clastic material prograded into the basins.

Sedimentation Sequence IV

The last recorded major tectonic event was a large-scale epeirogenic uplift of the whole Atlantic Coastal Plain. This movement was greatest in the south, where if all peninsular Florida is considered, a part of the shelf more than 550 mi wide was elevated. The continental block probably began to rise sometime during the late(?)

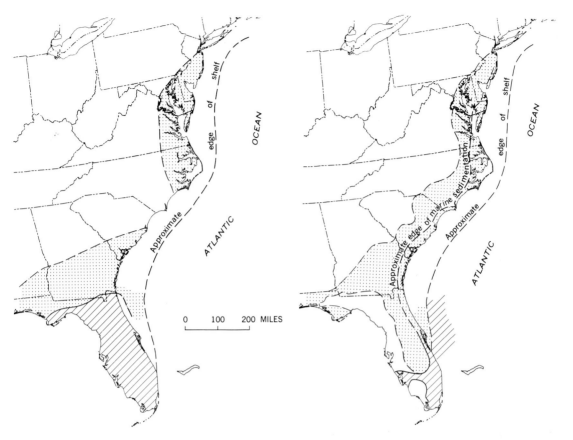

FIGURE 6. Distribution of carbonate (hachured) and clastic (dotted) facies along part of the eastern seaboard of the United States. (A) Miocene to early Plio-cene(?) time. The clastic wedge has been greatly enlarged and the carbonate facies has retreated to near the Florida–Georgia boundary. Dominantly clastic-rich sediments were found on shelf off Florida. (B) Pliocene to Quaternary time. Carbonate deposition confined to southern Florida. Carbonate deposits were en-countered during drilling on the shelf off Florida. Both carbonate and clastic rocks are very thin.

Pliocene, but the incomplete understanding of this epoch makes the establishment of the precise time difficult. The withdrawal of the sea at this time was initially accompanied by the outpour-ing of widespread very gravelly deposits from the central and southern Appalachians onto the Coastal Plain. Uplift continued well into the Pleistocene, and these older gravel plains and the underlying Coastal Plain formations were ex-tensively dissected. Continued tectonic emergence was accompanied by large eustatically controlled rises and falls in sea level because of continental glaciations. Deep valleys were entrenched in the emerged Coastal Plain, some of which extended as much as 200 ft (60 m) below present sea level. During a subsequent eustatically controlled rise, these deep valleys were aggraded. The spatial arrangement of the glaciofluviatile deposits in the northern Coastal Plain indicates continued tec-tonic uplift during this entire period. The

youngest Pleistocene deposits are topographically the lowest, and the older deposits are found at progressively higher elevations. The highest de-posits crop out at elevations of 400 ft (120 m) or more, well beyond any calculated eustatically con-trolled rise in sea level. Nearly all the deposits at elevations of 50 ft (15 m) or more are fluviatile in character, although considerable debate exists as to the origin of these higher level "terraces." In this review, only the lower level (Talbot–Pamlico–Cape May Formation) terraces are con-sidered marine.

The carbonate rocks are restricted to southern-most Florida, in the South Florida embayment of Applin (1951) (Fig. 6B). In this region a thin sequence of very calcareous sediments was de-posited. Other highly calcareous sediments of Pleistocene age were also found in borings from the Blake Plateau (lat. 28°30' and 30°33' long. 77°31' and 81°) east of Florida (Bunce and others, 1965). The time relationships between these two calcareous bodies is still unknown, and the two were grouped as a single unit in this report.

The Pliocene to Quaternary deposits of the Atlantic Coastal Plain, therefore, are mainly clastic rocks except in Southern Florida and on the submerged shelf. The widespread and coarse clastic nature of these deposits indicate uplift along the entire central and southern Appalachians during this period. The deposits, however, are thin (typically 50 ft (15 m) or less) and do not suggest uplift of the intensity or duration of that indicated by the Early Cretaceous deposits (sedi-

mentation sequence I). In addition, no marked accumulations in centers of deposition, so common in all the other sedimentation sequences, has been recognized. This suggests a general uniform uplift of the entire Appalachians at this time.

SUMMARY

As has been shown, the irregularities in the basement beneath the Atlantic Coastal Plain resemble those beneath the Gulf Coastal Plain. These irregularities are interpreted as tectonic and indicate that the two regions have undergone similar crustal deformations although the intensities have not been the same. The general tectonic

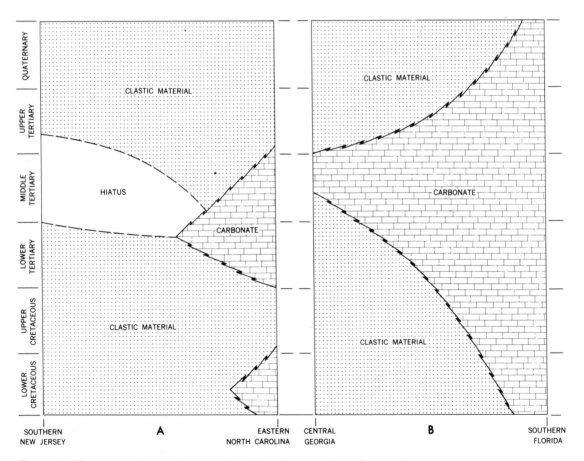

FIGURE 7. The clastic-carbonate boundary throughout the depositional history of the Atlantic Coastal Plain. The northern Coastal Plain (A) includes the area from North Carolina to New Jersey, whereas the southern Coastal Plain (B) includes the area from southern Florida to central Georgia.

style is a series of troughs and arches which are generally oriented perpendicular to the older structures in the adjacent craton. Downdip, these cross-warps are flanked by a series of disconnected, deeper troughs which have a similar orientation to the older structures in the craton. These deep furrows contain at least 25,000 ft (7,500 m) of sediment off the northeastern coast of the United States. Both the Atlantic and Gulf Coastal Plains have shapes like orthogeosynclines.

The Atlantic Coastal Plain has been subjected to periodic diastrophic movements and is not a "stable shelf" region. Prograding of clastic wedges into the depositional basin reflects structural activity in the sourceland. Recycling of material from the older Coastal Plain formations is minor. The movements of the carbonate–clastic boundary in the northern and southern Coastal Plain through time is shown in Figure 7. The carbonate facies is much more widespread in the southern Atlantic Coastal Plain than in the northern. This suggests less tectonic activity (uplift) in the southern than in the central Appalachians during the formation of this sedimentary prism. As shown, the carbonate facies progressively transgressed across the southern Coastal Plain from Early Cretaceous through the middle Tertiary; then rapidly regressed to its present position.

In the north, the carbonate facies only appeared twice during the depositional history of the Coastal Plain, in the Early Cretaceous and early and middle Tertiary. Diastrophic movements in the central Appalachians were greater than in the southern Appalachians, and appear to have been more irregular.

REFERENCES

Anderson, J. L., 1948, Cretaceous and Tertiary subsurface geology (Md.): Maryland Dept. Geology, Mines and Water Resources Bull. 2, p. 1–113, app., p. 385–441.

Applin, P. L., 1951, Preliminary report on buried pre-Mesozoic rocks in Florida and adjacent states: U. S. Geol. Survey Circ. 91, 28 p.

Bunce, E. T., Emery, K. O., Gerard, R. D., Knott, S. T., Lidz, L., Saito, T., and Schlee, J., 1965, Ocean drilling on the continental margin: Science, v. 150, no. 3697, p. 709–716.

Cederstrom, D. J., 1945, Structural geology of southeastern Virginia: Am. Assoc. Petroleum Geologists Bull., v. 29, no. 1, p. 71–95.

Cooke, C. W., 1936, Geology of the Coastal Plain of South Carolina: U. S. Geol. Survey Bull. 867, 196 p.

Drake, C. L., Ewing, M., and Sutton, G. H., 1959, Continental margins and geosynclines—The east coast of North America north of Cape Hatteras, in Ahrens, L. H., and others, eds., Physics and chemistry of the earth, v. 3: New York, Pergamon Press, p. 110–198.

Dryden, A. L., Jr., and Dryden, Clarissa, 1956, Coastal plains and heavy minerals (abs.): Internat. Geol. Cong., 20th, Mexico City, 1956, Resumenes de los trabajos presentados, p. 279–280.

Hardin, G. C., Jr., 1962, Notes on Cenozoic sedimentation in the Gulf Coast geosyncline, U.S.A., in Geology of the Gulf Coast and central Texas and guidebook of excursions, Geol. Soc. America, 1962 Ann. Meeting: Houston, Tex., Houston Geol. Soc., p. 1–15.

Herrick, S. M., and Vorhis, R. C., 1963, Subsurface geology of the Georgia Coastal Plain: Georgia Dept. Mines, Mining and Geology Inf. Circ. 25, 78 p.

Hersey, J. B., Bunce, E. T., Wyrick, R. F., and Dietz, F. T., 1959, Geophysical investigations of the continental margin between Cape Henry, Virginia, and Jacksonville, Florida: Geol. Soc. America Bull., v. 70, no. 4, p. 437–466.

Kay, G. H., 1951, North American geosynclines: Geol. Soc. America Mem. 48, 143 p.

LeGrand, H. E., 1961, Summary of geology of Atlantic Coastal Plain: Am. Assoc. Petroleum Geologists Bull. v. 45, no. 9, p. 1557–1571.

Murray, G. E., 1961, Geology of the Atlantic and Gulf coastal province of North America: New York, Harper & Bros., 692 p.

Puri, H. S., and Vernon, R. O., 1964, Summary of the geology of Florida and a guidebook to the classic exposures: Florida Geol. Survey Spec. Pub. no. 5 (rev.), 312 p.

Richards, H. G., 1945, Subsurface stratigraphy of Atlantic Coastal Plain between New Jersey and Georgia: Am. Assoc. Petroleum Geologists Bull., v. 29, no. 7, p. 885–955.

Stetson, H. C., 1949, The sediments and stratigraphy of the east coast continental margin, Georges Bank to Norfolk Canyon: Massachusetts Inst. Technology and Woods Hole Oceanog. Inst. Papers in Physics, Oceanography and Meteorology, v. 11, no. 2, p. 1–60.

Swain, F. M., 1947, Two recent wells in the Coastal Plain of North Carolina: Am. Assoc. Petroleum Geologists Bull., v. 31, no. 11, p. 2054–2060.

————, 1952, Ostracoda from wells in North Carolina: U. S. Geol. Survey Prof. Paper 234-A, p. 1–58; 234-B, p. 59–73, 9 pls.

————, and Brown, P. M., 1964, Cretaceous Ostracoda from wells in the southeastern United States: North Carolina Div. Mineral Resources Bull. 78, 55 p., 5 pls.

Tipsword, H. L., 1962, Tertiary Foraminifera in Gulf Coast petroleum exploration and development, in Geology of the Gulf Coast and central Texas and guidebook of excursions, Geol. Soc. America, 1962 Ann. Meeting: Houston, Tex. Houston Geol. Soc., p. 16–57.

U. S. Geological Survey and American Association of Petroleum Geologists, 1961, Tectonic map of the United States, exclusive of Alaska and Hawaii: 2 sheets, scale 1:2,500,000 [1962].

Vernon, R. O., and Puri, H. S., 1964, Geologic map of Florida: Florida Geol. Survey Map Ser. no. 18.

Chapter 29

Zircon Age Measurements in the Maryland Piedmont, with Special Reference to Baltimore Gneiss Problems*

G. R. TILTON, B. R. DOE AND C. A. HOPSON

INTRODUCTION

THE BALTIMORE GNEISS, which forms the basement rock in the Maryland Piedmont, is exposed in a series of seven domes mantled by metamorphosed sediments of the Glenarm Series. The lower three formations of the Glenarm are the Setters Formation, Cockeysville Marble and Wissahickon Formation. The gneiss is one of the first rocks for which geochronological studies clearly indicated two different ages for minerals from a single rock. Tilton, Wetherill, Davis, and Hopson (1958) showed that zircon in gneiss from two of the domes was 1,100 m.y. old, whereas biotite from the same rocks was 300 m.y. old. These authors advanced an argument, based on 1,100 m.y. Rb–Sr ages of potassium feldspars, that the gneiss has been a crystalline rock for the past 1100 m.y., that is, that the zircon was not relict grains from a clastic sediment that was metamorphosed into a crystalline gneiss during Paleozoic metamorphism. The discovery that radiogenic strontium may be exchanged between minerals in metamorphic processes (Compston and Jeffery, 1959; Fairbairn, Pinson, and Hurley, 1961) subsequently cast some doubts on the interpretation of the Baltimore Gneiss feldspar ages. However, Wetherill, Davis, and Lee-Hu (1968) have recently found a Rb–Sr age of 1,050 m.y.

* Publication authorized by the director, U. S. Geological Survey.

for total rock samples of Baltimore Gneiss and associated Hartley Augen Gneiss (Knopf and Jonas, 1929). Their total rock measurements show that the 1100-m.y. age given by Tilton et al. (1958) is still valid in the light of current knowledge.

Because the earlier studies established Precambrian mineral ages in gneiss from only two of the seven domes, additional observations are desirable. There is a particular need for data from the gneiss of the so-called Baltimore dome, shown on most maps as an outcrop of Baltimore Gneiss in the southwest part of the city of Baltimore. This outcrop is, in fact, the type locality for the Baltimore Gneiss, as first described by Williams (1892). Hopson (1964) pointed out that the lower two formations of the Glenarm Series are lacking at the margins of the gneiss body at Baltimore and questioned whether it is a dome at all. He also noted important contrasts in chemical composition and degree of deformation between gneiss at Baltimore, which he termed paragneiss, and the veined or migmatitic gneiss of the mantled domes. For instance, the light layers of the veined gneiss are rich in potassium feldspar but the light-colored layers in the paragneiss contain little or none. The veined gneiss shows plastic deformation, whereas the layers in the paragneiss are straight or gently warped. He presented evidence supporting the view that the gneiss at Baltimore is a metamorphosed sequence of either epiclastic or pyroclastic rocks of quartz keratophyric to basaltic composition, or, less likely, graywacke. Regarding the first possibility, Hopson drew attention to the metavolcanic rocks in Cecil County, which are situated along strike to the northeast of the paragneiss at Baltimore. More recently Southwick (1966) commented on the similarity in bedding characteristics and chemical composition between the paragneiss at Baltimore, paragneiss in Harford County northeast of Baltimore along strike, and metavolcanic rocks of Cecil County still farther to the northeast, and he suggested that these rocks may be correlative. The study described in this paper provides data pertinent to the above discussions.

WOODSTOCK DOME

The $^{207}Pb-^{206}Pb$ age of zircon from the Woodstock dome (Table 1) is greater than the $^{207}Pb-^{206}Pb$ ages of the zircons of the gneiss from the Towson and Phoenix domes, reported previously by Tilton et al. (1958). The $^{207}Pb-^{235}U$ and $^{206}Pb-^{238}U$ ratios are shown in a concordia diagram (Wetherill, 1956) in Figure 1. We interpret the discordances as being due to loss of lead by continuous diffusion at a constant value of the

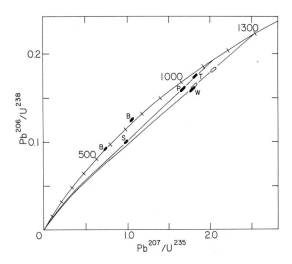

FIGURE 1. Concoria diagram showing the results of analyses of zircon from rocks identified as Baltimore Gneiss. (B) paragneiss at Baltimore; (P) Phoenix dome; (T) Towson dome; (W) Woodstock dome. Also included is (S) Setters Formation.

Open hexagons are zircons from gneisses in western North Carolina and eastern Tennessee. Hexagons represent estimated error limits of analyses. These describe hexagons because the $^{207}Pb-^{206}Pb$ ratios for lead in the samples can be determined more accurately than can the uranium–lead ratios.

parameter, D/a^2, where D is the diffusion coefficient and the crystals are assumed to be spheres having a radius, a (Tilton, 1960). Since the data are limited, interpretations involving episodic loss of lead as a result of Appalachian metamorphism must also be considered. For example, the Woodstock zircon point lies on a chord connecting 450 and 1,450 m.y. on the concordia curve, that is, the results can be explained if the sample is 1,450 m.y. old and if it lost lead episodically 450 million years ago and remained a closed system for lead at all other times. We do not favor such an interpretation because zircon from the youngest Paleozoic granites in the Maryland Piedmont also give discordant ages, the $^{238}U-^{206}Pb$ ages being about 20 percent less than the $^{207}Pb-^{206}Pb$ ages. These are the quartz monzonite of the Woodstock pluton and granodiorite of the Ellicott City pluton for which results were reported by Wetherill et al. (1966). If the age discordances are due mainly to Paleozoic metamorphism, the zircons from the youngest Paleozoic intrusives might be expected to be nearly concordant. After considering various possibilities, Wetherill et al. (1966) concluded that continuous diffusion was the most likely cause of the discordant zircon ages in the young rocks. It is then reasonable to assume that the same mechanism applies to the Baltimore Gneiss zircons.

Figure 1 shows that the age of the Woodstock dome zircon is 1,300 m.y. if lead has been lost by continuous diffusion at a constant value of D/a^2. This is greater than the 1,140-m.y. age found for the Towson and Phoenix dome zircons, and is greater than any isotopic ages previously found in the central and northern Appalachians. However, very similar ages have been found for zircons from banded gneisses from western North Carolina and eastern Tennessee in the southern Appalachians (Davis, Tilton, and Wetherill, 1962). The southern Appalachian zircon data are plotted as open hexagons in Figure 1 to permit comparison of results from the two regions.

From the data at hand it is impossible to say why the Woodstock dome zircon gives an age so different from zircon of the other two domes. If the zircons in the gneisses are detrital as suggested by the very great abundance of zircon in all three gneisses, different ages for the source rocks might be indicated. Still another interpretation is that metamorphism 1,140 million years ago was sufficiently intense to cause zircon in the gneiss at the Phoenix and Towson domes to lose essentially all of their lead, but affected the zircons in gneiss at the Woodstock dome to a lesser, perhaps negligible extent.

TABLE 1.   Ages of Zircon from Rocks of the Maryland Piedmont

Sample no.	Rock	$\dfrac{^{206}Pb}{^{238}U}$	$\dfrac{^{207}Pb}{^{235}U}$	$\dfrac{^{207}Pb}{^{206}Pb}$	$\dfrac{^{208}Pb}{^{232}Th}$
B-2*	Baltimore Gneiss, Towson dome	1040	1070	1120	940
B-4*	Baltimore Gneiss, Phoenix dome	960	1020	1120	1100
B-5	Setters Formation	620	700	915	615
B-58	Baltimore Gneiss, Woodstock dome	960	1050	1250	—
B-70	Paragneiss, Baltimore dome	575	570	540	530
GF-1b	Paragneiss, Baltimore dome	765	735	650	—
GF-2	Paragneiss, Baltimore dome	740	—	—	—
B-83	James Run Gneiss	325	355	550	420
B-90	James Run Gneiss	420	435	490	—

* From Tilton, Wetherill, Davis, and Hopson (1958).

PARAGNEISS AT BALTIMORE

Table 1 lists ages for three zircon samples of the paragneiss from the Campbell Quarry in Baltimore. In contrast to the granitic gneisses in the other domes, zircon is not abundant in the paragneiss. The amount of sample recovered from B-70 and GF-1b, mafic phases of the gneiss, was rather limited as was also the case with the James Run Gneiss, described below. GF-2, a granitic phase, gave only sufficient sample to permit measurement of the ^{238}U–^{206}Pb age. We attempted to separate zircon from paragneiss at two other localities, but were unable to obtain sufficient material for analysis.

The zircon ages of the paragneiss are unusual in that the ^{238}U–^{206}Pb ages are greater than the ^{207}PB–^{206}Pb ages, although the discrepancy is rather small for B-70. The ^{238}U–^{206}Pb ages reported from zircons to date have all been equal to or less than the ^{207}Pb–^{206}Pb ages, although cyrtolite at Hybla, Ontario, has given the same type of discordance found at Baltimore (Tilton, unpublished data). To check further on the discordant ages given by sample GF-1b, sample GF-2, an acidic phase of the gneiss at the same locality, was analyzed. Although insufficient material was obtained for a complete analysis, the ^{238}U–^{206}Pb age agrees closely with that found for GF-1b. We find no reason to question these measurements. The purity of the zircon separates is comparable to that of the other samples reported here and to those in other papers. The analytical data (Table 2) reveal nothing unusual in the uranium or thorium contents of the samples. Finally the analytical work was performed at approximately the same time, using the same techniques employed for samples B-5 and B-58, which yield "normal" patterns of age discordance.

The best estimate of the age of the zircon in the paragneiss is about 550 m.y., a value based on the nearly concordant age of sample B-70. The cause of the discordant ages for sample GF-1b is difficult to specify, but it could be due in part to the metamorphism that produced the hornblende in the paragneiss, for which Wetherill et al. (1966) reported K–Ar ages of 297 and 301 m.y.

The most interesting feature of the results is the difference between the ages for the paragneiss at Baltimore and for the veined gneiss of the Phoenix, Towson, and Woodstock domes (Table 1, Fig. 1). As can be seen from Figure 1, zircon from the Phoenix and Towson domes has an age of 1,140 m.y., and zircon from the Woodstock dome an age of 1,300 m.y., when loss of lead is interpreted as due to continuous diffusion. In any case the Phoenix, Towson, and Woodstock dome zircons appear to be distinctly older than the zircons in the paragneiss at Baltimore, which could be of Paleozoic age. This supports the conclusions of Hopson (1964), who separated the paragneiss of the Baltimore dome from the veined gneisses of the mantled gneiss domes on the basis of geological and petrological criteria.

JAMES RUN GNEISS

Ages were measured from two samples of zircon from paragneiss in the Gatch Quarry at Churchville, midway between Baltimore and the Susquehanna River in Harford County. The gneiss lies northeast along strike from the paragneiss at Baltimore. The Cecil County metavolcanic rocks, which may be less metamorphosed equivalents of the James Run Gneiss, are located 12 mi still farther to the northeast at the Susquehanna River. There are marked similarities in chemical composition, layering, and stratigraphic position between all of these rocks, as noted by Hopson (1964) and Southwick (1966). In Cecil County the rocks are the least metamorphosed;

TABLE 2.   Analytical Data for Zircon

Sample no.	Concentration, ppm			Relatve isotopic abundance			
	U	Th	^{206}Pb (Radiogenic)	^{204}Pb	^{206}Pb	^{207}Pb	^{208}Pb
B-5	928	133.6	82.4	0.103	100	8.35	8.80
B-58	892	—	121.7	0.049	100	8.83	6.06
B-70	526.5	251.7	41.7	0.237	100	9.25	22.37
GF-1b	643.1	—	68.8	0.194	100	8.92	22.15
GF-2*	626.7	—	65.0	(0.237)	100	(9.25)	(22.37)
B-83	514.8	441.3	22.8	0.134	100	7.78	40.12
B-90	705.0	411	40.6	0.0872	100	6.94	(0.190)†

* Insufficient material for isotope analysis. Isotopic composition assumed to be the same as that for B-70.

† Calculated from U and Th concentrations. See text for explanation. Uncertainty in lead concentration about twice that of the other samples.

the metamorphic grade apparently increases southwestward along strike. None of the rocks are known to underlie the Setters Formation and Cockeysville Marble, but they underlie or are interlayered with pelitic schists strongly resembling the Wissahickon Formation.

Zircon is sparse in the James Run Gneiss at Churchville. For sample B-90 lead data were obtained only from a sample spiked with very pure ^{208}Pb. The ^{208}Pb is sufficiently pure to permit accurate determination of the ^{207}Pb–^{206}Pb age of the sample, but the lead concentration data are somewhat uncertain since the ^{208}Pb–^{206}Pb ratio of the lead in the zircon had to be estimated from the uranium and thorium data. We estimate that the lead concentration for this sample has about twice the uncertainty of the remaining lead values in Table 2. A complete analysis is available for a second zircon (sample B-83) from the same quarry. The two samples yield discordant, but analytically different age values, as shown in Figure 2. Once more we interpret the age discordances as due to loss of lead by continuous diffusion. It would be difficult to ascribe the discordancy of B-83 to episodic loss of lead during Paleozoic metamorphism. The youngest age found for minerals in the Maryland Piedmont is the 300-m.y. age of biotite (Wetherill et al., 1966). It is obvious from Figure 2 that an attempt to ascribe the B-83 result to episodic loss of lead 300 m.y. ago would yield an impossibly old age for the zircon and the situation is even worse if any of the older ages in the area such as 550 m.y. (this paper) are assumed to be the time of lead loss.

Figure 2 shows that the zircon data are compatible with an age of 550 m.y., assuming loss of lead by continuous diffusion. Similar ages are

also found for zircon in the lower Paleozoic intrusive rocks in the area, for example, the Norbeck and Kensington Quartz Diorite of Wetherill et al. (1966) and the Relay and Port Deposit Quartz Diorites of Steiger, Hopson and Fisher (manuscript in preparation). This age is also in agreement with the probable age of zircon from the paragneiss at Baltimore. There are additional similarities which indicate a relation between the paragneiss at Baltimore and the paragneiss in Harford County. First, the abundance of zircon in both rocks is much lower in the paragneiss than in the veined gneiss. Approximately 2-kg samples of the specimens from the Phoenix, Towson, and Woodstock domes yielded 1–2 g of zircon, more than enough for isotopic analyses, whereas some samples of paragneiss at Baltimore failed to yield sufficient zircon, even from 20-kg samples. The same situation holds for the Harford County gneisses. Zircon recovery from specimens B-83 and B-90 was marginal for isotopic work. We also performed two mineral separations on samples from the Cecil County volcanics at the Susquehanna River but were unable to obtain zircon in any appreciable quantity. Second, the ^{208}Pb–^{206}Pb and thorium–uranium ratios of the zircons (Table 2) are higher by a factor of 3–5 in the paragneiss zircons than in the Woodstock dome zircon. The Phoenix and Towson dome zircons have ratios that closely resemble that for the Woodstock dome zircon (Tilton et al., 1958).

SETTERS FORMATION

Most geologists believe that the Setters Formation (the basal unit of the Glenarm Series) lies unconformably on the Baltimore Gneiss. Hopson (1964) has summarized the evidence for an un-

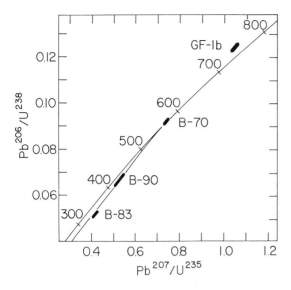

FIGURE 2. Concordia diagram showing zircon analyses for samples from paragneisses of Harford County (B-83, B-90) and Baltimore City (B-70, GF-1b). Hexagons represent error limits.

conformity. Because the Wissahickon Formation (the uppermost unit of the Glenarm Series) is certainly older than 500 m.y. and very likely older than 550 m.y. (Wetherill et al., 1966; Steiger, Hopson and Fisher, manuscript in preparation), Paleozoic rocks are unlikely sources of zircons in the Setters Formation.

Zircon from the Setters Formation near its contact with the Baltimore Gneiss in the Phoenix dome was analyzed. The isotopic ages (Table 1, Fig. 1) are discordant. When interpreted according to loss of lead by continuous diffusion, the data are compatible with an age of 1,140 m.y. for the zircon from the quartzite, the same age found for zircon from the Phoenix and Towson domes. Alternative explanations of the discordant ages would be that the sample consists of a mixture of crystals of Paleozoic and Precambrian ages, or some kind of episodic loss of lead. Because the Phoenix, Towson, and Setters zircons closely fit a chord connecting 300 m.y. and 1,150 m.y. (Fig. 1), the data in themselves do not exclude an episodic loss mechanism. The main reason for favoring a continuous diffusion mechanism is that zircons from some of the younger rocks in the area do not fit an episodic loss mechanism satisfactorily. For example, samples B-83 and B-90 cannot be reconciled to episodic loss of lead at any time in the Paleozoic Era if, as is likely, the two zircon populations have the same age. As mentioned above, the data for B-83 do not yield a reasonable age for episodic loss of lead at any

time in the Paleozoic. Neither is it likely, for reasons already mentioned, that zircons of Paleozoic age form part of the zircon population.

Stern, Goldich, and Newell (1966) have suggested that weathering is a significant cause of discordant ages for zircons in clays derived from gneiss at Morton, Minnesota. We do not favor such a process as a cause of the discordant ages for the Setters zircon for the following reason: the zircons in the Setters quartzite must have eroded from their parent rocks prior to approximately 550 m.y. ago, the minimum age of the quartzite. The time of erosion would be the time of major loss of lead according to the weathering model. If the loss occurred 550 m.y. ago, the age of the zircon in the source rocks must have been about 2,700 m.y., a value older than any yet found in Appalachian rocks. Certainly there is no evidence for such an age for the zircon from any samples of the Baltimore Gneiss, which is a likely source rock for the zircon in the Setters quartzite.

CONCLUSIONS

The results at hand show that zircon from paragneiss in the Baltimore Gneiss at its type locality has a distinctly younger age than do zircons from the veined gneisses of the Towson, Phoenix, and Woodstock domes, across strike to the west and north. A revision of the geographic extent of the name Baltimore Gneiss may be required. It is especially critical to obtain reliable geochronological data from the metavolcanic rocks in Cecil County to ascertain whether they are correlative with the paragneisses of Harford County and Baltimore, as suggested by geological and petrological observations.

ACKNOWLEDGMENTS

We thank D. L. Southwick of the U. S. Geological Survey for providing the zircon separate, B-90, for the James Run Gneiss.

APPENDIX 1, ANALYTICAL DATA

The concentration data were all obtained by stable isotope dilution, using methods previously described (Tilton et al., 1957). All concentrations given in Table 2 are accurate to about ± 1½ percent unless otherwise noted. These limits yield uncertainties of approximately ±15 m.y. for all age values given in Table 1. Prior to analysis all samples were washed in hot 1:1 nitric acid to remove pyrite.

The decay constants and isotopic abundance for uranium used in the age calculations are:

$$^{238}U: \quad \lambda = 1.54 \times 10^{-10} \ yr^{-1}$$
$$^{235}U: \quad \lambda = 9.72 \times 10^{-10} \ yr^{-1}$$
$$^{232}Th: \quad \lambda = 4.99 \times 10^{-11} \ yr^{-1}$$
$$^{238}U/^{235}U \quad = 137.8 \ (atom \ ratio)$$

APPENDIX 2, SAMPLE DESCRIPTIONS

B-5, quartzite. Middle quartzite member of Setters Formation, 50 ft from contact with Baltimore Gneiss of the Phoenix dome at the change in direction of the roadside channel along U. S. Highway 111 (Interstate Route 83), 0.5 mi southwest of Verona. Location is 250 ft north of location for sample B-4 (Wetherill et al., 1966). Coordinates: lat. 39°31.1′ N. long. 76°40.2′ W. Zircon morphology is very similar to that noted for Baltimore Gneiss at B-4. Large subhedral crystals, length/breadth ratio averaging about 2.5, zoned, some anhedral inclusions.

B-58, Baltimore Gneiss, Woodstock dome. From high rock bluff above Baltimore and Ohio railroad tracks 200 yd west of Woodstock. Coordinates: lat. 39°19.5′ N. long. 76°52.5′ W. Banded oligoclase–quartz–biotite–microcline gneiss. Accessory apatite, sphene, zircon, clinozoisite, allanite, garnet, and magnetite. Seriate texture, conspicuous foliation due to segregated and weakly aligned biotite. Zircon clear, few inclusions, rounded, subhedral habit, length/breadth ratios range from 1 to 4, averaging about 2.5. Zoned, a few with cores.

B-70, GF-1b, GF-2, paragneiss, Baltimore. Campbell Corporation quarry, Gwynns Falls at W. Baltimore St., Baltimore, Maryland. Hornblende–quartz–oligoclase granofels phase of gneiss complex. Texture of rock is granoblastic, with very little preferred orientation or segregation of mineral grains. Rock composed chiefly of plagioclase (An_{24}), quartz, green hornblende, with minor microcline, sphene, biotite, zircon, and apatite. The subhedral zircon, GF-2, is from a less mafic phase of the gneiss. This zircon contains many inclusions that appear to be apatite. Morphology is similar to zircon from B-58.

B-83, B-90, James Run Gneiss. Gatch Quarry, 2.5 mi south of Churchville, Maryland. Coordinates: lat. 39°31.5′ N. long. 76°15.5′ W. Fine-grained biotite gneiss from sequence of thinly interlayered gneiss and amphibolite. Individual layers range in thickness from several inches to 15 ft and are generally in sharp contact with each other. Rock is a fine-grained grano-

blastic mosaic of plagioclase (An_{22}), quartz, olive-brown biotite, and epidote, with accessory magnetite, sphene, muscovite, apatite, allanite, and zircon. Zircon clear, with a few inclusions. Subhedral crystals. Length/breadth ratios variable, mostly around 1.5. A few have ratios of 1, and appear to be rounded.

REFERENCES

Compston, W., J., P. M., 1959, Anomalous "Common strontium" in granite: Nature, v. 184, p. 179–181.

Davis, G. L., Tilton, G. R., and Wetherill, G. W., 1962, Mineral ages from the Appalachian Province in North Carolina and Tennessee: J. Geophy. Res., v. 67, p. 1987–1996.

Fairbairn, H. W., Pinson, W. H., and Hurley, P. M., 1961, The relation of discordant Rb–Sr mineral and whole rock ages in an igneous rock to its time of crystallization and to the time of subsequent Sr^{87}/Sr^{86} metamorphism: Geochim. et Cosmochim. Acta, v. 23, p. 135–147.

Hopson, C. A., 1964, The crystalline rocks of Howard and Montgomey Counties: Md. Geol. Survey, The geology of Howard and Montgomery Counties, p. 27–215.

Knopf, E. B., and Jonas, A. I., 1929, Geology of the crystalline rocks, Baltimore County: Maryland Geological Survey, p. 97–199.

Southwick, D. L., 1966, Paragneisses of the Northeast Piedmont—some facts and speculations (abs.): annual meeting of Northeastern Section, Geol. Soc. America, Philadelphia, Program, p. 43.

Stern, T. W., Goldich, S. S., and Newell, M. F., 1966, Effects of weathering on U-Pb ages of zircon from the Morton gneiss, Minnesota: Earth and Planetary Sci. Letters, v. 1, p. 369–371.

Tilton, G. R., 1960, Volume diffusion as a mechanism for discordant lead ages: J. Geophys. Res., v. 65, p. 2933–2945.

————, Davis, G. L., Wetherill, G. W., and Aldrich, L. T., 1957, Isotopic ages of zircon from granites and pegmatites: Trans. Am. Geophys. Union, v. 38, p. 360–371.

————, Wetherill, G. W., Davis, G. L., and Hopson, C. A., 1958, Ages of minerals from the Baltimore gneiss near Baltimore, Maryland: Geol. Soc. America Bull., v. 69, p. 1469–1474.

Wetherill, G. W., 1956, Discordant uranium–lead ages: Trans. Am. Geophys. Union, v. 37, p. 320–326.

————, Davis, G. L., and Lee-Hu, C., 1968, Rb–Sr Measurements on whole rocks and separated minerals from the Baltimore gneiss, Maryland: Geol. Soc. America Bull., v. 79, p. 757–762.

————, Tilton, G. R., Davis, G. L., Hart, S. R., and Hopson, C. A., 1966, Age measurements in the Maryland Piedmont: J. Geophys. Res., v. 71, p. 2139–2155.

Williams, G. H., 1892, Guide to Baltimore with an account of the geology of its environs: Am. Inst. Mining Engineers, guidebook prepared by local committee, J. Murphy and Co., Baltimore, Maryland, 139 p.

EPILOGUE

Epilogue

PHILIP B. KING

SOMETIME DURING the late Thirties, Ernst Cloos reported to the Geological Society of Washington on the field work he was doing in next-door Montgomery County, Maryland, and he included a special little sting for us Survey geologists in the audience. His structural map of the county was bordered by arrows, pointing southwest, west, and northwest—"To Texas," "To California," "To Wyoming," "To Alaska." Ernst's reproof was deserved, for all of us in those days looked on Washington as our "bedroom," our "winter quarters," where we could hibernate until the arrival of spring or summer, and then hasten away to our real homes in the mountains and deserts of the Great West and Northwest. Who cared about Eastern geology? Who wanted to wear out his life looking for outcrops? Hadn't it all been done, anyway? So Ernst went it alone, bringing new insight into old problems, and inspiring a few students to follow in his footsteps.

Many of our Survey geologists never reformed, but some of us perforce had to learn about eastern geology the hard way, during the exigencies of World War II. The Survey then undertook to explore for and to appraise strategic mineral deposits all over the country, and the Eastern deposits and their geologic settings were assigned to some of us. It wasn't easy for us to make a new start in strange country. In the west, we could tell dolomite from limestone by the weathering, but here the weathering was quite different. And the topographic forms were strangely reversed—carbonates made the valleys, shale made the ridges! During my own first field season in the Blue Ridge I fought residuum over the carbonate formations in the valleys, and brush and boulder fields over the quartzites and greenstones in the mountains, and was as useless as a fish out of water.

But by the end of the season I was beginning to feel at home, and I wanted more. The outcrops were there, if one made the effort to find them, and some of them were very good outcrops. And to my surprise I found that the country I was in had never really been mapped at all; bands of outcrop and major faults had been projected through it, but the formations and the faults didn't go where the maps said they should. And—well, all of us found we were falling in love with this very beautiful country, and with the rustic people who made their homes there.

Historically, all this has a strangely familiar ring. After the U. S. Geological Survey was organized in 1879, its second Director, John Wesley Powell, decided to break with the tradition of the earlier Territorial Surveys and to extend Federal geological work from the West into the East. He assigned his loyal henchman, Grove Karl Gilbert, to develop a program of Eastern work. Gilbert and the men he recruited were at first uncertain as to what to do. Hadn't all the geology been done anyway, during the more than half a century of effort of the excellent State Surveys? The first field work of the new program was therefore devoted to coordinating the earlier results of the State Surveys, by means of a series of carefully surveyed geological profiles across the Appalachian Chain. But so many unanswered problems became evident during this profiling that within a few years the effort was shifted to the preparation of folio quadrangle maps—resulting in some of the finest geologic work of its time, parts of which have not been superseded even today.

The Appalachian Chain is the most elegant on earth, so regularly arranged that its belts of formations and structures persist virtually from one end to the other—from its first appearance from beneath the sea in Newfoundland, to its final disappearance under the Gulf Coastal Plain in Alabama. What a contrast to the twisted and contorted mountain chains of middle Europe, or to the confusion of superposed rocks and structures in our own western Cordillera! No wonder is it that the Appalachians have been the birthplace of many of the great principles of North American geology and of World geology—to the theory of geosynclines and to theories of folding and faulting, to name only a few! But the apparent simplicity of the Appalachian Chain is deceiving; actually, it is full of guile, and its geology has aroused controversies as acrimonious as any of those in our science.

To these controversies and problems Ernst Cloos has brought a new and refreshing insight by his special brand of investigation, whose principles were set forth in 1937 in his "Application of recent structural methods in the interpretation of the crystalline rocks of Maryland." He says, "A large portion of the original evidence for deciphering the geological history is now lost forever. It is therefore the more necessary to recover all the evidence available in order to piece those fragments together into a picture of former conditions and from this to deduce the history of the region and its component rocks." The way in which this is done is as follows:

(1) "Accurate topographic maps are the basis of all geologic work."

(2) "Geologic maps are the basis for all other geological work. Mapping is slow work because the geologist in the field must find all rock exposures in order to attain accuracy. Fortunately rocks occur together in certain formations which can be recognized, grouped together conveniently, and then be inserted on topographic maps."

(3) "Besides rock mapping it is essential to map the distribution of rock structures. These have to be measured in the field and then also platted on maps. Detailed structural data are essential everywhere, and in many regions a general geologic map may be insufficient."

(4) "Microscopic investigations have to be made of the composition of the rocks, their mineral content, and also of the orientation of these minerals within the rocks."

Is this elementary? Certainly his statement is a strong affirmation of the old tried and true methods of field geology, but note that more sophisticated touches have been added which raise the method above the ranks of mere surveying, and which give the results greater strength— the meticulous recording of structural detail, supported finally by microscopic investigation of the rock fabric. An important supplement to Ernst's recording of structural detail has been supplied by his colleague Francis Pettijohn, who has led the way in an equally detailed recording of the primary features of the sedimentary rocks from which many of the structures in the region were built. It is true that other techniques have come into flower later, such as radiometric dating which has provided greater support for correlations based on geological evidence, but they are handmaidens to the main endeavor, and the fundamentals remain as Ernst propounded them more than a quarter of a century ago.

The effects of the philosophy of Ernst Cloos, and of his colleague Francis Pettijohn are evident in the contributions to this Volume—many by their former students, but not all (as their teaching has been widely disseminated). Field work based on the methods of Cloos and Pettijohn, some of which is reported on in these contributions, has modified or overturned many of our long-held ideas regarding the structure of the Blue Ridge and Piedmont provinces, and regarding the manner of deposition and the environment of the Paleozoic geosynclinal sequence in the Valley and Ridge province.

Nevertheless, field work in the Appalachian Chain is really a study in two dimensions, from which a three-dimensional picture must be deduced. These ancient, worn-down mountains now have a relief so slight that they afford nothing comparable to the cross-sectional picture that can be seen, for example, in the European Alps, in the Greenland fiords, or in our own western mountains. Even where cross sections are available in the Appalachians, they are small in proportion to the magnitude of the structures involved. This is not so vital in the crystalline southeastern part of the chain, where all the structures plunge steeply to great depths, but it is crucial in the folds and faults of the stratified northwestern part, or Valley and Ridge province. With all due regard to the value of field work, we also need to know more about what lies beneath the surface, in order to unravel the secrets of the Appalachian Chain.

Are the structures of the Valley and Ridge province rooted in basement ridges beneath, or have they formed over one or more surfaces of décollement in the sedimentary column? Have the structures grown secularly with time under the influence of excessive sedimentation and the gradual upthrust of basement highs; or are they an edifice of gigantic transported sheets, propelled by lateral thrust, by gravity sliding, or by some other mechanism? Please observe that there is a fundamental difference between the two questions just propounded. The first question deals with items of fact which, if not available now, will probably become available later. The second question deals with items of theory as to what happened during a far-distant orogenic time. The two questions, and the two kinds of items which they involve should never be confused; a basement core in the structures does not automatically mean that they formed by secular growth, nor does a décollement automatically mean that they formed by gravity sliding.

Many of the questions about items of fact will be dispelled in our own lifetimes. It is a sobering thought that the geologically less complex but economically more attractive Gulf Coast area in Texas and Louisiana is intimately known to depths of four or five miles beneath the surface, whereas very little of this kind of information is now available in the geologically more intriguing Appalachian Chain. But this thought demonstrates that such information would be within our grasp if there were the economic urge to find it; perhaps the urge will come. Even the few deep wells that have been drilled are enough to demonstrate the geological surprises that are in store for us. For example, the Nittany Arch in Pennsylvania appears from surface geology to be a great, unbroken culmination of folds, yet wells recently drilled there reveal many structures within 5,000 or 10,000 feet of the surface; one well passed through two or three thrust planes, and ended in a stratigraphically higher formation than the one in which it started!

Perhaps the reader will permit me to close these comments by asking some even more speculative questions about the larger relations about the Appalachian Chain:

What happens underneath? Data presently available suggest that the chain is underlain by a rather broad, shallow "root" at the base of the crust, quite in contrast with the very "bumpy" subcrustal topography beneath the western mountains. Is the apparent simplicity of the subcrustal topography beneath the Appalachian Chain genuine, or is it due merely to insufficiency of instrumental determinations of crustal thickness? If genuine, has an originally "bumpy" subcrustal topography like that beneath the western mountains been removed by some subterranean process during the ages since the Appalachian foldbelt was created?

What happens to the Appalachian Chain at its ends? It plunges beneath the sea in full strength along the east coast of Newfoundland, yet no trace of it exists beyond the continental shelf, on the floor of the present Atlantic Ocean. Was it broken off from its logical continuation in the British Isles across the ocean by some process of continental separation? And does the chain connect southwestward with the very similar Ouachita Chain after it disappears beneath the Gulf Coastal Plain? There seems to be every geological reason why it should do so, yet there is surprisingly little proof, aside from the indication of a few drill holes. The Appalachian gravity and magnetic trends fade away only a little after the chain passes under cover, to be replaced by trends in other directions, as though some drastic structural change has occurred just where we are unable to see it.

And last of all, what happens on the southeastern flank of the Appalachian Chain, and what was its southeastern side during its orogenic history? Present geophysical evidence suggests that rocks much like those in the southeastern part of the Appalachians extend beneath the coastal plain cover and nearly, if not quite, to the edge of the continental shelf. How far southeast does the Precambrian sialic basement extend? The last we see of such a basement is in the mantled gneiss domes near Baltimore. Does it continue to the edge of the shelf, or does it give way to a more simatic basement? And was the southeastern flank of the chain always open ocean, as it is today, or did it once adjoin other lands beyond—the vanished borderland of the legendary Appalachia, now foundered and oceanized— or perhaps even the northwest coast of Africa?

Perhaps we should not try to answer such questions now, and perhaps they will always remain elusive. But surely even a few clues more tangible than those available to us today would go far toward an understanding of the past history and the present tectonics of this most elegant folded mountain chain on earth.

Author Index

Numbers in italics indicate the first page of an article which the individual authored or coauthored. The index does not include citations of articles in this volume.

Junior authors included by the expression "et al." are cited, even when their names do not appear on the page listed.

Editors of guidebooks are cited where the title of the guidebook is given under "References," but not when mentioned in text.

Aaron, J. M., 271, 274, 284, 286, 287, 290
Adams, R. W., *83*, 98, 99
Aldrich, L. T., 433, 434
Allen, J. R. L., 61, 66, 95, 99, 110, 112, 119, 120
Allen, R. M., Jr., 201, 202, 204, 205, 208, 210, 337, 338, 340, 348
Alling, H. L., 102, 107, 120
Alvord, D. C., 271, 277, 290
American Association of Petroleum Geologists, 7, 19, 39, 43, 142, 145, 243, 245, 281, 286, 289, 291, 353, 367, 418, 427
American Geophysical Union, 218, 220, 223, 225, 261, 269, 280, 281, 289, 365
Amsden, T. W., 30, 39, 57, 66, 99, 104, 120
Anderson, J. L., 299, 305, 314, 420, 423, 424, 426
Anderson, T. H., 325, 331
Andreasen, G. E., 281, 291
Applin, P. L., 425, 426
Arndt, H. H., 46, 147, 160
Ashley, G. H., 164, 172
Austin, R. S., 385, 390, 395
Averitt, P., 34, 39, 46
Ayrton, W. G., 36, 39, 74, 80

Bailey, E. D., 286, 289
Bain, G. L., 351, 353, 355, 357, 359, 360, 364, 366 369, 381
Baker, D. R., 271, 272, 278, 289, 290
Balk, Robert, 273, 280, 284, 289
Ball, M. M., 96, 97, 99
Baragar, W. R. A., 401, 414
Barrell, J., 3
Bascom, Florence, 129f, 130, 141, 146, 205, 210, 401, 402, 405, 414
Bass, M. N., 202, 208, 211, 213, 225, 295, 298
Bassett, W. A., 314, 315, 325, 329, 331
Bassler, R. S., 103, 121
Bates, R. L., 164, 166, 172
Bates, T. F., 277, 289
Battey, M. H., 405, 414
Baum, J. L., 282, 289
Bayley, W. S., 249, 258, 271, 273, 282, 284, 290, 391, 395
Becker, F. G., 365
Behre, C. H., Jr., 164, 169, 172, 277, 290
Bell, Henry, III, 215, 225, 353-355, 357, 363, 364, 366, 369-374, 376, 377, 379-382, 392, 397
Bennison, A. P., 300, 308, 314
Bentley, R. D., 313, 315, 320, 325, 326, 331
Berg, G., 118, 120
Bergin, M. J., *147*
Berkey, C. P., 273, 290
Berman, B. L., 116, 120

Bernard, H. A., 61, 66
Berryhill, H. L., Jr., 46
Bersten, J. M., 36, 41
Beven, A. C., 104, 120
Billings, M. P., 227, 244
Blatt, H., 116, 120
Bloomer, R. O., 17, 19, 41, 200-202, 204-206, 208, 210, 338, 340, 344, 348
Bloomer, R. R., 205, 210
Bofinger, V. M., 223, 225
Bolton, E. E., 103, 120
Bottino M. L., 222, 224
Boucot, A. J., 3, 103, 121
Bouma, A. H., 301, 314
Bowen, B. M., Jr., 385, 395
Bowen, N. L., 411, 414
Brannock, W. W., 413, 415
Branson, C. C., 57, 66
Branson, E. R., 103, 121
Brent, W. B., 163, 172, 385, 395
Brobst, D. A., 214, 215, 223, 224, 228, 232, 236, 244, 253, 254, 258
Brock, M. R., 297, 298
Broedel, C. H., 257, 258
Brokaw, A. L., 164, 169, 172
Bromery, R. W., 281, 282, 290, 318, 331, 401, 402, 414 and others, 280, 283, 290
Brosgé, W. P., 161-163, 165, 169, 173
Broughton, J. G., and others, 271, 273, 276, 278, 282, 290
Brown, C. Q., 369, 381
Brown, P. M., 420, 427
Brown, W. R., 200-202, 205, 206, 209, 210, *235*, 244, 257, 258, 337, 340, 344, 345, 347, 348
Brückner, W., 177, 178
Bryant, B., *213*, 213-216, 218, 219, 223, 228-233, 238, 239, 244, 251, 253, 258, *262*, 263, 267-269, 369, 382, 392, 396
Bucher, W., 142, 145, 195, 197
Buckwalter, T. V., 271, 274, 278-280, 284, 285, 290
Buddington, A. F., 210, 271, 274, 290, 291
Bunce, E. T., 418, 425, 426
Burchfiel, B. C., 262, 268, 269, 288, 290, 297, 298
Burst, J., F., 114, 118, 120
Burtner, R. L., 57, 59, 66
Butler, J. R., 358, 361-363, 365, 369-378, 381, 382
Butts, Charles, 23, 28, 39, 41, 45-47, 84, 96, 99, 103, 120, 162, 164, 172, 179, 180, 191, 218, 224, 256, 258, 340, 348
Buxtorf, A., 179, 191

Calver, J. L., 385, 395
Calvert, W. L., 23, 41, 45

General Subject Index

Names of specific geologic features are indexed under appropriate major headings such as: Anticlines; Anticlinoria; Faults, major; Synclines; Synclinoria.

Names of States are subheadings under the following primary headings: names of geologic periods, geologic structure; Maps, aeromagnetic; Maps, geologic.

Discussion of certain processes, concepts, theories, etc. are indexed even though the author may have used a different term on the page cited. However, the index generally follows the authors' terminology.

Index to Stratigraphic Units

Names of igneous and metamorphic units are included, but those of series, stages and zones are not. Informal or obsolete names are generally in quotation marks. Authors were not required to follow uniform nomenclature, so names in the index may differ slightly from those used on a specific page cited.